Wangzikun

王梓坤文集 | 李仲来 主编

03
论 文（上卷）

王梓坤 著

北京师范大学出版集团
BEIJING NORMAL UNIVERSITY PUBLISHING GROUP
北京师范大学出版社

前　言

　　王梓坤先生是中国著名的数学家、数学教育家、科普作家、中国科学院院士。他为我国的数学科学事业、教育事业、科学普及事业奋斗了几十年，做出了卓越贡献。他是中国概率论研究的先驱者，是将马尔可夫过程引入中国的先行者，是新中国教师节的提出者。作为王先生的学生，我们非常高兴和荣幸地看到我们敬爱的老师8卷文集的出版。

　　王老师于1929年4月30日（农历3月21日）出生于湖南省零陵县（今湖南省永州市零陵区），7岁时回到靠近井冈山的老家江西省吉安县枫墅村，幼时家境极其贫寒。父亲王肇基，又名王培城，常年在湖南受雇为店员，辛苦一生，受教育很少，但自学了许多古书，十分关心儿子的教育，教儿子背古文，做习题，曾经凭记忆为儿子编辑和亲笔书写了一本字典。但父亲不幸早逝，那年王老师才11岁。母亲郭香娥是农村妇女，勤劳一生，对人热情诚恳。父亲逝世后，全家的生活主要靠母亲和兄嫂租种地主的田地勉强维持。王老师虽然年幼，但帮助家里干各种农活。他聪明好学，常利用走路、放牛、车水的时间看书、算题，这些事至今还被乡亲们传为佳话。

　　王老师幼时的求学历程是坎坷和充满磨难的。1940年念完初小，村里没有高小。由于王老师成绩好，家乡父老劝他家长送他去固江镇县立第三中心小学念高小。半年后，父亲不幸去世，家境更为贫困，家里希望他停学。但他坚决不同意并做出了他人生中的第一大决策：走读。可是学校离家有十里之遥，而且翻山越岭，路上有狼，非常危险。王老师往往天不亮就起床，黄昏才回家，好不容易熬到高小毕业。1942年，王老师考上省立吉安中学（现江西省吉安市白鹭洲中学），只有第一个学期交了学费，以后就再也交不起了。在班主任高克正老师的帮助下，王老师申请缓交学费获批准，可是初中毕业时却因欠学费拿不到毕业证，更无钱报考高中。幸而学长王寄萍出资帮助，才拿到了毕业证并且去县城考取了国立十三中（现江西省泰和中学）的公费生。这事发生在1945年。他以顽强的毅力、勤奋的天性、优异的成绩、诚朴的品行，赢得了老师、同学和亲友的同情、关心、爱护和帮助。母亲和兄嫂在经济极端困难的情况下，也尽力支持他，终于完成了极其艰辛的小学、中学学业。

1948年暑假，在长沙有5所大学招生。王老师同样没有去长沙的路费，幸而同班同学吕润林慷慨解囊，王老师才得以到了长沙。长沙的江西同乡会成员欧阳伯康帮王老师谋到一个临时的教师职位，解决了在长沙的生活困难。王老师报考了5所学校，而且都考取了。他选择了武汉大学数学系，获得了数学系的两个奖学金名额之一，解决了学费问题。在大学期间，他如鱼得水，在知识的海洋中遨游。1952年毕业，他被分配到南开大学数学系任教。

王老师在南开大学辛勤执教28年。1954年，他经南开大学推荐并考试，被录取为留学苏联的研究生，1955年到世界著名大学莫斯科大学数学力学系攻读概率论。三年期间，他的绝大部分时间是在图书馆和教室里度过的，即使在假期里有去伏尔加河旅游的机会，他也放弃了。他在莫斯科大学的指导老师是近代概率论的奠基人、概率论公理化创立者、苏联科学院院士柯尔莫哥洛夫（А. Н. Колмогоров）和才华横溢的年轻概率论专家杜布鲁申（Р. Л. Добрушин），两位导师给王老师制订

前言

了学习和研究计划,让他参加他们领导的概率论讨论班,指导也很具体和耐心。王老师至今很怀念和感激他们。1958年,王老师在莫斯科大学获得苏联副博士学位。

学成回国后,王老师仍在南开大学任教,曾任概率信息教研室主任、南开大学数学系副主任、南开大学数学研究所副所长。他满腔热情地投身于教学和科研工作之中。当时在国内概率论学科几乎还是空白,连概率论课程也只有很少几所高校能够开出。他为概率论的学科建设奠基铺路,向概率论的深度和广度进军,将概率论应用于国家经济建设;他辛勤地培养和造就概率论的教学和科研队伍,让概率论为我们的国家造福。1959年,时年30岁还是讲师的王老师就开始带研究生,主持每周一次的概率论讨论班,为中国培养出一些高水平的概率论专家。至今他已指导了博士研究生和博士后22人,硕士研究生30余人,访问学者多人。他为本科生、研究生和青年教师开设概率论基础及其应用、随机过程等课程。由于王老师在教学、科研方面的突出成就,1977年11月他就被特别地从讲师破格晋升为教授,这是"文化大革命"后全国高校第一次职称晋升,只有两人(另一位是天津大学贺家李教授)。1981年国家批准第一批博士生导师,王老师是其中之一。

1965年,他出版了《随机过程论》,这是中国第一部系统论述随机过程理论的著作。随后又出版了《概率论基础及其应用》(1976)、《生灭过程与马尔可夫链》(1980)。这三部书成一整体,从概率论的基础写起,到他的研究方向的前沿,被人誉为概率论三部曲,被长期用作大学教材或参考书。1983年又出版专著《布朗运动与位势》。这些书既总结了王老师本人、他的同事、同行、学生在概率论的教学和研究中的一些成果,又为在中国传播、推动概率论学科发展,培养中国概率论的教学和研究人才,起到了非常重要的作用,哺育了中国的几代概率论学人(这4部著作于1996年由北京师范大学出版社再版,书名分别是：《概率论基础及其应用》，即本8卷文集的第5卷；《随机过程通论》上、下卷，即本8卷文集的第6卷和第7卷）。1992年《生灭过程与马尔可夫链》的扩大修订版（与杨向群合作）被译成英文，由德国的施普林格（Springer）出版社和中国的科学出版社出版。1999年由湖南科技出版社出版的《马尔可夫过程与今日数学》，则是将王老师1998年底以前发表的主要论文进行加工、整理、编辑而成的一本内容系统、结构完整的书。

1984年5月，王老师被国务院任命为北京师范大学校长，这一职位自1971年以来一直虚位以待。王老师在校长岗位上工作了5年。王老师常说："我一辈子的理想，就是当教师。"他一生都在实践做一位好教师的诺言。任校长后，就将更多精力投入到发展师范教育和提高教师地位、待遇上来。1984年12月，王老师与北京师范大学的教师们提出设立"教师节"的建议，并首次提出了"尊师重教"的倡议，提出"百年树人亦英雄"，以恢复和提高人民教师在社会上的光荣地位，同时也表达了全国人民对教师这一崇高职业的高度颂扬、崇敬和爱戴。1985年1月，全国人民代表大会常务委员会通过决议，决定每年的9月10日为教师节。王老师任校长后明确提出北京师范大学的办学目标：把北京师范大学建成国内第一流的、国际上有影响力的、高水平、多贡献的重点大学。对于如何处理好师范性和学术性的问题，他认为两者不仅不能截然分开，而且是相辅相成的；不搞科研就不能叫大学，如果学术水平不高，培养的老师一般水平不会太高，所以必须抓学术；但师范性也不能丢，师范大学的主要任务就是干这件事，更何况培养师资是一项光荣任务。对师范性他提出了三高：高水平的专业、高水平的师资、高水平的学术著作。王老师也特别关心农村教育，捐资为农村小学修建教学楼，赠送书刊，设立奖学金。王老师对教育事业付出了辛勤的劳动，做出了重要贡献。正如著名教育家顾明远先生所说："王梓坤是教育实践家，他做成的三件事情：教师节、抓科研、建大楼，对北京师范大学的建设意义深远。"2008年，王老师被中国几大教育网站授予改革开放30年"中国教育时代人物"称号。

1981年，王老师应邀去美国康奈尔（Cornell）大学做学术访问；1985年访问加拿大里贾纳（Regina）大学、曼尼托巴（Manitoba）大学、温尼伯（Winnipeg）大学。1988年，澳大利亚悉尼麦考瑞（Macquarie）大学授予他荣誉科学博士学位和荣誉客座学者称号，王老师赴澳大利亚参加颁授仪式。该校授予他这一荣誉称号是由于他在研究概率论方面的杰出成就和在提倡科学教育和研究方法上所做出的贡献。

1989年，他访问母校莫斯科大学并作学术报告。

1993年，王老师卸任校长职务已数年。他继续在北京师范大学任职的同时，以极大的勇气受聘为汕头大学教授。这是国内的大学第一次高薪聘任专家学者。汕头大学的这一举动横扫了当时社会上流行的"读书无用论""搞导弹的不如卖茶叶蛋的"等论调，证明了掌握科学技术的人员是很有价值的，为国家改善广大知识分子的待遇开启了先河。但此事引起极大震动，一时引发了不少议论。王老师则认为：这对改善全国的教师和科技人员的待遇、对发展教育和科技事业，将会起到很好的作用。果然，开此先河后，许多单位开始高薪补贴或高薪引进人才。在汕头大学，王老师与同事们创办了汕头大学数学研究所，并任所长6年。汕头大学的数学学科有了很大的发展，不仅获得了数学学科的硕士学位授予权，而且聚集了一批优秀的数学教师，为后来获得数学学科博士学位授予权打下了坚实的基础。

王老师担任过很多兼职：天津市人民代表大会代表，国家科学技术委员会数学组成员，中国数学会理事，中国科学技术协会委员，中国高等教育学会常务理事，中国自然辩证法研究会常务理事，中国人才学会副理事长，中国概率统计学会常务理事，中国地震学会理事，中国高等师范教育研究会理事长，

《中国科学》《科学通报》《科技导报》《世界科学》《数学物理学报》等杂志编委，《数学教育学报》主编，《纯粹数学与应用数学》《现代基础数学》等丛书编委。

王老师获得了多种奖励和荣誉：1978年获全国科学大会奖，1982年获国家自然科学奖，1984年被中华人民共和国人事部授予"国家有突出贡献中青年专家"称号，1986年获国家教育委员会科学技术进步奖，1988年获澳大利亚悉尼麦考瑞大学荣誉科学博士学位和荣誉客座学者称号，1990年开始享受政府特殊津贴，1993年获曾宪梓教育基金会高等师范院校教师奖，1997年获全国优秀科技图书一等奖，2002年获何梁何利基金科学与技术进步奖。王老师于1961年、1979年和1982年3次被评为天津市劳动模范，1980年获全国新长征优秀科普作品奖，1990年被全国科普作家协会授予"新中国成立以来成绩突出的科普作家"称号。

1991年，王老师当选为中国科学院院士，这是学术界对他几十年来在概率论研究中和为这门学科在中国的发展所做出的突出贡献的高度评价和肯定。

王老师是将马尔可夫过程引入中国的先行者。马尔可夫过程是以俄国数学家 А. А. Марков 的名字命名的一类随机过程。王老师于1958年首次将它引入中国时，译为马尔科夫过程。后来国内一些学者也称为马尔可夫过程、马尔柯夫过程、Markov过程，甚至简称为马氏过程或马程。现在统一规范为马尔可夫过程，或直接用 Markov 过程。生灭过程、布朗运动、扩散过程都是在理论上非常重要、在应用上非常广泛、很有代表性的马尔可夫过程。王老师在马尔可夫过程的理论研究和应用方面都做出了很大的贡献。

随着时代的前进，特别是随着国际上概率论研究的进展，王老师的研究课题也在变化。这些课题都是当时国际上概率论研究前沿的重要方向。王老师始终紧随学科的近代发展步伐，力求在科学研究的重要前沿做出崭新的、开创性的成果，以带

前言

动国内外一批学者在刚开垦的原野上耕耘。这是王老师一生中数学研究的一个重大特色。

20世纪50年代末,王老师彻底解决了生灭过程的构造问题,而且独创了马尔可夫过程构造论中的一种崭新的方法——过程轨道的极限过渡构造法,简称极限过渡法。王老师在莫斯科大学学习期间,就表现出非凡的才华,他的副博士学位论文《全部生灭过程的分类》彻底解决了生灭过程的构造问题,也就是说,他找出了全部的生灭过程,而且用的方法是他独创的极限过渡法。当时,国际概率论大师、美国的费勒(W. Feller)也在研究生灭过程的构造,但他使用的是分析方法,而且只找出了部分的生灭过程(同时满足向前、向后两个微分方程组的生灭过程)。王老师的方法的优点在于彻底性(构造出了全部生灭过程)和明确性(概率意义非常清楚)。这项工作得到了苏联概率论专家邓肯(Е. Б. Дынкин,E. B. Dynkin,后来移居美国并成为美国科学院院士)和苏联概率论专家尤什凯维奇(А. А. Юшкевич)教授的引用和好评,后者说:"Feller构造了生灭过程的多种延拓,同时王梓坤找出了全部的延拓。"在解决了生灭过程构造问题的基础上,王老师用差分方法和递推方法,求出了生灭过程的泛函的分布,并给出此成果在排队论、传染病学等研究中的应用。英国皇家学会会员肯德尔(D. G. Kendall)评论说:"这篇文章除了作者所提到的应用外,还有许多重要的应用……该问题是困难的,本文所提出的技巧值得仔细学习。"在王老师的带领和推动下,对构造论的研究成为中国马尔可夫过程研究的一个重要的特色之一。中南大学、湘潭大学、湖南师范大学等单位的学者已在国内外出版了几部关于马尔可夫过程构造论的专著。

1962年,他发表了另一交叉学科的论文《随机泛函分析引论》,这是国内较系统地介绍、论述、研究随机泛函分析的第一篇论文。在论文中,他求出了广义函数空间中随机元的极限定理。此文开创了中国研究随机泛函的先河，并引发了吉林大学、武汉大学、四川大学、厦门大学、中国海洋大学等高校的不少学者的后继工作，取得了丰硕成果。

20世纪60年代初，王老师将邓肯的专著《马尔可夫过程论基础》译成中文出版，该书总结了当时的苏联概率论学派在马尔可夫过程论研究方面的最新成就，大大推动了中国学者对马尔可夫过程的研究。

20世纪60年代前期，王老师研究了一般马尔可夫过程的通性，如0-1律、常返性、马丁（Martin）边界和过分函数的关系等。他证明的一个很有趣的结果是：对于某些马尔可夫过程，过程常返等价于过程的每一个过分函数是常数，而过程的强无穷远0-1律成立等价于过程的每一个有界调和函数是常数。

20世纪60年代后期和70年代，由于众所周知的原因，王老师停下理论研究，应海军和国家地震局的要求，转向数学的实际应用，主要从事地震统计预报和在计算机上模拟随机过程。他带领的课题小组首创了"地震的随机转移预报方法"和"利用国外大震以预报国内大震的相关区方法"，被地震部门采用，取得了实际的效果。在这期间，王老师也发表了一批实际应用方面的论文，例如《随机激发过程对地极移动的作用》等，还有1978年出版的专著《概率与统计预报及在地震与气象中的应用》（与钱尚玮合作）。

20世纪70年代，马尔可夫过程与位势理论的关系是国际概率论界的热门研究课题。王老师研究布朗运动与古典位势的关系，求出了布朗运动、对称稳定过程的一些重要分布。如对球面的末离时、末离点、极大游程的精确分布。他求出的自原点出发的 d（不小于3）维布朗运动对于中心是原点的球面的末离时分布，是一个当时还未见过的新分布，而且分布的形式很简单。美国数学家格图（R. K. Getoor）也独立地得到了同样的结果。王老师还证明了：从原点出发的布朗运动对于中心是原点的球面的首中点分布和末离点分布是相同的,都是球面上的均匀分布。

20世纪80年代后期,王老师研究多参数马尔可夫过程。他于1983年在国际上最早给出多参数有限维奥恩斯坦-乌伦贝克(OU,Ornstein-Uhlenbeck)过程的严格数学定义并得到了系统的研究成果。如三点转移、预测问题、多参数与单参数的关系等。次年,加拿大著名概率论专家瓦什(J. B. Walsh)也给出了类似的定义,其定义是王老师定义的一种特殊情形。1993年,王老师在引进多参数无穷维布朗运动的基础上,给出了多参数无穷维OU过程定义,这是国际上最早提出并研究多参数无穷维OU过程的论文,该文发现了参数空间有分层性质。王老师关于多参数马尔可夫过程的开创性工作,推动和引发了国内对于多参数马尔可夫过程的研究,如中山大学、武汉大学、南开大学、杭州大学、湘潭大学、湖南师范大学等的后继研究。湖南科学技术出版社1996年出版的杨向群、李应求的专著《两参数马尔可夫过程论》,就是在王老师开垦的原野上耕耘的结果。

20世纪90年代至今,王老师带领同事和研究生研究国际上的重要新课题——测度值马尔可夫过程(超过程)。测度值马氏过程理论艰深,但有很明确的实际意义。粗略地说,如果普通马尔可夫过程是刻画"一个粒子"的随机运动规律,那么超过程就是刻画"一团粒子云"的随机飘移运动规律。王老师带领的集体在超过程理论上取得了丰富的成果,特别是他的年轻的同事和学生们,做了许多很好的工作。

2002年,王老师和张新生发表论文《生命信息遗传中的若干数学问题》,这又是一项旨在开拓创新的工作。1953年沃森(J. Watson)和克里克(F. Crick)发现DNA的双螺旋结构,人们对生命信息遗传的研究进入一个崭新的时代,相继发现了"遗传密码字典"和"遗传的中心法则"。现在,人类基因组测序数据已完成,其数据之多可以构成一本100万页的书,而且书中只有4个字母反复不断地出现。要读懂这本宏厚的巨著，需要数学和计算机学科的介入。该文首次向国内学术界介绍了人类基因组研究中的若干数学问题及所要用到的数学方法与模型，具有特别重要的意义。

除了对数学的研究和贡献外，王老师对科学普及、科学研究方法论，甚至一些哲学的基本问题，如偶然性、必然性、混沌之间的关系，也有浓厚兴趣，并有独到的见解，做出了一定的贡献。

在"文化大革命"的特殊年代，王老师仍悄悄地学习、收集资料、整理和研究有关科学发现和科学研究方法的诸多问题。1977年"文化大革命"刚结束，王老师就在《南开大学学报》上连载论文《科学发现纵横谈》（以下简称《纵横谈》），次年由上海人民出版社出版成书。这是"文化大革命"后中国大陆第一本关于科普和科学方法论的著作。这本书别开生面，内容充实，富于思想，因而被广泛传诵。书中一开始就提出，作为一个科技工作者，应该兼备德识才学，德是基础，而且德识才学要在实践中来实现。王老师本人就是一位成功的德识才学的实践者。《纵横谈》是十年"文化大革命"后别具一格的读物。数学界老前辈苏步青院士作序给予很高的评价："王梓坤同志纵览古今，横观中外，从自然科学发展的历史长河中，挑选出不少有意义的发现和事实，努力用辩证唯物主义和历史唯物主义的观点，加以分析总结，阐明有关科学发现的一些基本规律，并探求作为一名自然科学工作者，应该力求具备一些怎样的品质。这些内容，作者是在'四人帮'①形而上学猖獗、唯心主义横行的情况下写成的，尤其难能可贵……作者是一位数学家，能在研究数学的同时，写成这样的作品，同样是难能可贵的。"《纵横谈》以清新独特的风格、简洁流畅的笔调、扎实丰富的内容吸引了广大读者，引起国内很大的反响。书中不少章节堪称

---

① 指王洪文、张春桥、江青、姚文元.

优美动人的散文，情理交融回味无穷，使人陶醉在美的享受中。有些篇章还被选入中学和大学语文课本中。该书多次出版并获奖，对科学精神和方法的普及起了很大的作用。以至19年后，这本书再次在《科技日报》上全文重载（1996年4月4日至5月21日）。主编在前言中说："这是一组十分精彩、优美的文章。今天许许多多活跃在科研工作岗位上的朋友，都受过它的启发，以至他们中的一些人就是由于受到这些文章中阐发的思想指引，决意将自己的一生贡献给伟大的科学探索。"1993年，北京师范大学出版社将《纵横谈》进一步扩大成《科学发现纵横谈（新编）》。该书收入了《科学发现纵横谈》、1985年王老师发表的《科海泛舟》以及其他一些文章。2002年，上海教育出版社出版了装帧精美的《莺啼梦晓——科研方法与成才之路》一书，其中除《纵横谈》外，还收入了数十篇文章，有的论人才成长、科研方法、对科学工作者素质的要求，有的论数学学习、数学研究、研究生培养等。2003年《莺啼梦晓——科研方法与成才之路》获第五届上海市优秀科普作品奖之科普图书荣誉奖（相当于特等奖）。2009年，北京师范大学出版社出版的《科学发现纵横谈》（第3版）于同年入选《中国文库》（第四辑）（新中国60周年特辑）。《中国文库》编辑委员会称：该文库所收书籍"应当是能够代表中国出版业水平的精品""对中国百余年来的政治、经济、文化和社会的发展产生过重大积极的影响，至今仍具有重要价值，是中国读者必读、必备的经典性、工具性名著。"王老师被评为"新中国成立以来成绩突出的科普作家"，绝非偶然。

　　王老师不仅对数学研究、科普事业有突出的贡献，而且对整个数学，特别是今日数学，也有精辟、全面的认识。20世纪90年代前期，针对当时社会上对数学学科的重要性有所忽视的情况，王老师受中国科学院数学物理学部的委托，撰写了《今日数学及其应用》。该文对今日数学的特点、状况、应用，以及其在国富民强和提高民族的科学文化素质中的重要作用等做了全面、深刻的阐述。文章提出了今日数学的许多新颖的观点和新的认识。例如，"今日数学已不仅是一门科学，还是一种普适性的技术。""高技术本质上是一种数学技术。""某些重点问题的解决，数学方法是唯一的，非此'君'莫属。"对今日数学的观点、认识、应用的阐述，使中国社会更加深切地感受到数学学科在自然科学、社会科学、高新技术、推动生产力发展和富国强民中的重大作用，使人们更加深刻地认识到数学的发展是国家大事。文章中清新的观点、丰富的事例、明快的笔调和形象生动的语言使读者阅后感到是高品位的享受。

王老师在南开大学工作28年，吃食堂42年。夫人谭得伶教授是20世纪50年代莫斯科大学语文系的中国留学生，1957年毕业回国后一直在北京师范大学任教，专攻俄罗斯文学，曾指导硕士生、博士生和访问学者20余名。王老师和谭老师1958年结婚后育有两个儿子，两人两地分居26年。谭老师独挑家务大梁，这也是王老师事业成功的重要因素。

王老师为人和善，严于律己，宽厚待人，有功而不自居，有傲骨而无傲气，对同行的工作和长处总是充分肯定，对学生要求严格，教其独立思考，教其学习和研究的方法，将学生当成朋友。王老师有一段自勉的格言："我尊重这样的人，他心怀博大，待人宽厚；朝观剑舞，夕临秋水，观剑以励志奋进，读庄以淡化世纷；公而忘私，勤于职守；力求无负于前人，无罪于今人，无愧于后人。"

本8卷文集列入北京师范大学学科建设经费资助项目，由北京师范大学出版社出版。李仲来教授从文集的策划到论文的收集、整理、编排和校对等各方面都付出了巨大的努力。在此，我们作为王老师早期学生，谨代表王老师的所有学生向北京师范大学、北京师范大学出版社、北京师范大学数学科学学院和李仲来教授表示诚挚的感谢！

杨向群　吴　荣　施仁杰　李增沪

2016年3月10日

目 录

全部生灭过程的分类 ………………………………… (1)

On a Birth and Death Process ……………………… (12)

On Distributions of Functionals of Birth and Death Processes and Their Applications in the Theory of Queues ……… (18)

随机泛函分析引论 …………………………………… (32)

 §1. 引 言 ………………………………………… (32)

 §2. 随机元 ………………………………………… (35)

 §3. 随机变换 ……………………………………… (49)

 §4. 广义函数空间中的随机元 …………………… (60)

生灭过程构造论 ……………………………………… (76)

 §1. 引 言 ………………………………………… (76)

 §2. 基本特征数的概率意义 ……………………… (80)

 §3. Doob 过程的变换 …………………………… (84)

 §4. 连续流入不可能的充分必要条件 …………… (93)

 §5. 一般 Q 过程变换为 Doob 过程 …………… (98)

 §6. $S<+\infty$ 时 Q 过程的构造 ……………… (103)

 §7. 方程组的非负解与结果的深化 ……………… (116)

 §8. $S=+\infty$ 时 Q 过程的构造 ……………… (123)

§9. 进一步的问题 ……………………………………… (129)
　　附　录 ………………………………………………… (130)
波兰应用数学中若干结果的概述 ……………………… (139)
§1. 前　言 ………………………………………………… (139)
§2. 统计估值 ……………………………………………… (141)
§3. 量水计的最佳验收方法 ……………………………… (146)
§4. 估计到观察错误的产品质量的验收检查 …………… (152)
§5. 两个生产过程的比较与对偶性原则 ………………… (157)
§6. 度量分析的新代数理论及其在产品抽样
　　实验中的应用 ………………………………………… (170)
§7. 统计弹性理论中的一个奇论 ………………………… (189)
§8. 根据个体间已知距离，将个体所成的集
　　排序与分类的方法 …………………………………… (193)
§9. 平面上点的随机分布与类别存在的统计
　　判定法 ………………………………………………… (206)
§10. 电话总局最佳局址的选定 …………………………… (212)
§11. 关于个体（看成 n 维空间的点）集合的
　　几个注意 ……………………………………………… (221)
§12. 关于地质矿藏参数的估计 …………………………… (226)
§13. 关于统计参数的极大中极小估计 …………………… (237)
§14. 关于平面上经验曲线的长度 ………………………… (248)
§15. 某些生物问题中贝叶斯公式的应用 ………………… (251)
扩散过程在随机时间替换下的不变性 ………………… (254)
§1. 扩散过程的随机时间替换 …………………………… (254)
§2. Dirichlet 问题与特征算子方程 ……………………… (260)
生灭过程的遍历性与 0-1 律 …………………………… (264)
§1. 基本概念与特征数 …………………………………… (264)

§2. 常返性和遍历性 ·················· (267)

§3. 过分函数极限的存在性和 0-1 律 ········· (269)

On Zero-One Laws for Markov Processes ········· (273)

§1. Infinitely near zero-one laws ············ (275)

§2. Infinitely far zero-one laws ············ (285)

§3. The infinitely far zero-one law for homogeneous Markov processes ··········· (289)

The Martin Boundary and Limit Theorems for Excessive Functions ···················· (301)

§1. The martin boundary ·············· (302)

§2. The limit theorems for excessive functions ····· (308)

§3. Application to denumerable Markov processes ···················· (316)

Some Properties of Recurrent Markov Processes ······ (322)

§1. Introduction ·················· (322)

§2. First entrance and first contact times ········ (325)

§3. Necessary and sufficient conditions for recurrent processes ················ (333)

§4. Excessive functions and strong zero-one law ···· (340)

§5. Applications in differential equations ········ (347)

地震迁移的统计预报 ···················· (353)

§1. 引 言 ····················· (353)

§2. 中国中部南北地震带的地震迁移 ·········· (357)

§3. 全国 5 个主要地震区的地震迁移 ·········· (371)

§4. 存在问题 ···················· (374)

预测大地震的一种数学方法 ················ (376)

§1. 基本思想 ···················· (376)

§2. 相关区的选择 ·· (378)
§3. 判别量 X 的精确公式 ······························ (381)
§4. 举 例 ·· (383)
§5. 实践检验 ·· (384)
§6. 预报地区 ·· (385)

华北地区地震的统计预报（一） ······················ (389)

华北地区地震的统计预报（二） ······················ (399)
§1. 引 言 ·· (399)
§2. 预报因子提取的方法 ···································· (400)
§3. 分类预报的 Bayes（贝叶斯）聚合过程的
　　概率模型 ·· (402)
§4. 分类预报的具体过程 ···································· (404)
§5. 效果检验 ·· (406)

中断生灭过程的构造 ······································ (411)
§1. ·· (411)
§2. ·· (413)
§3. ·· (418)

随机激发过程对地极移动的作用 ······················ (421)
§1. 地极移动的随机微分方程模型 ······················ (421)
§2. 地极移动模型的概率性质 ···························· (425)
§3. 地极移动模型的预测问题 ···························· (431)
§4. 小 结 ·· (433)

中断生灭过程构造中的概率分析方法 ················ (435)
§1. 引 言 ·· (435)
§2. 两个引理 ·· (438)
§3. 中断过程的延拓过程 ···································· (446)
§4. Q 过程的变换 ·· (458)

§5. Q过程的构造 ································ (470)

Sojourn Times and First Passage Times for Birth and Death Processes ································ (476)
    §1. Introduction ································ (476)
    §2. Distributions of integral functionals ············ (479)
    §3. Distributions of Sojourn times and first passage times ································ (485)
    §4. The limit distribution ························ (489)

Last Exit Distributions and Maximum Excursion for Brownian Motion ································ (494)
    §1. Distributions of last exit place ················ (494)
    §2. Distributions of last exit time ················· (499)
    §3. Maximum excursion ···························· (502)
    §4. Time for first attaining maximum ·············· (505)

高维布朗运动的末遇时间与位置 ···················· (508)

概率论的若干新进展 ································ (513)
    §1. 关于马尔可夫过程 ···························· (514)
    §2. 关于鞅、随机积分与随机微分方程 ············· (517)
    §3. 关于随机场及概率论在统计物理中的应用 ······ (518)
    §4. 关于极限定理 ································ (519)
    §5. 关于点过程与排队论 ·························· (520)
    §6. 其 他 ······································ (521)

Stochastic Waves for Symmetric Stable Processes and Brownian Motion ································ (525)
    §1. ··· (525)
    §2. ··· (528)
    §3. ··· (535)

关于1976年四川松潘大地震的统计预报 ················ (538)
  §1. 预报意见及根据 ····························· (538)
  §2. 方法的基本思想 ····························· (540)
  §3. 结束语 ······································· (548)
后　记 ··· (550)

Классификация всех процессов размножния и гнбели' Науные доклады высшей школы, Физико-математические науки, 1958, (4)

全部生灭过程的分类

具有可数个状态 $E=\mathbf{N}$ 的齐次马尔可夫过程亦即转移概率 $P_{ij}(t)(i, j\in\mathbf{N}; 0\leqslant t<+\infty)$，它们是满足下列条件的一些实值函数：

$$P_{ij}(t)\geqslant 0, \qquad (1)$$

$$\sum_{j=0}^{+\infty}P_{ij}(t)=1, \qquad (2)$$

$$\sum_{j=0}^{+\infty}P_{ij}(t)P_{jk}(s)=P_{ik}(t+s). \qquad (3)$$

我们假定在零点满足连续性条件：

$$\lim_{t\to 0+}P_{ij}(t)=P_{ij}(0)=\delta_{ij}, \qquad (4)$$

其中 $\delta_{ij}=1, i=j; \delta_{ij}=0, i\neq j$. 从这些条件推出（见[6]）极限存在：

$$\lim_{t\to 0+}\frac{P_{ij}(t)-\delta_{ij}}{t}=q_{ij}, \qquad (5)$$

而且满足

$$0\leqslant q_{ij}<+\infty \quad (i\neq j),$$
$$\sum_{j\neq i}q_{ij}\leqslant -q_{ii}\leqslant +\infty. \qquad (6)$$

记 $q_i=-q_{ii}$. 数 q_i 称为从状态 i 流出的概率密度，q_{ij} 称为

从状态 i 转移到 j 的概率密度,矩阵

$$Q = (q_{ij}) \tag{7}$$

称为流出和转移的概率密度矩阵,满足(1)~(5)的族$\{P_{ij}(t)\}$称为 Q 过程.

Колмогоров[14]举出例子,说明(6)中的记号"$\leqslant$"不能用记号"$=$"代替. 但我们假定

$$\sum_{j \neq i} q_{ij} = q_i < +\infty \quad (i \in \mathbf{N}). \tag{8}$$

对给定的 Q,条件(8)等价于([1][6])每个 Q 过程$\{P_{ij}(t)\}$满足 Колмогоров 向后微分方程组

$$P'(t) = QP(t), \quad P(0) = I, \tag{9}$$

这里 $P(t)$ 是以 $P_{ij}(t)$ 为元素的矩阵, I 是单位矩阵.

反之,设给定满足(8)的形如(7)的矩阵,其中 $q_{ij} \geqslant 0$ 对 $i \neq j$,而 $q_i = -q_{ii} \geqslant 0$. 自然地提出下述问题:$Q$ 过程存在吗?即,是否存在满足(1)~(4)的$\{P_{ij}(t)\}$,使其流出和转移的概率密度矩阵与给定的矩阵 Q 重合.

[16][7]指出,对给定的 Q,Q 过程永远存在. Feller 给出了 Q 过程唯一的充分必要条件.

在[3]中 Добрушин 详细地研究了 Feller 条件. 满足 Feller 条件的矩阵 Q 称为规则的. 对于任意给定的非规则矩阵 Q,在[6]中 Doob 给出了无穷多个 Q 过程. 我们将称在[6]中构造的 Q 过程类为 Doob 过程类. 每个 Doob 过程由矩阵 Q 和一个概率分布 $\pi = (\pi_i)$ 唯一地给出([5,P267])[①],称之为(Q,π)Doob 过程. 但是,Doob 过程类并未穷尽全部的 Q 过程. 因此产生了第二个问题,对给定的 Q 矩阵,如何刻画所有的 Q 过程?此问

---

① 确切地说,$\pi_i = P(x(\tau) = i)$,其中 $\tau = \tau(\omega)$, $x(\tau) = x(\tau(\omega), \omega)$将在下面定义.

题等价于下述问题：对给定的 Q，如何求出方程(9)的满足条件(1)~(5)的所有解.

对于非规则的矩阵 Q，在[4]出现以前，此问题毫无解答.

在[4]中，对于充分广泛的 Q 矩阵类，问题得以解决. Feller 的论文[8][9]提出了相近的问题.

我们找出了全部生灭过程，即具有下列形式的矩阵 Q 的全部 Q 过程：

$$Q=\begin{bmatrix} -b_0 & b_0 & 0 & \cdots & 0 & 0 & 0 & \cdots \\ a_1 & -(a_1+b_1) & b_1 & \cdots & 0 & 0 & 0 & \cdots \\ \vdots & \vdots & \vdots & \vdots & \vdots & \vdots & \vdots & \\ 0 & 0 & 0 & \cdots & a_n & -(a_n+b_n) & b_n & \cdots \\ \vdots & \vdots & \vdots & & \vdots & \vdots & \vdots & \end{bmatrix},$$

这里 $a_i>0(i>0)$，$b_i>0(i\geqslant 0)$. 近些年来，生灭过程类为许多作者所研究[11][12][13][15].

现在假设 $\{P_{ij}(t)\}$ 是任意 Q 过程. 可以构造(见[2][16][19])概率空间 $(\Omega, \mathscr{B}, P)$(见[5])，及定义在其上的取值 0, 1, 2, … 及附加值 $+\infty$ 的随机变量族 $X=\{x(t,\omega), 0\leqslant t<+\infty\}(\omega\in\Omega)$[①]，使得有下述性质：

(i) 对任意指定的 t，$P(x(t)=+\infty)=0$；

(ii) 过程 X 是齐次马尔可夫过程，具有预先给定的转移概率 $P_{ij}(t)$；

(iii) 过程 X 是可测的，关于直线上的闭集类是可分的，而

---

① 在本卷第 132 页的参考文献[1]中称 $X=\{x(t), t\geqslant 0\}$ 为右下半连续修正，在本套书第 4 卷《生灭过程理论的若干新进展》的参考文献[1]中称之为典范链. 典范链是 Borel 可测的，完全可分的，轨道在 $(0, 1, 2, \cdots, +\infty)$ 中右下半连续(在一切 $q_i<+\infty$ 时成为右连续)，且有性质(i)(ii). 于是典范链具有性质(iii)(iv). 在 §1 中的 X，均可假定为典范链，而且当 $\omega\in\Omega$ 不特指时，ω 常略写，而记为 $X=\{x(t), t\geqslant 0\}$. ——编者

且可以任意取可数的处处稠密的 t 集作为满足可分性定义条件中的可数集;

(iv) 对任意指定的非负整数 k 和任意的 ω, 使 $x(t, \omega)=k$ 的 t 值的集合是一些不相交的左闭右开区间的并.

今后, 我们说样本函数的性质, 指的就是上面描述的函数 $x(t, \omega)$ 的性质. 我们也称过程 X 为 Q 过程, 它由 $P_{ij}(t)$ 唯一决定[①].

定义 $\tau(\omega)$ 为下述 t 值的最小上界, 使得函数 $x(s, \omega)$ 在 $[0, t]$ 中只有有限多个间断点[②].

引进矩阵 Q 的下列特征数

$$R = \sum_{i=0}^{+\infty} m_i, S = \sum_{i=1}^{+\infty} e_i,$$

其中

$$m_1 = \frac{1}{b_i} + \sum_{k=0}^{i-1} \frac{a_i a_{i-1} \cdots a_{i-k}}{b_i b_{i-1} \cdots b_{i-k} b_{i-k-1}},$$

$$e_i = \frac{1}{a_i} + \sum_{k=0}^{+\infty} \frac{b_i b_{i+1} \cdots b_{i+k}}{a_i a_{i+1} \cdots a_{i+k} a_{i+k+1}}.$$

在[13]中已证明: $R = E(\tau \mid x(0)=0)$; 当且仅当 $R = +\infty$ 时, 存在唯一的 Q 过程. 这个唯一的 Q 过程已被 Feller 找到[7]. 因此, 我们只需研究使 $R < +\infty$ 的矩阵 Q.

在[17]中首次引入量 S. 我们指出该量的概率意义. 设 $X_N = \{x_N(t), t \geq 0\}$ 是具有有限个状态 $(0, 1, 2, \cdots, N)$ 的 Q_N 过程, 其中

---

① 如果两个过程有相同的转移概率, 我们视它们为同一个过程, 不予区分.
② 在本卷第 5 篇的参考文献[1]中称 τ 为首次无穷, 在本卷第 8 篇的参考文献[1]中称 τ 为第一个飞跃点. ——编者

$$Q_N = \begin{bmatrix} -b_0 & b_0 & & & & 0 \\ a_1 & -(a_1+b_1) & b_1 & & & \\ & \vdots & \vdots & & & \vdots \\ & & a_{N-1} & -(a_{N-1}+b_{N-1}) & b_{N-1} & \\ & 0 & & & a_N+b_N & -(a_N+b_N) \end{bmatrix}$$

如果 $P(x_N(0)=N)=1$ 且 τ_N 是 X_N 首次到达状态 0 的时刻,那么
$$\lim_{N\to+\infty} E\tau_N = S.$$

现在,对于给定的非规则的矩阵 Q,我们着手寻找所有的 Q 过程.

情形 $R<+\infty$, $S=+\infty$. 设 $X=\{x(t), t\geq 0\}$ 是任意的 Q 过程. 设 $u>0$,如果对任意 $\varepsilon>0$,$x(t, \omega)$ 在区间 $[u-\varepsilon, u)$ 中取无穷多个不同的值且 $x(u, \omega)=j$,我们将称 u 是 $x(t, \omega)$ 飞跃到状态 j 的时刻,设 $\tau^{(n)}(\omega)=\inf\{u: u\geq\tau(\omega); u$ 是 $x(t, \omega)$ 飞跃到状态 $j(\leq n)$ 的时刻$\}$. 可以证明,对于从某个 n 后的一切 n,有 $P(\tau^{(n)}<+\infty)=1$,记
$$u_i^{(n)} = P(x(\tau^{(n)})=i) \quad (i=0, 1, 2, \cdots, n).$$

还可以证明,存在 k,使 $u_k^{(n)}>0$ 对某个 n 后的一切 n 成立,而且 $\dfrac{u_i^{(n)}}{u_k^{(n)}}$ 不依赖于 n,$n\geq\max(i, k)$ $(i\in \mathbf{N})$.

按如下方式定义非负数列 $s_0, s_1, s_2, \cdots$:
$$s_k = c(c \text{ 是任意的正数}),$$
$$s_i = \frac{u_i^{(n)}}{u_k^{(n)}} s_k \quad (n\geq\max(i, k), i\in \mathbf{N}).$$

我们称 $s_0, s_1, s_2, \cdots$ 为 Q 过程 X 的无穷小特征数列. 显然地,除常数因子不计外,特征数列被 Q 过程 X 唯一地决定. 令 $n_i = \sum_{j=i}^{+\infty} m_j$,我们有下面的定理.

定理 1 (i) 设给定满足 $R<+\infty$, $S=+\infty$ 的 Q 过程 X,

则其特征数列 s_i, $i \in \mathbf{N}$ 满足关系

$$0 < \sum_{i=0}^{+\infty} s_i n_i < +\infty. \tag{10}$$

(ii) 反之，给定满足(10)的非负数列 $\{s_i\}$，则存在唯一的 Q 过程 X，其特征数列与 $\{s_i\}$ ($i \in \mathbf{N}$) 重合。

定理 1 的 (ii) 中的 Q 过程 X 的转移概率 $P_{ij}(t)$ 可以按如下方式得到：用公式

$$\pi_i^{(n)} = \frac{s_i}{\sum_{j=0}^{n} s_j} \quad (i = 0, 1, 2, \cdots, n)$$

给出集中在状态 $(0, 1, 2, \cdots, n)$ 上的概率分布 $\pi^{(n)} = (\pi_0^{(n)}, \pi_1^{(n)}, \cdots, \pi_n^{(n)})$ ($n = k, k+1, \cdots$)。设 $P_{ij}^{(n)}(t)$ 是 $(Q, \pi^{(n)})$ Doob 过程的转移概率，则

$$\lim_{n \to +\infty} P_{ij}^{(n)}(t) = P_{ij}(t).$$

情形 $P < +\infty$, $S < +\infty$。设 X 是任意 Q 过程，且

$$\beta^{(n)}(\omega) = \inf\{t: t \geq \tau(\omega); x(t, \omega) \leq n\},$$

可以证明

$$P(\beta^{(n)} < +\infty) = 1 \quad (n \in \mathbf{N}^*),$$

记

$$z_0 = 0, \quad z_n = 1 + \sum_{k=1}^{n-1} \frac{a_1 a_2 \cdots a_k}{b_1 b_2 \cdots b_k},$$

$$z = \lim_{n \to +\infty} z_n \quad (\text{当 } R < +\infty \text{ 时}, z < +\infty).$$

定理 2 (i) 设给定满足 $R < +\infty$, $S < +\infty$ 的 Q 过程 X，则由

$$v_n = P(x(\beta^{(n)}) = n), \quad (n \in \mathbf{N}^*) \tag{11}$$

确定的数列 $\{v_n\}$ 满足条件

$$0 \leq v_1 \leq 1,$$

$$0 \leq \frac{v_{n+1}(z - z_{n+1})}{(z - z_n) - v_{n+1}(z_{n+1} - z_n)} \leq v_n \leq 1.$$

$$(n \in \mathbf{N}^*). \tag{12}$$

(ii) 反之，设给定满足(12)的数列$\{v_n\}$，则存在唯一的Q过程X满足(11).

定理 2 的(ii)中的 Q 过程 X 的转移概率 $P_{ij}(t)$ 可以按如下方式得到：用下面的递推公式给出集中在状态$(0, 1, 2, \cdots, n)$的分布 $\pi^{(n)} = (\pi_0^{(n)}, \pi_1^{(n)}, \cdots, \pi_n^{(n)})(n \in \mathbf{N}^*)$：

$$\pi_0^{(1)} = 1 - v_1, \quad \pi_1^{(1)} = v_1;$$

$$\pi_i^{(n+1)} = \pi_i^{(n)} \left(1 - v_{n+1} \frac{z_{n+1} - z_n}{z - z_n}\right) \quad (0 \leqslant i < n),$$

$$\pi_n^{(n+1)} = \pi_n^{(n)} \left(1 - v_{n+1} \frac{z_{n+1} - z_n}{z - z_n}\right) - v_{n+1} \frac{z - z_{n+1}}{z - z_n},$$

$$\pi_{n+1}^{(n+1)} = v_{n+1} \quad (n \in \mathbf{N}^*).$$

设 $P_{ij}^{(n)}(t)$ 是 $(Q, \pi^{(n)})$ Doob 过程的转移概率，则

$$\lim_{n \to +\infty} P_{ij}^{(n)}(t) = P_{ij}(t).$$

从定理 2 得出，所有的 Q 过程和所有满足(12)的序列$\{v_n\}$之间存在一一对应，对应关系由等式(11)给出.

现在设

$$v_i^{(n)} = P(x(\beta^{(n)}) = i) \quad (i = 0, 1, 2, \cdots, n; n \in \mathbf{N}^*),$$

$$c_{kj} = \begin{cases} \dfrac{z - z_k}{z - z_j}, & k > j, \\ 1, & k \leqslant j. \end{cases}$$

注意，c_{kj}是系统从状态k出发，经过有穷($\geqslant 0$)步到达状态j的概率(见[10]). 可以证明，存在极限

$$\lim_{n \to +\infty} \frac{\sum_{k=0}^{n-1} v_k^{(n)} c_{k0}}{\sum_{k=0}^{n} v_k^{(n)} c_{k0}} = p \quad (\geqslant 0);$$

$$\lim_{n \to +\infty} \frac{v_n^{(n)} c_{n0}}{\sum_{k=0}^{n} v_k^{(n)} c_{k0}} = q \quad (\geqslant 0).$$

我们按下面的方式定义非负数列 $\{r_n\}(n\in \mathbf{N})$：如果 $v_n^{(n)}=1(n\in \mathbf{N}^*)$，那么定义 $r_n=0(n\in \mathbf{N})$。如果存在 k 使 $v_j^{(j)}=1$ 对 $j\leqslant k$ 而 $v_{k+1}^{(k+1)}<1$，那么可以证明 $v_k^{(n)}>0$ 对一切 $n\geqslant k$，且 $\dfrac{v_i^{(n)}}{v_k^{(n)}}$ 与 $n>\max(i,k)(i\in \mathbf{N})$ 无关。定义

$$r_k = c \ (c\text{ 是任意正数}),$$
$$r_i = r_k \frac{v_i^{(n)}}{v_k^{(n)}} \ (n>\max(i,k),\ i\in \mathbf{N}).$$

显然地，除常数因子不计外，$r_0, r_1, r_2, \cdots$ 被 Q 过程决定，而 p, q 被 Q 过程唯一地决定。我们将称 $p, q, r_0, r_1, r_2, \cdots$ 为 Q 过程 X 的特征数列。

定理 3 (i) 设给定满足 $R<+\infty$，$S<+\infty$ 的 Q 过程 X，则其特征数列满足关系式

$$\begin{cases} p+q=1, \\ 0<\sum_{n=0}^{+\infty}r_n c_{n0}<+\infty,\ p>0; \\ r_n=0 \ (n\in \mathbf{N}), \quad p=0. \end{cases} \tag{13}$$

(ii) 反之，设给定满足 (13) 的非负数列 $p, q, r_0, r_1, r_2, \cdots$，则存在唯一的 Q 过程 X，其特征数列与给定的数列重合。

定理 3 的(ii)中的 Q 过程的转移概率 $p_{ij}(t)$ 可以按如下方式得到：按下面的公式给出集中在状态 $(0, 1, 2, \cdots, n)$ 上的分布

$$\pi^{(n)} = (\pi_0^{(n)}, \pi_1^{(n)}, \cdots, \pi_n^{(n)}) \quad (n\in \mathbf{N}^*),$$

如果 $p>0$，那么令

$$\pi_j^{(n)} = X_n \frac{r_j}{A_n},\ j=0, 1, 2, \cdots, n-1,$$

$$\pi_n^{(n)} = Y_n + X_n \frac{\sum_{m=n}^{+\infty} r_m c_{mn}}{A_n}.$$

其中

$$0 < A_n = \sum_{m=0}^{+\infty} r_m c_{mn} < +\infty,$$

$$X_n = \frac{pA_n(z-z_n)}{pA_n(z-z_n)+qA_0 z},$$

$$Y_n = \frac{qA_0 z}{pA_n(z-z_n)+qA_0 z};$$

如果 $p=0$，那么令

$$(\pi_0^{(n)}, \cdots, \pi_{n-1}^{(n)}, \pi_n^{(n)}) = (0, \cdots, 0, 1).$$

设 $P_{ij}^{(n)}(t)$ 是 $(Q, \pi^{(n)})$ Doob 过程的转移概率，则

$$\lim_{n \to +\infty} P_{ij}^{(n)}(t) = P_{ij}(t).$$

参考文献

[1] Austin D. On the existence of Markov transition probability functions. Proc. Nat. Acad. U.S.A., 1955, 41: 224-226.

[2] Chung K L. Foundations of the theory of continuous parameter Markov chains. Proc. Third Berkeley Symp. Math. Statistics and Prob., Berkeley: University of California Press, 1956, 2: 29-40.

[3] Добрушин Р Л. Об условиях регулярности однородных по времени Марковских процессов со счетным числом возможных состояний. УМН, 1952, 7(6): 185-191.

[4] Добрушин Р Л. Некоторые классы однородных счетных Марковскнх процессов. Теория Вероят. и ее Примен., 1957, 11(3): 377-380.

[5] Doob J L. Stochastic Process. New York: John Wiley & Sons, 1953.

[6] Doob J L. Markov chains—denumerable case. Trans. Am. Math. Soc., 1945, 58: 455-473.

[7] Feller W. On the integro-differential equations of purely discontinuous Markov processes. Trans. Am. Math. Soc. 1940, 48: 488-515. Errata. ibib., 1954, 58: 474.

[8] Feller W. Boundaries induced by non-negative matrices. Trans. Am. Math. Soc., 1956, 83: 19-54.

[9] Feller W. On boundaries and lateral conditions for the Kolmogorov differential equations. Ann. of Math., 1957, 65: 527-570.

[10] Harris T E. First passage and recurrence distribution. Trans. Am. Math. Soc., 1952, 73: 471-486.

[11] Karlin S, McGregor J. Representation of a class of stochastic processes. Proc. Nat. Acad. Sci., 1955, 41: 387-391.

[12] Karlin S, McGregor J. The differential equations of birth and death processes and the Stieltjies moment problem. Trans. Am. Math. Soc., 1957, 85: 489-546.

[13] Karlin S, McGregor J. The classification of birth and death processes. Trans. Am. Math. Soc., 1957, 86: 366-400.

[14] Колмогоров А Н. К вопросу о дпфференцируемости нереходиых вероятностей в однородных по времени процессах Маркова со счётным числом состояний. Учен. Зап. МГУ, Матем. 1951, 148(4): 53-59.

[15] Ledermann W. Reuter G E H. Spectral theory for the differential equations of simple birth and death

processes. Phil. Trans. Roy. Soc. London (Ser. A), 1954, 246: 321-369.

[16] Lévy P. Systèmes Markovients et stationnaires; cas dénumerable. Ann. Sci. Ecole Norm. Syp., 1951, 68(3): 327-381.

[17] Reuter G E H. Demumerable Markov processes and the associated contraction semigroups on I. Acta. Math., 1957, 97: 1-46.

[18] Reuter G E H. A note on contraction semigroups. Math. Scand., 1955, 3: 275-280.

[19] Юшкевич А А. О Дифференцируемости переходных вероятностей однородного марковского процесса со счетным числом состояний. Учен. Зап. Мгу, Матем., 1956, 186(9): 141-160.

杨向群译自俄文

On a Birth and Death Process[①]

Let (Ω, B, P) be a probability space, on which we define a separable (for definition of separability, see [3] Chap. 2) stationary Markov process $\{x(t, w), t \geq 0\}$ with denumerable states $0, 1, 2, \cdots$ and a fictive state $+\infty$, such that for any fixed $t \geq 0$, $P(x(t, w) = +\infty) = 0$. Its transition probabilities $P_{ij}(t) = P(x(t, w) = j \mid x(0, w) = i)$ satisfy the following conditions:

$$P_{ij}(t) \geq 0, \tag{1}$$

$$\sum_{j=0}^{+\infty} P_{ij}(t) = 1, \tag{2}$$

$$\sum_{j=0}^{+\infty} P_{ij}(t) P_{jk}(s) = P_{ik}(t+s), \tag{3}$$

$$\lim_{t \downarrow 0} P_{ij}(t) = P_{ij}(0) = \delta_{ij}, \tag{4}$$

where $t \geq 0$, $s \geq 0$, $\delta_{ij} = 1$, $\delta_{ij} = 0 (i \neq j)$. From (4) follows the

---

[①] Received: 1959-06-02.

All birth and death processes with a given density matrix Q are found in [6] without discovering what is the important unique Q-process satisfying [7][8].

existence of limits[2][3]:

$$\lim_{t\downarrow 0}\frac{P_{ij}(t)-\delta_{ij}}{t}=q_{ij}. \qquad (5)$$

Let $Q=(q_{ij})$ be a matrix with elements q_{ij}. We call Q the density matrix of this process. If Q has the form

$$Q=\begin{pmatrix} -b_0 & b_0 & 0 & \cdots & 0 & 0 & 0 & \cdots \\ a_1 & -(a_1+b_1) & b_1 & \cdots & 0 & 0 & 0 & \cdots \\ \vdots & \vdots & \vdots & \vdots & \vdots & \vdots & & \\ 0 & 0 & 0 & \cdots & a_n & -(a_n+b_n) & b_n & \cdots \\ \vdots & \vdots & \vdots & \vdots & \vdots & \vdots & & \end{pmatrix}, \qquad (6)$$

where $a_i>0$ $(i>0)$, $b_i>0$ $(i\geq 0)$ are constants, we say that $\{x(t, w), t\geq 0\}$ is a birth and death process. Put $R = \sum_{i=0}^{+\infty} m_i$;

$S=\sum_{i=1}^{+\infty}e_i$, where $\quad m_i = \dfrac{1}{b_i} + \sum_{k=0}^{i-1}\dfrac{a_i a_{i-1}\cdots a_{i-k}}{b_i b_{i-1}\cdots b_{i-k}b_{i-k-1}}$;

$$e_i = \frac{1}{a_i} + \sum_{k=0}^{+\infty}\frac{b_i b_{i+1}\cdots b_{i+k}}{a_i a_{i+1}\cdots a_{i+k}a_{i+k+1}}.$$

For the probability meaning of R, see [1]. Now we shall point out the probability meaning of S. Put

$$Q_N=\begin{pmatrix} -b_0 & b_0 & 0 & \cdots & 0 & 0 & 0 \\ a_1 & -(a_1+b_1) & b_1 & \cdots & 0 & 0 & 0 \\ \vdots & \vdots & \vdots & \vdots & \vdots & \vdots \\ 0 & 0 & 0 & \cdots & a_{N-1} & -(a_{N-1}+b_{N-1}) & b_{N-1} \\ 0 & 0 & 0 & \cdots & 0 & a_N+b_N & -(a_N+b_N) \end{pmatrix}$$

and let $x_N(t, w)$ be a stationary separable Markov process with states 0, 1, 2, $\cdots$, N and density matrix Q_N, $P(x_N(0, w) = N) = 1$. We have

Lemma If $\tau_N(w) = \inf(t : x_N(t, w) = 0)$, then when

$N \to +\infty$, the expectation of $\tau_N(w)$ tends to S.

Loosely speaking, for the process $\{x(t, w), t \geqslant 0\}$, S is the mean time of random motion from the fictive state $+\infty$ to the state 0; but R, by [1], is that from 0 to $+\infty$.

For a fixed w, the point $\tau(w)$ on the t-axis is called a transcendent jump point for the function $x(t, w)$, if for any given $\varepsilon > 0$, $x(t, w)$ takes infinite different values in the interval $[\tau(w) - \varepsilon, \tau(w)]$.

Now we turn to consider the opposite problem: if a matrix Q of form (6) is given, are there any birth and death processes, of which the $P_{ij}(t)$ satisfy (1)~(4), and of which the density matrix coincides with Q? An affirmative answer is given by [4], in which (also in [1]) it is proved that, if $R = +\infty$, such a process (we shall call it a Q process) is unique, and its $P_{ij}(t)$ satisfy also Kolmogorov's forward differential equations

$$\boldsymbol{P}'(t) = \boldsymbol{P}(t) \cdot \boldsymbol{Q}, \tag{7}$$

$$\boldsymbol{P}(0) = \boldsymbol{1}, \tag{8}$$

where $\boldsymbol{P}(t)$ ($\boldsymbol{P}'(t)$) is an infinite matrix with elements $P_{ij}(t)$ ($P'_{ij}(t)$), and $\boldsymbol{1}$ the identity. This unique Q process is found in [4]. But if $R < +\infty$, an infinite number of Q processes, but not all, are given in [2]. The following result is obtained in [5]: if $R < +\infty$, $S = +\infty$, there is no Q process satisfying (7)(8), but if $R < +\infty$, $S < +\infty$, there is a unique Q process, which satisfies (7)(8).

In this paper we want to construct this unique Q process, which satisfies (7)(8) under conditions $R < +\infty$, $S < +\infty$.

If $R < +\infty$, $S < +\infty$, we can prove that there exists a probability space (Ω, B, P) such that:

(i) On (Ω, B, P) we can define a sequence of Q processes $\{x_n(t, w), t \geqslant 0\}(n \in \mathbf{N}^*)$, where every $\{x_n(t, w), t \geqslant 0\}$ is defined by Q and by a discrete probability distribution $\Pi^{(n)} = \{\delta_{in}\}$ (i. e., the mass is concentrated on the state n) according to a method of Doob (see [3], P267, Theorem 2.5), and has transition probabilities $P_{ij}^{(n)}(t)$.

(ii) If $n > m$, then $\{x_n(t, w), t \geqslant 0\}$ is connected with $\{x_m(t, w), t \geqslant 0\}$ as follows: Let $\{\tau_k(w)\}$, $\{\beta_k(w)\}$ be two sequences of points on the t - axis, where $\tau_1(w)$ is the first transcendent jump point of $x_n(t, w)$, and $\beta_1(w) = \inf(t; t \geqslant \tau_1(w); x_n(t, w) \leqslant m)$; if $\tau_{k-1}(w)$, $\beta_{k-1}(w)$ are defined, we take $\tau_k(w)$ as the least transcendent jump point which exceeds $\beta_{k-1}(w)$, and $\beta_k(w) = \inf(t; t \geqslant \tau_k(w); x_n(t, w) \leqslant m)$. It is easy to prove that $P(\tau_k(w) < +\infty) = 1$, $P(\beta_k(w) < +\infty) = 1$. Now, if we cut off those pieces of curves of $x_n(t, w)$ (as a function of t) which correspond to the intervals $(\tau_k(w), \beta_k(w))$, transfer the remaining pieces but the first (which corresponds to $[0, \tau_1(w)]$) to the left so that their order remains as before, and the end point $\tau_k(w)$ of an interval coincides with the initial point $\beta_k(w)$ of the following interval, then in the result we obtain $x_m(t, w)$. This operation we denote by

$$g_{nm}[x_n(t, w)] = x_m(t, w). \tag{9}$$

Now we can state our main result:

Theorem If a matrix (6) satisfying conditions $R < +\infty$, $S < +\infty$ is given, then on (Ω, B, P) there is one and only one Q process $\{x(t, w), t \geqslant 0\}$, which has the following properties:

(i) there exists a measurable set C such that $P(C)=1$ and for any fixed $w\in C$, $x_n(t, w)$ tends to $x(t, w)$ almost everywhere (with respect to Lebesgue measures on the t - axis) as $n\to +\infty$;

(ii) for any k and any fixed t_1, t_2, $\cdots$, t_k, we have
$$P(x_n(t_i, w)\to x(t_i, w); i=1, 2, \cdots, k)=1;$$

(iii) let $P_{ij}(t)$ be the transition probabilities of $\{x(t, w), t\geqslant 0\}$, then $P_{ij}^{(n)}(t)\to P_{ij}(t)$ and $P_{ij}(t)$ satisfy Kolmogorov's forward differential equations (7)(8).

The construction of $\{x(t, w), t\geqslant 0\}$ can be understood intuitively as follows: Imagine that a particle A moves on the trajectory of $\{x_n(t, w), t\geqslant 0\}$; every time when it reaches the "boundary" $+\infty$, it goes immediately to the state n and continues its motion as before. Now put $n\to +\infty$, by the theorem we can imagine that, if A moves on the trajectory of $\{x(t, w), t\geqslant 0\}$, then when reaching $+\infty$, it goes back "continuously" to the finite states. The condition $R<+\infty$ ensures that it reaches $+\infty$ with probability 1, and the condition $S<+\infty$ ensures the possibility of its "continuous" return from $+\infty$ to the finite states.

References

[1] Добрушин Р Л. Усп. Матем. Наук, 1952, 7(6): 185-191.

[2] Doob J L. Trans. Am. Math. Soc., 1945, 58: 455-473.

[3] Doob J L. Stochastic Process. New York: John Wiley and Sons, 1953.

[4] Feller W. Trans. Am. Math. Soc., 1940, 48: 488-515; Trans. Am. Math. Soc., 1945, 58: 474.
[5] Reuter G E H. Acta Math., 1957, 97: 1-46.
[6] Ван Цзы-кун. Научные доклады высшей школы. физико-Матем Науки, 1958, (4): 19-25.

On Distributions of Functionals of Birth and Death Processes and Their Applications in the Theory of Queues[①]

(I) Let $x(t, w)$ ($0 \leqslant t < +\infty$) be a birth and death process (for definition see [1] or [2]) with the density matrix Q of transition probabilities:

$$Q = \begin{pmatrix} -b_0 & b_0 & 0 & 0 & \cdots \\ a_1 & -(a_1+b_1) & b_1 & 0 & \cdots \\ 0 & a_2 & -(a_2+b_2) & b_2 & \cdots \\ \vdots & \vdots & \vdots & \vdots & \vdots \end{pmatrix}, \quad (1)$$

where $b_i > 0$ ($i \geqslant 0$), $a_i > 0$ ($i > 0$). According to [3]~[5] we can suppose that this process is a separable and almost strong Markov process[②]. This hypothesis has no influence on the transition probabilities, nor therefore, on matrix Q, but warrants the legitimacy of some subsequent operations. Let 0, 1,

---

① Received: 1960-09-15.
② In [4] it is called "Почти строго марковский цроцесс".

2, ⋯ denote the states of this process, on which a non-negative function[1] $V(k)$ ($k \in \mathbf{N}$) is defined. The main aim of this note is to study the distributions and moments of the random functionals

$$\xi^{(n)}(w) = \int_0^{\tau^{(n)}} V[x(t)]dt, \qquad (2)$$

and

$$\xi(w) = \lim_{n \to +\infty} \xi^{(n)}(w), \qquad (3)$$

where $x(t) = x(t, w)$, and $\tau^{(n)} = \tau^{(n)}(w)$ is the first moment when the process reaches the state n, in other words,

$$\tau^{(n)}(w) = \inf(t: x(t, w) = n). \qquad (4)$$

Some results of this note may find applications in the theory of queues.

It will be seen that the following determinant of order n plays an important role:

$$\delta_n(\lambda) =$$

$$\begin{vmatrix} -(\lambda V(0)+a_0+b_0) & b_0 & 0 & 0 & \cdots & 0 & 0 \\ a_1 & -(\lambda V(1)+a_1+b_1) & b_1 & 0 & \cdots & 0 & 0 \\ 0 & a_2 & -(\lambda V(2)+a_2+b_2) & b_2 & \cdots & 0 & 0 \\ \vdots & \vdots & \vdots & \vdots & \vdots & \vdots & \vdots \\ 0 & 0 & 0 & \cdots & a_{n-2} & -(\lambda V(n-2)+a_{n-2}+b_{n-2}) & b_{n-2} \\ 0 & 0 & 0 & \cdots & 0 & a_{n-1} & -(\lambda V(n-1)+a_{n-1}+b_{n-1}) \end{vmatrix}$$

where λ is a real parameter and $a_0 = 0$. Expanding it according to the last row, we have

$$\delta_n(\lambda) = -(\lambda V(n-1)+a_{n-1}+b_{n-1})\delta_{n-1}(\lambda) - a_{n-1}b_{n-2}\,\delta_{n-2}(\lambda),$$
$$\delta_1(\lambda) = -(\lambda V(0)+a_0+b_0), \qquad (5)$$

---

[1] It is supposed that $V(k) \not\equiv 0$ unless specially stated otherwise.

$\delta_0(\lambda) = 1$ (say).

We denote by $\delta_n^{(k)}(\lambda)$ the determinant obtained by substituting the column vector $(0, 0, \cdots, 0, -b_{n-1})^T$ for the kth column in $\delta_n(\lambda)$ (A' represents the transposed matix of matix A).

Let us prove two lemmas first:

Lemma 1 There exists a constant $\theta > 0$ such that $\delta_n(\lambda)$ is not zero and has the same sign as $(-1)^n$, when $\lambda > -\theta$.

Proof The lemma is trivial for $\delta_0(\lambda)$ and $\delta_1(\lambda)$. Suppose that it holds for all $\delta_k(\lambda)$ ($k \leqslant n-1$), then we prove that it holds also for $\delta_n(\lambda)$. Since $\delta_n(\lambda)$ is a continuous function of λ, it is sufficient to show that $\delta_n(\lambda)$ has the same sign as $(-1)^n$ when $\lambda \geqslant 0$. Consider

$$\frac{d\delta_n(\lambda)}{d\lambda} = -V(0)\delta_{1,1}(\lambda) + V(1)(\lambda V(0) + a_0 + b_0)\tilde{\delta}_{2,2}(\lambda) -$$

$$V(2)\delta_{3,3}(\lambda) - V(3)\delta_{4,4}(\lambda) - \cdots - V(n-3)\delta_{n-2,n-2}(\lambda) +$$

$$V(n-2)[\lambda V(n-1) + a_{n-1} + b_{n-1}]\delta_{n-2}(\lambda) -$$

$$V(n-1)\delta_{n-1}(\lambda), \qquad (6)$$

where $\delta_{i,i}(\lambda)$ is a determinant of order $n-1$ obtained from $\delta_n(\lambda)$ by omitting the kth row and kth column, and $\tilde{\delta}_{2,2}$ is a determinant of order $n-2$ obtained from $\delta_n(\lambda)$ by omitting the first two rows and first two columns. Since the sign of $\delta_k(\lambda)$ ($\lambda \geqslant 0$) depends not on the exact values of its elements, but only on their signs, the determinants $\delta_{1,1}(\lambda)$, $\tilde{\delta}_{2,2}(\lambda)$, $\delta_{3,3}(\lambda)$, $\delta_{4,4}(\lambda)$, $\cdots$, $\delta_{n-2,n-2}(\lambda)$ have the same signs as those of $\delta_{n-1}(\lambda)$, $\delta_{n-2}(\lambda)$, $\delta_2(\lambda) \cdot \delta_{n-3}(\lambda)$, $\delta_3(\lambda) \cdot \delta_{n-4}(\lambda)$, $\cdots$, $\delta_{n-3}(\lambda) \cdot \delta_2(\lambda)$ respectively. By hypothesis of induction we see that every term on the right-hand side of (6) has the same sign as $(-1)^n$, and therefore, the sign of $\dfrac{d\delta_n(\lambda)}{d\lambda}$ coincides also with that of $(-1)^n$.

Since $\dfrac{d\delta_n(\lambda)}{d\lambda}$ is continuous, we see that when $\lambda>0$,

$$\delta_n(\lambda)=\int_0^\lambda \frac{d\delta_n(x)}{dx}dx,$$

also has the same sign as $(-1)^n$.

Finally, if $\lambda=0$, by (5) and induction it is easy to prove that

$$\delta_n(0)=(-1)^n b_{n-1}b_{n-2}\cdots b_1 b_0.$$

Lemma 2 If $f_i^{(n)}\geqslant 0$, $f_i^{(n)}\uparrow f_i$, then

$$\lim_{n\to+\infty}\left(\sum_{i=0}^{n-1}f_i^{(n)}\right)=\sum_{i=0}^{+\infty}f_i.$$

Proof Clearly we have

$$\lim_{n\to+\infty}\left(\sum_{i=0}^{n-1}f_i^{(n)}\right)\leqslant\sum_{i=0}^{+\infty}f_i.$$

Moreover, for an arbitrary positive integer m and for $\varepsilon>0$, if all f_i are finite, then we are able to find an $n>m$, such that for all $i=0, 1, \cdots, m$,

$$f_i^{(n)}>f_i-\frac{\varepsilon}{m+1},$$

and therefore

$$\lim_{n\to+\infty}\left(\sum_{i=0}^{n-1}f_i^{(n)}\right)\geqslant\sum_{i=0}^{n-1}f_i^{(n)}>\sum_{i=0}^{m}f_i-\varepsilon.$$

Since both m and ε are arbitrary, Lemma 2 holds when all f_i are finite. Finally, if there is at least one $f_i=\infty$, then the lemma is trivial.

(Ⅱ) Consider a fixed positive integer n. Put

$$\varphi_{k,n}(\lambda)=M_k e^{-\lambda\xi^{(n)}(w)}, \qquad (7)$$

where λ is a real parameter, and M_k denotes the conditional expectation when the initial distribution is concentrated on the state k. Let $c_k = a_k + b_k$, and

$$h = \min_{k \leq n-1, \theta} \left(\frac{c_k}{V(k)}; \theta \right) > 0,$$

where θ is defined by Lemma 1 $\left(\text{put } \frac{c}{0} = +\infty, \text{ if } c > 0 \right)$.

Theorem 1 (i) All $\varphi_{k,n}(\lambda)$ ($k \leq n$) are finite for $\lambda > -h$;

(ii) They are the unique solutions of the system of equations

$$a_k \varphi_{k-1,n}(\lambda) - (a_k + b_k) \varphi_{k,n}(\lambda) + b_k \varphi_{k+1,n}(\lambda) - \lambda V(k) \varphi_{k,n}(\lambda) = 0$$
$$(0 \leq k \leq n-1), \qquad (8)$$
$$\varphi_{n,n}(\lambda) = 1,$$

for $\lambda > -h$, that is,

$$\varphi_{k,n}(\lambda) = \frac{\delta_n^{(k+1)}(\lambda)}{\delta_n(\lambda)} \quad (0 \leq k \leq n, \ \delta_n^{(n+1)}(\lambda) = \delta_n(\lambda)). \qquad (9)$$

Proof We first prove (ii) by supposing that (i) holds. Let β be the sojourn time of the process at state k (β depends on k). It is well known that

$$P_k(\beta \leq t) = \begin{cases} 0, & t < 0, \\ 1 - e^{-c_k t}, & t \geq 0, \end{cases} \qquad (10)$$

where P_k denotes the conditional probability when the initial distribution is concentrated on state k. Moreover, we have

$$P_k(x(\beta+0) = k+1) = \frac{b_k}{c_k},$$
$$P_k(x(\beta+0) = k-1) = \frac{a_k}{c_k}. \qquad (11)$$

By (11), the strong Markov property, and (10), we have for $k < n$:

$$\varphi_{k,n} = M_k \left(e^{-\lambda \int_0^{\tau^{(n)}} V[x(t)] dt} \right)$$

$$= \frac{b_k}{c_k} M_k (e^{-\lambda \int_0^\beta V[x(t)]dt - \lambda \int_\beta^{\tau^{(n)}} V[x(t)]dt} \mid x(\beta+0) = k+1) +$$

$$\frac{a_k}{c_k} M_k (e^{-\lambda \int_0^\beta V[x(t)]dt - \lambda \int_\beta^{\tau^{(n)}} V[x(t)]dt} \mid x(\beta+0) = k-1)$$

$$= \frac{b_k}{c_k} M_k e^{-\lambda V(k)\beta} M_{k+1} e^{-\lambda \int_0^{\tau^{(n)}} V[x(t)]dt} +$$

$$\frac{a_k}{c_k} M_k e^{-\lambda V(k)\beta} M_{k-1} e^{-\lambda \int_0^{\tau^{(n)}} V[x(t)]dt}$$

$$= \frac{b_k}{\lambda V(k) + a_k + b_k} \varphi_{k+1,n}(\lambda) + \frac{a_k}{\lambda V(k) + a_k + b_k} \varphi_{k-1,n}(\lambda).$$

It is assumed that $\lambda > -\frac{c_k}{V(k)}$ in the last step of the calculation. This result verifies the first n equalities of (8). The last equality is obvious by the definition of $\varphi_{n,n}(\lambda)$. By Lemma 1, the coefficient determinant of equations (8) is not zero when $\lambda > -h$. Therefore, $\varphi_{k,n}(\lambda)$ ($k=0, 1, 2, \cdots, n$) furnishes the unique solution of (8). Then solving (8) we get (9).

Now we turn to prove (i). By (7), (i) is trivial when $\lambda \geq 0$. Let $0 > \lambda > -h$. In order to prove (i) we use an idea of [6]. We know from Lemma 1 that (8) has an unique solution $\varphi(k)$, $k=0, 1, 2, \cdots, n$. It can be proved that

$$\varphi(k) = -\lambda M_k \int_0^{\tau^{(n)}} \varphi(x_t) V(x_t) dt + 1, \qquad (12)$$

where $x_t = x(t) = x(t, \omega)$. In fact, consider the generalized infinitesimal operator A of the process defined by

$$A\varphi(k) = \frac{M_k \varphi(x_\beta) - \varphi(k)}{M_k \beta}. \qquad (13)$$

From (10) and (11) we obtain

$$A\varphi(k) = \left[\frac{b_k}{c_k} \varphi(k+1) + \frac{a_k}{c_k} \varphi(k-1) - \varphi(k) \right] \div \frac{1}{c_k}$$

$$= a_k\varphi(k-1) - (a_k+b_k)\varphi(k) + b_k\varphi(k+1).$$

Therefore, equations (8) can be rewritten as

$$A\varphi(k) - \lambda V(k)\varphi(k) = 0, \quad 0 \leqslant k \leqslant n-1, \tag{14}$$
$$\varphi(n) = 1.$$

Let $\Psi(k)$ denote the function on the right-hand side of (12). Clearly $\Psi(n)=1$. By (13) and (14) we have, when $k \leqslant n-1$,

$$A\Psi(k) = \frac{M_k \Psi(x_\beta) - \Psi(k)}{M_k \beta}$$

$$= \frac{-\lambda M_k M_{x_\beta} \int_0^{\tau^{(n)}} \varphi(x_t) V(x_t) dt + \lambda M_k \int_0^{\tau^{(n)}} \varphi(x_t) V(x_t) dt}{M_k \beta}$$

$$= \frac{\lambda M_k \int_0^{\beta} \varphi(x_t) V(x_t) dt}{M_k \beta} = \lambda V(k)\varphi(k) = A\varphi(k).$$

Therefore, by the linearity property of A, we have

$$A[\Psi(k) - \varphi(k)] = 0, \quad 0 \leqslant k \leqslant n-1.$$

The coefficient determinant of these linear equations is $\delta_n(0)$, which is not zero by Lemma 1, so that the system has only a trivial solution, i.e.,

$$\Psi(k) = \varphi(k), \quad 0 \leqslant k \leqslant n-1.$$

Moreover,

$$\Psi(n) = 1 = \varphi(n).$$

Hence the equality (12) is proved. In order to prove $M_k e^{-\lambda \int_0^{\tau^{(n)}} V(x_t) dt} < +\infty$, when $-h < \lambda < 0$, define

$$u_0(k) \equiv 1,$$
$$u_m(k) = -\lambda M_k \int_0^{\tau^{(n)}} V(x_t) u_{m-1}(x_t) dt + 1. \tag{15}$$

By (12) and induction, we have

$$u_m(k) \leqslant \varphi(k). \tag{16}$$

On the other hand, it can be proved that[6]

$$u_m(k) = \sum_{s=0}^{m} M_k \frac{\left[-\lambda \int_0^{\tau^{(n)}} V(x_t) dt\right]^s}{s!} \uparrow M_k e^{-\lambda \xi^{(n)}(\omega)}, \quad (17)$$

so that

$$M_k e^{-\lambda \xi^{(n)}(\omega)} \leqslant \varphi(k). \quad (18)$$

Corollary 1 Equations (8) have an unique positive solution when $\lambda > -h$; all determinants $\delta_n^{(k)}(\lambda)$ ($k=1, 2, \cdots, n$) have the same sign as $(-1)^n$.

Proof These results follow immediately from (18) and (9).

As the distribution of $\xi^{(n)}(\omega)$ is uniquely determined by $\varphi_{k,n}(\lambda)$ (see [8, P38] and Theorem 1. (i)) which we have obtained, the problem of finding the distribution of $\xi^{(n)}(\omega)$ has been already solved (with respect to the measure p_k, $k \leqslant n$).

Now we turn to study the moments $M_k [\xi^{(n)}(\omega)]^l$ (l is a positive integer). Their existence follows from Theorem 1 (i), and then

$$M_k [\xi^{(n)}(\omega)]^l = (-1)^l \varphi_{k,n}^{(l)}(0), \quad (19)$$

where $\varphi^{(l)}(0) = \dfrac{d^l}{d\lambda^l} \varphi(\lambda) \bigg|_{\lambda=0}$. We denote $M_k [\xi^{(n)}(\omega)]^l$ by $m_{k,n}^{(l)}$.

Theorem 2 When $k \leqslant n-1$,

$$m_{k,n}^{(l)} = \sum_{i=k}^{n-1} G_{i,n}^{(l)}, \quad m_{n,n}^{(l)} = 0, \quad (20)$$

where

$$G_{i,n}^{(l)} = \frac{lV(i) m_{i,n}^{(l-1)}}{b_i} + \sum_{k=0}^{i-1} \frac{a_i a_{i-1} \cdots a_{i-k} lV(i-k-1) m_{i-k-1,n}^{(l-1)}}{b_i b_{i-1} \cdots b_{i-k} b_{i-k-1}}.$$

$$(21)$$

Proof Differentiating (8) l times with respect to λ, putting $\lambda=0$, and multiplying by $(-1)^l$, we have from (19)

$$a_k m_{k-1,n}^{(l)} - (a_k+b_k) m_{k,n}^{(l)} + b_k m_{k+1,n}^{(l)} + lV(k) m_{k,n}^{(l-1)} = 0,$$
$$0 \leq k \leq n-1, \quad (22)$$
$$m_{n,n}^{(l)} = 0.$$

Solving these equations, we have

$$m_{k,n}^{(l)} = \frac{\widetilde{\delta}_n^{(k+1)}(0)}{\delta_n(0)}, \quad 0 \leq k \leq n-1, \quad (23)$$

where $\widetilde{\delta}_n^{(k+1)}$ is a determinant obtained from $\delta_n(0)$ by substituting the vector $-l(V(0) m_{0,n}^{(l-1)}, \cdots, V(n-1) m_{n-1,n}^{(l-1)})^{\mathrm{T}}$ for the $(k+1)$th column. Expanding the two determinants in (23), we have (20) and (21), as was to be proved.

Therefore, $m_{k,n}^{(l)}$ can be expressed by $m_{k,n}^{(l-1)}$. In the special case when $l=1$, $G_{i,n}^{(1)}$ does not depend on n. We denote it by G_i, then

$$G_i = \frac{V(i)}{b_i} + \sum_{k=0}^{i-1} \frac{a_i a_{i-1} \cdots a_{i-k} V(i-k-1)}{b_i b_{i-1} \cdots b_{i-k} b_{i-k-1}}. \quad (24)$$

If, moreover, $V \equiv 1$, then

$$m_{k,n}^{(1)} = M_k \tau^{(n)} = \sum_{i=k}^{n-1} \left(\frac{1}{b_i} + \sum_{k=0}^{i-1} \frac{a_i a_{i-1} \cdots a_{i-k}}{b_i b_{i-1} \cdots b_{i-k} b_{i-k-1}} \right). \quad (25)$$

Formula (25) was first found in [9].

(Ⅲ) Now consider the limiting case. Let

$$\tau(\omega) = \lim_{n \to +\infty} \tau^{(n)}(\omega), \quad (26)$$

$$\xi(\omega) = \lim_{n \to +\infty} \xi^{(n)}(\omega) = \int_0^\tau V[x(t)] dt. \quad (27)$$

These limits exist by the monotonic property, but they can equal $+\infty$ with positive probability. By the monotonic convergence theorem we have

$$m_k = M_k \xi(\omega) = \lim_{n \to +\infty} m_{k,n}^{(1)} = \sum_{i=k}^{+\infty} G_i. \quad (28)$$

For $\lambda > 0$ let

$$\varphi_k(\lambda) = \lim_{n \to +\infty} \varphi_{k,n}(\lambda) = M_k(e^{-\lambda \xi(\omega)}). \tag{29}$$

Theorem 3 For any integer $k \geqslant 0$, there are only two possibilities:

(i) either $P_k(\xi(\omega) = +\infty) = 1$, which holds if and only if

$$M_0 \xi(\omega) = \sum_{i=0}^{+\infty} G_i = +\infty;$$

(ii) or $P_k(\xi(\omega) < +\infty) = 1$, which holds if and only if

$$M_0 \xi(\omega) = \sum_{i=0}^{+\infty} G_i < +\infty.$$

In case (ii)

$$\varphi_k(\lambda) = \lim_{n \to +\infty} \frac{\delta_n^{(k+1)}(\lambda)}{\delta_n(\lambda)}, \quad k \geqslant 0, \tag{30}$$

which is the unique (with an unspecified constant multiplier), bounded, nontrivial solution of the following equations:

$$a_k \varphi_{k-1}(\lambda) - (a_k + b_k) \varphi_k(\lambda) + b_k \varphi_{k+1}(\lambda) - \lambda V(k) \varphi_k(\lambda) = 0,$$
$$k \geqslant 0. \tag{31}$$

Proof Since $\lambda > 0$, $0 \leqslant \varphi_k(\lambda) \leqslant 1$. As we have done in Theorem 1, it can be proved that $\varphi_k(\lambda)$ ($k \geqslant 0$) should satisfy (31).

Besides the trivial solution, (31) can have only one linearly independent solution, which may be unbounded. Take $\varphi_0(\lambda) \geqslant 0$ arbitrarily. By Lemma 7 of [10], this solution is bounded if and only if

$$\sum_{n=1}^{+\infty} \left[\lambda \left(\frac{V(n)}{b_n} + \frac{a_n V(n-1)}{b_n b_{n-1}} + \frac{a_n a_{n-1} V(n-2)}{b_n b_{n-1} b_{n-2}} + \cdots + \frac{a_n a_{n-1} \cdots a_2 V(1)}{b_n b_{n-1} \cdots b_2 b_1} \right) + \frac{a_n a_{n-1} \cdots a_1}{b_n b_{n-1} \cdots b_1} \right] < +\infty.$$

The last condition is equivalent to $\sum_{i=0}^{+\infty} G_i < +\infty$ when $V \not\equiv 0$.

If $\sum_{i=0}^{+\infty} G_i = +\infty$, $\{\varphi_k(\lambda)\}$ can only be the trivial solution since $\{\varphi_k(\lambda)\}$ is bounded, that is, $\varphi_k(\lambda) \equiv 0$ ($k \geqslant 0$, $\lambda > 0$). Therefore $P_k(\xi(\omega) = +\infty) = 1$.

If $\sum_{i=0}^{+\infty} G_i < +\infty$, $\{\varphi_k(\lambda)\}$ is either a trivial solution, or a nontrivial bounded solution. But is can not be the trivial solution, or otherwise, as we stated above, we would have $P_k(\xi(\omega) = +\infty) = 1$, $M_k \xi(\omega) = \sum_{i=k}^{+\infty} G_i = +\infty$, contradicting $\sum_{i=0}^{+\infty} G_i < +\infty$. Therefore, $\{\varphi_k(\lambda)\}$ must be the unique bounded nontrivial solution. In order to find this solution, we note that $\varphi_{k,n}(\lambda) \downarrow \varphi_k(\lambda)$ ($n \to +\infty$) when $\lambda > 0$, and then obtain (30) from (9).

Now consider the moments of $\xi(\omega)$:
$$m_k^{(l)} = M_k [\xi(\omega)]^l = \lim_{n \to +\infty} m_{k,n}^{(l)}.$$

Let
$$G_i^{(l)} = \lim_{n \to +\infty} G_{i,n}^{(l)}$$
$$= \frac{lV(i) m_i^{(l-1)}}{b_i} + \sum_{k=0}^{i-1} \frac{a_i a_{i-1} \ldots a_{i-k} lV(i-k-1) m_{i-k-1}^{(l-1)}}{b_i b_{i-1} \ldots b_{i-k} b_{i-k-1}}.$$
(32)

Theorem 4 For all integers $k \geqslant 0$, $l > 0$:

(i) $m_k^{(l)} = \sum_{i=k}^{+\infty} G_i^{(l)}$;

(ii) $m_k^{(n)} \leqslant n! \, (m_0)^n$;

(iii) all $m_k^{(l)}$ are either infinite or finite, and they are finite if and only if $m_0 = \sum_{i=0}^{+\infty} G_i < +\infty$.

Proof By Lemma 2 we have

On Distributions of Functionals of Birth and Death Processes and Their Applications in the Theory of Queues

$$m_k^{(l)} = \lim_{n \to +\infty} m_{k,n}^{(l)} = \lim_{n \to +\infty} \Big(\sum_{i=k}^{n-1} G_{i,n}^{(l)} \Big) = \sum_{i=k}^{+\infty} G_i^{(l)}. \quad (33)$$

Since $m_0 \geqslant m_1 \geqslant m_2 \geqslant \cdots$, and from (33) and (32), it follows that

$$m_k^{(2)} \leqslant 2m_0 \Big(\sum_{i=k}^{+\infty} G_i \Big) = 2(m_0)^2.$$

Suppose that $m_k^{(n-1)} \leqslant (n-1)! \, (m_0)^{n-1}$, and from (33), (32) and $m_0^{(n-1)} \geqslant m_1^{(n-1)} \geqslant \cdots$, it follows that

$$m_k^{(n)} \leqslant n m_0^{(n-1)} m_k \leqslant n! \, (m_0)^n. \quad (34)$$

(iii) follows immediately from (ii)(i) and (32).

Corollary 2 The distribution function $F_k(x) = P_k(\xi(\omega) \leqslant x)$ is uniquely determined by its moments $m_k^{(l)}$ ($l \in \mathbf{N}$; $m_k^{(0)} = 1$).

Proof By (34), when $r < \dfrac{1}{m_0}$, the series

$$\sum_{n=0}^{+\infty} \frac{m_k^{(n)}}{n!} r^n \leqslant \sum_{n=0}^{+\infty} (m_0 r)^n < +\infty,$$

hence from a theorem in § 15.4 of [11], the corollary holds if $m_k < +\infty$ (and therefore, $m_0 < +\infty$). If $m_k = +\infty$, then $P_k(\xi(\omega) = +\infty) = 1$ by Theorem 3 (i).

(Ⅳ) We give two examples to show how our results find applications in the theory of queues ($[m \backslash m \backslash n]$ system).

Example 1 Let $V(0) = 1$, $V(k) = 0 (k > 0)$. Then $\xi^{(n)}(\omega)$ is the total sojourn time of state 0 before reaching the state n (in terms of the theory of queues, this is the total free time before n service facilities being occupied). G_i in (24) reduces to g_i:

$$g_0 = \frac{1}{b_0}, \qquad g_i = \frac{a_i a_{i-1} \cdots a_1}{b_i b_{i-1} \cdots b_1 b_0}.$$

$$\varphi_{0,n}(\lambda) = M_0 \, e^{-\lambda \xi^{(n)}} = \frac{1}{\lambda \sum_{i=0}^{n-1} g_i + 1} \quad \text{(see (9))}.$$

By Laplace transform we have[①]

$$P_0(\xi^{(n)} \leqslant x) = \begin{cases} 0, & x<0, \\ 1-e^{-\frac{x}{\sum_{i=0}^{n-1} g_i}}, & x \geqslant 0. \end{cases}$$

$$M_0 \xi^{(n)} = \sum_{i=0}^{n-1} g_i \quad (\text{see } (20)).$$

From Theorem 3 it follows that $P_0(\xi < +\infty) = 1$ if and only if $M_0\xi = \sum_{i=0}^{+\infty} g_i < +\infty$; in this case

$$\varphi_0(\lambda) = \lim_{n \to +\infty} \varphi_{0,n}(\lambda) = \frac{1}{\lambda \sum_{i=0}^{+\infty} g_i + 1},$$

$$P_0(\xi \leqslant x) = \begin{cases} 0, & x<0, \\ 1-\exp\left\{-\frac{x}{\sum_{i=0}^{+\infty} g_i}\right\}, & x \geqslant 0. \end{cases}$$

Example 2 If we are interested in the total time before n service facilities being firstly occupied, we need only take $V(k)=1$ ($k<n$), $V(n)=0$. Applying the theorems above, we find the distribution of $\xi^{(n)}(\omega)$.

References

[1] Wang Tzu-kwen. Science Record, New Ser, 1959, 3: 266-268.

[2] Feller W. An Introduction to Probability Theory and Its Applications. 1957.

[3] Doob J L. Stochastic Processes. 1953.

---

① This result has also been obtained by Wu Li-de independently.

[4] Юшкевич А А. Теория Вероят и ее Nримен, 1960, 2: 187-213.

[5] Chung K L. Ann. of Math. , 1958, 68: 126-149.

[6] Хасьминский Р З. Теория Вероят и ее Nримен, 1959, 4: 332-341.

[7] Дыикин Е В. Теория Вероят и ее Nримен, 1956, 1: 38-60.

[8] Wilks S S. Mathematical Statistics. 1944.

[9] Добрушин Р. Л Усп. Мамем. Наун, 1952, 7: 185-191.

[10] Reuter G E H. Acta Math , 1957, 97: 1-46.

[11] Gramer H. Mathematical Methods of Statistics. 1946.

随机泛函分析引论

§1. 引 言

随机泛函分析是概率论与泛函分析交界的边缘学科．产生的原因主要有二：一是由于概率论研究对象的日益扩大，古典概率论主要研究随机变数，这远不能满足其他学科与技术的需要，应该研究一般的随机元，例如随机的曲线、连续函数、可积函数等，这些元素已不再是实数或复数了．因此，有必要建立一般的理论，以研究抽象空间中的随机元，这样，就必须用到泛函分析中的方法和成果．其次是由于实际中不断地提出随机方程，在这些方程的内部或边界条件中包含着随机的函数．正如泛函分析以一般的算子理论来研究方程一样，有必要建立一般的随机算子的理论，或者说，随机变换的理论，来研究随机方程，例如随机积分方程等．基于这些原因，研究随机泛函分析的人日益增多，然而由于它的历史很短（就笔者所知，1956年才正式提出随机泛函分析的名称），目前它还处于发展的早期阶段，即主要是用泛函分析来解决一些概率论中的问题的阶段，关于随机变换及其对随机方程的应用还研究得很少．因此，它的内容范围，也远没有定型，不过，如一般所设想的[14]，它至少应该包含 Banach 空间中随机元、随机变换与随机广义函数三个方面．作为对这一学科的介绍，本文就依此而分为三节．§2～§3 是捷克数学家们特别是 Otto Hanš 的工作，这里基本上包含了[15～18]中的主要结果与方法．§4 研究广义函数空间中的随机元，由于全体广义函数不构成 Banach 空间，故不能利用§2～§3 的主要结果．与§2～§3 不同，本章内所有结果均由笔者得到而系初次发表．现在分别介绍各章中主要内容与研究情况．

随机元的定义各家不同．E. Mourier[19] 称自可测空间 (Ω, σ) 到 Banach 空间 E 的映像 V 为一随机元，如果对每有界线性泛函 $f \in E^*$（E 的共轭空间），$f(V)$ 是一随机变数．当 E 为可分 Banach 空间时，此定义与 O. Hanš 的定义一致，但在一般的 Banach 空间则不同．А. Н. Колмогоров 和 Ю. В. Прохоров[20] 所引进的概念则是随机过程的一般化，把随机元看成可测空间中的测度，我国在此方面工作的有胡国定、郑曾同[21]等人．本文中考虑的是概率 1 收敛，在[20]中则研究可测距离空间中测度列的收敛，相当于随机变数列依分布收敛的一般化．因此，在关于极限定理的强弱这一点上，正好彼此补充．§2 在讨论随机元的一般性质后，主要研究随机元列的概率 1 的(χ)-收敛，然后利用所得结果以研究平衡序列．这里还应该提到 L. E. Dubins 关于随机元的工作[22]，他的定义与上面提到的均不同．

§3 讨论随机变换，证明了随机的不动点原理、Banach-Hahn 定理、$C[0,1]$ 中随机线性泛函的表现定理和逆变换、共轭变换的可测性，为研究随机泛函方程做了基本的准备工作[14]．

广义函数空间中的随机元是广义过程. 广义过程的定义也各家不同. [4]中定义广义过程为基本空间 Φ 上的线性连续随机泛函, 收敛性指依概率收敛; [5]中称广义过程为取值于 $L^2(\Omega)$ 中的线性连续随机泛函, 收敛性指均方收敛; 而依[6]中的定义, 则广义过程的每一现实是 J. Minkusinski, R. Sikorski 观点下的广义函数[8], 我国王寿仁在[23]、郑绍濂在[7]的工作中, 即采用这种定义; 本文采用[9][10][11]中提出的定义, 其特点在于使广义过程的每一现实是 L. Schwartz 意义下的广义函数[1][2][3]. §4 中, 引进了广义过程的条件数学期望的定义, 合理的定义应该具备两个条件: 既要保留随机变数的条件期望的基本性质, 又应满足极限定理的需要. 从这两点出发, 不期发现, 广义过程的条件期望应该这样定义, 它在形式上与随机变数的条件期望完全一样, 接着研究了它的存在与表现形式. 在这项工作完成以后, 自然地想到应该类似地研究 Banach 空间中随机元的条件数学期望, 后来知道这项工作已经在[24]中完成. 在§4.3 中, 得到了随机元列的概率 1 收敛定理, 然后应用它于平稳广义过程序列, 以证明对后者的加强大数定理.

为便于阅读, 证明的叙述相当详细, 文中所需的泛函分析知识, 可在任何一本泛函分析书如[25]中找到, 概率论知识则(至少形式上)不要求, 测度论知识见[27], 例如, 可测空间的定义见[27]§17.

最后, 关于随机泛函分析进一步的问题, 这里只提出随机积分方程的研究, 看来这是不可避免的课题, 但目前已有的工作还是非常初步的(参看[14]).

§2. 随机元

§2.1. 随机元的基本性质

定义 1 设 (Ω, σ),$(X, \mathscr{B})$ 为二非空可测空间,又 V 为自全 $\Omega=(\omega)$ 到 $X=(x)$ 的映像,如果对任意的 $B\in\mathscr{B}$,有 $(\omega: V(\omega)\in B)\in\sigma$,那么称 V 为界定于 Ω 上而取值于 X 中的随机元,简记为 (Ω,σ)-$(X,\mathscr{B})$(或 Ω-X)随机元,或称随机变量.

特别,若 $(X,\mathscr{B})$ 为一维 Borel 可测空间,即 $X=(-\infty, +\infty)$,$\mathscr{B}$ 为含 $(-\infty, +\infty)$ 中全体开集的最小 σ-代数,则此时 Ω-X 随机元即通常的随机变数.

有时为了在 X 中考虑收敛性,有必要引进拓扑. 称 $(X, C, \mathscr{B})$ 为可测拓扑空间,如果 (X, C) 为拓扑空间,$(X, \mathscr{B})$ 为可测空间,而且① $\mathscr{B}=\sigma(C)$,这里 C 是 X 中全体开集所成的集. 显然,$\mathscr{B}=\sigma(\mathscr{A})$,$\mathscr{A}$ 为 X 中全体闭集所成的集. 特别,如果 (X, C) 为距离空间,那么称 $(X, C, \mathscr{B})$ 为可测距离空间. 以后无必要时 C 及 $\mathscr{B}$ 常略去,而简记 $(X, C, \mathscr{B})$ 为 X.

由定义 1 容易推出随机元的下列基本性质.

(i) 设 V 为 (Ω, σ)-$(X, \mathscr{B})$ 随机元,而 τ 为 $(X, \mathscr{B})$-$(Y, \mathfrak{A})$ 随机元,则 τV 为 (Ω, σ)-$(Y, \mathfrak{A})$ 随机元.

(ii) 反之,为使 V 为 (Ω, σ)-$(X, \mathscr{B})$ 随机元,只要对任意 $(X, \mathscr{B})$-$(Y, \mathfrak{A})$ 随机元 τ,τV 为 (Ω, σ)-$(Y, \mathfrak{A})$ 随机元,但这里 Y 至少含两不同点 y_1,y_2,而且单点集 $(y_1)\in\mathfrak{A}$,$(y_2)\in\mathfrak{A}$.

(iii) 为使 V 为 (Ω, σ)-$(X, \mathscr{B})$ 随机元,充分必要条件是:

---

① 若 N 为一集族,则 $\sigma(N)$ 表含 N 的最小 σ-代数.

对任意界定在 $(X, \mathcal{B})$ 上的 Borel 可测函数 $g(x)$，gV 是随机变数.

实际上，对任 $A\in\mathfrak{A}$, $A^{-1}=(x: \tau(x)\in A)\in \mathcal{B}$, 故 $(\omega: \tau[V(\omega)]\in A)=(\omega: V(\omega)\in A^{-1})\in\sigma$. 得证 (i)；对任意 $B\in\mathcal{B}$, 令 $\tau(x)=y_1$ 或 y_2, 视 $x\in B$ 或 $x\overline{\in} B$ 而定，则 $(\omega: V(\omega)\in B)=(\omega: \tau[V(\omega)]=y_1)\in\sigma$. 得证 (ii)；由 (i) (ii) 即得 (iii).

如对 X 稍加条件，可以缩小 $g(x)$ 所在的类，并保持 (iii) 的正确性，如下：称可测拓扑空间 $(X, C, \mathcal{B})$ 为 N 空间，如对任一闭集 $B\in\mathcal{B}$, 存在界定于 X 上的连续（因之 $\mathcal{B}$-可测）函数 $h_B(x)$, 使当且仅当 $x\in B$ 时，$h_B(x)=0$. 易见可测距离空间是 N 空间①.

(iv) 为使 V 为 (Ω, σ)-N 空间随机元，充分必要条件是对任意闭集 $B\in\mathcal{B}$, $h_B(V)$ 是随机变数.

事实上，由 (iii) 得**必要性**. **充分性**则因：对任意闭集 $B\in\mathcal{B}$, $(\omega: V(\omega)\in B)=(\omega: h_B(V(\omega))=0)\in\sigma$；其次，易见使 $(\omega: V(\omega)\in B)\in\sigma$ 的一切 x-集 B 构成 X 中一 σ-代数 $\mathcal{F}$, 既然 $\mathcal{F}$ 包含一切闭集，故也包含 $\mathcal{B}=\sigma(\mathcal{A})$.

(v) 设 $\{V_n\}_{n=1}^{+\infty}$ 为一列 Ω-$X(N$ 空间$)$ 随机元，而且对每 $\omega\in\Omega$, $V_n(\omega)\to V(\omega)$, $V(\omega)\in X$, 则 V 也是 Ω-N 空间 X 的随机元.

因为，由 (iv), $\{h_B(V_n)\}_{n=1}^{+\infty}$ 是一列随机变数，由 $h_B(x)$ 的连续性，得 $h_B(V_n)\to h_B(V)(n\to +\infty)$, 故由普通概率论知 $h_B(V)$ 也是随机变数，再由 (iv) 即得 (v).

以下简称可测可分距离空间为 SM 空间，可测可分 Banach 空间为 SB 空间，可测距离空间为 M 空间，可测 Banach 空间为

---

① 例如，令 $h_B(x)=\inf\limits_{y\in B}\rho(x, y)$.

B 空间. 当空间可分时, 随机元具有更多的性质, 如下:

(vi) 为使 V 为 Ω - X(SM 空间)的随机元, **充分必要条件**是: 存在一列取有穷多个值的随机元列 $\{V_n(\omega)\}_{n=1}^{+\infty}$, 使对任意 $\omega \in \Omega$, $V_n(\omega) \to V(\omega)(n \to +\infty)$.

充分性由(v)推出. 设 $\{x_n\}_{n=1}^{+\infty}$ 为 X 中可列子集, 稠于 X. 对每 $n \in \mathbf{N}^*$, $i = 1, 2, \cdots, n-1$, 令 $B_{11} = X$, $B_{nn} = \{x: \rho(x, x_n) < \min\limits_{1 \leqslant j \leqslant n-1} \rho(x, x_j)\}$, $B_{in} = B_{i, n-1} \bigcap (X - B_{nn})$. 显然, 对每固定的 n, $\{B_{in}, i=1, 2, \cdots, n\}$ 是 X 的有穷分解. 定义
$$V_n(\omega) = x_i, \qquad \omega \in V^{-1}(B_{in}).$$
于是对每 $\omega \in \Omega$, $\rho(V_n(\omega), V(\omega)) \to 0$, ρ 表距离.

(vii) 为使 V 为 Ω - X(SM 空间)的随机元, 充分必要条件是: 存在取可列多个值的随机元列 $\{V_n(\omega)\}_{n=1}^{+\infty}$, 使 $\{V_n(\omega)\}_{n=1}^{+\infty}$ 均匀收敛于 $V(\omega)$.

只需证**必要性**. 设 $\{x_n\}_{n=1}^{+\infty}$ 为 X 中可列稠子集, 对每 $i \in \mathbf{N}^*$ 及每 $n \in \mathbf{N}^*$, 令
$$A_{in} = \left\{x: \rho(x, x_i) \leqslant \frac{1}{n}\right\} - \bigcup_{j=1}^{i-1} A_{jn},$$
显然, 对每固定的 n, $\{A_{in}, i \in \mathbf{N}^*\}$ 是 X 的可列分解. 定义
$$V_n(\omega) = x_i, \qquad \omega \in V^{-1}(A_{in}),$$
则对每 ω, $\rho(V_n(\omega), V(\omega)) \leqslant \dfrac{1}{n}$.

(viii) 为使 V 为 Ω - X(SB 空间)的随机元, **充分必要条件**是: 对每 $f \in X^*$(X 的共轭空间), $f(V)$ 是一随机变数.

必要性显然. 下证**充分性**. 由泛函分析中 Banach-Mazur 定理, 可视 X 为 $C[0,1]$ 的一子集, 这里 $C[0,1]$ 是 $[0,1]$ 上一切连续函数所成的 Banach 空间. $\|x\| = \max\limits_{0 \leqslant t \leqslant 1} |x(t)|$. 因为 $\mathscr{B}$ 为含一切集 $\{z: \|z - x\| \leqslant r\}$, $x \in X$, $z \in X$, $r \in [0, +\infty)$

的最小 σ-代数,故只要证
$$\{\omega:\|V(\omega)-x\|\leqslant r\}\in\sigma.$$
在 $C[0,1]$ 上定义线性泛函 $g_t: g_t(x)=x(t)(x\in C[0,1])$,则
$$\{\omega:\|V(\omega)-x\|\leqslant r\}=\bigcap_{t\in H}\{\omega:x(t)-r\leqslant g_t[V(\omega)]\leqslant x(t)+r\},$$
这里 H 为 $[0,1]$ 中任一可列稠集,由此即得证充分性.

(ix) 设 V_1,V_2 为两 Ω-X(SB 空间)的随机元,则 $V(\omega)=V_1(\omega)+V_2(\omega)$ 也是 Ω-X 随机元.

因为,对任意 $f\in X^*$,$f(V(\omega))=f(V_1(\omega))+f(V_2(\omega))$,由(viii),$f(V_1)$ 及 $f(V_2)$ 为两随机变数,故 $f(V(\omega))$ 也是,再由(viii),即得所欲证.

注 1 如 X 为 B 空间,如[26]中所证,(ix)中 $V(\omega)$ 未必是随机元. 这一点是定义 1 的弱点;但它具有性质(i),故又有优点.

§2.2. 概率 1 收敛

从泛函分析中的各种收敛性:强收敛(即依范收敛)、弱收敛等,抽象出一般的所谓(χ)-收敛,证明以概率 1 的(χ)-收敛定理,并将它运用到各种具体的收敛中去,这就是本节的主要目的.

设 X 为任一非空集,X_1 为 X 中一切序列 $\{x_n\}_{n=1}^{+\infty}$ 所成的集,X_2 为 X 的一切子集所成的集,设 χ 为 X_1 到 X_2 的映像,具有性质:

(i) 若 $x_n=x$,$n\in \mathbf{N}^*$,则 $x\in\chi(\{x_n\}_{n=1}^{+\infty})$;

(ii) $\chi(\{x_n\}_{n=1}^{+\infty})\subset\chi(\{x_{n_i}\}_{i=1}^{+\infty})$;

(iii) 若 $x_0\overline{\in}\chi(\{x_n\}_{n=1}^{+\infty})$,则存在子列 x_{n_1},x_{n_2},…,使 $x_0\overline{\in}\chi(\{x_{n_{i_k}}\}_{k=1}^{+\infty})$ 对一切 $x_{n_{i_1}}$,$x_{n_{i_2}}$,…成立.

定义 2 称 $\{x_n\}_{n=1}^{+\infty}$(χ)-收敛于 x_0,如果 $x_0\in\chi(\{x_n\}_{n=1}^{+\infty})$;

X 的子集 M 称为 χ-列紧的，如对 M 的每序列 $\{x_n\}_{n=1}^{+\infty}$，存在子列 $\{x_{n_i}\}_{i=1}^{+\infty}$，使 $\chi(\{x_{n_i}\}_{i=1}^{+\infty}) \neq \varnothing$；$X$ 的子集 N 称为 $M(\subset X)$ 的 χ-闭包，如果 N 由一切如下的 $x \in X$ 构成：存在 M 中的序列 $\{x_n\}_{n=1}^{+\infty}$，使 $x \in \chi(\{x_n\}_{n=1}^{+\infty})$；称 (χ_1)-收敛蕴含 (χ_2)-收敛，如对 X 的每一序列 $\{x_n\}_{n=1}^{+\infty}$，$\chi_1(\{x_n\}_{n=1}^{+\infty}) \subset \chi_2(\{x_n\}_{n=1}^{+\infty})$.

显然，强收敛与弱收敛都是特殊的 (χ)-收敛. 比弱收敛稍许一般化的 (χ)-收敛为 (Δ)-收敛. 设 Δ 为 X^* 的任一子集，(Δ)-收敛由下述映像定义：

$$\Delta(\{x_n\}_{n=1}^{+\infty}) = \bigcap_{f \in \Delta} \{x: x \in X, \lim_{n \to +\infty} f(x_n - x) = 0\}.$$

为叙述本节的基本定理(定理1)，还要引进一概念. 设 X 为 Banach 空间，集 $\Delta(\subset X^*)$ 称为在集 $M(\subset X)$ 上是全体的，如果由下列两条件就可推出 $x = y$：

(i) $x \in M$, $y \in M$；

(ii) 对任意 $f \in \Delta$，有 $f(x) = f(y)$.

定理 1 设 $\{V_n\}_{n=0}^{+\infty}$ 为概率空间 (Ω, σ, μ)—$X(B$ 空间$)$ 的随机元列，又 μ 为完全概率测度；设对每 $\omega \in \Omega$，存在 $\Delta(\omega) \subset X^*$，使 $\Delta(\omega)$ 在 X 的子集：$\{V_0(\omega)\}$ 与 $\{\bigcup_{n=1}^{+\infty}(V_n(\omega))$ 的 χ-闭包$\}$ 的和集上是全体的；最后，设 (χ)-收敛蕴含 $(\Delta(\omega))$-收敛，$(\omega \in \Omega)$. 则 $\{V_n\}_{n=1}^{+\infty}$ 以概率 $1(\chi)$-收敛于 V_0 的充分必要条件是：

(i) $\mu\{\omega: \bigcup_{n=1}^{+\infty}\{V_n(\omega)\}$ 为 χ-列紧的$\} = 1$；

(ii) $\mu\{\omega:$ 对每 $f \in \Delta(\omega)$, $\lim_{n \to +\infty} f(V_n(\omega) - V_0(\omega)) = 0\} = 1$.

证 由 $\{V_n\}_{n=1}^{+\infty}$ 以概率 $1(\chi)$-收敛于 V_0 的假定，存在 $E \in \sigma$，$\mu(E) = 1$，使至少对每 $\omega \in E$，$\{V_n(\omega)\}_{n=1}^{+\infty}(\Delta(\omega))$-收敛于 $V_0(\omega)$，由此及 μ 的完全性得证(ii). 其次，对任意 $\omega \in E$，任取 $\bigcup_{n=1}^{+\infty}\{V_n(\omega)\}$

中一序列 $\{x_i\}_{i=1}^{+\infty}$，则或者存在某 $x_0 \in \bigcup\limits_{n=1}^{+\infty}\{V_n(\omega)\}$，使 x_0 在 $\{x_i\}_{i=1}^{+\infty}$ 中出现无穷次，或者对任一 $x \in \bigcup\limits_{n=1}^{+\infty}\{V_n(\omega)\}$，$x$ 在 $\{x_i\}_{i=1}^{+\infty}$ 中只出现有穷次. 在前一情况下，利用性质(i)，在后一情况下，利用性质(ii)，都可以选取子列 x_{i_1}，x_{i_2}，…，使 $\chi(\{x_{i_j}\}_{j=1}^{+\infty}) \neq \emptyset$. 此事实既对任一 $\omega \in E$ 以及任一上述的序列 $\{x_n\}_{n=1}^{+\infty}$ 成立，故由 μ 的完全性即得证(i)，于是**必要性**证毕.

设使(i)(ii)同时成立的 ω-集为 F，$\mu(F) = 1$. 任取 $\omega \in F$，如果说 $\{V_n(\omega)\}_{n=1}^{+\infty}$ 不 (χ)-收敛于 $V_0(\omega)$，即设 $V_0(\omega) \overline{\in} \chi(\{V(\omega)\}_{n=1}^{+\infty})$，那么存在一子列 $\{V_{n_i}(\omega)\}_{i=1}^{+\infty}$，具有性质(iii). 但因集 $\bigcup\limits_{n=1}^{+\infty}\{V_n(\omega)\}$ 是 χ-列紧的，故存在后一子列的子列 $\{V_{n_{i_k}}(\omega)\}_{k=1}^{+\infty}$ 及某元 $x_0 \in X$，使 $x_0 \in \chi(\{V_{n_{i_k}}(\omega)\}_{k=1}^{+\infty})$. 由于 (χ)-收敛蕴含 $(\Delta(\omega))$-收敛，故 $x_0 \in \bigcap\limits_{f \in \Delta(\omega)} \{x: \lim\limits_{k \to +\infty} f(V_{n_{j_k}}(\omega) - x) = 0\}$. 由此及(ii)即知 $f(x_0) = f(V_0(\omega))$ 对每一有界线性泛函 $f \in \Delta(\omega)$ 成立. 既然 $x_0 \in \chi(\{V_{n_{j_k}}(\omega)\})$，故 x_0 属于 $\bigcup\limits_{n=1}^{+\infty}\{V_n(\omega)\}$ 的闭包. 根据 $\Delta(\omega)$ 在 $\{V_0(\omega)\}$ 与 $\{\bigcup\limits_{n=1}^{+\infty}\{V_n(\omega)\}$ 的 χ-闭包$\}$的和集上是全体的假定，得知 $V_0(\omega) = x_0 = \chi(\{V_{n_{i_k}}(\omega)\}_{k=1}^{+\infty})$，这与性质(iii)矛盾. 由 μ 的完全性即得证**充分性**. ∎

注 2 如 μ 未必是完全的，也已证明

$$\{\omega: \{V_n(\omega)\}_{n=1}^{+\infty} \ (\chi)\text{-收敛于} V_0(\omega)\}$$

$$= \{\omega: \bigcup\limits_{n=1}^{+\infty}\{V_n(\omega)\} \text{为}(\chi)\text{-列紧的}\} \cap$$

$$\{\omega: \lim\limits_{n \to +\infty} f(V_n(\omega) - V_0(\omega)) = 0 \text{ 对每 } f \in \Delta(\omega) \text{ 成立}\}.$$

在实际中如下运用定理 1：考虑 Banach 空间中某种具体的收敛，如果它是一种 (χ)-收敛，而且蕴含 $(\Delta(\omega))$-收敛，那么

可于定理 1 中换(χ)-收敛为此具体的收敛而得到相应的或更强的结果. 下定理用强收敛而例示一般.

定理 2 设 $\{V_n(\omega)\}_{n=0}^{+\infty}$ 为一列 $(\Omega,\sigma,\mu)-X(SB\text{ 空间})$ 随机元, 则 $\{V_n\}_{n=1}^{+\infty}$ 以概率 1 强收敛于 V_0 的充分必要条件是:

(i) $\mu\{\omega: \bigcup_{n=1}^{+\infty}\{V_n(\omega)\}$ 为强列紧的$\}=1$;

(ii) 对每 $f\in\Delta$, 这里 $\Delta(\subset X^*)$ 在 X 上是全体的, 有
$$\mu\{\omega: \lim_{n\to+\infty}f(V_n(\omega)-V_0(\omega))=0\}=1.$$

证 强收敛是一种(χ)-收敛, 而且蕴含(Δ)-收敛, 这里 $\Delta\subset X^*$ 任意. 注意此地未预设 μ 为完全的. 为证**必要性**, 首先指出, 由于使随机变数列收敛的 ω-集可测, 故 B_1 中的 ω-集可测. 由
$$\mu\{\omega: \lim_{n\to+\infty}f(V_n(\omega)-V_0(\omega))=0\}\geqslant\mu(E)=1$$
即得证(ii), 这里 $E=\{\omega: \{V_n\}_{n=1}^{+\infty}$ 强收敛于 $V_0\}$.

同样, 若能证(i)中的 ω-集 D 可测, 则由定理 1 的证明知 $D\supset E$, 故 $\mu(D)=1$. 像证§2.1(viii)一样, 可视 $X\subset C[0,1]$. 由 Arzela 定理, 有

$\{\omega: \bigcup_{n=1}^{+\infty}\{V_n(\omega)\}$ 为强列紧的$\}$
$=\bigcap_{m=1}^{+\infty}\bigcup_{k=1}^{+\infty}\bigcap_{\substack{t_1,t_2\in H \\ |t_1-t_2|\leqslant\frac{1}{k}}}\bigcap_{n=1}^{+\infty}(\{\omega: |g_{t_1}(V_n(\omega))|\leqslant k\}\cap$

$\left\{|g_{t_1}(V_n(\omega))-g_{t_2}(V_n(\omega))|\leqslant\dfrac{1}{m}\right\}),$

其中 H 为 $[0,1]$ 中任一可列稠集. 既然 $g_t(V_n(\omega))$ 是随机变数, 即知集 D 可测.

充分性如下证明. 由于 X 的可分性, 对每在 X 上为全体的集 $\Delta(\subset X^*)$, 存在 Δ 的可列子集 Δ_0, Δ_0 也在 X 上是全体的. 因此

$\{\omega: \lim_{n\to+\infty} f(V_n(\omega) - V_0(\omega)) = 0$ 对每 $f \in \Delta$ 成立$\}$
$= \bigcap_{f \in \Delta_0} \{\omega: \lim_{n\to+\infty} f(V_n(\omega) - V_0(\omega)) = 0\}$,

故等号前的 ω-集的测度为 1. 再由注 2 即得证充分性. ∎

自然希望建立对 (χ)-收敛的 Cauchy 收敛原则. 我们放弃一般情况而满足于强收敛, 因为一般情况不难类似处理.

定理 3 设 $\{V_n\}_{n=1}^{+\infty}$ 为一列 (Ω, σ, μ)-X(SB 空间) 随机元, 为使存在 Ω-X 随机元 V_0, 使 $\{V_n\}_{n=1}^{+\infty}$ 以概率 1 强收敛于 V_0, 充分必要条件是定理 2 (i) 及 (ii′) 满足:

(ii′) 对每 $f \in \Delta$, $\Delta \subset X^*$, Δ 在 X 上是全体的, 有
$$\mu\{\omega: \lim_{n,m\to+\infty} f(V_n(\omega) - V_m(\omega)) = 0\} = 1,$$

当定理 2 (i), (ii′) 满足时, V_0 在某 ω-集 E 上的值唯一, $\mu(E) = 1$.

证 必要性 由定理 2 的必要部分推出. **证充分性**. 令
$$E = \{\omega: \bigcup_{n=1}^{+\infty} \{V_n(\omega)\} \text{为强列紧的}\} \cap$$
$$\bigcap_{f \in \Delta_0} \{\omega: \lim_{n,m\to+\infty} f(V_n(\omega) - V_m(\omega)) = 0\},$$

则 $\mu(E) = 1$ (Δ_0 的意义同前). 任取 $\omega \in E$, 由于 $\bigcup_{n=1}^{+\infty} \{V_n(\omega)\}$ 为强列紧的, 故存在 $x(\omega) \in X$, 使 $\bigcup_{n=1}^{+\infty} \{V_n(\omega)\}$ 中某一子列 $\{V_{n_k}\}_{k=1}^{+\infty}$ 强收敛于 $x(\omega)$. 既然强收敛性蕴含任一 (Δ)-收敛性, 故对任意 $f \in \Delta$, $f(V_{n_k}(\omega)) \to f(x(\omega))$. 由此及
$$\lim_{n,m\to+\infty} [f(V_n(\omega)) - f(V_m(\omega))] = 0 \ (f \in \Delta_0),$$

可见对任一 $f \in \Delta_0$, $f(V_n(\omega)) \to f(x(\omega))$. 如果说具有这种性质的 $x(\omega)$ 有两个为 $x_1(\omega)$ 及 $x_2(\omega)$, 那么由于 Δ_0 在 X 上的全体性及 $f(x_1(\omega)) = \lim_{n\to+\infty} f(V_n(\omega)) = f(x_2(\omega))$ 即得 $x_1(\omega) = x_2(\omega)$. 今定义

$$V(\omega)=\begin{cases}x(\omega), & x\in E,\\ \theta(\text{零元}), & \omega\in\Omega-E,\end{cases}$$

则 $V(\omega)$ 具备定理 2 中 $V_0(\omega)$ 所应满足的条件,从而 $\{V_n\}_{n=1}^{+\infty}$ 以概率 1 强收敛于 V. 易见 V 是一随机元,因为对任意开集 $\Gamma\in X$,

$$(\omega: V(\omega)\in\Gamma)$$
$$=E\cap\bigcup_{m=1}^{+\infty}\bigcap_{k=m}^{+\infty}(V_n(\omega)\in\Gamma),\qquad \overline{\theta\in\Gamma},$$
$$=E\cap\bigcup_{m=1}^{+\infty}\bigcap_{k=m}^{+\infty}(V_n(\omega)\in\Gamma)\bigcup(\Omega-E),\quad \theta\in\Gamma. \blacksquare$$

§2.3. 平稳序列

试运用上节一般的收敛定理,来研究平稳序列的加强大数定理.

定义 3 设 $\{V_n\}_{n=1}^{+\infty}$ 为一列 (Ω, σ, μ)-$(X, \mathscr{B})$ 随机元,μ 为 σ 上概率测度. 称 $\{V_n\}_{n=1}^{+\infty}$ 为平稳序列,如对任意两正整数 n,k,任意 $B_i\in\mathscr{B}$,$i=1, 2, \cdots, n$,有

$$\mu\{\bigcap_{i=1}^{n}(\omega: V_i(\omega)\in B_i)\}=\mu\{\bigcap_{i=1}^{n}(\omega: V_{i+k}(\omega)\in B_i)\}.$$

注 3 由此立知,若 $\{V_n\}_{n=1}^{+\infty}$ 为平稳序列,又 τ 为 $(X, \mathscr{B})$-$(Y, \mathfrak{U})$ 随机元,则 $\{\tau(V_n)\}_{n=1}^{+\infty}$ 为 (Ω, σ, μ)-$(Y, \mathfrak{U})$ 平稳序列. 事实上,对任意 $B_i\in\mathfrak{U}$,$i=1, 2, \cdots, n$,

$$\mu\{\bigcap_{i=1}^{n}(\omega: \tau(V_i)\in B_i)\}=\mu\{\bigcap_{i=1}^{n}(\omega: V_i\in\tau^{-1}B_i)\}$$
$$=\mu\{\bigcap_{i=1}^{n}(\omega: V_{i+k}\in\tau^{-1}B_i)\}=\mu\{\bigcap_{i=1}^{n}(\omega: \tau(V_{i+k})\in B_i)\}.$$

定理 4 设 $\{V_n\}_{n=1}^{+\infty}$ 为 (Ω, σ, μ)-$X(SB\text{ 空间})$ 平稳序列,又 $\int_{\Omega}\|V_1(\omega)\|\mathrm{d}\mu(\omega)<+\infty$,则存在一 Ω-X 随机元 V_0,使随机元列 $\{U_s\}_{s=1}^{+\infty}$,

$$U_s(\omega)=\frac{1}{s}\sum_{k=1}^{s}V_k(\omega),$$

以概率 1 强收敛于 V_0,又 V_0 的值在某 E 上唯一,$\mu(E)=1$.

证 只要验证条件定理 2 中(i)与定理 3 中(ii′)成立. 对任意有界线性泛函 $f \in X^*$，由注 3 知 $\{f(V_n)\}_{n=1}^{+\infty}$ 是平稳随机变数列. 由 Birkhoff-Хинчин 定理，可见随机变数列 $\{f(U_n)\}_{n=1}^{+\infty}$ 以概率 1 收敛. 剩下只要证

$$\mu\left\{\omega: \bigcup_{s=1}^{+\infty}\left\{\frac{1}{s}\sum_{k=1}^{s}V_k(\omega)\right\} \text{为强列紧的}\right\}=1.$$

设 $\{x_n\}_{n=1}^{+\infty}$ 为不含 θ 的稠于 X 的可列集. 对每 $i \in \mathbf{N}^*$ 及每 $n \in \mathbf{N}^*$，令

$$A_{in}=\left\{x: \|x-x_i\| \leqslant \frac{1}{n}\right\} - \bigcup_{j=1}^{i-1} A_{jn}.$$

对每 $n \in \mathbf{N}^*$ 及 $m \in \mathbf{N}^*$，定义 X-X 的映像 τ_n，$_1\sigma_{mn}$，$_2\sigma_{mn}$ 如下：

$$\tau_n^{-1}(x_i)=A_{in}, \quad i \in \mathbf{N}^*$$

$$_1\sigma_{mn}^{-1}(\theta)=\bigcup_{j=1}^{m} A_{jn},$$

$$_1\sigma_{mn}^{-1}(x_i)=A_{in}-\bigcup_{j=1}^{m} A_{jn}, \quad i \in \mathbf{N}^*$$

$$_2\sigma_{mn}^{-1}(\theta)=\bigcup_{j=m+1}^{+\infty} A_{jn},$$

$$_2\sigma_{mn}^{-1}(x_i)=A_{in}-\bigcup_{j=m+1}^{+\infty} A_{jn}, \quad i \in \mathbf{N}^*$$

显然，对每 $m \in \mathbf{N}^*$，$n \in \mathbf{N}^*$ 及 $x \in X$，有

$$_1\sigma_{mn}(x)+_2\sigma_{mn}(x)=\tau_n(x), \tag{2.1}$$

$$\|x-\tau_n(x)\| \leqslant \frac{1}{n}. \tag{2.2}$$

又

$$\sum_{i=1}^{+\infty}\|x_i\|\mu\{\omega: V_1(\omega) \in A_{in}\}$$

$$=\int_{\Omega}\|V_k(\omega)+\tau_n(V_k(\omega))-V_k(\omega)\|\,d\mu(\omega)$$

$$\leqslant \int_{\Omega}\|V_1(\omega)\|\,d\mu(\omega)+\frac{1}{n}<+\infty,$$

因此

$$\int_\Omega \|\,_1\sigma_{mn}(V_k(\omega))\|\,\mathrm{d}\mu(\omega)$$
$$= \sum_{i=m+1}^{+\infty} \|x_i\|\mu\{\omega: V_1(\omega) \in A_{in}\} < +\infty,$$
$$\int_\Omega \|\,_2\sigma_{mn}(V_k(\omega))\|\,\mathrm{d}\mu(\omega)$$
$$= \sum_{i=1}^{m} \|x_i\|\mu\{\omega: V_1(\omega) \in A_{in}\} < +\infty.$$

由注 3 知 $\{\|\,_1\sigma_{mn}(V_k)\|\}_{k=1}^{+\infty}$ 及 $\{\chi_{\{x_i\}}(\,_2\sigma_{mn}(V_k))\}_{k=1}^{+\infty}$ 为两平稳随机变数序列(这里 $\{x_i\}$ 为单点集,而 $\overline{\chi_A(y)}=0$,如 $y\overline{\in}A$,$=1$,如 $y\in A$)。前后两序列中每随机变数的数学期望分别等于

$$\sum_{i=m+1}^{+\infty} \|x_i\|\mu\{\omega: V_1(\omega) \in A_{in}\},$$

及

$$\mu\{\omega: V_1(\omega) \in A_{in}\}, \quad i\leqslant m,$$
$$0, \qquad\qquad\qquad i>m.$$

因此,存在非负随机变数 $_1\xi_{mn}$ 及 $_2^i\xi_{mn}$,使

$$\frac{1}{s}\sum_{k=1}^{s} \|\,_1\sigma_{mn}(V_k)\|, \tag{2.3}$$

$$\frac{1}{s}\sum_{k=1}^{s} \chi_{\{x_i\}}(\,_2\sigma_{mn}(V_k)) \quad (s \in \mathbf{N}^*),$$

分别以概率 1 收敛于 $_1\xi_{mn}$ 与 $_2^i\xi_{mn}$。

在前一情况下,因

$$\int_\Omega {}_1\xi_{mn}\mathrm{d}\mu(\omega) = \sum_{i=m+1}^{+\infty} \|x_i\|\mu\{\omega: V_1(\omega) \in A_{in}\} < +\infty,$$

故 $\lim\limits_{m\to+\infty}\int_\Omega {}_1\xi_{mn}(\omega)\mathrm{d}\mu(\omega) = 0$,于是存在子列

$$_1\xi_{m_1 n}, \quad _1\xi_{m_2 n}, \quad \cdots \tag{2.4}$$

以概率 1 收敛于 0;在后一情况下,因为平稳随机元列

$\{{}_2\sigma_{mn}(V_k)\}_{k=1}^{+\infty}$ 中之通项可写为

$${}_2\sigma_{mn}(V_k) = \sum_{i=1}^{m} x_i \chi_{\{x_i\}}({}_2\sigma_{mn}(V_k)),$$

故知序列

$${}_2\sigma_{mn}(V_1),\quad \frac{1}{2}\sum_{k=1}^{2}{}_2\sigma_{mn}(V_k),\cdots \qquad (2.5)$$

在概率 1 强收敛于随机元

$${}_2\xi_{mn} = \sum_{i=1}^{m} x_i\, {}_2^i\xi_{mn}.$$

使序列 (2.3)～(2.5) 收敛的 ω-集分别记为 ${}_1E_{mn}$，E_n 及 ${}_2E_{mn}$，令

$$E = \bigcap_{m=1}^{+\infty}\bigcap_{n=1}^{+\infty}({}_1E_{mn}\cap E_n\cap {}_2E_{mn})\in\sigma,\quad \mu(E)=1.$$

对任意 $\varepsilon>0$，取正整数 $N\geqslant\dfrac{4}{\varepsilon}$，又对每 $\omega\in E$，存在正整数 p_ω，使 $p\geqslant p_\omega$ 时，

$$|{}_1\xi_{m_p N}(\omega)|\leqslant\frac{\varepsilon}{4}. \qquad (2.6)$$

对此相同的 $\omega\in E$，存在正整数 s_ω，使 $s\geqslant s_\omega$ 时，下两不等式同时成立：

$$\left|\frac{1}{s}\sum_{k=1}^{s}\|{}_1\sigma_{m_{p_\omega}N}(V_k(\omega))\| - {}_1\xi_{m_{p_\omega}N}(\omega)\right|\leqslant\frac{\varepsilon}{4}, \qquad (2.7)$$

$$\left\|\frac{1}{s}\sum_{k=1}^{s}{}_2\sigma_{m_{p_\omega}N}(V_k(\omega)) - {}_2\xi_{m_{p_\omega}N}(\omega)\right\|\leqslant\frac{\varepsilon}{4}. \qquad (2.8)$$

于是对每 $\omega\in E$，集

$$\bigcup_{s=1}^{s_\omega}\left\{\frac{1}{s}\sum_{k=1}^{s}V_k(\omega)\right\}\cup\{{}_2\xi_{m_{p_\omega}N}(\omega)\}$$

形成一有穷 ε-网。实际上，对每 $\omega\in E$ 及每 $s>s_\omega$，由 (2.2) (2.1) (2.8) (2.7) (2.6) 及 $\dfrac{1}{N}\leqslant\dfrac{\varepsilon}{4}$ 有

$$\left\| \frac{1}{s}\sum_{k=1}^{s} V_k(\omega) - {}_2\xi_{m_{p_\omega}N}(\omega) \right\|$$

$$= \left\| \left[\frac{1}{s}\sum_{k=1}^{s} V_k(\omega) - \frac{1}{s}\sum_{k=1}^{s}\tau_N(V_k(\omega))\right] + \right.$$

$$\left[\frac{1}{s}\sum_{k=1}^{s}\tau_N(V_k(\omega)) - \frac{1}{s}\sum_{k=1}^{s}{}_2\sigma_{m_{p_\omega}N}(V_k(\omega))\right] +$$

$$\left.\left[\frac{1}{s}\sum_{k=1}^{s}{}_2\sigma_{m_{p_\omega}N}(V_k(\omega)) - {}_2\xi_{m_{p_\omega}N}(\omega)\right] \right\|$$

$$\leqslant \frac{\varepsilon}{4} + \left\|\frac{1}{s}\sum_{k=1}^{s}{}_1\sigma_{m_{p_\omega}N}(V_k(\omega))\right\| + \frac{\varepsilon}{4}$$

$$\leqslant \frac{\varepsilon}{4} + \frac{1}{s}\sum_{k=1}^{s}\left\|{}_1\sigma_{m_{p_\omega}N}(V_k(\omega))\right\| + \frac{\varepsilon}{4}$$

$$\leqslant \frac{\varepsilon}{4} + \left|\frac{1}{s}\sum_{k=1}^{s}\left\|{}_1\sigma_{m_{p_\omega}N}(V_k(\omega))\right\| - {}_1\xi_{m_{p_\omega}N}(\omega)\right| + {}_1\xi_{m_{p_\omega}N} + \frac{\varepsilon}{4}$$

$$\leqslant \varepsilon. \blacksquare$$

定义 4 设 $\{V_n\}_{n=1}^{+\infty}$ 为 (Ω, σ, μ)-$(X, \mathcal{B})$ 随机元列. 称 $\{V_n\}_{n=1}^{+\infty}$ 为独立序列, 如果对任意正整数 n, 任意 $B_i \in \mathcal{B}$, $i=1, 2, \cdots, n$, 有

$$\mu\left(\bigcap_{i=1}^{n}\{\omega: V_i(\omega)\in B_i\}\right) = \bigcap_{i=1}^{n}\mu(\omega: V_i(\omega)\in B_i);$$

称 $\{V_n\}_{n=1}^{+\infty}$ 同分布, 如对任意正整数 i, k, $B\in\mathcal{B}$, 有

$$\mu(\omega: V_i(\omega)\in B) = \mu(\omega: V_k(\omega)\in B);$$

对任一 (Ω, σ, μ)-$X(B$ 空间$)$ 随机元 $V(\omega)$, 如果它关于 μ 的 Bochner 积分 $\int_\Omega V(\omega)\mathrm{d}\mu(\omega)$ 存在, 那么称此积分为 $V(\omega)$ 的数学期望, 记为 EV.

由泛函分析知, 对可分 Banach 空间, $EV = \int_\Omega V(\omega)\mathrm{d}\mu(\omega)$ 存在的充分必要条件是: 勒贝格积分 $\int_\Omega \|V(\omega)\|\mathrm{d}\mu(\omega) < +\infty$.

定理 5(独立同分布随机元列的加强大数定理) 设 $\{V_n\}_{n=1}^{+\infty}$ 为 (Ω, σ, μ)- $X(SB$ 空间$)$的独立同分布随机元列，为使随机元列 $\{U_s\}_{s=1}^{+\infty}$，

$$U_s(\omega) = \frac{1}{s}\sum_{k=1}^{s} V_k(\omega)$$

以概率 1 收敛于 EV_1 的充分必要条件是 EV_1 存在.

证 必要性不必证.

反之，既然 EV_1 存在，故 $\int_\Omega \|V_1(\omega)\| \,\mathrm{d}\mu(\omega) < +\infty$，又因 $\{V_n\}_{n=1}^{+\infty}$ 为平稳序列，故由定理 4 之证知

$$\mu\{\omega: \bigcup_{n=1}^{+\infty}\{U_n(\omega)\} \text{是强列紧的}\} = 1.$$

既然对每 $f \in X^*$，随机变数列 $\{f(U_n)\}_{n=1}^{+\infty}$ 以概率 1 收敛于 $f(EV_1)$，故由定理 2 即得证. ∎

§3. 随机变换

§3.1. 随机变换的定义

定义 1 设 $(\Omega, \sigma)(E, \mathcal{B})$ 为两可测空间，Γ 为任一非空集. 自 $\Omega \times \Gamma$ 到 E 的变换 $T(\omega, \gamma)$ 称为随机变换，如对任意固定的 $\gamma \in \Gamma$, $B \in \mathcal{B}$,

$$(\omega: T(\omega, \gamma) \in B) \in \sigma, \quad (3.1)$$

换言之，若对任意固定的 $\gamma \in \Gamma$, $T(\omega, \gamma)$ 是 Ω-E 随机元，则 T 是一随机变换.

如果把 $T(\omega, \gamma)$ 记成 $T(\omega)\gamma$，那么更易看出它与泛函分析中算子 T 的类似性. 因为当 $\omega \in \Omega$ 固定时，$T(\omega)$ 将 Γ 变到 E，故有时也称 $T(\omega, \gamma) = T(\omega)\gamma$ 为随机算子.

设 $(E, \mathcal{B})(\Gamma, \mathfrak{U})$ 为两可测距离空间，各有距离为 δ, ρ. 又 $f(\omega)$ 为 (Ω, σ)-$(\Gamma, \mathfrak{U})$ 随机元，称随机变换 $T(\omega, \gamma)$ 在 $f(\omega)$ 为局部压缩的，如存在随机变数 $C(\omega) < 1$，使对每 $\omega \in \Omega$ 及 $\gamma \in \Gamma$，有

$$\delta(T(\omega, \gamma), T(\omega, f(\omega))) \leqslant C(\omega)\rho(\gamma, f(\omega)), \quad (3.2)$$

称 $T(\omega, \gamma)$ 为压缩的，如存在随机变数 $C(\omega) < 1$，使对每 $\omega \in \Omega$ 及 $\gamma_1, \gamma_2 \in \Gamma$，有

$$\delta(T(\omega, \gamma_1), T(\omega, \gamma_2)) \leqslant C(\omega)\rho(\gamma_1, \gamma_2). \quad (3.3)$$

如果 (3.2)(3.3) 中的 $C(\omega) \leqslant C_0 < 1$，$C_0$ 为一常数，那么分别称 $T(\omega, \gamma)$ 在 $f(\omega)$ 为均匀局部压缩的及均匀压缩的. 称 $T(\omega, \gamma)$ 为连续的，如对任意 $\omega \in \Omega$，任意 $\gamma_0 \in \Gamma$，当 $\gamma_n \in \Gamma$, $\gamma_n \to \gamma_0 (n \to +\infty)$ 时，有

$$T(\omega, \gamma_n) \to T(\omega, \gamma_0). \quad (3.4)$$

称 $T(\omega, \gamma)$ 为有界的，如对任意 $\omega \in \Omega$，$\gamma_1 \in \Gamma$，$\gamma_2 \in \Gamma$，有
$$\delta(T(\omega, \gamma_1), T(\omega, \gamma_2)) \leqslant C(\omega)\rho(\gamma_1, \gamma_2)$$
$$(+\infty > C(\omega) \geqslant 0). \tag{3.5}$$

设 $(E, \mathscr{B})(\Gamma, \mathfrak{A})$ 为两 B 空间，称 $T(\omega, \gamma)$ 为线性的，如对每 $\omega \in \Omega$，$\gamma_1 \in \Gamma$，$\gamma_2 \in \Gamma$ 及任两个实数 α，β，有
$$T(\omega, \alpha\gamma_1 + \beta\gamma_2) = \alpha T(\omega, \gamma_1) + \beta T(\omega, \gamma_2), \tag{3.6}$$
称随机变换 $T(\omega, \gamma)$ 为随机泛函，如果 $(E, \mathscr{B})$ 为一维 Borel 可测空间.

如果 (Ω, σ) 为概率空间 (Ω, σ, μ)，那么可类似地定义以概率 1 压缩的、以概率 1 连续的随机变换等.

引理 1 设 $T(\omega, \gamma)$ 为 $\Omega \times \Gamma \text{-}(E, \mathscr{B})$ 连续随机变换，其中 $(\Gamma, \mathfrak{A})$ 为 SM 空间，又 V 为 $\Omega\text{-}\Gamma$ 随机元，则 $U(\omega) = T(\omega, V(\omega))$ 为 $\Omega\text{-}E$ 随机元.

证 由 §2.1(vii)，存在一列取可列多个值的 $\Omega\text{-}\Gamma$ 随机元 $\{V_n\}_{n=1}^{+\infty}$，均匀收敛于 V. 定义
$$U_n(\omega) = T(\omega, V_n(\omega)),$$
则对每 $B \in \mathscr{B}$，有
$$(\omega: U_n(\omega) \in B) = \bigcup_{i=1}^{+\infty}(\omega: T(\omega, x_i) \in B)$$
$$\bigcap(\omega: V_n(\omega) = x_i) \in \sigma,$$
这里的 $\{x_i\}\{V_n\}$ 的选取如 §2.1(vii)，因而 $\{U_n(\omega)\}_{n=1}^{+\infty}$ 为一列 $(\Omega, \sigma)\text{-}(E, \mathscr{B})$ 随机元. 由于 T 的连续性即得所欲证. ∎

§3.2. 随机不动点原理

不动点原理在研究各类方程的存在与唯一性问题中，起着重要作用. 类似地，为了研究随机方程，有必要建立随机不动点原理，作为初步，有

定理 1 设 T 为 $(\Omega, \sigma, \mu) \times E\text{-}(E, \mathscr{B})$ 的随机变换，E 为完备的 SM 空间，又 T 以概率 1 为压缩的，则存在 $\Omega\text{-}E$ 随机

元 $\varphi(\omega)$，它在 ω-集 G 上唯一，$\mu(G)=1$，使
$$\mu(\omega: T(\omega, \varphi(\omega))=\varphi(\omega))=1;$$
又 $\varphi(\omega)$ 可以由某随机元列 $\{V_n\}_{n=1}^{+\infty}$ 逼近，其中 V_1 可任意选取.

证 令 $G=(\omega: T$ 为压缩的$)$，则 $\mu(G)=1$. 对任意固定的 $\omega\in G$，由泛函分析知存在唯一 $e(\omega)\in E$，它满足 $T(\omega, e(\omega))=e(\omega)$. 定义
$$\varphi(\omega)=\begin{cases} e(\omega), & \omega\in G, \\ e, & \omega\in\Omega-G, \end{cases}$$
其中 e 为 E 中任一定点. 因为对每固定 $\omega\in G$，T 是连续的，所以可以用 §3.1 引理 1. 任意选取随机元 V_1，并定义
$$V_{n+1}(\omega)=\begin{cases} T(\omega, V_n(\omega)), & \omega\in G, \\ e, & \omega\in\Omega-G, \end{cases}$$
显然随机元列 $\{V_n\}_{n=1}^{+\infty}$ 收敛于 $\varphi(\omega)$. 由 §2.1(v) 即知 $\varphi(\omega)$ 也是随机元而且具有所需的性质. ∎

下面考虑一列连续随机变换的情形，作为随机逼近，定理 2 甚为有用.

定理 2 设 φ, f_1, f_2, $\cdots$ 为 (Ω, σ, μ)-$E(SM$ 空间$)$ 的随机元，μ 为完全的概率测度，$C(\omega)$ 为随机变数，$\{T_n\}_{n=1}^{+\infty}$ 为一列 $\Omega\times E$-E 的连续随机变换，使下列条件满足：
$$0\leq C(\omega)<1, \omega\in\Omega,$$
$$\mu\{\omega: \lim_{n\to+\infty}\rho(f_n(\omega), \varphi(\omega))=0\}=1,$$
$$\mu\{\omega: \lim_{n\to+\infty}\rho(T_n(\omega, \varphi(\omega)), \varphi(\omega))=0\}=1,$$
$$\mu\{\omega: \rho(T_n(\omega, x), T_n(\omega, f_n(\omega)))\leq C(\omega)\rho(x, f_n(\omega))\}=1,$$
$$n\in\mathbf{N}^*, x\in E,$$
取 $V_1(\omega)$ 为任意 Ω-E 随机元，令
$$V_{n+1}(\omega)=T_n(\omega, V_n(\omega)),$$
则 $\{V_n\}_{n=1}^{+\infty}$ 为一列 Ω-E 随机元，以概率 1 收敛于 φ.

证 由 §3.1 引理 1 知 $V_n(\omega)$ 为随机元. 令

$$G = \{\omega : \lim_{n \to +\infty} f_n(\omega) = \varphi(\omega)\} \cap \{\omega : \lim_{n \to +\infty} T_n(\omega, \varphi(\omega)) = \varphi(\omega)\}$$

$$\cap \bigcap_{n=1}^{+\infty} \bigcap_{x \in H} \{\omega : \rho(T_n(\omega, x), T_n(\omega, f_n(\omega)))$$

$$\leqslant C(\omega)\rho(x, f_n(\omega))\},$$

其中 H 为 E 中可列稠集, 由假定知 $\mu(G)=1$. 由随机变换的连续性, 有

$$G \subset \bigcap_{n=1}^{+\infty} \bigcap_{x \in E} \{\omega : \rho(T_n(\omega, x), T_n(\omega, f_n(\omega)))$$

$$\leqslant C(\omega)\rho(x, f_n(\omega))\}.$$

今任意固定 $\omega \in G$, 反复用三角不等式, 即得

$$\rho(V_{n+1}(\omega), \varphi(\omega)) = \rho(T_n(\omega, V_n(\omega)), \varphi(\omega))$$

$$\leqslant \rho(T_n(\omega, V_n(\omega)), T_n(\omega, f_n(\omega))) + \rho(T_n(\omega, f_n(\omega)),$$

$$T_n(\omega, \varphi(\omega))) + \rho(T_n(\omega, \varphi(\omega)), \varphi(\omega))$$

$$\leqslant C(\omega)\rho(V_n(\omega), f_n(\omega)) + C(\omega)\rho(f_n(\omega), \varphi(\omega)) +$$

$$\rho(T_n(\omega, \varphi(\omega)), \varphi(\omega))$$

$$\leqslant C(\omega)\rho(V_n(\omega), \varphi(\omega)) + 2C(\omega)\rho(f_n(\omega), \varphi(\omega)) +$$

$$\rho(T_n(\omega, \varphi(\omega)), \varphi(\omega)).$$

对 $\varepsilon > 0$, 存在正整数 N, 使 $n \geqslant N$ 时, 下两式均成立:

$$C(\omega)\rho(f_n(\omega), \varphi(\omega)) \leqslant \frac{1-C(\omega)}{8}\varepsilon,$$

$$\rho(T_n(\omega, \varphi(\omega)), \varphi(\omega)) \leqslant \frac{1-C(\omega)}{4}\varepsilon.$$

因此, 对每 $n \geqslant N$ 有

$$\rho(V_{n+1}(\omega), \varphi(\omega)) \leqslant C(\omega)\rho(V_n(\omega), \varphi(\omega)) + \frac{1-C(\omega)}{2}\varepsilon,$$

由归纳法, 即得对任意 $k \in \mathbf{N}^*$, 有

$$\rho(V_{N+k}(\omega), \varphi(\omega))$$

$$\leqslant C^k(\omega)\rho(V_N(\omega), \varphi(\omega)) + \frac{1-C(\omega)}{2}\varepsilon\sum_{i=1}^{k}C^{i-1}(\omega).$$

因 $0\leqslant C(\omega)<1$, 故当 k 大于某正整数 M 时, 上式右方小于 ε, 于是得证对任意 $\omega\in G$, $V_n(\omega)\to\varphi(\omega)$. 由 μ 之完全性即知定理结论正确. ∎

作为上述定理的应用, 试讨论下列问题: 设 $T(\omega, x)$ 为均匀压缩随机变换, 而且对每固定的 $x\in E$, Bochner 积分

$$S(x) = \int_\Omega T(\omega, x)\mathrm{d}\mu(\omega) \tag{3.7}$$

存在. 易见 $S(x)$ 是 E-E 压缩变换, 因而存在唯一不动点 x_0, 使 $S(x_0)=x_0$. 换言之, x_0 是 $T(\omega, x)$ 的数学期望 $S(x)$ 的不动点. 试问如何根据对 $T(\omega, x)$ 的一列独立的观察以求 x_0? 与概率论中加强大数定理类似, 有

定理 3 设 $\{T_n\}_{n=1}^{+\infty}$ 为一列 $(\Omega, \sigma, \mu)\times E$-$E$ 的连续随机变换, 这里 μ 为完全的概率测度, 又 $E=(E, \mathscr{B})$ 为 SB 空间. 设

(i) 对任意固定的 $x\in E$, $\{T_n(\omega, x)\}_{n=1}^{+\infty}$ 是一列独立同分布随机元;

(ii) 对任意固定的 $x\in E$, $T_1(\omega, x)$ 的 Bochner 积分 (3.7) 存在;

(iii) 存在常数 $c<1$, 使对每对 $x, y\in E$, 每 $n\in\mathbf{N}^*$

$$\mu\{\omega: \|T_n(\omega, x)-T_n(\omega, y)\|\leqslant c\|x-y\|\}=1,$$

则存在随机元列 $\{V_n\}_{n=1}^{+\infty}$, 以概率 1 收敛于 $S(x)$ 的唯一不动点 x_0. 可如下取 $\{V_n\}_{n=1}^{+\infty}$: $V_1(\omega)$ 为任意随机元, $V_{n+1}(\omega)=S_n(\omega, V_n(\omega))$, 而

$$S_n(\omega, x) = \frac{1}{n}\sum_{i=1}^{n}T_i(\omega, x).$$

证 不妨把 $S(x)$ 也看成随机变换 $S(\omega, x)(\equiv S(x))$. 由于 S, $\{S_n\}_{n=1}^{+\infty}$ 都是以概率 1 为均匀压缩随机变换, 故由本节定理

1，存在随机元列 $\varphi(\equiv x_0)$，$\{f_n\}_{n=1}^{+\infty}$ 分别是它们的随机不动点（概率 1）. 由 §2.3 定理 5 及假设(i)(ii)，对每 $x \in E$，随机元列 $\{S_n(\omega, x)\}_{n=1}^{+\infty}$ 以概率 1 收敛于 $S(x)$. 既然每个 $f_n(\omega)$ 都是随机不动点，故对每 $\omega \in \bigcap_{i=1}^{+\infty} \bigcap_{x \in E} \{\|T_i(\omega, x) - T_i(\omega, x_0)\| \leqslant c \|x - x_0\|\}$，有

$$\|f_n(\omega) - x_0\|$$
$$\leqslant \|f_n(\omega) - S_n(\omega, x_0)\| + \|S_n(\omega, x_0) - x_0\|$$
$$= \left\| \frac{1}{n} \sum_{i=1}^{n} T_i(\omega, f_n(\omega)) - \frac{1}{n} \sum_{i=1}^{n} T_i(\omega, x_0) \right\| + \|S_n(\omega, x_0) - x_0\|$$
$$\leqslant c \|f_n(\omega) - x_0\| + \|S_n(\omega, x_0) - x_0\|.$$

因此，$\|f_n(\omega) - x_0\| \leqslant \dfrac{1}{1-c} \|S_n(\omega, x_0) - x_0\|$，从而

$$\mu\{\omega: \lim_{n \to +\infty} f_n(\omega) = \varphi(\omega)\} = 1.$$

这样，定理 2 中一切条件满足而本定理得证. ∎

§3.3. 若干定理的随机化

泛函分析中许多重要的定理，如 Tietze 定理、Banach-Hahn 定理以及具体空间中线性泛函的表现定理等，在随机情况，自应有对应的定理. 本节中即研究此问题. 结果发现，问题的实质归结为可测性是否成立. 由于基本思想容易掌握，这里只讨论随机的 Banach-Hahn 定理及 $C[0, 1]$ 中线性泛函的表现.

以 $(\mathbf{R}, \mathscr{B}_1)$ 表一维 Borel 可测空间，$\varGamma$ 为可分实线性赋范空间，$M \subset \varGamma$，M 是一线性流型.

定理 4 设 V 为 $(\varOmega, \sigma) \times M$-$(\mathbf{R}, \mathscr{B}_1)$ 的随机变换，$\varGamma$，M 满足上述条件. 如果

(i) 对每 $\omega \in \varOmega$，$\alpha \in \mathbf{R}$，$\beta \in \mathbf{R}$，$x \in M$，$y \in M$，
   $V(\omega, \alpha x + \beta y) = \alpha V(\omega, x) + \beta V(\omega, y)$；

(ii) 对每对 $(\omega, x) \in \varOmega \times M$，有 $|V(\omega, x)| \leqslant S(\omega) \|x\|$，

其中 $S(\omega) = \sup\limits_{x \in O \cap M} |V(\omega, x)|$,而 $O = \{x: \|x\| = 1\}$.

于是存在 $(\Omega, \sigma) \times \Gamma\text{-}(\mathbf{R}, \mathscr{B}_1)$ 的随机变换 T,使

(i) 对每对 $(\omega, x) \in \Omega \times M$,$T(\omega, x) = V(\omega, x)$;

(ii) 对每 $\omega \in \Omega$,$\alpha \in \mathbf{R}$,$\beta \in \mathbf{R}$,$x \in \Gamma$,$y \in \Gamma$,
$$T(\omega, \alpha x + \beta y) = \alpha T(\omega, x) + \beta T(\omega, y);$$

(iii) 对每对 $(\omega, x) \in \Omega \times \Gamma$,$|T(\omega, x)| \leqslant S(\omega) \|x\|$.

证 如下造 T. 因 Γ 可分,存在可列集 $\{x_n\}_{n=1}^{+\infty}$,它是 Γ-M 的稠密子集. 令 M_n 为 $M \cup \bigcup\limits_{k=1}^{n} \{x_k\}$ 所产生的线性流型($n \in \mathbf{N}$),又 $\Gamma_0 = \bigcup\limits_{n=0}^{+\infty} M_n$. 令

$V_0(\omega, x) = V(\omega, x)$ $(\omega, x) \in \Omega \times M_0$,

$\qquad\qquad = 0$, $(\omega, x) \in \Omega \times (\Gamma_0 - M_0)$.

$V_n(\omega, x) = V_{n-1}(\omega, x) = 0$, $(\omega, x) \in \Omega \times (\Gamma_0 - M_n)$,

$V_n(\omega, x + t x_n) = V_{n-1}(\omega, x) + t \cdot \sup\limits_{x \in M_{n-1}} (V_{n-1}(\omega, x) -$
$\qquad\qquad S(\omega) \|x - x_n\|)$, $\omega \in \Omega$,$x \in M_{n-1}$,$t \in \mathbf{R}$.

再对每一对 $(\omega, x) \in \Omega \times \Gamma_0$,令
$$T(\omega, x) = T_0(\omega, x) = \lim_{n \to +\infty} V_n(\omega, x),$$

对 $y \in \Gamma - \Gamma_0$,$y = \lim\limits_{n \to +\infty} y_n$,$y_n \in \Gamma_0$,令
$$T(\omega, y) = \lim_{n \to +\infty} T_0(\omega, y_n).$$

由泛函分析知,对每固定的 $\omega \in \Omega$,$T(\omega, y)$ 是一有界线性泛函,它是 $V(\omega, y)$ 自 M 到全 Γ 上的扩张,而且保留范不变. 因此,剩下只要证可测性. 由 Γ 的可分性,
$$\{\omega: S(\omega) \leqslant c\} = \bigcap_{x \in \tilde{o}} \{\omega: |V_n(\omega, x)| \leqslant c\},$$

其中 $\tilde{o} \subset O$ 为某可列稠集. 因 V 为随机变换,故 V_0 为 $\Omega \times \Gamma_0$-R_1 的随机变换. 对每个 $c \in \mathbf{R}$,有
$$\{\omega: \sup_{x \in M_{n-1}} (V_{n-1}(\omega, x) - S(\omega) \|x - x_n\|) \leqslant c\}$$

$$= \bigcap_{x \in \widetilde{M}_{n-1}} \{\omega: (V_{n-1}(\omega, x) - S(\omega) \| x - x_n \|) \leqslant c\},$$

其中 $\widetilde{M}_{n-1}(\subset M_{n-1})$ 为可列并稠于 M_{n-1} 之集, 故 T_0 是 $\Omega \times \Gamma_0$-R 的随机变换, 因而 T 是 $\Omega \times \Gamma$-$\mathbf{R}$ 的随机变换. ∎

如果除去 Γ 为可分空间的假定, 定理 4 是否正确? 尚不可知, 但如 Γ 是 Hilbert 空间(不必可分), M 为 Γ 的子 Hilbert 空间时, 定理 4 的结论正确, 因为每 $x \in \Gamma$ 可唯一地展为 $x = y + z$, 其中 $y \in M$ 而 $z \perp M$, 故只要令 $T(\omega, x) = V(\omega, y)$ 即可.

定理 5(M. Ullrich) 设 $T(\omega, x)$ 为 $(\Omega, \sigma) \times C[0, 1]$-$(\mathbf{R}, \mathscr{B}_1)$ 线性连续随机变换, 则存在 $(\Omega, \sigma) \times [0, 1]$-$(\mathbf{R}, \mathscr{B}_1)$ 随机变换 $g(\omega, t)$, 当 $\omega \in \Omega$ 固定时, $g(\omega, t)$ 是 t 的有界变差函数, 而且对每 $\omega \in \Omega$, $x \in C[0, 1]$, 有

$$T(\omega, x) = \int_0^1 x(t) \mathrm{d}g(\omega, t).$$

证 对每固定的 $\omega \in \Omega$, $T(\omega, x)$ 是 $C[0, 1]$ 上线性泛函, 故由泛函分析知存在有界变差函数 $g(\omega, t)$ 使上式成立. 不失一般性可设 $g(\omega, 1) = 0$ $(\omega \in \Omega)$, 而且 $g(\omega, t)$ 在 $(0, 1)$ 中关于 t 右连续. 剩下只要证 $g(\omega, t)$ 是一随机变换, 即只要证对每固定的 $t \in [0, 1]$, $g(\omega, t)$ 是一随机变数. 当 $t = 0$ 时, 因①

$$T(\omega, 1) = \int_0^1 \mathrm{d}g(\omega, t) = -g(\omega, 0),$$

而结论成立. 令 $t_0 \in (0, 1)$, 定义一列连续函数 $\{a_n(t)\}_{n=1}^{+\infty}$ 如下:

$$a_n(t) = \begin{cases} 1, & 0 \leqslant t \leqslant t_0, \\ -n(t - t_0) + 1, & t_0 < t < t_0 + \frac{1}{n}, \\ 0, & t \geqslant t_0 + \frac{1}{n}. \end{cases}$$

---

① 下式 $T(\omega, 1)$ 中, 1 表恒等于 1 的函数.

显然

$$g(\omega,t_0) = g(\omega,0) + \lim_{n\to+\infty}\int_0^1 a_n(t)\mathrm{d}g(\omega,t)$$
$$= g(\omega,0) + \lim_{n\to+\infty}T(\omega,a_n),$$

因而 $g(\omega,t_0)$ 是一随机变数. 最后, 因 $g(\omega,1)$ 是一常数, 故是一随机变数. ∎

§3.4. 随机逆变换与共轭变换

在泛函分析中,逆算子与共轭算子起着重要作用. 自然地提出问题:对已给随机变换 $T(\omega,x)$,试问其逆变换(依赖于 ω)是否存在?如存在,是否是随机变换?前一问题由普通泛函分析解决. 事实上,对固定的 ω,$T(\omega,x)=T(\omega)x$ 化为普通的算子,有一般的处理方法,因此,只要研究后一问题. 此外,我们还研究共轭变换是否随机的. 结果发现,问题的实质都归结为可测性,然而,证明可测性并不常常容易.

本节中,以 X, Z 表 Banach 空间,X^*, Z^* 为其共轭空间,它们的闭集全体分别记为 C_x, C_z, C_x^*, C_z^*,包含全体闭集的最小 σ-代数分别记为,$\mathcal{B}_x$, $\mathcal{B}_z$, $\mathcal{B}_x^*$, $\mathcal{B}_z^*$. 又若 A 是闭集,则以 $\widetilde{A}$ 表 A 的如下可列稠子集,使若零元 $\theta\in A$,则 $\theta\in\widetilde{A}$ (如果这种 $\widetilde{A}$ 存在).

设 T 为 $\Omega\times Z\text{-}X$ 的线性有界随机变换,称 $\Omega\times X^*\text{-}Z^*$ 变换 T^* 为 T 的共轭变换,如对每 $\omega\in\Omega$,下二等式等价:

$$z^* = T^*(\omega, x^*),$$
$$x^*(T(\omega, z)) = z^*(z) \quad (一切 z\in Z).$$

作为初步,利用随机不动点原理,容易证明下列简单事实:若 $T(\omega,x)$ 为 $(\Omega,\sigma)\times X\text{-}X$($SB$ 空间)上的线性压缩随机变换,则存在线性有界随机变换 S,它是随机变换 $T\text{-}I$ 的逆变换.

事实上,由泛函分析知 S 的存在,故只要证 S 的可测性. 为此,注意对固定的 $z\in X$,$\Omega\text{-}X$ 变换 $S(\omega,z)$ 是线性压缩随

机变换 $T_z(\omega, x)(\omega\in\Omega, x\in X)$ 的随机不动点，其中
$$T_z(\omega, x)=T(\omega, x)-z \quad (\omega\in\Omega, x\in X).$$
由§3.2定理1, 此随机不动点是一随机元.

然而，下述更一般的定理6却不能由随机不动点原理推出：

定理6 设 T 为 $\Omega\times Z$-$X(B$ 空间) 上的线性有界随机变换，这里 Z 是 SB 空间.

(i) 如果 T 可逆, 那么其 $\Omega\times X$-Z 上的逆变换 S 也是线性有界随机变换；

(ii) 如果 Z^* 为可分的，那么共轭变换 T^* 也是线性有界随机变换.

证 S 及 T^* 的线性及有界性均由泛函分析推出. 先证 S 为随机变换. 对每 $x\in X, A\in C_z$,
$$\{\omega: S(\omega, x)\in A\}=\bigcup_{z\in A}\{\omega: S(\omega, x)=z\}$$
$$=\bigcup_{z\in A}\{\omega: T(\omega, z)=x\}. \quad (3.8)$$
此 ω-集等于
$$\bigcap_{n=1}^{+\infty}\bigcup_{z\in\widetilde{A}}\left\{\omega: T(\omega, x)\in O\left(x, \frac{1}{n}\right)\right\}\in\sigma, \quad (3.9)$$
其中 $O(x, r)$ 表以 x 为 0, $r>0$ 为半径的闭球. 实际上, 若 $\omega\in\bigcup_{z\in A}\{\omega: T(\omega, z)=x\}$, 则存在 $z_0\in A$, 使 $T(\omega, z_0)=x$, 由 T 的有界性,
$$\|T(\omega, z)-x\|=\|T(\omega, z)-T(\omega, z_0)\|\leq C(\omega)\|z-z_0\|$$
对一切 $z\in Z$ 成立. 既然 $\widetilde{A}$ 稠于 A, 故对任意正整数[①]n, 存在 $z\in\widetilde{A}$, 使 $T(\omega, z)\in O\left(x, \frac{1}{n}\right)$, 从而
$$\omega\in\bigcap_{n=1}^{+\infty}\bigcup_{z\in\widetilde{A}}\left\{\omega: T(\omega, z)\in O\left(x, \frac{1}{n}\right)\right\}.$$

---

① 例如，任取 $z\in O\left(z_0, \frac{1}{(C(\omega)+1)n}\right)\cap\widetilde{A}=\varnothing$ 即可.

反之，由上式知对每正整数 n，存在 $z_n \in \widetilde{A}$，使 $\omega \in \left\{\omega: T(\omega, z_n) \in O\left(x, \frac{1}{n}\right)\right\}$，故

$$\|z_n - z_m\| = \|S(\omega, T(\omega, z_n)) - S(\omega, T(\omega, z_m))\|$$
$$\leqslant \|S(\omega, \cdot)\| \cdot \|T(\omega, z_n) - T(\omega, z_m) + x - x\|$$
$$\leqslant \|S(\omega, \cdot)\| \cdot \left(\frac{1}{n} + \frac{1}{m}\right),$$

这里 $\|S(\omega, \cdot)\|$ 表 ω 固定时算子 $S(\omega, \cdot)$ 的范数，故存在 $z_0 \in Z$ 使 $z_n \to z_0$. 既然 $z_n \in \widetilde{A}$，故 $z_0 \in A$，因此

$$\omega \in \{\omega: T(\omega, z_0) = x\} \subset \bigcup_{z \in A} \{\omega: T(\omega, z) = x\}.$$

这样，证明了(3.8)式左端集与(3.9)中集的相等性，既然(3.9)中的集可测，于是得证对任意 $A \in C_z$，$\{\omega: S(\omega, x) \in A\} \in \sigma$，由此易知此关系式对任意 $A \in \mathcal{B}_z$ 也成立，因为 $\mathcal{B}_z = \sigma(C_z)$.

次证 T^* 为随机变换. 类似地，对每 $x^* \in X^*$，$A \in C_{z^*}$，有

$$\{\omega: T^*(\omega, x^*) \in A\}$$
$$= \bigcap_{n=1}^{+\infty} \bigcup_{z^* \in \widetilde{A}} \left\{\omega: T^*(\omega, x^*) \in O\left(z^*, \frac{1}{n}\right)\right\}.$$

由共轭变换的定义，对每 $z^* \in Z^*$，$x^* \in X^*$，及正整数 n，

$$\left\{\omega: T^*(\omega, x^*) \in O\left(z^*, \frac{1}{n}\right)\right\}$$
$$= \bigcap_{z \in Z} \left\{\omega: |x^*(T(\omega, z)) - z^*(z)| \leqslant \frac{1}{n}\|z\|\right\},$$

这等价于 $\bigcap_{z \in z} \left\{\omega: |x^*(T(\omega, z)) - z^*(z)| \leqslant \frac{1}{n}\|z\|\right\}$. 故对每 $x^* \in X^*$，$A \in C_{z^*}$，有

$$\{\omega: T^*(\omega, x^*) \in A\}$$
$$= \bigcap_{n=1}^{+\infty} \bigcup_{z^* \in \widetilde{A}} \bigcap_{z \in z} \left\{\omega: x^*(T(\omega, z)) \in O\left(z^*(z), \frac{1}{n}\|z\|\right)\right\},$$

因此，若 T 为随机变换，则 T^* 也是随机变换. ∎

§4. 广义函数空间中的随机元

§4.1. 基本概念

设 $\mathbf{R}^k = (x)$ 为 k 维实数空间，$[a, b]$ 表 $\mathbf{R}^k$ 中子集 $\{x: a_i \leqslant x_i \leqslant b_i, i=1, 2, \cdots, k\}$. 对实数 $m > 0$，以 $[-m, m]$ 表点集 $\{x: -m \leqslant x_i \leqslant m, i=1, 2, \cdots, k\}$. 以 $\widetilde{P}$ 表全体如下的 k 维向量 $\boldsymbol{p}$ 的集：$\boldsymbol{p} = (p_1, p_2, \cdots, p_k)$, $p_i \geqslant 0$ 为整数. Φ 表界定在 $\mathbf{R}^k$ 上的全体基本函数的集，称为基本空间（亦可采用其他基本空间，只要对它下面的定理 2 成立）. 函数 $\varphi(x)$ 称为基本的，如果它取复值；对任意 $p \in \widetilde{P}$，导数

$$D^p \varphi(x) = \frac{\partial^{p_1 + p_2 + \cdots + p_k}}{\partial x_1^{p_1} \partial x_2^{p_2} \cdots \partial x_k^{p_k}} \varphi(x)$$

存在；而且 $\varphi(x)$ 的支集是有界的. 所谓 $\varphi(x)$ 的支集是指点集 $(x: \varphi(x) \neq 0)$ 的闭包. 支集含于 $[-m, m]$ 中的全体基本函数的集记作 Φ_m.

对每 $\varphi(x) \in \Phi$，令 $\|\varphi\| = \sup\limits_{x \in \mathbf{R}^k} |\varphi(x)|$. 称基本函数列 $\{\varphi_n\}_{n=1}^{+\infty}$ 收敛于 $\varphi(x) \in \Phi$，并记为 $\varphi_n \to \varphi$，如果存在实数 $m > 0$，使每 $\varphi_n(x)$ 及 $\varphi(x)$ 的支集均含于 $[-m, m]$ 中，而且对每 $p \in \widetilde{P}$，有 $\|D^p(\varphi_n - \varphi)\| \to 0 (n \to +\infty)$. 集 $B \subset \Phi$ 称为有界的，如存在实数 $m > 0$，使 $B \subset \Phi_m$，而且对每 $\varphi(x) \in B$，$\|D^p \varphi\| \leqslant C_p$，这里 $C_p > 0$ 为不依赖于 $\varphi(x) (\in B)$ 的常数.

界定在 Φ 上的复值、线性、连续泛函 $F(\varphi)$ 称为广义函数. Φ 上全体广义函数所成的空间记作 T. 显然，关于常用的泛函（与函数）的加法运算和对复数的乘法运算，空间 T（及 Φ）是线性的.

称广义函数列$\{F_n\}_{n=1}^{+\infty}$收敛于泛函F，并记为$F_n \to F$，如对每$\varphi \in \Phi$，数列$F_n(\varphi) \to F(\varphi)(n \to +\infty)$.

我们要用到广义函数论中两个重要定理：

定理 1（[2]，卷2，P91 及 P76） 设$\{F_n\}_{n=1}^{+\infty} \subset T$，$F_n \to F$，则$F \in T$，而且此收敛在每一有界集$B$上是均匀的.

定理 2（[9]） 在Φ中存在可列子集H，H关于常用的函数加法成群，而且对每$m > 0$及每$\varphi \in \Phi_m$，存在$\{\varphi_n\}_{n=1}^{+\infty} \subset \Phi_m \cap H$，使$\varphi_n \to \varphi(n \to +\infty)$.

本章以后恒固定概率空间为(Ω, σ, ρ)，$\Omega = (\omega)$，并简称(Ω, σ, ρ)-$(E, \mathscr{B})$随机元为$(E, \mathscr{B})$中的随机元.

特别，可测复值函数空间$(A, \mathscr{A})$中的随机元$\eta(=\eta_x(\omega))$称为随机过程. 其中$A = (a(x))$为界定在$\mathbf{R}^k$上全体复值函数$a(x)$的集，而$\mathscr{A}$为含下型a-集的最小σ-代数：
$$(a: \operatorname{Re} a(x) < c_1, \operatorname{Im} a(x) < c_2),$$
其中$x \in \mathbf{R}^k$，$c_1 \in \mathbf{R}$，$c_2 \in \mathbf{R}$任意固定，Re 及 Im 分别表实、虚部分. 容易看出，为使η为随机过程，必须也只需：

(i) 对每固定的$\omega \in \Omega$，η是$x \in \mathbf{R}^k$的复值函数；

(ii) 对每固定的$x \in \mathbf{R}^k$，η是复值随机变量.

实际上，条件(i)是显然的；其次，由σ-代数$\mathscr{A}$的定义，条件(ii)等价于映像η的可测性.

类似地，可测广义函数空间$(T, \mathscr{F})$中的随机元$\xi(=\xi_\varphi(\omega))$称为广义过程. 这里$\mathscr{F}$为含下型的$F(\in T)$-集的最小$\sigma$-代数：
$$F_{\varphi,c} = (F: \operatorname{Re} F(\varphi) < c_1, \operatorname{Im} F(\varphi) < c_2),$$
其中$\varphi \in \Phi$，$c_1 \in \mathbf{R}$，$c_2 \in \mathbf{R}$任意固定. 同样可证，为使ξ为广义过程，必须也只需：

(i) 对每固定的$\omega \in \Omega$，ξ是$\varphi \in \Phi$的广义函数；

(ii) 对每固定的$\varphi \in \Phi$，ξ是复值随机变量.

称广义函数 $E\xi$ 为广义过程 ξ 的数学期望，如果对每 $\varphi \in \Phi$，$[E\xi](\varphi) = E\xi(\varphi)$（这里 $[E\xi](\varphi)$ 表 $E\xi$ 在 φ 上的值，而 $E\xi(\varphi)$ 则表随机变量 $\xi(\varphi) = \xi_\varphi(\omega)$ 的数学期望）。

容易证明：为使广义过程 ξ 的数学期望 $E\xi$ 存在，必须也只需：对每 $\varphi \in \Phi$，$E\xi(\varphi)$ 存在，而且对每列 $\{\varphi_n\}_{n=1}^{+\infty} \subset \Phi$，$\varphi_n \to 0$，有 $\lim\limits_{n \to +\infty} E\xi(\varphi_n) = 0$。

实际上，如果 $E\xi$ 存在，由 $E\xi$ 的定义 $E\xi(\varphi)$ 存在；既然 $E\xi$ 是广义函数，那么 $E\xi(\varphi_n) = [E\xi](\varphi_n) \to 0$ $(\varphi_n \to 0)$。

为证**充分性**. 定义泛函 $E\xi$：$[E\xi](\varphi) = E\xi(\varphi)$，则
$$[E\xi](\varphi_1 + \varphi_2) = E\xi(\varphi_1 + \varphi_2) = E(\xi(\varphi_1) + \xi(\varphi_2))$$
$$= E\xi(\varphi_1) + E\xi(\varphi_2) = [E\xi](\varphi_1) + [E\xi](\varphi_2).$$

其次，如 $\{\psi_n\}_{n=1}^{+\infty} \subset \Phi$，$\psi_n \to \psi \in \Phi$，$\varphi_n = \psi_n - \psi \to 0$。由 $E\xi(\varphi_n) \to 0$，立得 $E\xi(\psi_n) \to E\xi(\psi)$，即 $[E\xi](\psi_n) \to [E\xi](\psi)$。从而得证 $E\xi$ 为广义函数。

§4.2. 广义过程的条件数学期望

考虑概率空间 (Ω, σ, P) 及广义过程 ξ. 又 $\mathfrak{A}$ 为 σ 的某一子 σ-代数. 本文以后恒设 P 为完全的.

定义 1 界定在 $\omega \in \Omega$，$\varphi \in \Phi$ 上的二元函数 $E(\xi \mid \mathfrak{A})$ $(= E(\xi \mid \mathfrak{A}) = E(\xi \mid \mathfrak{A})(\omega, \varphi))$ 称为广义过程 ξ 关于 $\mathfrak{A}$ 的条件数学期望，如果

(i) 它是一广义过程；关于 $\mathfrak{A}'$ 可测，即对任意 $D \in \mathscr{F}$，有 $(\omega : E(\xi \mid \mathfrak{A}) \in D) \in \mathfrak{A}'$；

(ii) 对任意固定的 $A \in \mathfrak{A}$，$\varphi \in \Phi$，有[①]
$$\int_A \xi(\varphi) P(\mathrm{d}\omega) = \int_A E(\xi \mid \mathfrak{A})(\varphi) P(\mathrm{d}\omega),$$

这里及今后以 $\mathfrak{A}'$ 表 σ-代数，它由一切如下的集 A 构成：A 与

---

① 积分按虚实部分别进行，以后同此。

𝔄 中某一元的对称差是一测度为零的集(简称零测集).

定理 1 为使界定在 $\omega \in \Omega$，$\varphi \in \Phi$ 上的二元函数 $E(\xi \mid \mathfrak{A})$ 是 ξ 关于 $\mathfrak{A}$ 的条件数学期望，必须也只需：

(i) 对任意固定的 $\omega \in \Omega$，$E(\xi \mid \mathfrak{A})$ 是一广义函数；

(ii) 对任意固定的 $\varphi \in \Phi$，存在随机变量 $\xi(\varphi)$ 关于 $\mathfrak{A}$ 的条件数学期望 $E(\xi(\varphi) \mid \mathfrak{A})$([12，13])，使以概率 1，

$$E(\xi \mid \mathfrak{A})(\varphi) = E(\xi(\varphi) \mid \mathfrak{A}). \tag{4.1}$$

证 必要性 设 $E(\xi \mid \mathfrak{A})$ 为 ξ 关于 $\mathfrak{A}$ 的条件数学期望. 由定义 1(i)得(i). 于定义 1(i)中取 $D = F_{\varphi,c}$，由定义 1(i)得

$$(\omega: \operatorname{Re} E(\xi \mid \mathfrak{A})(\varphi) < c_1, \operatorname{Im} E(\xi \mid \mathfrak{A})(\varphi) < c_2)$$
$$= (\omega: E(\xi \mid \mathfrak{A}) \in F_{\varphi,c}) \in \mathfrak{A}',$$

故 $E(\xi \mid \mathfrak{A})(\varphi)$ 关于 $\mathfrak{A}'$ 可测. 再由定义 1(ii)立知 $E(\xi \mid \mathfrak{A})(\varphi)$ 为 $\xi(\varphi)$ 关于 $\mathfrak{A}$ 的一条件数学期望，故(ii)成立.

充分性 设满足条件(i)(ii)的函数 $E(\xi \mid \mathfrak{A})$ 存在. 由(4.1)立得定义 1(ii)，并知 ω-集

$$(\omega: E(\xi \mid \mathfrak{A}) \in F_{\varphi,c})$$
$$= (\omega: \operatorname{Re} E(\xi \mid \mathfrak{A})(\varphi) < c_1, \operatorname{Im} E(\xi \mid \mathfrak{A})(\varphi) < c_2)$$

与 ω-集

$$(\omega: \operatorname{Re} E(\xi(\varphi) \mid \mathfrak{A}) < c_1, \operatorname{Im} E(\xi(\varphi) \mid \mathfrak{A}) < c_2)$$

最多相差一零测集①，由 $E(\xi(\varphi) \mid \mathfrak{A})$ 的定义知后一集属于 $\mathfrak{A}'$，故 $(\omega: E(\xi \mid \mathfrak{A}) \in F_{\varphi,c}) \in \mathfrak{A}'$. 一切满足关系式 $(\omega: E(\xi \mid \mathfrak{A}) \in D) \in \mathfrak{A}'$ 的集 $D(\subset T)$ 构成集 k，显然，k 是一 σ-代数. k 既含一切 $F_{\varphi,c}$，而 $\mathscr{F}$ 又是含一切 $F_{\varphi,c}$ 的最小 σ-代数，故 $k \supset \mathscr{F}$，这表示 $E(\xi \mid \mathfrak{A})$ 关于 $\mathfrak{A}'$ 可测. 由此及(i)即得定义 1(i). ∎

设 $E(\xi \mid \mathfrak{A})$ 为 ξ 关于 $\mathfrak{A}$ 的条件数学期望，又设广义过程

---

① 即对称差为零测集(注意 P 的完全性).

$\widetilde{E}(\xi\mid\mathfrak{A})$ 具有下列性质：存在 ω-集 A，$P(A)=1$，使 $\omega\in A$ 固定时，两广义函数 $E(\xi\mid\mathfrak{A})$ $\widetilde{E}(\xi\mid\mathfrak{A})$ 恒等（即对一切 $\varphi\in\Phi$，$E(\xi\mid\mathfrak{A})$ 与 $\widetilde{E}(\xi\mid\mathfrak{A})$ 在 φ 上值相同），则 $\widetilde{E}(\xi\mid\mathfrak{A})$ 也是 ξ 关于 $\mathfrak{A}$ 的条件数学期望. 这由定义 1（本节）直接推出. 反之，设 $E(\xi\mid\mathfrak{A})$ 及 $\widetilde{E}(\xi\mid\mathfrak{A})$ 均是 ξ 关于 $\mathfrak{A}$ 的条件数学期望，则存在 ω-集 A，$P(A)=1$，使 $\omega\in A$ 固定时，两广义函数 $E(\xi\mid\mathfrak{A})$，$\widetilde{E}(\xi\mid\mathfrak{A})$ 恒等. 事实上，这里以及今后永以 H 表 Φ 中某一可列、稠密子集（不妨设 H 关于常用的函数加法成群，这并不影响可列性），它的存在由定理 B 保证. 由定理 1，对每 $\varphi\in\Phi$，$E(\xi\mid\mathfrak{A})(\varphi)$ 及 $\widetilde{E}(\xi\mid\mathfrak{A})(\varphi)$ 均为 $\xi(\varphi)$ 关于 $\mathfrak{A}$ 的条件数学期望. 故存在可测集 A_φ，使 $P(A_\varphi)=1$，当 $\omega\in A_\varphi$ 时，有

$$E(\xi\mid\mathfrak{A})(\varphi)=\widetilde{E}(\xi\mid\mathfrak{A})(\varphi), \tag{4.2}$$

令 $A=\bigcap_{\varphi\in H}A_\varphi$，$P(A)=1$. 当 $\omega\in A$ 固定时，上式对一切 $\varphi\in H$ 成立. 由广义函数的连续性及 H 的稠密性，可见对每一固定的 $\omega\in A$，(4.2) 对一切 $\varphi\in\Phi$ 成立.

下面研究条件数学期望的存在性. 由定理 1，自然地联想到 $E(\xi\mid\mathfrak{A})$ 非常类似于普通过程论中的条件分布（[12][13]）. 既然后者的存在问题较为复杂，故易想到 $E(\xi\mid\mathfrak{A})$ 也不会常常存在.

定理 2 为使 $E(\xi\mid\mathfrak{A})$ 存在，必须也只需对每 $\varphi\in\Phi$，存在 $E(\xi(\varphi)\mid\mathfrak{A})$，满足下面两条件：

(i) 对任一列 $\{\varphi_n\}_{n=1}^{+\infty}\subset\Phi$，$\varphi_n\to 0$，有

$$P(\omega: \lim_{n\to+\infty} E(\xi(\varphi_n)\mid\mathfrak{A})=0)=1;$$

(ii) 存在 ω-集 A，$P(A)=1$，使 $\omega\in A$ 固定时，对任一列 $\{\varphi_n\}_{n=1}^{+\infty}\subset H$，$\varphi_n\to 0$，有 $\lim_{n\to+\infty} E(\xi(\varphi_n)\mid\mathfrak{A})=0$.

证 充分性 固定一族满足 (i)(ii) 的 $\{E(\xi(\varphi)\mid\mathfrak{A}),\varphi\in\Phi\}$. 因以概率 1，

$$E(\xi(\varphi_1+\varphi_2) \mid \mathfrak{A})$$
$$=E(\xi(\varphi_1) \mid \mathfrak{A})+E(\xi(\varphi_2) \mid \mathfrak{A})(\varphi_1, \varphi_2 \in \Phi),$$

故由 H 的可数性,存在 ω-集 C,$p(C)=1$,使 $\omega \in C$ 固定时,作为 φ 的函数,$E(\xi(\varphi) \mid \mathfrak{A})$ 在 H 上是线性的,因而,当 $\omega \in AC$($p(AC)=1$)固定时,$E(\xi(\varphi) \mid \mathfrak{A})$ 在 H 上是线性连续的. 于是,可依连续性将它拓展到全 Φ 上,同时保持在 Φ 上的线性与连续性. 今定义函数

$$g(\omega, \varphi) = \begin{cases} 0, & \omega \in \overline{AC}, \\ E(\xi(\varphi) \mid \mathfrak{A}), & \omega \in AC, \varphi \in H, \\ \lim_{i=m} E(\xi(\varphi_i) \mid \mathfrak{A}), & \omega \in AC, \varphi \in \Phi, \varphi \in \overline{H} \end{cases} \quad (4.3)$$

$$(\{\varphi_i\}_{n=1}^{+\infty} \subset H, \varphi_i \to \varphi).$$

后一极限显然不依赖于 $\{\varphi_i\}_{n=1}^{+\infty}$ 的选择. 由定义立知:对每固定的 $\omega \in \Omega$,$g(\omega, \varphi)$ 是一广义函数. 若能再验证定理 1 中条件 (ii) 成立,则可取 $g(\omega, \varphi)$ 为 $E(\xi \mid \mathfrak{A})$.

如 $\varphi \in H$ 固定,由 (4.3) 及 $P(CD)=1$,可见

$$g(\omega, \varphi) = E(\xi(\varphi) \mid \mathfrak{A}) \quad \text{a.s.},$$

a.s. 表 "关于 P 几乎处处成立". 故 $g(\omega, \varphi)$ 是 $\xi(\varphi)$,$\varphi \in H$,关于 $\mathfrak{A}$ 的条件数学期望([12]).

如 $\varphi \in \Phi$ 但 $\varphi \in \overline{H}$ 固定,仍由 (4.3) 及定理 2 条件 (i) 有

$$g(\omega, \varphi) = \lim_{i \to +\infty} E(\xi(\varphi_i) \mid \mathfrak{A}) = \lim_{i \to +\infty} E(\xi(\varphi_i - \varphi) \mid \mathfrak{A}) +$$
$$E(\xi(\varphi) \mid \mathfrak{A}) = E(\xi(\varphi) \mid \mathfrak{A}) \quad \text{a.s..}$$

必要性 如 $E(\xi \mid \mathfrak{A})$ 存在,取 $E(\xi(\varphi) \mid \mathfrak{A}) = E(\xi \mid \mathfrak{A})(\varphi)$. 由于对每固定的 $\omega \in \Omega$,广义函数 $E(\xi \mid \mathfrak{A})$ 关于 $\varphi \in \Phi$ 连续,故 (i)(ii) 均成立. ∎

系 1 为使 $E(\xi \mid \mathfrak{A})$ 存在,必须而且只需存在 $\{E(\xi(\varphi) \mid \mathfrak{A}), \varphi \in \Phi\}$ 以及 ω-集 A,$p(A)=1$,使当 $\omega \in A$ 固定时,对一切 $\{\varphi_j\}_{j=1}^{+\infty} \subset \Phi$,$\varphi_j \to 0$,有 $\lim_{j \to +\infty} E(\xi(\varphi_j) \mid \mathfrak{A}) = 0$.

事实上，由系 1 之条件立知定理 2 (i)(ii) 两条件满足. 必要性之证完全与定理 2 中必要性之证相同.

系 1 中的充分必要条件，形式上虽较简单，运用时却不方便，此由下系 2 之证可见.

系 2 设存在广义函数 F，使对每 $\varphi \in \Phi$，有 $|\xi(\varphi)| \leq |F(\varphi)|$ a.s.，则 $E(\xi | \mathfrak{A})$ 存在.

证 只要验证定理 2 中 (i)(ii) 满足. 首先，由 $|\xi(\varphi)| \leq |F(\varphi)|$ a.s.，$E\xi(\varphi)$ 存在，故 $E(\xi(\varphi) | \mathfrak{A})$ 有意义，而且对任意 $\{\varphi_n\}_{n=1}^{+\infty} \subset \Phi$，$\varphi_n \to 0$，有

$$|E(\xi(\varphi_n) | \mathfrak{A})| \leq E(|\xi(\varphi_n)| | \mathfrak{A}) \leq E(|F(\varphi_n)| | \mathfrak{A})$$
$$= |F(\varphi_n)| \to 0 \quad \text{a.s.}$$

其次，存在 ω-集 A，$p(A)=1$，使 $\omega \in A$ 时，对一切 $\varphi_n \in H$，下三式同时成立：

$$E(|F(\varphi_n)| | \mathfrak{A}) = |F(\varphi_n)|;$$
$$E(|\xi(\varphi_n)| | \mathfrak{A}) \leq E(|F(\varphi_n)| | \mathfrak{A});$$
$$|E(\xi(\varphi_n) | \mathfrak{A})| \leq E(|\xi(\varphi_n)| | \mathfrak{A}),$$

故当 $\omega \in A$ 固定时，数列

$$|E(\xi(\varphi_n) | \mathfrak{A})| \leq E(|\xi(\varphi_n)| | \mathfrak{A}) \leq E(|F(\varphi_n)| | \mathfrak{A})$$
$$= |F(\varphi_n)| \to 0 \quad (\varphi_n \in H, \varphi_n \to 0). \blacksquare$$

今研究 $E(\xi | \mathfrak{A})$ 的表现问题. 记 $H = \{\varphi_j\}_{j=1}^{+\infty}$. 以 B_H 表含下型 ω-集：$(\omega: \operatorname{Re} \xi(\varphi_j) < C_1, \operatorname{Im} \xi(\varphi_j) < C_2)$，$\varphi_j \in H$，$C_1 \in \mathbf{R}$，$C_2 \in \mathbf{R}$ 的最小 σ-代数.

定理 3 设 $E(\xi | \mathfrak{A})$ 存在，而且 $\{\xi(\varphi_j)\}_{j=1}^{+\infty}$ 的值域为 Borel 集 (可列无穷维)，则存在 ω-集 A，$p(A)=1$，使 $\omega \in A$ 固定时，对一切 $\varphi \in \Phi$ 有

$$E(\xi | \mathfrak{A})(\varphi) = \lim_{j \to +\infty} \int_\Omega \xi(\psi_j, \omega') p(d\omega', \omega), \quad (4.4)$$

这里 $p(M, \omega)$ 当 $\omega \in \Omega$ 固定时是 B_H 上的概率测度；$\{\psi_j\}_{j=1}^{+\infty} \subset$

H, $\psi_j \to \varphi$.

证 取 $p(M, \omega)$ 为 $\{\xi(\varphi_j)\}_{j=1}^{+\infty}$ 关于 $\mathfrak{A}$ 的条件分布([12][13]),则对 $\varphi_i \in H$,有

$$E(\xi \mid \mathfrak{A})(\varphi_i) = E(\xi(\varphi_i) \mid \mathfrak{A})$$
$$= \int_\Omega \xi(\varphi_i, \omega') p(\mathrm{d}\omega', \omega) \quad \text{a.s.},$$

故存在 ω-集 A,$p(A)=1$,当固定 $\omega \in A$ 时,对一切 $\varphi_i \in H$ 同时有

$$E(\xi \mid \mathfrak{A})(\varphi_i) = \int_\Omega \xi(\varphi_i, \omega') p(\mathrm{d}\omega', \omega).$$

因此,由广义函数 $E(\xi \mid \mathfrak{A})(\varphi)$($\omega \in A$ 固定)关于 φ 的连续性及 H 的稠密性即得证(4.4). ∎

若利用广义条件分布([12][13]),则无须 $\{\xi(\varphi_j)\}_{j=1}^{+\infty}$ 的值域为 Borel 集的假定,仍可得到 $E(\xi \mid \mathfrak{A})$ 的积分表现,但那时积分将在无穷维空间中进行.

$E(\xi \mid \mathfrak{A})$ 具有下列性质,证明甚易,故从略.

(i) 若 $\mathfrak{A} = (\varnothing, \Omega)$,则 $E(\xi \mid \mathfrak{A}) = E(\xi)$ a.s.;

(ii) 若对每 $\varphi \in H$,$\xi(\varphi) \geqslant 0$ a.s.,则 $E(\xi \mid \mathfrak{A}) \geqslant 0$ a.s.;

(iii) 若 C_i 为常数,ξ_i 为广义过程,$i=1, 2, \cdots, n$,则

$$E\left(\sum_{i=1}^n C_i \xi_i \mid \mathfrak{A}\right) = \sum_{i=1}^n C_i E(\xi_i \mid \mathfrak{A}) \quad \text{a.s.};$$

(iv) 若 ξ 关于 $\mathfrak{A}'$ 可测,则 $E(\xi \mid \mathfrak{A}) = \xi$ a.s.;

(v) 若 $\mathfrak{A}_1 \subset \mathfrak{A}_2 \subset \mathscr{B}$,则

$$E(E(\xi \mid \mathfrak{A}_1) \mid \mathfrak{A}_2) = E(\xi \mid \mathfrak{A}_1)$$
$$= E(E(\xi \mid \mathfrak{A}_2) \mid \mathfrak{A}_1) \quad \text{a.s.}.$$

§4.3. 极限定理

可测广义函数空间 $(T, \mathscr{F})$ 中一切随机元 ξ(亦即一切广义

过程)构成集 ε[①]. 称随机元列 $\{\xi_n\}_{n=1}^{+\infty}$ 收敛于 $\xi\in\varepsilon$,并记为 $\xi_n\to\xi$ a.s.,如存在 ω-集 A,$p(A)=1$ 使 $\omega\in A$ 固定时,广义函数列 $\{\xi_n(\omega)\}_{n=1}^{+\infty}$ 收敛于广义函数 $\xi(\omega)$.

定义 2 广义函数列 $\{F_n\}_{n=1}^{+\infty}$ 称为在原点拟同等连续,如对任意 $\varepsilon>0$ 及任一列 $\{\varphi_j\}_{j=1}^{+\infty}\subset\Phi$,$\varphi_j\to 0$,存在两正整数 N 及 M,使对一切 $n>N$,$j>M$,有 $|F_n(\varphi_j)|<\varepsilon$.

定理 4 为使随机元列 $\{\xi_n\}_{n=1}^{+\infty}$ 以概率 1 收敛于某随机元 ξ,必须也只需:

(i) 对任意固定的 $\varphi_j\in H$,随机变量列 $\{\xi_n(\varphi_j)\}_{n=1}^{+\infty}$ 以概率 1 收敛;

(ii) $\{\xi_n\}_{n=1}^{+\infty}$ 以概率 1 在原点拟同等连续.

证 充分性 令

$A=\bigcap_{\varphi_j\in H}(\omega:\{\xi_n(\varphi_j)\}_{n=1}^{+\infty}\text{收敛})\bigcap(\omega:\{\xi_n\}_{n=1}^{+\infty}\text{在原点拟同等连续})$,

则 $p(A)=1$. 令任意固定 $\omega\in A$,对每 $\varphi\in\Phi$,存在 $\varphi_h\in H$,$h\in \mathbf{N}^*$,$\varphi_h\to\varphi$,或 $\varphi_h-\varphi\to 0(h\to+\infty)$,而且

$|\xi_n(\varphi)-\xi_m(\varphi)|$
$\leqslant|\xi_n(\varphi)-\xi_n(\varphi_h)|+|\xi_n(\varphi_h)-\xi_m(\varphi_h)|+|\xi_m(\varphi_h)-\xi_m(\varphi)|.$
(4.5)

由于 $\omega\in A$ 及 D_2,对任意 $\varepsilon>0$,存在 $M>0$,$N>0$,使 n,$m>N$,$h>M$ 时,上式右方第一项

$$|\xi_n(\varphi)-\xi_n(\varphi_h)|=|\xi_n(\varphi-\varphi_h)|<\frac{\varepsilon}{3},$$

同时也使第三项小于 $\frac{\varepsilon}{3}$. 固定 $h>M$. 由 $\omega\in A$,存在 $k>0$,使 n,$m>k$ 时,第二项也小于 $\frac{\varepsilon}{3}$. 故当 n,m 均大于 $\max(k,N)$

[①] 请勿与正数 ε 混淆.

时，$|\xi_n(\varphi)-\xi_m(\varphi)|<\varepsilon$. 这表示数列 $\{\xi_n(\varphi)\}_{n=1}^{+\infty}$ 收敛，其中 $\varphi\in\Phi$ 任意固定. 从而当 $\omega\in A$ 固定时，广义函数列 $\{\xi_n(\omega)\}_{n=1}^{+\infty}$ 收敛于某极限 $\xi_0(\omega)$. 定义 $\xi(\omega)$ 为

$$\xi(\omega)=\begin{cases}\xi_0(\omega),&\omega\in A,\\ 0,&\omega\overline{\in}A,\end{cases}$$

由定理 A，可见当 $\omega\in\Omega$ 固定时，$\xi(\omega)$ 是一广义函数. 其次，既然随机变量列 $\{\xi_n(\varphi)\}_{n=1}^{+\infty}$ 以概率 1 收敛于 $\xi(\varphi)$，故 $\xi(\varphi)$ 也是一随机变量. 由 §4.1 立知 $\xi\in\varepsilon$，从而 $\xi_n\to\xi$ a.s..

必要性 (i) 必要显然. 下证 (ii) 必要. 设 $\xi_n\to\xi$ a.s.. 由定义知存在 A，$p(A)=1$ 当 $\omega\in A$ 固定时，广义函数列 $\{\xi_n(\omega)\}_{n=1}^{+\infty}$ 收敛于广义函数 $\xi(\omega)$，由定理 A，此收敛性关于 $\varphi\in B$ ($B\subset\Phi$ 为任一有界集) 是均匀的. 取 $B=\{\varphi_j\}_{j=1}^{+\infty}\subset\Phi$，$\varphi_j\to 0$ ($j\to+\infty$)，则 B 必为有界集，故对任意 $\varepsilon>0$，存在 $N>0$，当 $n>N$ 时，对一切 $\varphi_j\in B$ 有

$$|\xi_n(\varphi_j)-\xi(\varphi_j)|<\frac{\varepsilon}{2},$$

其次，由广义函数 $\xi(\omega)$ 的连续性，存在 $M>0$，使 $j>M$ 时

$$|\xi(\varphi_j)|<\frac{\varepsilon}{2}.$$

因此，对一切 $n>N$，$j>M$，有 $|\xi_n(\varphi_j)|<\varepsilon$. 既然此性质对每 $\omega\in A$ 成立，故得证 (ii). ∎

现在应用定理 4 来研究独立随机元列与平稳随机元列的强大数定理.

称 $\{\xi_n\}_{n=1}^{+\infty}\subset\varepsilon$ 为独立的，如对任意正整数 m，任意 $E_j\in\mathscr{F}$，$j=1,2,\cdots,m$，任一组正整数 $i_1<i_2<\cdots<i_m$，有

$$p(\bigcap_{i=1}^{m}\{\omega:\xi_{i_j}(\omega)\in E_j\})=\bigcap_{j=1}^{m}p(\omega:\xi_{i_j}(\omega)\in E_j). \quad (4.6)$$

称 $\{\xi_n\}_{n=1}^{+\infty}\subset\varepsilon$ 为平稳的，如对任意正整数 τ，有

$$p(\bigcap_{i=1}^{m}\{\omega: \xi_{i_j+\tau}(\omega) \in E_j\}) = p(\bigcap_{j=1}^{m}\{\omega: \xi_{i_j}(\omega) \in E_j\}). \quad (4.7)$$

以上独立性与平稳性的定义均已在[10]中给出.

今定义随机元 $\xi \in \varepsilon$ 的方差为

$$D\xi = \int_{\Omega} \sup_{\varphi \in \Phi} |\xi(\varphi) - M\xi(\varphi)|^2 p(\mathrm{d}\omega)$$

$$= \int_{\Omega} \sup_{\varphi \in H} |\xi(\varphi) - M\xi(\varphi)|^2 p(\mathrm{d}\omega),$$

只要上式中所用到的积分收敛.

定理 5 设 $\{\xi_n\}_{n=1}^{+\infty} \subset \varepsilon$ 为独立随机元列,而且 $E\xi_n = 0$,$n \in \mathbf{N}^*$,则为使随机元列

$$\zeta_j = \frac{1}{j}\sum_{n=1}^{j} \xi_n \to 0 \text{ a.s.} \quad (j \to +\infty),$$

只需 (i) $\sum_{n=1}^{+\infty} \dfrac{D\xi_n}{n^2} < +\infty$;

(ii) $\{\xi_n\}_{n=1}^{+\infty}$ 以概率 1 在原点拟同等连续.

证 对任意固定的 $\varphi \in \Phi$,令 $E_j = F_{\varphi, c_j}$,代入(4.6)即得随机变量列 $\{\xi_n(\varphi)\}_{n=1}^{+\infty}$ 的独立性. 由(i)知 $\sum_{n=1}^{+\infty} \dfrac{D\xi_n(\varphi)}{n^2} < +\infty$,故由独立随机变量列的 Колмогоров 强大数定理得 $\zeta_j(\varphi) \to 0$ a.s.. 如能证明 $\{\zeta_j\}_{j=1}^{+\infty}$ 以概率 1 在原点拟同等连续,则由定理 4 即得 $\zeta_j \to 0$ a.s.. 由条件(ii),存在 ω-集 A,$p(A) = 1$,使 $\omega \in A$ 固定时,广义函数列 $\{\xi_n(\omega)\}_{n=1}^{+\infty}$ 在原点拟同等连续. 今证对此 ω,$\{\zeta_j(\omega)\}_{j=1}^{+\infty}$ 也在原点拟同等连续. 任取 $\varepsilon > 0$ 及 $\{\varphi_m\} \subset \Phi$,$\varphi_m \to 0$,由假设存在两整数 $N > 0$,$M > 0$,使 $n > N$,$m > M$ 时,

$$|\xi_n(\varphi_m)| < \frac{\varepsilon}{2}. \quad (4.8)$$

其次,由于 $\xi_n(\varphi)(n=1, 2, \cdots, N)$ 都是广义函数而且 $\varphi_m \to 0$,故存在整数 $L > 0$,当 $m > L$ 时

$$|\xi_n(\varphi_m)| < \frac{\varepsilon}{2} \quad (n=1, 2, \cdots, N). \tag{4.9}$$

因此,当 $j > N$, $m > \max(L, M)$ 时,由(4.8)(4.9)得

$$|\zeta_j(\varphi_m)| = \frac{1}{j} \left| \sum_{n=1}^{j} \xi_n(\varphi_m) \right|$$

$$\leqslant \frac{1}{j} \sum_{n=1}^{N} |\xi_n(\varphi_m)| + \frac{1}{j} \sum_{n=N+1}^{j} |\xi_n(\varphi_m)| < \varepsilon. \blacksquare$$

两个 ω-集 A_1, A_2 称为对等的,并记为 $A_1 \approx A_2$,如果它们最多相差一零测集.

设 $\{\xi_n\}_{n=1}^{+\infty} \subset \varepsilon$ 为平稳随机元列,它所产生的 σ-代数记为 $\mathscr{B}_\xi$,换言之,$\mathscr{B}_\xi$ 为含下型 ω-集

$$(\omega: \xi_n(\omega) \in E) \quad (n \in \mathbf{N}^*, E \in \mathscr{F})$$

的最小 σ-代数. 由平稳性,$\{\xi_n\}$ 在 $\mathscr{B}_\xi'$ 上决定唯一具有下性质的保测度集变换 T_ξ:对任意正整数 n,任意 $E^{(n)} \in \mathscr{F}^n$ ($\mathscr{F}^n = \mathscr{F} \times \mathscr{F} \times \cdots \times \mathscr{F}$, n 次),

$$T_\xi(\omega: \{\xi_1(\omega), \xi_2(\omega), \cdots, \xi_n(\omega)\} \in E^{(n)})$$

$$\approx (\omega: \{\xi_2(\omega), \xi_3(\omega), \cdots, \xi_{n+1}(\omega)\} \in E^{(n)}). \tag{4.10}$$

其证明完全与平稳随机变量序列情况一样([12]).

集 $A \in \mathscr{B}_\xi'$ 称为关于 $\{\xi_n\}_{n=1}^{+\infty}$ 为不变的,如 $T_\xi A \approx A$. 全体不变集构成一 σ-代数 U_ξ,$U_\xi \subset \mathscr{B}_\xi'$.

固定 $\varphi \in \Phi$. 于(4.7)中取 $E_i = F_{\varphi, c_i}$,$i=1, 2, \cdots, m$,立见随机变量列 $\{\xi_n(\varphi)\}_{n=1}^{+\infty}$ 是平稳的,以 $\mathscr{B}_\varphi$ 表含下型 ω-集

$$(\omega: \operatorname{Re} \xi_n(\varphi) < C_1, \operatorname{Im} \xi_n(\varphi) < C_2) \quad (C_1 \in \mathbf{R}, C_2 \in \mathbf{R})$$

的最小 σ-代数,由通常的关于平稳随机变量序列的理论([12]),$\{\xi_n(\varphi)\}_{n=1}^{+\infty}$ 在 $\mathscr{B}_\varphi'$ 上决定唯一具有下性质的保测度集变换 T_φ:

$$T_\varphi(\omega: \{\operatorname{Re} \xi_1(\varphi), \operatorname{Im} \xi_1(\varphi), \cdots,$$

$$\operatorname{Re} \xi_n(\varphi), \operatorname{Im} \xi_n(\varphi)\} \in A_{2n})$$

$$\approx (\omega: \{\operatorname{Re}\xi_2(\varphi), \operatorname{Im}\xi_2(\varphi), \cdots,$$
$$\operatorname{Re}\xi_{n+1}(\varphi), \operatorname{Im}\xi_{n+1}(\varphi)\} \in A_{2n}), \quad (4.11)$$

其中 A_{2n} 为任意 $2n$ 维 Borel 集，$n \in \mathbf{N}^*$。

另一方面，因 $\mathcal{B}_\varphi' \subset \mathcal{B}_\xi'$，故 T_ξ 在 $\mathcal{B}_\xi'$ 上派生一保测度集变换 $\widetilde{T}_\varphi: \widetilde{T}_\varphi A = T_\xi A, (A \in \mathcal{B}_\varphi')$。由(4.10)易见 $\widetilde{T}_\varphi$ 也具有(4.11)中的性质(因由 $\mathcal{F}$ 的定义，易证(4.11)左方的 ω-集可表为(ω: $\{\xi_1(\omega), \xi_2(\omega), \cdots, \xi_n(\omega)\} \in E_\varphi^{(n)}), E_\varphi^{(n)} \in \mathcal{P}^n$, 之形)。既然具此性质的保测度集变换唯一，故 T_φ 与 $\widetilde{T}_\varphi$ 重合。

关于 T_φ 不变的集所成的 σ-代数记为 U_φ。可证 $U_\varphi = U_\xi \cap \mathcal{B}_\varphi'$。实际上，若 $A \in U_\xi \cap \mathcal{B}_\varphi'$，则 $T_\varphi A = \widetilde{T}_\varphi A = T_\xi A \approx A$，故 $A \in U_\varphi$。反之，若 $A \in U_\varphi$，则 $A \in \mathcal{B}_\varphi'$，而且 $T_\xi A = \widetilde{T}_\varphi A = T_\varphi A \approx A$，故 $A \in U_\xi$，因此 $A \in U_\xi \cap \mathcal{B}_\varphi'$。

定理 6[①] 设 $\{\xi_n\}_{n=1}^{+\infty} \subset \varepsilon$ 为平稳随机元列，$E\xi_1$ 存在，而且以概率 1 在原点拟同等连续，则

$$\zeta_j = \frac{1}{j}\sum_{n=1}^{j}\xi_n \to E(\xi_1 \mid U_\xi) \quad \text{a.s.}$$

证 对每 $\varphi \in \Phi$，由假设 $E\xi_1(\varphi)$ 存在，又 $\{\xi_n(\varphi)\}_{n=1}^{+\infty}$ 为平稳随机变量序列，故由 Brikhoff-Хинчин 定理([12])

$$\zeta_j(\varphi) \to E(\xi_1(\varphi) \mid U_\varphi) \quad \text{a.s.} \quad (4.12)$$

其次，在证定理 5 时已证明 $\{\zeta_j\}_{j=1}^{+\infty}$ 以概率 1 在原点拟同等连续，故由定理 4，存在 $\xi \in \varepsilon$，使

$$\zeta_j \to \xi \quad \text{a.s.}$$

固定 $\varphi \in \Phi$。由(4.12)可见 $\xi(\varphi) = E(\xi_1(\varphi) \mid U_\varphi)$ a.s.，既然 $U_\varphi \subset U_\xi$，故 $\xi(\varphi)$ 关于 U_ξ' 可测。若能再证明对任意 $A \in U_\xi$，有

$$\int_A \xi(\varphi) p(\mathrm{d}\omega) = \int_A \xi_1(\varphi) p(\mathrm{d}\omega),$$

[①] [10]中证明了 $\{\zeta_j\}_{j=1}^{+\infty}$ 有极限而未求出。

则由定理 1 即得证本定理(附带证明了 $E(\xi_1 \mid U_\xi)$ 存在).

为此，以 $\tilde{\eta}$ 表复数 η 的实部(或虚部)，注意 A 关于 T_ξ 的不变性以及 $T_\xi C = T_\varphi C$ ($C \in \mathscr{B}_\varphi'$) 得

$$\int_A \tilde{\xi}_1(\varphi) p(\mathrm{d}\omega) = \lim_{n \to +\infty} \sum_{k=-\infty}^{+\infty} \frac{k}{2^n} p\left(\left\{\frac{k}{2^n} \leqslant \tilde{\xi}_1(\varphi) < \frac{k+1}{2^n}\right\} \cap A\right)$$

$$= \lim_{n \to +\infty} \sum_{k=-\infty}^{+\infty} \frac{k}{2^n} p\left(T_\xi^j\left\{\frac{k}{2^n} \leqslant \tilde{\xi}_1(\varphi) < \frac{k+1}{2^n}\right\} \cap T_\xi^j A\right)$$

$$= \lim_{n \to +\infty} \sum_{k=-\infty}^{+\infty} \frac{k}{2^n} p\left(\left\{\frac{k}{2^n} \leqslant \tilde{\xi}_{j+1}(\varphi) < \frac{k+1}{2^n}\right\} \cap A\right)$$

$$= \int_A \tilde{\xi}_{j+1}(\varphi) p(\mathrm{d}\omega) = \int_A \tilde{\zeta}_j(\varphi) p(\mathrm{d}\omega) \to \int_A \tilde{\xi}(\varphi) p(\mathrm{d}\omega).$$

这里可在积分号下取极限是由于 $\tilde{\zeta}_j(\varphi)$, $j \geqslant 1$ 的均匀可积性 [12]. ∎

参考文献

[1] Schwartz L. Théorie des Distributions. Hermann et Cie, Paris, 1950.

[2] Гельфанд Н М и Шилов Г Е. Обобшенные функцни. Вып. 1, 1960；Вып. 2, 1958.

[3] 冯康. 广义函数论. 数学进展, 1955, 1(3)：405-590.

[4] Гелвфанд И М. Обобшенные случайные процессы. ДАН СССР, 1955, 100：853-856.

[5] Ito K. Stationary random distributions. Mem. Col. Sic. Univ. Kyoto, Ser. A, 1954, 28：209-223.

[6] Урбаник К. Случайные процессы, реализации которых являются обобшенными функциями. Теор. Вероят. и ее Npим., Вып, 1956, 1(1)：146-148.

[7] 郑绍濂. 正则与奇异的平稳广义随机过程. 科学记录,

1959, 3(8): 281-286.

[8] Mikusiński J and Sikorski. The elementary theory of distributions. Inventiones Mathematicae, 1957, 38(2): 187-206.

[9] Winkelbauer K. К теории обобшенных случайных процессов. Чехослиовский Мат. Лсурнап. 1956, 6: 517-521.

[10] Ullrich M. Some theorems on random Schwartz distributions. Trans. of the 1st Prague Conference on Information Theory. etc., 1956: 273-291.

[11] 伊藤清. 确率过程(中译本, 随机过程. 刘璋温, 译. 上海: 上海科学技术出版社, 1961).

[12] Doob J L. Stochastic Processes. 1953.

[13] Loève M. Probability Theory. 1955.

[14] Bharucha-Reid A T. On random solutions of integral equations in Banach space. Trans. of the Second Prague Conference on Information Theory, etc., 1960: 27-48.

[15] Hanš O. Generalized random variables. Trans. of the 1st Prague Conference on Information Theory, etc., 1957: 61-103.

[16] Hanš O. Random fixed point theorems. Trans. of the 1st Prague Conference on Information Theory, etc., 1957: 105-125.

[17] Hanš O. Invers and adjoint transforms of linear bounded random transforms. Trans. of the 1st Prague Conference on Information Theory, etc.,

1957：127-133.

[18] Hanš O. Measurability of extensions of continuous random transforms. Annals of Math. Statistics，1959，30：1 152-1 157.

[19] Mourier E. Éléments aléatoires dans un espace de Banach. Annales de l'Institut Henri Poincaré，1953，13：161-244.

[20] Прохоров Ю В. Сходимость случайных процессов н предельные теоремы теории вероятностей. Теор. Вероят. и ее Ирим, 1956, 1(2)：177-238.

[21] 郑曾同. 测度的弱收敛与强马氏过程. 数学学报，1961，11：126-132.

[22] Dubins L E. Generalized random variables. Trans. of the Amer. Math. Society，1957，84：273-309.

[23] 王寿仁. 关于广义随机过程的一个注记. 科学记录，1958，2(1)：15-17.

[24] Driml M and Hanš O. Conditional expectations for generalized random variables. Trans. of the Second Prague Conference on Information Theory，etc.，1960：123-144.

[25] 关肇直. 泛函分析讲义. 北京：高等教育出版社，1958.

[26] Nedoma J. Note on generalized random variables. Trans. of the 1st Prague Conference on Information Theory. etc.，1957：139-141.

[27] Halmos P R. Measure Theory. 1950(中译本，测度论. 王建华，译. 北京：科学出版社，1980).

生灭过程构造论

§1. 引 言

设 $x(t, \omega)$, $t \geqslant 0$ 为定义在概率空间 $(\Omega, \mathscr{F}, P)$ 上的具可列多个状态的马氏过程，其相空间为 $E = \mathbf{N}$，转移概率为 $P_{ij}(t)$, $i, j \in E$, $t \geqslant 0$，它们是一组满足下列条件的实值函数：

$$P_{ij}(t) \geqslant 0, \tag{1.1}$$

$$\sum_j P_{ij}(t) = 1, \tag{1.2}$$

$$\sum_k P_{ik}(t) P_{kj}(s) = P_{ij}(t+s), \tag{1.3}$$

$$\lim_{t \to 0} P_{ij}(t) = P_{ij}(0) = \delta_{ij}, \tag{1.4}$$

其中 $\sum_j$ 等表对一切 $j \in E$ 求和，又 $\delta_{ii} = 1$, $\delta_{ij} = 0$, $(i \neq j)$。由此可证明[1,P126~127]存在极限

$$\lim_{t \to 0} \frac{P_{ij}(t) - \delta_{ij}}{t} = q_{ij}. \tag{1.5}$$

---

① 收稿日期：1962-01-09；收修改稿日期：1962-03-20.

令 $q_i = -q_{ii} \geqslant 0$，以后恒设对一切 $i \in E$，有
$$0 < \sum_{j \neq i} q_{ij} = -q_{ii} = q_i < +\infty, \quad (1.6)$$
称 $\boldsymbol{Q} = (q_{ij})$ 为密度矩阵，而马氏过程 $x(t, \omega)$，$t \geqslant 0$，则简称为 Q 过程，以表示它与 Q 有(1.5)的关系. 如果两个 Q 过程有相同的 $P_{ij}(t)$，我们就把它们看作同一 Q 过程. 故有时也称一组满足(1.1)~(1.4)的 $(P_{ij}(t))$ 为一 Q 过程. 特别当
$$\begin{cases} q_{ii+1} = b_i (>0), & q_{ii-1} = a_i (>0), \\ q_{ii} = -(a_i + b_i); & q_{ij} = 0 \quad (|i-j| > 1) \end{cases} \quad (1.7)$$
(补定义 $a_0 = 0$)时，称如的 Q 过程为生灭过程.

反之，设已给矩阵 $\boldsymbol{Q} = (q_{ij})$，$q_{ij} \geqslant 0 (i \neq j)$ 并使(1.6)成立[1]，一个重要的问题是求出一切 Q 过程，即求出一切矩阵 $\boldsymbol{P}(t) = (P_{ij}(t))$，使满足条件(1.1)~(1.5). 可以证明[1,P226]，这问题在假定(1.6)下，等价于求下列向后微分方程组的满足(1.1)~(1.5)的全部解：
$$\boldsymbol{P}'(t) = \boldsymbol{Q}\boldsymbol{P}(t), \quad \boldsymbol{P}(0) = \boldsymbol{1}, \quad (1.8)$$
这里 $\boldsymbol{P}'(t) = (P'_{ij}(t))$，$\boldsymbol{I} = (\delta_{ij})$.

关于此问题的研究情况如下：1945 年 Doob[1,P241] 证明，Q 过程总是存在的，而且只有两种可能，或者只存在一个，或者有无穷多个；早在 1940 年，Feller 在[7]中甚至对更一般情况找到了 Q 过程唯一的充分必要条件，此条件后为 Добрущин 所详细研究[2]，满足 Feller 条件的矩阵 $\boldsymbol{Q}$ 称为规则的，于是 Q 过程的存在与唯一问题得以完满解决. 剩下是求出全部解(Q 过程)的问题. 对规则矩阵，这问题早已解决[7]；对任一不规则矩阵，Doob 在[4,P267]中虽找出了无穷多个 Q 过程(以后称这种 Q 过程为 Doob 过程)，但却远未穷尽一切 Q 过程. 于是求出

---

① 本文以后所用的 Q 均满足此两个条件，不再声明.

全部 Q 过程的问题，引起了广泛的注意，虽然经过不少人的努力，距离彻底解决，似乎还要做许多工作. 目前试图解决此问题的方法至少有三种，各均取得一定成果. Feller，孙振祖，Reuter，Karlin 与 Mcgregor 分别用各种分析方法[8][10][11][13]对某种 Q 求出了全部$(P_{ij}(t))$，但[8]与[10]中所求出的$(P_{ij}(t))$除满足(1.8)外，还满足向前微分方程组

$$P'(t)=P(t)Q, \quad P(0)=1; \quad (1.9)$$

Дынкин 与 Reuter 用半群方法[6][11]，找出了部分 Q 过程的无穷小算子的定义域；第三种方法见 Добрущин[3]及作者[14][15]，这种方法的基本思想类似于函数构造论：根据样本函数（或称轨道）的性质，可以看出 Doob 过程的结构较为简单，然后通过这种较简单的过程来逼近任一 Q 过程. 这种方法目前虽然只用来研究几种特殊的 Q 矩阵，但估计它的潜力尚未全部发挥[参看 3]，本文以前此法的逻辑基础尚未发表.

求出全部 Q 过程的问题之所以重要至少是由于：一方面，在现实中遇到的马氏过程，容易求出的是 Q 而不是 $P_{ij}(t)$，例如排队论中许多例子[12, §20, §34]就是如此，因此，Q 是否能唯一决定过程、如何决定以及如不能唯一决定时，尚需补充些什么数字特征才能唯一决定等问题，就具有重要的理论与实际的意义；另一方面，如上所述，此问题紧密联系于微分方程论（解方程组(1.8)）及半群理论（求出某已给无穷小算子表达式的全体可能的定义域，或等价地，求出全体以此算子为无穷小算子的半群，这些半群由马尔可夫转移概率产生）. 因此，如能以概率方法求出全部 Q 过程，就等价于用概率方法求出了(1.8)的全部解或全体半群，对微分方程及半群理论均有一定影响. 从这种观点看来，上述第三种方法也许更合乎要求.

本文的主要目的是：研究生灭过程样本函数的性质；并在

此基础上，应用第三种方法来求出全体生灭过程（对已给 Q）；同时就这种过程来叙述第三种方法的逻辑基础．因此，它的目的是结果性的，也是方法性的．文中部分结果已预告于[14, 15]中，此地给予证明．在这一部分成果的获得过程中，作者得到 Р. Л. Добрущин 先生指导，谨致以衷心的谢意．

§2～§5 研究生灭过程样本函数的性质与过程的变换；§6～§8 对已给的 Q 求出全部 Q 过程并讨论其构造；§9 中叙述若干尚待进一步研究的问题；最后，为便于阅读起见，在附录中证明了本文用到的关于生灭过程的性质．阅读本文还需要过程论中一些最基本的知识，如过程的可分性等，在附录中不可能补齐，且幸所需的全部预备知识不多，它们都可在[1. §Ⅱ.2～§Ⅱ.9]或相应的书中找到．

§2. 基本特征数的概率意义

设已给形如(1.7)的密度矩阵 Q. 对这类矩阵,重要的是下列基本特征数[①]:

$$m_i = \frac{1}{b_i} + \sum_{k=0}^{i-1} \frac{a_i a_{i-1} \cdots a_{i-k}}{b_i b_{i-1} \cdots b_{i-k} b_{i-k-1}} \quad (i \geqslant 0), \qquad (2.1)$$

$$e_i = \frac{1}{a_i} + \sum_{k=0}^{+\infty} \frac{b_i b_{i+1} \cdots b_{i+k}}{a_i a_{i+1} \cdots a_{i+k} a_{i+k+1}} \quad (i > 0), \qquad (2.2)$$

$$R = \sum_{i=0}^{+\infty} m_i, \quad S = \sum_{i=0}^{+\infty} e_i, \qquad (2.3)$$

以及

$$z_0 = 0, \quad z_n = 1 + \sum_{k=1}^{n-1} \frac{a_1 a_2 \cdots a_k}{b_1 b_2 \cdots b_k}, \quad z = \lim_{n \to +\infty} z_n. \qquad (2.4)$$

为了叙述这些数字的概率意义,考虑任一 Q 过程 $x(t, \omega)$, $t \geqslant 0$. 不影响转移概率及 Q,可设此过程为完全可分的[4,P57],即关于一维闭集可分,而且可分 t-集可取为 $[0, +\infty)$ 中任一可列稠集 $R = (r_i)$. 由于可分性的需要,有时须引入虚状态 $+\infty$,但对任一固定的 $t \geqslant 0$, $P(x(t, \omega) = +\infty) = 0$. 以后记 $\bar{E} = E \cup \{+\infty\}$. 定义随机变量

$$\xi_i(\omega) = \inf(t : x(t, \omega) = i) \qquad (2.5)$$

并以 P_j 及 E_j 表由过程的转移概率及集中在状态 j 上的开始分布所产生的概率测度及对此测度而取的数学期望,[2]中证明了:$m_i = E_i \xi_{i+1}(\omega)$,因而 $R = \lim_{n \to +\infty} E_0 \xi_n(\omega)$;并且还证明了,当且仅当 $R = +\infty$ 时,Q 过程是唯一的(见附录定理1),因此,我

[①] 特征数 R 等的一般化见[16].

们以后只考虑 $R<+\infty$ 的情形. 如果令

$$\tau(\omega)=\inf(t: \lim_{s\uparrow t} x(s,\omega)=+\infty) \tag{2.6}$$

（即[1，P235]所称的第一个无穷），则由积分的单调收敛定理，$R=E_0\tau(\omega)$. 换言之，R 是自 0 出发，沿着过程的轨道而运动的质点初次到达 $+\infty$ 的平均时间.

S 的概率意义会预告于[14，15]. 考虑矩阵

$$Q_N = \begin{Bmatrix} -b_0 & b_0 & 0 & \cdots & 0 & 0 & 0 \\ a_1 & -(a_1+b_1) & b_1 & \cdots & 0 & 0 & 0 \\ \vdots & \vdots & \vdots & \vdots & \vdots & \vdots & \vdots \\ 0 & 0 & 0 & \cdots & a_{N-1} & -(a_{N-1}+b_{N-1}) & b_{N-1} \\ 0 & 0 & 0 & \cdots & 0 & a_N+b_N & -(a_N+b_N) \end{Bmatrix}, \tag{2.7}$$

它由 Q 中前 $n+1$ 横列与前 $n+1$ 直行的元素构成，但需将其中第 $n+1$ 横列与第 n 直行上的元 a_N 换成 a_N+b_N. 考虑可分 Q_N 过程 $x_N(t,\omega)$，$t\geq 0$. 定义

$$\sigma_N(\omega)=\inf(t: x_N(t,\omega)=0). \tag{2.8}$$

若设 $P(x_N(0,\omega)=N)=1$，则有

引理 2.1 $\lim_{N\to+\infty} E\sigma_N(\omega)=S.$

证　引进

$$\xi_i^{(N)}(\omega)=\inf(t: x_N(t,\omega)=i), \tag{2.9}$$

并令 $e_i^{(N)}=E_i\xi_{i-1}^{(N)}(\omega)$，即 $e_i^{(N)}$ 为关于 $x_N(t,\omega)$ 自 i 出发初次到达 $i-1$ 的平均时间. 众所周知[1，P148，P215]，对 Q_N 过程，在状态 k 上的逗留时间 $\beta(=\beta_k)$ 有指数分布

$$P_k(\beta\leq t)=\begin{cases} 1-e^{-c_k t}, & t\geq 0, \\ 0, & t<0, \end{cases} \quad k\geq 0, \tag{2.10}$$

其中 $c_k = a_k + b_k$，故 $E_k\beta = \dfrac{1}{c_k}$，而且

$$\begin{cases} P_k(x_N(\beta+0) = k+1) = \dfrac{b_k}{c_k}, \\ P_k(x_N(\beta+0) = k-1) = \dfrac{a_k}{c_k}, \end{cases} \quad 0 < k < N, \quad (2.11)$$

$$P_0(x_N(\beta+0) = 1) = P_N(x_N(\beta+0) = N-1) = 1 \quad (2.12)$$

（为使(2.11)(2.12)有意义，应补设此过程是 Borel 可测的，见[1, P141]）. (2.12) 表示，状态 O 与 N 都是 Q_N 过程的反射壁. 由此可见，$e_i^{(N)}$ 应满足方程组

$$e_i^{(N)} = \dfrac{a_i}{a_i+b_i} \cdot \dfrac{1}{a_i+b_i} + \dfrac{b_i}{a_i+b_i}\left(\dfrac{1}{a_i+b_i} + e_{i+1}^{(N)} + e_i^{(N)}\right)$$

$$(i = 1, 2, \cdots, N-1),$$

$$e_N^{(N)} = \dfrac{1}{a_N + b_N}.$$

解出后得

$$e_i^{(N)} = \dfrac{1}{a_i} + \sum_{k=0}^{N-2-i} \dfrac{b_i b_{i+1} \cdots b_{i+k}}{a_i a_{i+1} \cdots a_{i+k} a_{i+k+1}} + \dfrac{b_i b_{i+1} \cdots b_{N-1}}{a_i a_{i+1} \cdots a_{N-1}(a_N + b_N)}$$

$$(i = 1, 2, \cdots, N-2), \quad (2.12')$$

从而 $\lim\limits_{N \to +\infty} e_i^{(N)} = e_i$，又

$$\lim_{N \to +\infty} E\sigma_N(\omega) = \lim_{N \to +\infty} \sum_{i=1}^{N} e_i^{(N)} = S. \quad \blacksquare$$

直观地说，上引理表示，当 $+\infty$ 是"反射壁"时，e_i 是 Q 过程自 i 初次到 $i-1$ 的平均时间，而 S 则是此过程自 $+\infty$ 初次到 0 的平均时间.

根据 $S < +\infty$ 或 $S = +\infty$，可将全体不规则的（或等价地，使 $R < +\infty$ 的）(1.7)形的矩阵 Q 分为 S_1 与 S_2 两类. 以后（见定理 4.2 及其系）会看到，S_1 类中的 Q 所对应的 Q 过程，结构上要比 S_2 类中的复杂. 对于 S_2 类中的 Q，质点不可能自 $+\infty$ "连续地"回到有穷状态上来.

利用[9]中的一个结果，容易证明(2.4)中的数具有下列概率意义：任取 E 中三状态 $n>k>j$，并考虑 Q 过程的嵌入[①]马氏链 $x_n(\omega)$，$n\geqslant 0$，即具有下列转移概率的马氏链

$$\begin{cases} P(x_n=i-1 \mid x_{n-1}=i)=\dfrac{a_i}{c_i}, \\ P(x_n=i+1 \mid x_{n-1}=i)=\dfrac{b_i}{c_i}. \end{cases} \quad (2.13)$$

关于此链，质点自 k 出发，在到达 n 以前先到达 j 的概率为 $\dfrac{z_n-z_k}{z_n-z_j}$（见附录定理2）；当 n 上升时，此概率不下降．今令自 k 出发，终于要到达 j 的概率为 d_{kj}，并理解 $\dfrac{+\infty}{+\infty}=1$，则当 $j<k$ 时，

$$d_{kj}=\lim_{n\to+\infty}\dfrac{z_n-z_k}{z_n-z_j}=\dfrac{z-z_k}{z-z_j}. \quad (2.14)$$

考虑任一可分 Q 过程 $x(t,\omega)$，$t\geqslant 0$，仍以 c_{kj} 表自 k 出发，经有穷步转移到达 j 的概率，或者等价地，记

$$c_{kj}=P_k(\xi_j(\omega)<\tau(\omega)), \quad (2.15)$$

设 $R<+\infty$．由于 $z\leqslant b_0 R$，故 $d_{kj}<1$．显然，当 $j<k$ 时，$c_{kj}=d_{kj}$；如果 $j>k$，那么由(2.6)得 $P_k(\xi_j(\omega)<+\infty)\geqslant P_k(\tau(\omega)<+\infty)=1$（这里还用到生灭过程的特性：自 k 出发，下一步落到 $k-1$ 或 $k+1$ 的概率为1；因而当自 k 到达另一状态时，必历经一切中间状态．此特性将多次用到而不另说明），故 $d_{kj}=1$．因此如把 0 步也算作有穷步，总结上述便得

引理 2.2 对可分 Q 过程 $x(t,\omega)$，$t\geqslant 0$，若 $R<+\infty$，则自 k 出发，经有穷步到达 j 的概率

$$c_{kj}=\begin{cases} 1, & k\leqslant j, \\ \dfrac{z-z_k}{z-z_j}, & k>j. \end{cases} \quad (2.16)$$

[①] 参看定义 4.1，那里考虑一般情况．

§3. Doob 过程的变换

最简单的一类 Q 过程是 Doob 过程，它的定义见附录[或 4，第 267 页]. 此类过程的样本函数是所谓 T 跳跃函数，后者如下定义：

定义 3.1 设 $y(t)$，$t \geqslant 0$ 为取值于 $\bar{E}$ 的函数，称点 t_0 为它的跳跃点，如它在 t_0 不连续，而且存在 $\varepsilon > 0$，使在 $[t_0 - \varepsilon, t_0)$ 及 $[t_0, t_0 + \varepsilon)$ 中，它的值分别为两不相等的常数；称点 τ 为它的飞跃点，如对任意 $\varepsilon > 0$，在 $[\tau - \varepsilon, \tau)$ 中，它有无穷多个跳跃点.

定义 3.2 值域为 E 的函数 $y(t)$，$t \geqslant 0$ 称为 T 跳跃的，如果

(i) 在任一有穷区间中，只有有穷多个飞跃点 $\tau_i (\tau_0 = 0, \tau_i < \tau_{i+1})$;

(ii) 在任一飞跃区间 $[\tau_i, \tau_{i+1})$ 中，一切不连续点都是跳跃点 τ_{ij}，其数可列 $(\tau_i = \tau_{i0} < \tau_{i1} < \tau_{i2} < \cdots)$, $i \in \mathbf{N}$;

(iii) 在任两相邻的不连续点上，有
$$|y(\tau_{ij}) - y(\tau_{ij+1})| = 1 \quad (i, j \in \mathbf{N}).$$

T 跳跃函数称为 T_n 跳跃的，如在任一飞跃点 $\tau_i (i > 0)$ 上，$y(\tau_i) \leqslant n$.

注意 T 跳跃函数右连续，不以 $+\infty$ 为值.

对于 Doob 过程 $x(t, \omega)$，$t \geqslant 0$，由于 $R < +\infty$，一切随机变量 $\tau_{ij}(\omega)$ 均以概率 1 有穷. 此过程由 Q 及分布 $\pi = (\pi_j)$ 决定，这里 $\pi_j = P(x(\tau_i, \omega) = j)$ 与之无关 $(i \in \mathbf{N}^*)$，故称它为 (Q, π) 过程.

以后常要用到过程的一种变换.

定义 3.3 称函数 $y(t)$, $t \geq 0$, 自 $x(t)$, $t \geq 0$ 经 $c(\alpha_k, \beta_k)$ 变换得来, 如果存在两列正数 (α_k), (β_k), 使

$$0(=\beta_0) < \alpha_1 \leq \beta_1 < \alpha_2 \leq \beta_2 < \cdots < \alpha_k \leq \beta_k < \cdots, \sum_{i=0}^{\infty}(\alpha_{i+1}-\beta_i)=+\infty,$$

而且 $y(t)$ 如下定义: 令

$$\gamma_1 = \alpha_1; \ y(t) = x(t), \ 0 \leq t \leq \gamma_1,$$

若 $y(t)$ 已在 $[0, \gamma_k]$ 上定义, 则令

$$\gamma_{k+1} = \gamma_k + (\alpha_{k+1} - \beta_k); \ y(\gamma_k + t) = x(\beta_k + t), \ 0 \leq t < \alpha_{k+1} - \beta_k.$$

直观地说, 抛去 $x(t)$ 对应于 $[\alpha_i, \beta_i)$ 的那些段, 剩下的第一段 $[0, \alpha_1)$ 保留不动, 其余的段向左移动, 使 $[0, \alpha_1)$, $[\beta_i, \alpha_{i+1})$ $(i \in \mathbf{N}^*)$ 按原序联结而不相交, 所得函数即 $y(t)$.

今以 $x_n(t)$ 表某 T_n 跳跃函数, 用下列方法定义两列正数, 这种迭代定义方法将多次引用. 以

$$\tau_1 \ \text{表} \ x_n(t) \text{的第一个飞跃点}, \tag{3.1}$$

$$\tau_{k_1} = \inf(\tau: \tau \geq \tau_1, \tau \text{ 是飞跃点, 而且 } x_n(\tau) < n); \tag{3.2}$$

如果已定义 τ_{k_i}, 那么令

$$\tau_{k_i+1} = \inf(\tau: \tau > \tau_{k_i}, \tau \text{ 是飞跃点}), \tag{3.3}$$

$$\tau_{k_{i+1}} = \inf(\tau: \tau \geq \tau_{k_i+1}, \tau \text{ 是飞跃点}, x_n(\tau) < n), \tag{3.4}$$

于是

$$0 < \tau_1 \leq \tau_{k_1} < \tau_{k_1+1} \leq \tau_{k_2} < \cdots \leq \tau_{k_i+1} \leq \tau_{k_{i+1}} < \cdots$$

设以上诸数均有穷, 而且

$$\sum_{i=0}^{\infty}(\tau_{k_i+1} - \tau_{k_i}) = +\infty, \ k_0 = 0, \ \tau_0 = 0, \tag{3.5}$$

对 $x_n(t)$ 施行 $c(\tau_{k_i+1}, \tau_{k_{i+1}})$ 变换后, 得一 T_{n-1} 跳跃函数 $x_{n-1}(t)$, 记此关系为

$$f_{n,n-1}(x_n(t)) = x_{n-1}(t), \tag{3.6}$$

故 $f_{n,n-1}$ 表 T_n 跳跃函数到 T_{n-1} 跳跃函数的变换, 注意 (3.6) 并

不表示对固定的 t 双方相等.

现在考虑 $(Q, S^{(n)})$ 过程 $x_n(t, \omega)$，$t \geq 0 (\omega \in \Omega)$，这里 $S^{(n)} = (\mathscr{S}_0^{(n)}, \mathscr{S}_1^{(n)}, \cdots, \mathscr{S}_n^{(n)})$ 是集中在前 $n+1$ 个状态 $(0, 1, 2, \cdots, n)$ 上的分布，使 $P(x_n(\tau_i, \omega) = j) = \mathscr{S}_j^{(n)}$. 为简单计，设 $\mathscr{S}_0^{(n)} > 0$. 利用 (3.1) 至 (3.4) 定义随机变量列 $\tau_{k_i+1}(\omega)$，$\tau_{k_{i+1}}(\omega) (i \geq 0)$，则由于 $R < +\infty$ 及 $\mathscr{S}_0^{(n)} > 0$，它们均以概率 1 有穷而且 (3.5) 成立. 对过程 $x_n(t, \omega)$，$t \geq 0$ 施行 $C(\tau_{k_i+1}(\omega), \tau_{k_{i+1}}(\omega))$ 变换后，得二元函数 $x_{n-1}(t, \omega)$，$t \geq 0 (\omega \in \Omega)$，即

$$f_{n, n-1}(x_n(t, \omega)) = x_{n-1}(t, \omega). \tag{3.7}$$

引理 3.1 $x_{n-1}(t, \omega)$，$t \geq 0$ 是 $(Q, S^{(n-1)})$ 过程，这里

$$\mathscr{S}_i^{(n-1)} = \frac{\mathscr{S}_i^{(n)}}{\sum_{j=0}^{n-1} \mathscr{S}_j^{(n)}} \quad (0 \leq i < n). \tag{3.8}$$

证 对固定的 ω，由定义知 $x_{n-1}(t, \omega)$ 是 T_{n-1} 跳跃函数. 今证对每固定的 t，$x_{n-1}(t, \omega)$ 是随机变量. 以 $\sigma_l(\omega)$ 表 $x_{n-1}(t, \omega)$ 的第 l 个飞跃点 $(\sigma_0 = 0)$，并令

$$\eta_l(\omega) = \sum_{i=1}^{l} (\tau_{k_i}(\omega) - \tau_{k_{i-1}+1}(\omega)) \quad (k_0 = 0) \tag{3.9}$$

(换言之，$\eta_l(\omega)$ 是在 $\tau_{k_l}(\omega)$ 以前，自 $x_n(t, \omega)$ 所抛去的区间的总长). 注意 $x_n(t, \omega)$ 是右连续过程，故是 Borel 可测的，因而

$$(x_{n-1}(t, \omega) = i, \sigma_l(\omega) < t \leq \sigma_{l+1}(\omega))$$
$$= (x_n(t + \eta_l, \omega) = i, \tau_{k_l}(\omega) < t + \eta_l(\omega) < \tau_{k_{l+1}}(\omega))$$

是可测集，故 $(x_{n-1}(t, \omega) = i) = \sum_{l=0}^{+\infty}(x_{n-1}(t, \omega) = i, \sigma_l(\omega) < t \leq \sigma_{l+1}(\omega))$ 也可测，再留意 $x_{n-1}(0, \omega) = x_n(0, \omega)$，即得证 $x_{n-1}(t, \omega)$，$t \geq 0$ 是一随机过程. 它还是 $(Q, S^{(n-1)})$ 过程，因为对任意 $l \geq 1$，令 τ_m 为 $x_n(t, \omega)$ 的第 m 个飞跃点，由 (3.8) 得

$$P(x_{n-1}(\sigma_l) = j) = P(x_n(\tau_{k_l}) = j)$$

$$= \sum_{m=l}^{+\infty} P(x_n(\tau_{k_l}) = j \mid \tau_{k_l} = \tau_m) \cdot P(\tau_{k_l} = \tau_m)$$

$$= \sum_{m=l}^{+\infty} \frac{(1-\mathscr{S}_n^{(n)})^{l-1} \cdot \mathscr{S}_j^{(n)} \cdot [\mathscr{S}_n^{(n)}]^{m-l}}{(1-\mathscr{S}_n^{(n)})^l \cdot [\mathscr{S}_n^{(n)}]^{m-l}} P(\tau_{k_l} = \tau_m)$$

$$= \mathscr{S}_j^{(n-1)} \sum_{m=l}^{+\infty} P(\tau_{k_l} = \tau_m).$$

因为 $\mathscr{S}_0^{(n)} > 0$,故 $P(\tau_l \leqslant \tau_{k_l} < +\infty) = 1$,即 $\sum_{m=l}^{+\infty} P(\tau_{k_l} = \tau_m) = 1$,从而

$$P(x_{n-1}(\sigma_l) = j) = \mathscr{S}_j^{(n-1)} \quad (0 \leqslant j \leqslant n-1).$$

最后,根据 Doob 过程的定义[1,P241 定理 4],还要证明 $x_{n-1}(t, \omega)$ 是由相互独立的最小链[1,P232]组成. $x_{n-1}(t, \omega)$ 是由最小链组成是显然的,故只要证独立性,以 τ_{ij},σ_{ij} 分别表 $x_n(t, \omega)$ 及 $x_{n-1}(t, \omega)$ 第 i 个飞跃点后第 j 个跳跃点($j \geqslant 0$),$f_u(x_1, y_1, x_2, y_2, \cdots)$ 表任意无穷维 Borel 可测函数,$u \in \mathbf{N}^*$. 令

$$F_{uv}^{(n-1)}(\omega) = f_u(x_{n-1}(\sigma_{v0}), \sigma_{v1}-\sigma_{v0}, x_{n-1}(\sigma_{v1}), \sigma_{v2}-\sigma_{v1}, \cdots),$$
$$F_{uv}^{(n)}(\omega) = f_u(x_n(\tau_{v0}), \tau_{v1}-\tau_{v0}, x_n(\tau_{v1}), \tau_{v2}-\tau_{v1}, \cdots).$$

设 l 为任意正整数,$c_1, c_2, \cdots, c_l$ 为任意 l 个实数,则

$$P(x_{n-1}(\sigma_v) = j_v, F_{v,v}^{(n-1)} < c_v, v=1, 2, \cdots, l)$$
$$= P(x_n(\tau_{k_v}) = j_v, F_{vk_v}^{(n)} < c_v, v=1, 2, \cdots, l)$$
$$= \sum P(k_v(\omega) = m_v, x_n(\tau_{m_v}) = j_v, F_{vm_v}^{(n)} < c_v, v=1,2,\cdots,l)$$
$$= \sum P(x_n(\tau_{m_v}) = j_v, F_{vm_v}^{(n)} < c_v, v=1,2,\cdots,l, x_n(\tau_i) = n,$$
$$i \neq m_1, \neq m_2, \cdots, \neq m_l, i < m_l),$$

这里及以下的 $\sum$ 表对正整数 $m_l > m_{l-1} > \cdots > m_1 \geqslant 1$ 求和. 由于 $x_n(t, \omega)$ 是 Doob 过程,故构成 $x_n(t, \omega)$ 的最小链是相互独立的. 因此,如以 $\bar{P}_i$ 表开始分布集中在 i 上时最小链所产生的

测度,即得上式最右项

$$= \sum [\mathscr{S}_n^{(n)}]^{m_1-1} \cdot [\mathscr{S}_n^{(n)}]^{m_2-(m_1+1)} \cdots$$

$$[\mathscr{S}_n^{(n)}]^{m_l-(m_{l-1}+1)} \prod_{v=1}^{l} \overline{P}_{j_v}(F_{v,m_v}^{(n)} < c_v)$$

$$= \prod_{v=1}^{l} \left\{ \frac{\overline{P}_{j_v}(F_{vm_v}^{(n)} < c_v)}{1 - \mathscr{S}_n^{(n)}} \right\}$$

$$= \prod_{v=1}^{l} P(x_n(\tau_{k_v}) = j_v, F_{v,k_v}^{(n)} < c_v)$$

$$= \prod_{v=1}^{l} P(x_{n-1}(\sigma_v) = j_v, F_{v,v}^{(n-1)} < c_v). \quad (3.10)$$

然后对j_v自 0 到 $n-1$ 求和($v=1,2,\cdots,l$),即得

$$P(F_{vv}^{(n-1)} < c_v, v = 1, 2, \cdots, l) = \prod_{v=1}^{l} P(F_{vv}^{(n-1)} < c_v),$$

此即表诸构成 $x_{n-1}(t, \omega)$ 的最小链的独立性。∎

类似于 $f_{n,n-1}$,定义另一种变换 $g_{n,n-1}$ 如下:

对 T_n 跳跃函数 $x_n(t)$,仿(3.1)~(3.4),令

$$\tau_1 \text{ 为 } x_n(t) \text{ 的第一个飞跃点,} \quad (3.1')$$

$$\beta_{k_1} = \inf(t: t \geqslant \tau_1, x_n(t) < n); \quad (3.2')$$

$$\tau_{k_i+1} \text{ 为 } \beta_{k_i} \text{ 后的第一个飞跃点,} \quad (3.3')$$

$$\beta_{k_i+1} = \inf(t: t \geqslant \tau_{k_i+1}, x_n(t) < n). \quad (3.4')$$

仍设此诸数皆有穷而且

$$\sum_{i=0}^{\infty}(\tau_{k_i+1} - \beta_{k_i}) = +\infty, \quad k_0 = 0, \beta_0 = 0 \quad (3.5')$$

对 $x_n(t)$ 施以 $C(\tau_{k_i+1}, \beta_{k_i+1})$ 变换后,得一 T_{n-1} 跳跃函数 $x_{n-1}(t)$,记此关系为

$$g_{n,n-1}(x_n(t)) = x_{n-1}(t), \quad (3.11)$$

或者,为以后方便,记成

$$g_{n+1,n}(x_{n+1}(t)) = x_n(t). \quad (3.12)$$

这表示变换 $g_{n+1,n}$ 把 T_{n+1} 跳跃函数变为 T_n 跳跃函数.

今考虑 $(Q, V^{(n+1)})$ 过程 $x_{n+1}(t, \omega)$, $t \geq 0$, 这里 $V^{(n+1)} = (v_0^{(n+1)}, v_1^{(n+1)}, \cdots, v_{n+1}^{(n+1)})$ 表某集中在 $(0, 1, 2, \cdots, n+1)$ 上的分布, 它的样本函数是 T_{n+1} 跳跃函数. 由 (3.12), 令
$$g_{n+1,n}(x_{n+1}(t, \omega)) = x_n(t, \omega), \tag{3.13}$$
则类似地得

引理 3.2 $x_n(t, \omega)$, $t \geq 0$ 是 $(Q, V^{(n)})$ 过程, 其中
$$\begin{cases} v_j^{(n)} = \dfrac{v_j^{(n+1)}}{\sum\limits_{i=0}^{n} v_i^{(n+1)} + v_{n+1}^{(n+1)} c_{n+1,n}} & (j < n), \\ v_n^{(n)} = \dfrac{v_n^{(n+1)} + v_{n+1}^{(n+1)} c_{n+1,n}}{\sum\limits_{i=0}^{n} v_i^{(n+1)} + v_{n+1}^{(n+1)} c_{n+1,n}}, \\ \sum\limits_{i=0}^{n} v_i^{(n)} = \sum\limits_{i=0}^{n+1} v_i^{(n+1)} = 1. \end{cases} \tag{3.14}$$

而 c_{kj} 由 (2.16) 定义, $n \in \mathbf{N}^*$.

证 证明仿引理 3.1, 不同处在于证 (3.14). 分别以 $\sigma_l(\omega)$, $\tau_l(\omega)$ 表 $x_n(t, \omega)$ 与 $x_{n+1}(t, \omega)$ 的第 l 个飞跃点,
$$P(x_n(\sigma_l) = j) = P(x_{n+1}(\beta_{k_l}) = j)$$
$$= \sum_{m=l} P(x_{n+1}(\beta_{k_l}) = j \mid \tau_m \leq \beta_{k_l} < \tau_{m+1}) \cdot P(\tau_m \leq \beta_{k_l} < \tau_{m+1}). \tag{3.15}$$

由于 $R < +\infty$, 对 $x_{n+1}(t, \omega)$, 自 k 出发, 经有穷步到达 $j (\geq k)$ 的概率为 1, 到达 $k-1$ 的概率为 $c_{k,k-1}$, 故 $\Delta = \sum\limits_{i=1}^{n} v_i^{(n+1)} + v_{n+1}^{(n+1)} c_{n+1,n} > 0$ 是自任一飞跃点出发经有穷步[1]到达 $(0, 1, 2, \cdots, n)$ 的概率. 因而

---

[1] 0 步也算作有穷步.

$$P(x_{n+1}(\beta_{k_l})=j \mid \tau_m \leqslant \beta_{k_l} < \tau_{m+1})$$
$$=\frac{(1-\Delta)^{m-l}\Delta^{l-1}v_j^{(n+1)}}{(1-\Delta)^{m-l}\Delta^l}=v_j^{(n)}, \quad (0\leqslant j<n).$$

类似地有

$$P(x_{n+1}(\beta_{k_l})=n \mid \tau_m \leqslant \beta_{k_l} < \tau_{m+1})=v_n^{(n+1)}+v_{n+1}^{(n+1)}c_{n+1,n},$$

以此代入(3.15)，并注意易证 $P(\tau_l \leqslant \beta_{k_l} < +\infty)=1$，$P(\lim_{i\to+\infty}\tau_i=+\infty)=1$，即得证(3.14)中前两式；最后一式是显然的. ∎

更一般地，对 $n>m$，定义两变换

$$f_{nm}=f_{m+1,m}\cdots f_{n-1,n-2}f_{n,n-1}, \tag{3.16}$$
$$g_{nm}=g_{m+1,m}\cdots g_{n-1,n-2}g_{n,n-1}. \tag{3.17}$$

它们都是把 T_n 跳跃函数变为 T_m 跳跃函数的单值变换，逆变换 f_{nm}^{-1}，g_{nm}^{-1}，则把 T_m 跳跃函数变为 T_n 跳跃函数，但后者一般是多值的.

仍旧考虑 $(Q, S^{(n)})$ 过程 $x_n(t, \omega)$，$t\geqslant 0$. 根据随机过程的表现理论[4,§6]，可以取基本事件空间 $\Omega=\Omega_n$，这里 $\Omega_n=(\omega_n)$ 是全体 T_n 跳跃函数的集合，而且基本事件 ω_n 与样本函数 $x_n(t, \omega_n)$ 重合，即 $x_n(t, \omega_n)=\omega_n(t)$ $(t\geqslant 0)$. 这样取定的概率空间记为 $(\Omega_n, \mathscr{F}_n, P_n)$，$P_n$ 完全由 Q，$S^{(n)}$ 及一开始分布决定. 今如取由(3.8)定义的分布 $S^{(n-1)}$，则由(3.7)及引理 3.1，定义在 $(\Omega_n, \mathscr{F}_n, P_n)$ 上的过程 $f_{n,n-1}(x_n(t, \omega_n))$ 是 $(Q, S^{(n-1)})$ 过程. 由此易见 $f_{nm}(x_n(t, \omega_n))$ 是定义在 $(\Omega_n, \mathscr{F}_n, P_n)$ 上的 $(Q, S^{(m)})$ 过程 $(m<n)$，这里

$$\mathscr{S}_i^{(m)}=\frac{\mathscr{S}_i^{(n)}}{\sum_{j=0}^m \mathscr{S}_j^{(n)}} \quad (0\leqslant i\leqslant m), \tag{3.18}$$

此式是(3.8)的推广.

今设已给一列非负数 $(\mathscr{S}_i)$，使

$$0 < \sum_{i=0}^{+\infty} \mathscr{S}_i \leqslant +\infty \tag{3.19}$$

(注意此级数可以发散), 故至少有一 $\mathscr{S}_i > 0$. 不失以下讨论的一般性, 设 $\mathscr{S}_0 > 0$. 由 $(\mathscr{S}_i)$ 作集中在 $(0, 1, 2, \cdots, n)$ 上的分布 $S^{(n)} = (\mathscr{S}_0^{(n)}, \mathscr{S}_1^{(n)}, \cdots, \mathscr{S}_n^{(n)})$, 其中

$$\mathscr{S}_i^{(n)} = \frac{\mathscr{S}_i}{\sum_{j=0}^{n} \mathscr{S}_j}, \tag{3.20}$$

显然, 分布列 $(S^{(n)})$ 满足关系 (3.18).

引理 3.3 存在概率空间 $(\Omega, \mathscr{F}, P)$, 在其上可以定义一列 $(Q, S^{(n)})$ 过程 $x_n(t, \omega)$, $t \geqslant 0 (n \in \mathbf{N})$, 使满足关系 (3.7), 这里 $S^{(n)}$ 由 (3.20) 决定.

证 固定一分布 (v_i) 作为开始分布. 如上所述, 对每一 $n \geqslant 0$, 存在 $(\Omega_n, \mathscr{F}_n, P_n)$ 及定义于其上的 $(Q, S^{(n)})$ 过程 $x_n(t, \omega_n)$, $t \geqslant 0$. 对任意 $k (\geqslant 1)$ 个非负整数 $n_1, n_2, \cdots, n_k$, 任取 $n \geqslant \max(n_1, n_2, \cdots, n_k)$, 定义在 $(\Omega_n, \mathscr{F}_n, P_n)$ 上的过程 $z_{n_i}(t, \omega_n) = f_{nn_i}(x_n(t, \omega_n))$ 也是 $(Q, S^{(n_i)})$ 过程, 故与 $x_{n_i}(t, \omega_{n_i})$, $t \geqslant 0$ 有相同的有穷维分布. 今对 $t_i \in [0, +\infty)$ 及 $j_i \in E$, $(i = 1, 2, \cdots, k)$ 定义 k 维分布

$$F_{n_1 t_1, \cdots, n_k t_k}(j_1, j_2, \cdots, j_k)$$
$$= P_n(z_{n_i}(t_i, \omega_n) = j_i, i = 1, 2, \cdots, k), \tag{3.21}$$

易见此分布不依赖于 n 的选择, 而且有穷维分布族 $\{F_{n_1 t_1, n_2 t_2, \cdots, n_k t_k}\}$ 是相容的, 故根据柯尔莫哥洛夫定理[4,§6] 存在概率空间 $(\Omega, \mathscr{F}, P)$, 及定义于其上的过程列 $x_n(t, \omega)$, $t \geqslant 0$, $(n \in \mathbf{N})$, 使

$$P(x_{n_i}(t_i, \omega) = j_i, i = 1, 2, \cdots, k)$$
$$= F_{n_1 t_1, n_2 t_2, \cdots, n_k t_k}(j_1, j_2, \cdots, j_k). \tag{3.22}$$

由此及 (3.21), 特别地知 $x_n(t, \omega)$ 与 $z_n(t, \omega_n)$ 有相同的有穷维

分布. 其次, 按上引柯氏定理, 可取 $\Omega = (\omega)$, 其中 $\omega = \omega(n, t)$ 是取值于 E 的二元函数 ($n \in \mathbf{N}$, $t \in [0, +\infty)$), 并且 $x_n(t, \omega) = \omega(n, t)$. 由于对一切 $n \geqslant m \geqslant 0$, $P_n(f_{nm}(z_n(t, \omega_n)) = z_m(t, \omega_n)) = 1$, $P_n(z_n(t, \omega_n)$ 是 T_n 跳跃函数$) = 1$. 故可清洗 Q [参看 5, P53], 以使对每 ω, $x_n(t, \omega)$ 是 T_n 跳跃函数, 而过程 $x_n(t, \omega)$, $t \geqslant 0$ 则成为 $(Q, S^{(n)})$ 过程, 并且使 (3.7) 成立. 清洗 (缩小) 后的概率空间仍记为 $(\Omega, \mathscr{F}, P)$, 则此空间符合要求. ■

逐句重复引理 3.3 的证明, 作显然的记号上及字面上的修改后, 即可证明下面的

引理 3.4 存在概率空间 $(\Omega, \mathscr{F}, P)$, 在其上可以定义一列 $(Q, V^{(n)})$ 过程 $x_n(t, \omega)$, $t \geqslant 0$ ($n \in \mathbf{N}^*$), 使满足关系 (3.13). 这里 $V^{(n)} = (v_0^{(n)}, v_1^{(n)}, \cdots, v_n^{(n)})$ ($n \in \mathbf{N}^*$) 是 (3.14) 的任一列非负解.

注 引理 3.3 对一般的满足 (1.6) 的 Q 也成立, 证明不需作任何修改.

§4. 连续流入不可能的充分必要条件

设 Q 满足(1.6)而 $x(t, \omega)$，$t \geqslant 0$ 为可分 Q 过程. 可以证明：不影响转移概率，对每 $i \in E$，可设 t-集 $S_i(\omega) = (t: x(t, \omega) = i)$ 以概率 1(记为 a. a.)是有穷或可列多个左闭右开的不相交的区间的和，而且在任一有界区间中，只含有穷多个如此的区间，以后称为 i 区间[1,P149]；还可证明：在任一定点 t 后有第一个断点，它是跳跃点(a. a.)[1,P227].

定理 4.1 对任意满足(1.6)的可分 Q 过程 $x(t, \omega)$，$t \geqslant 0$，t-集 $\Gamma(\omega) = (t: t$ 是 $x(s, \omega)$，$s \geqslant 0$ 的飞跃点)是闭集(a. a.).

证 对固定的 ω，称 a 是 $\Gamma(=\Gamma(\omega))$ 的左极限点，如 a 是 Γ 的极限点，但存在 $\varepsilon > 0$，使 $x(t)$ 在 $[a-\varepsilon, a)$ 中为常数. 记 Γ 的左极限点集为 A，并令 $B = (b: x(t)$ 在 b 不连续，而且在某 $[b-\delta, b)(\delta > 0)$ 为常数). 显然 $A \subset B$. 但另一方面，因 $[b-\delta, b)$ 必含于某 i 区间之中，而且 B 中不同的 i 不能含于同一 i 区间之中，故 B 是可列集(a. a.). 记 $B = (b_n)$，则 b_n 不是跳跃点的概率等于 0. 否则，存在 $r \in R$(可分 t-集)，使 $P(r \in [b_n - \delta, b_n)$ 而且 b_n 非跳跃点$) > 0$，于是 r 后第一个断点以正概率不是跳跃点，此如上述由(1.6)不可能. 故 B 由跳跃点构成；然而由 A 的定义，A 中的点均非跳跃点，故 $AB = \varnothing$，从而
$$A = \varnothing (a.\ a.).$$
若点 γ 是 Γ 的极限点，但 $\gamma \bar{\in} A$，则在任一 $[b-\varepsilon, b)$ 中必有无穷多个跳跃点[参看 1，P160 的系]，故 $\gamma \in \Gamma$. 因而得证 $\Gamma(\omega)$ 是闭集(a. a.). ∎

任意固定 $\mathscr{S} \geqslant 0$. 由定理 4.1，可以定义

$$\tau_{\mathscr{S}}(\omega) = \max(\gamma: \gamma \leqslant \mathscr{S}, \gamma \in \Gamma(\omega)), \quad (4.1)$$

换言之，$\tau_{\mathscr{S}}(\omega)$ 是 $\mathscr{S}$ 前的最后一个飞跃点(若右方括号中集是空的，则令 $\tau_{\mathscr{S}}(\omega) = 0$). 它是随机变量. 易见存在(a. a.)极限 $\lim\limits_{t \downarrow \tau_{\mathscr{S}}(\omega)} x(t, \omega)$. 实际上，如说不然，必存在 $i \in E$ 使 $P(\overline{\lim\limits_{t \downarrow \tau_{\mathscr{S}}(\omega)}} x(t, \omega) > \underline{\lim\limits_{t \downarrow \tau_{\mathscr{S}}(\omega)}} x(t, \omega) = i) > 0$，由于 $P(\tau_{\mathscr{S}}(\omega) \leqslant \mathscr{S}) = 1$，故上式表示以正的概率在 $[0, \mathscr{S}]$ 中有无穷多个 i 区间，此不可能.

定义 $x(\tau_{\mathscr{S}}, \omega) = \lim\limits_{t \downarrow \tau_{\mathscr{S}}} x(t, \omega)$. 若对任意 $\mathscr{S} \geqslant 0$，有 $P(x(\tau_{\mathscr{S}}, \omega) = +\infty) = 0$，则说质点不能自 $+\infty$"连续地"流入有穷状态. 不久可证，对满足(1.7)的 Q 过程，其充分必要条件是 $S = +\infty$.

设 $x(t, \omega)$，$t \geqslant 0$ 是取值于 E 的齐次 Borel 可测马氏过程，对 $[0, +\infty)$ 中任一子集 B，以 $\mathscr{B}_B$ 表含 ω-集 $(x(t, \omega) = j)(t \in B, j \in E)$ 的最小 σ-代数. 称随机变量 $\zeta(\omega)(\leqslant +\infty)$ 为马尔可夫时间(M. T.)，若对任一 $s \geqslant 0$，$(\zeta(\omega) \leqslant s) \in \mathscr{B}_{[0,s]}$. 记 $\Omega_{\zeta} = (\zeta(\omega) < +\infty)$，令

$$\mathscr{B}_{[0,\zeta]} = (A: A \subset \Omega_{\zeta}, 对任 t \geqslant 0, A \cap (\zeta \leqslant t) \in \mathscr{B}_{[0,t]}), \quad (4.2)$$

则 $\mathscr{B}_{[0,\zeta]}$ 是 Ω_{ζ} 中一 σ-代数. 定义 $\mathscr{B}_{[0,+\infty)}$ 到 $\mathscr{B}_{[0,+\infty)}$ 中的集变换 θ_{ζ}，使保持和交与补集运算并且使

$$\theta_{\zeta}(x(t, \omega) \in \Gamma) = (x(t+\zeta, \omega) \in \Gamma) \quad (\Gamma \subset E), \quad (4.3)$$

可以证明[5,P133]，若(1.4)成立，而且 $x(t, \omega)$ 在 $\zeta(\omega)$ 右连续(a. a.)，则在 $\zeta(\omega)$ 强马氏性成立：对任 $B \in \mathscr{B}_{[0,+\infty)}$，在 Q_{ζ} 上，除去某零测度集外，有

$$P(\theta_{\zeta} B \mid \mathscr{B}_{[0,\zeta]}) = P_{x(\zeta)}(B). \quad (4.4)$$

下面引理 4.1 基本上属于 Дынкин.

引理 4.1 设 $\xi(\omega)$ 为随机变量，满足条件：

(i) 对任意 $s \geqslant 0$，$t \geqslant 0$，ω-集 $A_s = (\xi > s) \in \mathscr{B}_{[0,s]}$，而且

$A_{s+t} \subseteq A_s \cap \theta_s A_t$;

(ii) 存在 $T>0$, $\alpha>0$, 使对一切 $k \in E$, 有 $P_k(A_T) < 1-\alpha$, 则 $E\xi < +\infty$.

证 因 $A_s \in \mathscr{B}_{[0,s]}$, $\theta_s A_T \in \mathscr{B}_{[s,+\infty)}$, 由马氏性得
$$P_k(A_{s+T}) \leqslant P_k(A_s \cap \theta_s A_T)$$
$$= \int_{A_s} P_{x(s)}(A_T) P_k(\mathrm{d}\omega) \leqslant (1-\alpha) P_k(A_s),$$

从而 $P_k(A_{nT}) \leqslant (1-\alpha)^n$, 并且
$$E_k\xi = \int_0^{+\infty} P_k(\xi > s)\mathrm{d}s = \sum_{n=0}^{+\infty} \int_{nT}^{(n+1)T} P_k(\xi > s)\mathrm{d}s$$
$$\leqslant T \sum_{n=0}^{+\infty} P_k(\xi > nT) = T \sum_{n=0}^{+\infty} P_k(A_{nT}) \leqslant \frac{T}{\alpha} < +\infty.$$

由于 $k \in E$ 任意, 故 $E\xi < +\infty$. ∎

定理 4.2 设 Q 满足 (1.7) 而且 $S=+\infty$, 则对任意的可分、Borel 可测 Q 过程 $x(t, w)$, $t \geqslant 0$, 有 $P(x(\tau_{\mathscr{S}}, \omega)=+\infty)=0$, 这里 $\mathscr{S} \geqslant 0$ 任意.

证 若 $R=+\infty$, 则因第一个飞跃点 $\tau(\omega)=+\infty$ (a. a.), 故 $\tau_{\mathscr{S}}(\omega)=0$ (a. a.), 而定理显然正确.

设 $R<+\infty$, 令 $\eta_i(\omega) = \inf(t: x(t, \omega) = i)$, 则 $P_0(\eta_i < +\infty)=1$.

引进随机变量
$$\xi_k(\omega) = \inf(t: x(t, \omega) = k, x(\tau_t, \omega) = +\infty) \quad (k \in E)$$
(4.5)

(若右方括号中集是空的, 则令 $\xi_k(\omega)=+\infty$), 试证 $P(\xi_k(\omega)=+\infty)=1$.

先证 $P(\xi_0(\omega)=+\infty)=1$. 若说不然, 则 $P(\xi_0<+\infty)>0$, 故至少有一 $i \in E$, 使 $P_i(\xi_0<+\infty)>0$. 由于 (1.7)(2.13) 成立, 既然 $P(\xi_0(\omega) \geqslant \tau(\omega))=1$, 故

$$P_0(\xi_0 < +\infty) = P_0(\eta_i < +\infty, \xi_0 - \eta_i < +\infty),$$

对 η_i 用强马氏性即得

$$P_0(\xi_0 < +\infty) = \int_{(\eta_i < +\infty)} P_0(\xi_0 - \eta_i < +\infty \mid \mathscr{B}_{[0,\eta_i]}) P_0(d\omega)$$

$$= \int_{(\eta_i < +\infty)} P_{x(\eta_i)}(\xi_0 < +\infty) P_0(d\omega).$$

因为 $x(\eta_i) = i$, $P_0(\eta_i < +\infty) = 1$, 故由上式得 $P_0(\xi_0 < +\infty) = P_i(\xi_0 < +\infty) > 0$, 于是存在 $T > 0$, $\alpha > 0$, 使 $P_0(\xi_0 \leqslant T) \geqslant \alpha$. 既然对任意 $k \in E$, 有

$$P_0(\xi_0 \leqslant T) \leqslant P_0(\eta_k \leqslant T, \xi_0 - \eta_k \leqslant T)$$
$$= \int_{(\eta_k \leqslant T)} P_0(\xi_0 - \eta_k \leqslant T \mid \mathscr{B}_{[0,\eta_k]}) P_0(d\omega)$$
$$= P_k(\xi_0 \leqslant T) P_0(\eta_k \leqslant T),$$

故 $P_k(\xi_0 \leqslant T) \geqslant P_0(\xi_0 \leqslant T) \geqslant \alpha$, 即 $P_k(\xi_0 > T) < 1 - \alpha (k \in E)$, 从而引理 4.1 条件 (ii) 满足; 由 $\xi_0(\omega)$ 的定义易见条件 (i) 也满足, 故得 $E\xi_0 < +\infty$. 按引理 2.1 及假设,

$$\lim_{n \to +\infty} E\sigma_n (= E_n\sigma_n) = S = +\infty,$$

故存在 N, 使

$$E\sigma_N > 2E\xi_0. \qquad (4.6)$$

另一方面, 用迭代法定义

$$\alpha_1(\omega) = \inf(t: x(t, \omega) = N),$$
$$\beta_1(\omega) = \inf(t: t > \alpha_1(\omega), x(t, \omega) = N+1),$$
$$\alpha_k(\omega) = \inf(t: t > \beta_{k-1}(\omega), x(t, \omega) = N-1),$$
$$\beta_k(\omega) = \inf(t: t > \alpha_k(\omega), x(t, \omega) = N+1). \quad k > 1.$$

由于 $R < +\infty$, 易见[①] $P(\alpha_k < +\infty, \beta_k < +\infty, k \in \mathbf{N}^*) = 1$. 今保存区间 $[\alpha_k(\omega), \beta_k(\omega))(k \in \mathbf{N}^*)$ 而抛去其他区间, 并将保留区

---

① 这也可从定理 5.2 证 (i) 推出.

间向左按原序平移，使 $a_1(\omega)$ 重合于 0，并使各区间相连而不相交，所得为 Q_N 过程 $x_N(t,\omega)$，$t\geq 0$(见(2.7))，$P(x_N(0,\omega)=N)=1$. 用(2.8)定义 $\sigma_N(\omega)$，显然 $\sigma_N(\omega)\leq \xi_0(\omega)$，故 $E\sigma_N\leq E\xi_0$. 此与(4.6)矛盾，故 $P(\xi_0(\omega)=+\infty)=1$.

其次，由 $P(\xi_0<+\infty)\geq P(\xi_k<+\infty)\prod_{i=1}^{k}\dfrac{a_i}{a_i+b_i}$，得

$$P(\xi_k(\omega)=+\infty)=1 \quad (k\in E). \tag{4.7}$$

今若说定理不真，即对某 $\mathscr{S}\geq 0$，有 $P(x(\tau_\mathscr{S})=+\infty)>0$，则必存在 $k\in E$，使 $P(x(\tau_\mathscr{S})=+\infty, x(\mathscr{S})=k)>0$，故 $P(\xi_k<+\infty)\geq P(x(\tau_\mathscr{S})=+\infty, x(\mathscr{S})=k)>0$，此与(4.7)矛盾. ∎

系 4.1 若 $S=+\infty$，则存在 Ω_0，$P(\Omega_0)=1$，使 $\omega\in\Omega_0$ 时，t-集 $H_\omega=(t:\lim\limits_{s\downarrow t}x(s,\omega)=+\infty)$ 是空集.

证 只要令 $\Omega_0=\bigcap\limits_{k}(\xi_k(\omega)=+\infty$ 且存在 $\varepsilon=\varepsilon(\omega)$，使 $(t,t+\varepsilon)$ 中无飞跃点). ∎

反之，如 $S<+\infty$，由下面的系 7.1，知存在 Q 过程使定理 4.2 及系 4.1 不成立，故 $S=+\infty$ 是不可能"连续"流入有穷状态的充分必要条件.

§5. 一般 Q 过程变换为 Doob 过程

是否可变任一 Q 过程为 Doob 过程？本节给出一般方法. 此时 Q 不变而转移概率的变化则可控制得很小[①].

定义 5.1 使 (1.6) 成立的矩阵 Q 称为原子的，如满足

$$P(\xi_n = j \mid \xi_{n-1} = i) = \frac{q_{ij}}{q_i} \quad (i, j \in E) \tag{5.1}$$

的马氏链 $(\xi_n)(n \geqslant 0)$ 具有性质：对任一集 $R \subset E$，不论开始分布如何，存在正整数 $N(=N(\omega))$，使

$$P(\xi_n \in R, \ n \geqslant N) = 0 \ \text{或} \ 1. \tag{5.2}$$

称 Q 过程为原子的，若 Q 是原子的；而由 (5.1) 定义的马氏链 (ξ_n)，则称为此 Q 过程的嵌入马氏链.

易见生灭过程是原子的 [1, P113 的系].

今设 $x(t, \omega)$, $t \geqslant 0$ 为可分的 Borel 可测齐次马氏过程，τ 为其第一个飞跃点；又设 ζ 为停时，$P(\zeta < +\infty) = 1$，$P(x(\zeta) = +\infty) = 0$，$\tau_\zeta^{(n)}$ 为 ζ 后的第 n 个跳跃点，$\tau_\zeta = \lim\limits_{n \to +\infty} \tau_\zeta^{(n)}$ 是 ζ 后的第一个飞跃点，易见 $\tau_\zeta^{(n)}$ 为停时.

定理 5.1 若 Q 过程 $x(t, \omega)$，$t \geqslant 0$ 是原子的，且 $P(\tau < +\infty) = 1$，则

$$P(\theta_{\tau_\zeta} B \mid \mathscr{B}_{[0, \tau_\zeta)}) = C, \tag{5.3}$$

这里 $B \in \mathscr{B}_{[0, +\infty)}$，$\mathscr{B}_{[0, \tau_\zeta)}$ 是含一切 $\mathscr{B}_{[0, \tau_\zeta^{(n)}]}$ $(n \geqslant 0)$ $(\tau_\zeta^{(0)} = \zeta)$ 的最小 σ-代数，又 C 为不依赖于 ζ 的常数.

证 除去一零测集后，$\tau_\zeta^{(n)} < \tau_\zeta^{(n+1)}$，$\Omega_{\tau_\zeta^{(n)}} = \Omega_{\tau_\zeta^{(n+1)}}$.

[①] 参看定理 6.4 及 8.2 的证明.

先证 $\mathscr{B}_{[0,\tau_\zeta^{(n)}]} \subset \mathscr{B}_{[0,\tau_\zeta^{(n+1)}]}$. 任取 $A \in \mathscr{B}_{[0,\tau_\zeta^{(n)}]}$，则 $A \subset \Omega_{\tau_\zeta^{(n+1)}}$. 又

$$(A, \tau_\zeta^{(n+1)} \leqslant t) = (A, \tau_\zeta^{(n)} \leqslant t) \cap (\tau_\zeta^{(n+1)} \leqslant t, \tau_\zeta^{(n)} \leqslant t)$$

$$= (A, \tau_\zeta^{(n)} \leqslant t) \cap \bigcup_{v=1}^{+\infty} \bigcap_{u=v}^{+\infty} \bigcup_{k=1}^{2^n-1} \left[\left(\frac{(k-1)t}{2^n} < \tau_\zeta^{(n)} \leqslant \frac{kt}{2^n} \right) \cap \right.$$

$$\left. \left(存在 j, k < j \leqslant 2^n, 使 x\left(\frac{jt}{2^n}\right) \neq x\left(\frac{kt}{2^n}\right) \right) \right] \in \mathscr{B}_{[0,t]},$$

故 $A \in \mathscr{B}_{[0,\tau_\zeta^{(n+1)}]}$. 于是 $\mathscr{B}_{[0,\tau_\zeta^{(n)}]} \uparrow \mathscr{B}_{[0,\tau_\zeta]}$. 由 [4, 定理 4.3] 及关于 $\tau_\zeta^{(n)}$ 的强马氏性 (4.4)，得

$$P(\theta_{\tau_\zeta} B \mid \mathscr{B}_{[0,\tau_\zeta]}) = \lim_{n \to +\infty} P(\theta_{\tau_\zeta} B \mid \mathscr{B}_{[0,\tau_\zeta^{(n)}]}) = \lim_{n \to +\infty} P_{x(\tau_\zeta^{(n)})}(\theta_\tau B).$$

(5.4)

任意取实数 $\alpha > 0$，令 $R = (i: P_i(\theta_\tau B) > \alpha)$，由 (5.4)

$$P(\omega: P(\theta_{\tau_\zeta} B \mid \mathscr{B}_{[0,\tau_\zeta]}) > \alpha)$$

$= P(\omega: 存在 N(=N(\omega))$，使 $n \geqslant N$ 时，$x(\tau_\zeta^{(n)}) \in R)$.

然而 $(x(\tau_\zeta^{(n)}))(n \in \mathbf{N})$ 是 Q 过程的嵌入马氏链，开始分布为 $r_i = P(x(\zeta) = i)(i \geqslant 0)$，由原子性得

$$P(\omega: P(\theta_{\tau_\zeta} B \mid \mathscr{B}_{[0,\tau_\zeta]}) > \alpha) = 0 \text{ 或 } 1.$$

由于 α 为任一正数，故以概率 1

$$P(\theta_{\tau_\zeta} B \mid \mathscr{B}_{[0,\tau_\zeta]}) = C(\zeta),$$

其中 $C(\zeta)$ 表一常数，它可能依赖于 ζ，更精确些，可能依赖于开始分布 $r_i = P(x(\zeta) = i)(i \in E)$. 利用马氏链理论中下列简单事实：设 (x_n) 为马氏链，f 为实值函数，若对任意开始分布，$f(x_n)$ 当 $n \to +\infty$ 时以概率 1 收敛于常数，则此常数与开始分布无关. 因此，在此事实中取 $x_n = x(\tau_\zeta^{(n)})$，$f(x_n) = P_{x_n}(\theta_\tau B)$，即得 $C(\zeta)$ 与 ζ 无关. ∎

现在进一步设 Q 满足 (1.7) 而且 $R < +\infty$. 用迭代法定义

$\tau_1(\omega)(=\tau(\omega))$ 为 $x(t, \omega)$ 的第一个飞跃点， (5.5)

$\beta_1^{(n)}(\omega) = \inf(t: t \geqslant \tau_1(\omega), x(t, \omega) \leqslant n);$ (5.6)

若 $\tau_{m-1}(\omega)$，$\beta_{m-1}^{(n)}(\omega)$ 已定义，则令

$$\tau_m(\omega) \text{ 为 } \beta_{m-1}^{(n)}(\omega) \text{ 后的第一个飞跃点}, \quad (5.7)$$

$$\beta_m^{(n)}(\omega) = \inf(t: t \geqslant \tau_m(\omega), \ x(t, \omega) \leqslant n). \quad (5.8)$$

此外，令 $\beta_{mk}^{(n)}(\omega)$ 为 $\beta_m^{(n)}(\omega)$ 后的第 k 个跳跃点.

对 Q 过程 $x(t, \omega)$，$t \geqslant 0$ 施行 $C(\tau_m(\omega), \beta_m^{(n)}(\omega))$ 变换后，所得过程记为 $x_n(t, \omega)$，$t \geqslant 0$.

定理 5.2 若 $R < +\infty$，则 $x_n(t, \omega)$ 为 $(Q, V^{(n)})$ 过程，其中 $V^{(n)} = (v_0^{(n)}, v_1^{(n)}, \cdots, v_n^{(n)})$ 满足 (3.14).

证 为证此只需要证明下列结论：

(i) 对一切 n，$m = 1, 2, \cdots$，$P(\beta_m^{(n)} < +\infty) = 1$;

(ii) 对任意固定的 $n \geqslant 1$，$x(\beta_m^{(n)})(m \in \mathbf{N}^*)$ 独立同分布，而且 $v_i^{(n)} = P(x(\beta_m^{(n)}) = i)$ 满足 (3.14);

(iii) 对任意固定的 $n \geqslant 1$，随机变量族

$$(x(\beta_{mk}^{(n)}); (\beta_{m,k+1}^{(n)} - \beta_{mk}^{(n)}), \ k \in \mathbf{N}) \quad (\beta_{m0}^{(n)} = \beta_m^{(n)})$$

不依赖于随机变量族

$$(x(\beta_{jk}^{(n)}); \beta_{jk}^{(n)}, \ j = 0, 1, 2, \cdots, m-1, \ k \in \mathbf{N}).$$

如果这些结论得以证明，那么由于 $x_n(t, \omega)$，$t \geqslant 0$ 的密度矩阵是 Q（这由 Q 中元的概率意义 (2.10)(2.11) 推出），而且在飞跃点上的分布为 $P(x(\beta_m^{(n)}) = i) = v_i^{(n)}$ $(0 \leqslant i \leqslant n)$，故它是 $(Q, V^{(n)})$ 过程.

(i)～(iii) 的证明分成四步：

i) 由 $R < +\infty$，存在 $s > 0$ 及 $l \in E$，使

$$P(\beta_1^{(l)} < +\infty) \geqslant P(\tau_1 < s, \ x(s) \leqslant l) > 0. \quad (5.9)$$

由 (1.7) 可见①存在 $T > 0$，$\alpha > 0$，使对任一 $k \in E$，有 $P_k(\beta_1^{(l)} \leqslant$

---

① 实际上，由 (5.9) 至少有一 $j \in E$ 使 $P_j(\beta^{(l)} < +\infty) > 0$. 由 $R < +\infty$ 知 $P_0(\beta^{(l)} < +\infty) > 0$. 故有 T，α 使 $P_0(\beta_1^{(l)} \leqslant T) \geqslant \alpha$，于是 $P_k(\beta_1^{(l)} \leqslant T) \geqslant P_0(\beta_1^{(l)} \leqslant T) \geqslant \alpha$.

$T) \geqslant \alpha$. 从而根据引理 4.1 立得 $E\beta_1^{(l)} < +\infty$, 故有 $P(\beta_1^{(l)} < +\infty) = 1$. 今考虑 $y(t, \omega) = x(\beta_1^{(l)}(\omega) + t, \omega)$, $t \geqslant 0$, 由关于 $\beta_1^{(l)}$ 的强马氏性它也是 Q 过程. 仿 (5.6) 对此 $y(t, \omega)$ 定义的 $\beta_1^{(l)}(\omega)$ 记为 $\beta_{1y}^{(l)}(\omega)$, 则由上知 $P(\beta_{1y}^{(l)}(\omega) < +\infty) = 1$. 于是由 $\beta_2^{(l)}(\omega) = \beta_1^{(l)}(\omega) + \beta_{1y}^{(l)}(\omega)$ 得 $P(\beta_2^{(l)} < +\infty) = 1$. 如此继续, 得证 $P(\beta_m^{(l)} < +\infty) = 1 \, (m \geqslant 1)$. 既然当 $n \geqslant l$ 时, $\beta_m^{(n)}(\omega) \leqslant \beta_m^{(l)}(\omega)$, 故 (i) 对 $n \geqslant l$ 正确.

ii) 固定 $n(\geqslant l)$. 令 $\tau_0 \equiv 0$. 取 $B_i = (x(\beta_0^{(n)}) = i)$, 并于定理 5.1 中令 $\zeta(\omega) = \beta_{m-1}^{(n)}(\omega) \, (m = 2, 3, \cdots)$, 得知诸事件

$$\theta_{\tau_m} B_i = (x(\beta_m^{(n)}) = i) \quad (i = 0, 1, 2, \cdots, n)$$

与 $\mathscr{B}_{[0, \tau_m]}$ 中的事件独立, 特别与事件 $x(\beta_j^{(n)}) = i$, $j < m$, $i = 0$, $1, 2, \cdots, n$ 及其交独立, 从而诸 $x(\beta_m^{(n)}) \, (m \geqslant 1)$ 相互独立. 再在定理 5.1 中顺次令 $\zeta(\omega) = 0$, $\zeta(\omega) = \beta_1^{(n)}(\omega)$, $\zeta(\omega) = \beta_2^{(n)}(\omega), \cdots$, 可见诸事件 $\theta_{\tau_1} B_i = (x(\beta_1^{(n)}) = i)$, $\theta_{\tau_2} B_i = (x(\beta_2^{(n)}) = i)$, $\cdots$ 有相同的概率.

iii) 为证定理 5.2(iii) 对 $n(\geqslant l)$ 成立, 只需证 $(x(\beta_{mk}^{(n)}),$ $\beta_{m,k+1}^{(n)} - \beta_{mk}^{(n)}, k \in \mathbf{N})$ 与 $\mathscr{B}_{[0, \tau_m]}$ 中的事件独立. 对任两组整数 $k_1 < k_2 < \cdots < k_u$, $r_1 < r_2 < \cdots < r_v$, 有

$$(x(\beta_{mk_1}^{(n)}) = i_1, \cdots, x(\beta_{mk_u}^{(n)}) = i_u; \beta_{mr_1+1}^{(n)} - \beta_{mr_1}^{(n)}$$
$$> t_1, \cdots, \beta_{mr_v+1}^{(n)} - \beta_{mr_v}^{(n)} > t_v)$$
$$= \theta_{\tau_m} B,$$

其中 $B = (x(\beta_{0k_1}^{(n)}) = i_1, \cdots, x(\beta_{0k_u}^{(n)}) = i_u; \beta_{0r_1+1}^{(n)} - \beta_{0r_1}^{(n)} > t_1, \cdots,$ $\beta_{0r_v+1}^{(n)} - \beta_{0r_v}^{(n)} > t_v)$. 然后仿 ii) 利用定理 5.1 即可.

iv) 证定理 5.2(i)(ii)(iii) 中的结论对任一 $n(\geqslant 1)$ 成立. 只要证 $P(\beta_k^{(1)} < +\infty) = 1$ 即可. 若 $v_1^{(l)} = P(x(\beta_1^{(l)}) = 1) > 0$, 则由 i) 及 ii) 得 $P(\beta_1^{(1)} < +\infty) = v_1^{(l)} \sum_{n=0}^{+\infty} (1 - v_1^{(l)})^n = 1$; 若 $v_1^{(l)} = 0$,

则取 $k(\leqslant l)$，使 $v_k^{(l)} > 0$，于是

$$P(\beta_1^{(l)} < +\infty) = v_k^{(l)} c_{k_1} \cdot \sum_{n=0}^{+\infty} (1 - v_k^{(l)} c_{k_1})^n = 1.$$

然后利用 i) 中之方法得证 $P(\beta_k^{(1)} < +\infty) = 1$ $(k \geqslant 1)$，故 (i) 完全得证.

为完全证明 (ii)(iii)，只要在 ii) iii) 中以 1 换 l；最后，注意 $g_{n+1,n}(x_{n+1}(t, \omega)) = x_n(t, \omega)$，故 $(v_i^{(n)})$ 满足 (3.14). ∎

对于 $R < +\infty$，$S = +\infty$ 的 Q，尚可如下把 Q 过程变为 Doob 过程. 代替 (5.5)～(5.8)，令

$$\tau_1(\omega) \text{ 为 } x(t, \omega) \text{ 的第一个飞跃点}, \tag{5.10}$$

$$\alpha_1^{(n)}(\omega) = \inf(t: t \geqslant \tau_1(\omega), t \text{ 为飞跃点}, x(t, \omega) \leqslant n); \tag{5.11}$$

$$\tau_m(\omega) \text{ 为 } \alpha_{m-1}^{(n)}(\omega) \text{ 后的第一个飞跃点}, \tag{5.12}$$

$$\alpha_m^{(n)}(\omega) = \inf(t: t \geqslant \tau_m(\omega), t \text{ 为飞跃点}, x(t, \omega) \leqslant n), \tag{5.13}$$

此外，令 $\alpha_{mk}^{(n)}(\omega)$ 为 $\alpha_m^{(n)}(\omega)$ 后的第 k 个跳跃点.

对 Q 过程 $x(t, \omega)$，$t \geqslant 0$，施行 $c(\tau_m(\omega), \alpha_m^{(n)}(\omega))$ 变换后，所得过程也记为 $x_n(t, \omega)$，$t \geqslant 0$.

定理 5.3 设 $R < +\infty$，$S = +\infty$，则 $x_n(t, \omega)$，$t \geqslant 0 (n \geqslant l$，$l$ 为某非负数) 为 $(Q, S^{(n)})$ 过程，其中 $S^{(n)} = (\mathscr{S}_0^{(n)}, \mathscr{S}_1^{(n)}, \cdots, \mathscr{S}_n^{(n)})$ 满足 (3.8).

证 取 $\mathscr{S} > 0$，使 $P(\tau_1 < \mathscr{S}) > 0$. 由定理 4.2，$P(x(\tau_{\mathscr{S}}) \neq +\infty) = 1$，这里 $\tau_{\mathscr{S}}(\omega)$ 是 $\mathscr{S}$ 以前的最后一个飞跃点，故存在 $l \in E$，使 $P(\tau_1 \leqslant \tau_{\mathscr{S}}, x(\tau_{\mathscr{S}}) = l) = P(\tau_1 < \mathscr{S}, x(\tau_{\mathscr{S}}) = l) > 0$，从而

$$P(\alpha_1^{(l)} < +\infty) \geqslant P(\tau_1 \leqslant \tau_{\mathscr{S}}, x(\tau_{\mathscr{S}}) = l) > 0,$$

故得到了与 (5.9) 类似的式子，然后只要逐句重复上定理的证明至 iv) 以前，并作显然的改变即可. ∎

§6. $S<+\infty$ 时 Q 过程的构造

固定任一满足 (1.7) 的 Q，使 $R<+\infty$. 考虑方程组 (3.14)，任意给定 $v_1^{(1)}$，$0\leqslant v_1^{(1)}\leqslant 1$，则 $v_0^{(1)}$ 唯一决定；如果 $(v_0^{(n)}, v_1^{(n)}, \cdots, v_n^{(n)})$ 已求出，那么任意给定 $v_{n+1}^{(n+1)}$，$0\leqslant v_{n+1}^{(n+1)}\leqslant 1$ 后，可唯一决定 $(v_0^{(n+1)}, v_1^{(n+1)}, \cdots, v_{n+1}^{(n+1)})$. 因此给出一数列 $(v_n^{(n)})$ 后（以后简记为 (v_n)，即 $v_n=v_n^{(n)}$），可唯一决定 (3.14) 的一组解. 为使此组解中每 $(v_0^{(n)}, v_1^{(n)}, \cdots, v_n^{(n)})$ $(n\in \mathbf{N}^*)$ 均是一概率分布，不难看出，充分必要条件是 (v_n) 满足条件

$$1\geqslant v_1\geqslant 0,$$
$$1\geqslant v_n\geqslant \frac{v_{n+1}(z-z_{n+1})}{(z-z_n)-v_{n+1}(z_{n+1}-z_n)}\geqslant 0 \quad (n\geqslant 1). \quad (6.1)$$

由定理 5.2(ii)，立得

引理 6.1 设已给 Q 过程 $x(t,\omega)$，$t\geqslant 0$，使 $R<+\infty$，则序列 (v_n)

$$v_n=P(x(\beta_1^{(n)})=n), \quad n\geqslant 1 \quad (6.2)$$

满足 (6.1).

本节中，以下恒设 $R<+\infty$，$S<+\infty$.

重要的是，以后会证明：设已给满足 (6.1) 的序列 (v_n)，则必存在 Q 过程 $x(t,\omega)$，$t\geqslant 0$，满足 (6.2)，而且此过程是唯一的[1]. 为此要做相当准备.

设已给一列满足 (6.1) 的 (v_n)，它决定 (3.14) 的一组解记为 $(V^{(n)})$，$V^{(n)}=(v_0^{(n)}, v_1^{(n)}, \cdots, v_n^{(n)})$. 由引理 3.4，可在某空间

---

[1] 有相同转移概率的 Q 过程算作同一 Q 过程. 见 §1.

$(\Omega, \mathscr{F}, P)$ 上作一列 $(Q, V^{(n)})$ 过程 $x_n(t, \omega)$, $t \geqslant 0$, $n \geqslant 1$, 并且

$$g_{nm}(x_n(t, \omega)) = x_m(t, \omega) \quad (n \geqslant m), \quad (6.3)$$

换言之，对 $x_n(t, \omega)$ 施行 $C(\tau_i^{(n,m)}(\omega), \beta_i^{(n,m)}(\omega))$ 变换后即得 $x_m(t, \omega)$，这里 $\tau_i^{(n,m)}(\omega)$, $\beta_i^{(n,m)}(\omega)$ 仿 $(5.5) \sim (5.8)$ 对过程 $x_n(t, \omega)$ 定义，即

$$\tau_1^{(n,m)}(\omega) \text{ 为 } x_n(t, \omega) \text{ 的第一个飞跃点}, \quad (6.4)$$

$$\beta_1^{(n,m)}(\omega) = \inf(t: t \geqslant \tau_1^{(n,m)}(\omega), x_n(t, \omega) \leqslant m); \quad (6.5)$$

$$\tau_i^{(n,m)}(\omega) \text{ 为 } \beta_{i-1}^{(n,m)}(\omega) \text{ 后第一个飞跃点}, \quad (6.6)$$

$$\beta_i^{(n,m)}(\omega) = \inf(t: t \geqslant \tau_i^{(n,m)}(\omega), x_n(t, \omega) \leqslant m). \quad (6.7)$$

$$\tau_i^{(n)}(\omega) \text{ 为 } x_n(t, \omega) \text{ 的第 } i \text{ 个飞跃点}, \quad (6.8)$$

$$\tau_{ij}^{(n)}(\omega) \text{ 为 } \tau_i^{(n)}(\omega) \text{ 后的第 } j \text{ 个跳跃点}. \quad (6.9)$$

先考虑一特殊情况，即 $v_n = 1(n \geqslant 1)$ 时，此 (v_n) 所决定的 (3.14) 的解记为 $(\pi^{(n)})$，显然

$$\pi^{(n)} = (\pi_0^{(n)}, \cdots, \pi_{n-1}^{(n)}, \pi_n^{(n)}) = (0, \cdots, 0, 1). \quad (6.10)$$

引理 6.2 对 $(Q, \pi^{(n)})$ 过程 $x_n(t, \omega)$, $t \geqslant 0$, 令 $\xi^{(n,m)}(\omega)$ 表使 $\beta_i^{(n,m)}(\omega) < \beta_1^{(n,0)}(\omega)$ 的 i 的个数，则

$$E\xi^{(n,m)}(\omega) = \sum_{i=0}^{+\infty} P(\beta_i^{(n,m)} < \beta_1^{(n,0)}) = \frac{z}{z - z_m}. \quad (6.11)$$

证 定义 $\eta_i(\omega) = 1$ 或 0, 视 $\beta_i^{(n,m)} < \beta_1^{(n,0)}$ 与否而定，则

$$E\xi^{(n,m)} = \sum_{i=1}^{+\infty} E\eta_i = \sum_{i=1}^{+\infty} P(\beta_i^{(n,m)} < \beta_1^{(n,0)}).$$

但 $P(\beta_i^{(n,m)} < \beta_1^{(n,0)}) = (1 - c_{m0})^{i-1}$, 故由 (2.16) 得

$$E\xi^{(n,m)} = \sum_{i=1}^{+\infty} (1 - c_{m0})^{i-1} = \frac{1}{c_{m0}} = \frac{z}{z - z_m}. \blacksquare$$

对 $(Q, \pi^{(n)})$ 过程 $x_n(t, \omega)$, $t \geqslant 0$, 定义

$$T^{(n,r)}(\omega) = \beta_1^{(n,r)}(\omega) - \tau_1^{(n)}(\omega)(n > r), \quad (6.12)$$

故 $T^{(n,r)}$ 是自 $\tau_1^{(n)}$ 算起，初次到达 r 的时间. 注意 $P(x_n(\tau_1^{(n)}) =$

$n)=1$,故 $ET^{(n,r)}$ 是在"质点自 n 出发,到达 $+\infty$ 后立刻回到 n"的条件下,质点初次到达 r 的平均时间,因此,它不超过在"自 n 出发,到达 $+\infty$ 后'连续地'回到有穷状态"的条件下,此时间的平均值 $\sum_{k=r+1}^{+\infty} e_k$(见(2.2)).此直观上显然正确的事实可如下严格证明.

引理 6.3 对 $(Q,\pi^{(n)})$ 过程,$ET^{(n,r)} \leqslant \sum_{k=r+1}^{+\infty} e_k \leqslant S$ ($n > r \geqslant 0$).

证 首先注意一简单事实:设有差分方程

$$\mathscr{D}_k = \frac{1}{c_k}(1-e^{-c_k\varepsilon}) + \frac{a_k}{c_k}\mathscr{D}_{k-1} + \frac{b_k}{c_k}\mathscr{D}_{k+1},\ 0<n<k<N,$$
(6.13)

其中 $c_k = a_k + b_k$,$a_k > 0$,$b_k > 0$,$0 < \varepsilon \leqslant +\infty$(若 $\varepsilon = +\infty$,则令 $e^{-c_k\varepsilon} = 0$),在边值条件 $\mathscr{D}_n = 0$,$\mathscr{D}_N = c$($c \geqslant 0$)下其解记为 $(\mathscr{D}_n^{(c)},\mathscr{D}_{n+1}^{(c)},\cdots,\mathscr{D}_N^{(c)})$,则 $\mathscr{D}_k^{(c)} \geqslant \mathscr{D}_k^{(0)}$ ($n \leqslant k \leqslant N$).

今以 f_i 表对 $(Q,\pi^{(n)})$ 过程自 i 出发初次回到 $i-1$ 的平均时间,则像导出 (2.12′) 一样,得

$$f_i = \frac{a_i}{c_i} \cdot \frac{1}{c_i} + \frac{b_i}{c_i}\left(\frac{1}{c_i} + f_{i+1} + f_i\right),\ 1 \leqslant i \leqslant n \quad (6.14)$$

或 $f_i = \frac{1}{a_i} + \frac{b_i}{a_i}f_{i+1}$,$1 \leqslant i \leqslant n$.

故若已知 f_{n+1},则

$$f_i = \frac{1}{a_i} + \sum_{k=0}^{n-1-i} \frac{b_i b_{i+1} \cdots b_{i+k}}{a_i a_{i+1} \cdots a_{i+k} a_{i+k+1}} + \frac{b_i b_{i+1} \cdots b_n}{a_i a_{i+1} \cdots a_n} f_{n+1},$$
(6.15)

故若能证 $f_{n+1} \leqslant e_{n+1}$,则由 (6.15) 及 (2.2) 立得 $f_i \leqslant e_i$,而

$$ET^{(n,r)} \leqslant \sum_{i=r+1}^{+\infty} f_i \leqslant \sum_{i=r+1}^{+\infty} e_i = S.$$

为证 $f_{n+1} \leqslant e_{n+1}$，考虑 $N > k > n$. 以 $\overline{\mathcal{D}}_k^{(N)}$ 表自 k 出发初次到达 n 的平均时间，但当到达 N 时，立刻回到 n（更精确些，$\overline{\mathcal{D}}_k^{(N)}$ 为对 (Q, π_n) 过程，自 k 出发初次到达含两点之集 $\{n, N\}$ 的平均时间）；而 $\overline{\overline{\mathcal{D}}}_k^{(N)}$ 表对 Q_N-过程[见(2.7)]自 k 出发初次到达 n 的平均时间. 易见 $(\overline{\mathcal{D}}_k^{(N)})$ 与 $(\overline{\overline{\mathcal{D}}}_k^{(N)})$ $(n \leqslant k \leqslant N)$ 分别是 (6.13)，当 $\varepsilon = +\infty$ 时在边界条件 $\mathcal{D}_n = 0$，$\mathcal{D}_N = 0$ 及 $\mathcal{D}_n = 0$，$\mathcal{D}_N = c\left(\geqslant \dfrac{1}{a_N + b_N}\right)$ 下的解，故由上述事实 $\overline{\mathcal{D}}_{n+1}^{(N)} \leqslant \overline{\overline{\mathcal{D}}}_{n+1}^{(N)}$. 但 $\overline{\mathcal{D}}_{n+1}^{(N)} \uparrow f_{n+1}$；$\overline{\overline{\mathcal{D}}}_{n+1}^{(N)} \uparrow e_{n+1}$，$N \to +\infty$[注意 $\overline{\overline{\mathcal{D}}}_{n+1}^{(N)} = e_{n+1}^{(N)}$，见(2.12′)]，故 $f_{n+1} \leqslant e_{n+1}$. ∎

对任 ε，$0 < \varepsilon \leqslant +\infty$，考虑函数

$$f_\varepsilon(x) = x, \text{ 如 } 0 \leqslant x < \varepsilon; = \varepsilon, \text{ 如 } x \geqslant \varepsilon. \qquad (6.16)$$

设 $\xi(\geqslant 0)$ 是具有分布密度 ce^{-cy} $(c > 0)$ 的随机变量，则易见 $Ef_\varepsilon(\xi) = \dfrac{1}{c}(1 - e^{-c\varepsilon})$，特别，$Ef_{+\infty}(\xi) = \dfrac{1}{c}$.

对 $(Q, \pi^{(n)})$ 过程及 $n > r$，定义

$$T_\varepsilon^{(n,r)}(\omega) = \sum_{\tau_1^{(n)} \leqslant \tau_{ij}^{(n)} < \beta_1^{(n,r)}} f_\varepsilon(\tau_{ij+1}^{(n)}(\omega) - \tau_{ij}^{(n)}(\omega)), \qquad (6.17)$$

特别，$T_{+\infty}^{(n,r)} = T^{(n,r)}$. 直觉地说，将 $(Q, \pi^{(n)})$ 过程 $x_n(t, \omega)$，$t \geqslant 0$ 的常数区间如下变形：若其长不小于 ε，则缩短之使其长变为 ε；若长小于 ε，则保留不变. 于是 $T_\varepsilon^{(n,r)}(\omega)$ 是 $[\tau_1^{(n)}(\omega), \beta_1^{(n,r)}(\omega))$ 中变形后的区间的总长.

引理 6.4 对 $(Q, \pi^{(n)})$ 过程，

$ET_\varepsilon^{(n,r)}$

$$\leqslant \sum_{k=r+1}^{+\infty} \left[\dfrac{1}{a_k}(1 - e^{-c_k\varepsilon}) + \sum_{l=0}^{+\infty} \dfrac{b_k b_{k+1} \cdots b_{k+l}}{a_k a_{k+1} \cdots a_{k+l}} \times \dfrac{(1 - e^{-c_{k+l+1}\varepsilon})}{a_{k+l+1}} \right] \leqslant S.$$

$$(6.18)$$

证 $(Q, \pi^{(n)})$ 过程的 k 区间经如上变形后，有平均长度为 $E_k f_\varepsilon(\beta) = \frac{1}{c_k}(1-e^{-c_k\varepsilon})$ [见(2.10)]，然后重复引理 6.3 的证明，只要换 $\frac{1}{c_k}$，$\frac{1}{a_k}$ 为 $\frac{1}{c_k}(1-e^{-c_k\varepsilon})$ 及 $\frac{1}{a_k}(1-e^{-c_k\varepsilon})$ 即可. ∎

由于(6.3)，对固定 $\varepsilon>0$，有 $T_\varepsilon^{(n,0)}(\omega) \leqslant T_\varepsilon^{(n+1,0)}(\omega)$，令 $T_\varepsilon(\omega) = \lim\limits_{n\to+\infty} T_\varepsilon^{(n,0)}(\omega)$.

引理 6.5 对 $(Q, \pi^{(n)})$ 过程，
$$\lim_{\varepsilon\to 0} ET_\varepsilon(\omega)=0; \qquad P(\lim_{\varepsilon\to 0} T_\varepsilon(\omega)=0)=1.$$

证 由 $ET_\varepsilon^{(n,0)} \leqslant S_\varepsilon =$
$$\sum_{k=1}^{+\infty}\left[\frac{1}{a_k}(1-e^{-c_k\varepsilon}) + \sum_{l=0}^{+\infty} \frac{b_k b_{k+1}\cdots b_{k+l}}{a_k a_{k+1}\cdots a_{k+l}} \frac{(1-e^{-c_{k+l+1}\varepsilon})}{a_{k+l+1}}\right] \leqslant S < +\infty$$
得 $ET_\varepsilon \leqslant S_\varepsilon$，$\lim\limits_{\varepsilon\to 0} ET_\varepsilon \leqslant \lim\limits_{\varepsilon\to 0} S_\varepsilon = 0$. 又由 $T_\varepsilon(\omega)$ 关于 ε 的单调性，存在 $\lim\limits_{\varepsilon\to 0} T_\varepsilon(\omega) = T(\omega) \geqslant 0$. 根据积分单调定理 $ET(\omega) = \lim\limits_{\varepsilon\to 0} ET_\varepsilon(\omega)=0$，从而 $P(T(\omega)=0)=1$. ∎

令 $L_{nm}^{(i)}(\omega) = \begin{cases} \beta_i^{(n,m)}(\omega) - \tau_i^{(n,m)}(\omega), & \beta_i^{(n,m)}(\omega) < \beta_1^{(n,0)}(\omega), \\ 0, & \beta_i^{(n,m)}(\omega) > \beta_1^{(n,0)}(\omega). \end{cases}$

$$(6.17')$$

故 $\sum\limits_{i=1}^{+\infty} L_{nm}^{(i)}(\omega)$ 是经(6.3)自 $x_n(t, \omega)$ 得 $x_m(t, \omega)$ 时，在 $[\tau_1^{(n)}(\omega), \beta_1^{(n,0)}(\omega)]$ 中所抛去区间之总长. 由(6.3)可见

$$\sum_{i=1}^{+\infty} L_{nm}^{(i)}(\omega) \leqslant \sum_{i=1}^{+\infty} L_{n+1,m}^{(i)}(\omega),$$

故存在 $L_m(\omega) = \lim\limits_{n\to+\infty} \sum\limits_{i=1}^{+\infty} L_{nm}^{(i)}(\omega)$.

引理 6.6 对 $(Q, \pi^{(n)})$ 过程，
$$\lim_{m\to+\infty} EL_m(\omega)=0, \quad 又 P(\lim_{m\to+\infty} L_m(\omega)=0)=1.$$

证 由 $(Q, \pi^{(n)})$ 过程 $x_n(t, \omega)$，$t \geqslant 0$ 的构造，$\beta_i^{(n,m)} -$

$\tau_i^{(n,m)}$ 不依赖①于事件 $(\beta_1^{(n,m)} < \beta_1^{(n,0)})$，而且 $\beta_i^{(n,m)} - \tau_i^{(n,m)}$ $(i \in \mathbf{N}^*)$ 同分布，故由引理 6.2 及 $\tau_1^{(n,m)}(\omega) = \tau_1^{(n)}(\omega)$，得

$$\sum_{i=1}^{+\infty} EL_{nm}^{(i)}$$
$$= \sum_{i=1}^{+\infty} E(\beta_i^{(n,m)} - \tau_i^{(n,m)} \mid \beta_1^{(n,m)} < \beta_1^{(n,0)}) P(\beta_i^{(n,m)} < \beta_1^{(n,0)})$$
$$= \sum_{i=1}^{+\infty} E(\beta_i^{(n,m)} - \tau_i^{(n,m)}) P(\beta_i^{(n,m)} < \beta_1^{(n,0)})$$
$$= E(\beta_1^{(n,m)} - \tau_1^{(n,m)}) \sum_{i=1}^{+\infty} P(\beta_1^{(n,m)} < \beta_1^{(n,0)})$$
$$= E(\beta_1^{(n,m)} - \tau_1^{(n)}) \frac{z}{z - z_m}. \tag{6.18'}$$

再由引理 6.3 及 (2.2)(2.4)，经简单计算后

$$\sum_{i=1}^{+\infty} EL_{nm}^{(i)} \leqslant \sum_{k=m+1}^{+\infty} e_k \cdot \frac{z}{z - z_m} \leqslant z \sum_{k=m}^{+\infty} \frac{b_1 b_2 \cdots b_k}{a_1 a_2 \cdots a_k a_{k+1}}. \tag{6.19}$$

注意 $z < b_0 R < +\infty$，$\sum_{k=1}^{+\infty} \frac{b_1 b_2 \cdots b_k}{a_1 a_2 \cdots a_k a_{k+1}} < S < +\infty$，故

$$\lim_{m \to +\infty} EL_m(\omega) = \lim_{m \to +\infty} \Big(\lim_{n \to +\infty} \sum_{i=1}^{+\infty} EL_{nm}^{(i)} \Big) = 0.$$

再由 $L_m(\omega) \geqslant L_{m+1}(\omega)$ 即得 $P(\lim_{m \to +\infty} L_m(\omega) = 0) = 1$. ■

由 (6.7) 及 (6.3)，可见 $\beta_i^{(n,0)}(\omega) \leqslant \beta_i^{(n+1,0)}(\omega)$，故存在极限 $\beta_i^{(0)}(\omega) = \lim_{n \to +\infty} \beta_i^{(n,0)}(\omega)$. 由定义 $P(\lim_{i \to +\infty} \beta_i^{(n,0)}(\omega) = +\infty) = 1$ $(n \geqslant 1)$，既然 $\beta_i^{(n,0)}(\omega) \leqslant \beta_i^{(0)}(\omega)$，故得

$$P(\lim_{i \to +\infty} \beta_i^{(0)}(\omega) = +\infty) = 1. \tag{6.20}$$

以下"几乎一切 t"系对 Lebesgue 测度而言.

定理 6.1 以概率 1，$(Q, \pi^{(n)})$ 过程的样本函数 $x_n(t, \omega)$,

---

① 即对任意实数 a，事件 $(\beta_i^{(n,m)} - \tau_i^{(n,m)} < a)$ 与 $(\beta_i^{(n,m)} < \beta_1^{(n,0)})$ 独立.

当 $n \to +\infty$ 时几乎对一切 t 收敛.

证 除去一零测度集后，可设对每一 $\omega \in \Omega$，$x_n(t, \omega)$，在任一有限区间中只有有限多个 i 区间 ($n \geq 0$，$i \in E$). 若向左平移使每 $x_n(t, \omega)$ 为常数的区间，而且每区间平移的距离不大于 ε，则在 $[0, \beta_1^{(n,0)}(\omega))$ 中，使 $x_n(t, \omega)$ 不等于 $x_m(t, \omega)$ ($n > m$) 的点 t 所成区间的总长不超过 $\varepsilon + T_\varepsilon^{(n)}(\omega) < \varepsilon + T_\varepsilon(\omega)$. 固定 k，取 $n > m > l (> k)$. 由于 $\beta_1^{(n,0)}(\omega) \geq \beta_1^{(k,0)}(\omega)$，得

$$L(t: t \in [0, \beta_1^{(k,0)}(\omega)), x_n(t, \omega) \neq x_m(t, \omega)) \leq L_l(\omega) + T_{L_l}(\omega).$$
(6.21)

令 $\Omega_0 = (L_l(\omega) + T_{L_l}(\omega) \downarrow 0)(l \to +\infty)$，由引理 6.5，引理 6.6 得 $P(\Omega_0) = 1$. 固定 $\omega \in \Omega_0$，由 (6.21) 知 $x_n(t, \omega)(= x_n(t))$ 在 $[0, \beta_1^{(k,0)})$ 中依测度 L 收敛，故存在一列 $n_i \to +\infty$，使 $x_{n_i}(t)$ 在 $[0, \beta_1^{(k,0)})$ 中对几乎一切 t 收敛. 固定一收敛点 t_0，由于 Doob 过程的相空间为 E，故存在 $M \in E$，使 $x_{n_i}(t_0) \to M (i \to +\infty)$. 由于 E 离散，有正整数 L，使

$$x_{n_i}(t_0) = M \quad (i \geq L).$$
(6.22)

今证存在正整数 L'，使 $n > L'$ 时，$x_n(t_0) = M$，从而 $\{x_n(t_0)\}$ 收敛. 否则必存在一列 $m_i \to +\infty$，使

$$x_{m_i}(t_0) \neq M.$$

由此式及 (6.22)，并根据 $g_{nm}(x_n(t, \omega)) = x_m(t, \omega)$，即知在 $[0, t_0]$ 中，$x_M(t, \omega)$ 有无穷多个不同的 M 区间，此与证明开始时所说的矛盾.

于是得证在 $[0, \beta_1^{(k,0)}(\omega))$ 中，定理结论成立；令 $k \to +\infty$ 即得在 $[0, \beta_1^{(0)}(\omega))$ 中也成立；同样得证在 $[0, \beta_i^{(0)}(\omega))$ 中成立；再由 (6.20) 即得证定理. ∎

今考虑一般情况. 取 (3.14) 的任一非负解 $V^{(n)} = (v_0^{(n)}, v_1^{(n)}, \cdots, v_n^{(n)})$，$n \geq 1$. 下面看到，在证 $(Q, V^{(n)})$ 过程列的收敛

时，$(Q, \pi^{(n)})$ 过程列将在一定意义下起控制作用.

定理 6.2 以概率 1，$(Q, V^{(n)})$ 过程的样本函数 $x_n(t, \omega)$，当 $n \to +\infty$ 时几乎对一切 t 收敛.

证 若能证引理 6.5 及 6.6 对 $(Q, V^{(n)})$ 过程也成立，则只需逐字重复定理 6.1 的证明即可.

像对 $(Q, \pi^{(n)})$ 过程列定义 $L_{nm}^{(i)}(\omega)$，$T_\varepsilon^{(n,r)}(\omega)$ 等一样，对 $(Q, V^{(n)})$ 过程列定义 $\bar{L}_{nm}^{(i)}(\omega)$，$\bar{T}_\varepsilon^{(n,r)}(\omega)$ 等，只于其上加一短横线以表区别.

与推导 (6.18) 同样，得

$$E\Big(\sum_{i=1}^{+\infty} \bar{L}_{nm}^{(i)}\Big) = E(\bar{\beta}_1^{(n,m)} - \bar{\tau}_1^{(n,m)}) \cdot \sum_i P(\bar{\beta}_i^{(n,m)} < \bar{\beta}_1^{(n,0)}).$$

(6.23)

然而

$$P(\bar{\beta}_i^{(n,m)} < \bar{\beta}_1^{(n,0)})$$
$$= \sum_{\substack{d_j = 1 \\ (j=1,2,\cdots,i-1)}}^{m} P(\bar{\beta}_i^{(n,m)} < \bar{\beta}_1^{(n,0)} \mid x_n(\bar{\beta}_j^{(n,m)}) = d_j, j = 1, 2, \cdots, i-1) \cdot$$
$$P(x_n(\bar{\beta}_j^{(n,m)}) = d_j, j = 1, 2, \cdots, i-1)$$
$$= \sum_{\substack{d_j = 1 \\ (j=1,2,\cdots,i-1)}}^{m} \bigcap_{j=1}^{i-1} (1 - c_{d_j,0}) P(x_n(\bar{\beta}_j^{(n,m)}) = d_j, j = 1, 2, \cdots, i-1)$$
$$\leqslant (1 - c_{m0})^{i-1},$$

$$\sum_i P(\bar{\beta}_i^{(n,m)} < \bar{\beta}_1^{(n,0)}) \leqslant \sum_{i=1}^{+\infty} (1 - c_{m0})^{i-1} = \frac{z}{z - z_m}.$$

(6.24)

如果能够证明下列直觉上显然正确的事实：质点自 $(0, 1, 2, \cdots, n)$ 中的状态出发，每当到达 $+\infty$ 时，立即回到 $(0,$

$1, 2, \cdots, n$),这样运动直到初次[①]到达$(0, 1, 2, \cdots, m)$ ($m<n$)的平均时间,不大于它自 n 出发,每当到达$+\infty$时,立即回到 n,如是运动直到初次到达$(0, 1, 2, \cdots, m)$的平均时间.换言之,即

$$E(\bar{\beta}_1^{(n,m)} - \bar{\tau}_1^{(n,m)}) \leqslant E(\beta_1^{(n,m)} - \tau_1^{(n,m)}), \quad (6.25)$$

则由$(6.23) \sim (6.25)$立得

$$E\Big(\sum_{i=1}^{+\infty} \bar{L}_{nm}^{(i)}\Big) \leqslant E(\beta_1^{(n,m)} - \tau_1^{(n,m)}) \frac{z}{z - z_m}, \quad (6.26)$$

从而$(6.18')$对$(Q, V^{(n)})$过程正确,故引理 6.6 对$(Q, V^{(n)})$过程也正确.

为严格证明上列事实,用联结基本事件空间的技巧[4,P71],可造 Ω,使在其上同时定义$(Q, \pi^{(n)})$过程及一列独立随机变量$(y_n(\omega))$,有相同的分布 $P(y_n(\omega) = i) = v_i^{(n)}$ ($i=0, 1, 2, \cdots, n$).将$(Q, \pi^{(n)})$过程的样本函数自第一个飞跃点起至初次出现状态 $y_1(\omega)$ 的时刻止的那一段抛去,再将自下一飞跃点(即出现 $y_1(\omega)$ 的时刻后的第一飞跃点)起至以后初次出现状态 $y_2(\omega)$ 的时刻[②]止的那一段抛去,如此继续,经平移后所得即$(Q, V^{(n)})$过程,因而已将$(Q, V^{(n)})$过程嵌入于$(Q, \pi^{(n)})$过程之中.对此两过程,易见对几乎一切 $\omega \in \Omega$,有

$$\bar{\beta}^{(n,m)}(\omega) - \bar{\tau}_1^{(n,m)}(\omega) \leqslant \beta_1^{(n,m)}(\omega) - \tau_1^{(n,m)}(\omega),$$

甚至更一般的有

$$\bar{T}_\varepsilon^{(n,m)}(\omega) \leqslant T_\varepsilon^{(n,m)}(\omega),$$

因而得证(6.25),并且 $\bar{T}_\varepsilon^{(n,m)}(\omega) \leqslant T_\varepsilon^{(n,m)}(\omega)$;$\bar{T}_\varepsilon(\omega) \leqslant T_\varepsilon(\omega)$.由于 $ET_\varepsilon(\omega) \to 0$ ($\varepsilon \to 0$),$P(\lim_{\varepsilon \to 0} T_\varepsilon(\omega) = 0) = 1$,即知引理 6.5 对$(Q, V^{(n)})$过程成立. ∎

[①] 若自$(0, 1, 2, \cdots, m)$出发,则认为初次回到$(0, 1, 2, \cdots, m)$的时间为 0.
[②] 易见这些时刻以概率 1 有穷.

以 $x(t, \omega)$ 表 $(Q, V^{(n)})$ 过程列的极限. 由定理 6.2, 对几乎一切 ω, 存在 L(Lebesgue)零测集 T_ω, 当 $t \in T_\omega$ 时, $x(t, \omega)$ 无定义. 补定义

$$x(t, \omega) = +\infty, \quad t \in T_\omega, \tag{6.27}$$

从而以概率 1, $x(t, \omega)$ 在 $[0, +\infty)$ 有定义. 由于 $(Q, V^{(n)})$ 过程 $x_n(t, \omega), t \geq 0$ Borel 可测, 故对 $i \in E$, $((t, \omega): x(t, \omega) = i) \in \overline{\mathscr{B} \times \mathscr{F}}$, 这里 $\mathscr{B}$ 表 $[0, +\infty)$ 中 Borel 集族, 而 $\overline{\mathscr{B} \times \mathscr{F}}$ 则表 $\mathscr{B} \times \mathscr{F}$ 关于 $L \times P$ 的完全化 σ 代数. 因此

$$((t, \omega): x(t, \omega) = +\infty) \in \overline{\mathscr{B} \times \mathscr{F}}.$$

由定理 6.2, $L(t: x(t, \omega) = +\infty) = 0$ 对几乎一切 ω 成立, 故由 Fubini 定理, 存在 L 测度为 0 的集 T, 使 $t \in T$ 时,

$$P(\omega: x(t, \omega) = +\infty) = 0. \tag{6.28}$$

试证 (6.28) 对一切 $t \in [0, +\infty)$ 成立.

实际上, 由 (6.3) 可见对 $\omega \in \Omega$ 及一切 n

$$x(t, \omega) = x_n(t, \omega), \quad t < \tau(\omega), \tag{6.29}$$

这里 $\tau(\omega) = \tau_1^{(n)}(\omega)$ 是第一个飞跃点. 由此易见

$$P(x(t) = i \mid x(0) = i) \geq e^{-(a_i + b_i)t} \to 1, \quad t \to 0, \tag{6.30}$$

而且对任意给定的 $t > 0, \eta > 0$, 存在 $\delta > 0$, 使 $\varepsilon < \delta$ 时, 下两式成立:

$$|P(x(t) \neq +\infty \mid x(0) = i, x(\varepsilon) = i) - P(x(t) \neq +\infty \mid x(\varepsilon) = i)| < \eta, \tag{6.31}$$

$$|P(x(t) \neq +\infty \mid x(\varepsilon) = i) - P(x(t-\varepsilon) \neq +\infty \mid x(0) = i)| < \eta, \tag{6.32}$$

只要用到的条件概率有意义. 对 $t \in T$, 有

$$P(x(t) = j \mid x(0) = i) \geq P(x(t) = j \mid x(0) = i, x(\varepsilon) = i) \cdot$$
$$P(x(\varepsilon) = i \mid x(0) = i). \tag{6.33}$$

$$P(x(t) \neq +\infty \mid x(0) = i) = \sum_j P(x(t) = j \mid x(0) = i)$$

$$\geqslant P(x(\varepsilon)=i \mid x(0)=i)P(x(t)\neq+\infty \mid x(0)=i,$$
$$x(\varepsilon)=i). \tag{6.34}$$

今因 $L(T)=0$，$0\overline{\in}T$，故对 $\eta>0$，存在 ε_1，使 $t-\varepsilon_1\overline{\in}T$，同时使(6.31)(6.32)对 ε_1 成立，而且 $P(x(\varepsilon_1)=i \mid x(0)=i)>1-\eta$。于是由(6.34)立得 $P(x(t)\neq+\infty|x(0)=i)>(1-\eta)(1-2\eta)$。由 η 的任意性，$P(x(t)\neq+\infty|x(0)=i)=1$，或 $P(x(t)=+\infty)=0$。

定理 6.3 (i) $x(t,\omega)$，$t\geqslant 0$ 是 Q 过程；

(ii) 以 $P_{ij}^{(n)}(t)$ 及 $P_{ij}(t)$ 分别表 $(Q,V^{(n)})$ 过程 $x_n(t,\omega)$ 及 $x(t,\omega)$ 的转移概率，则 $\lim\limits_{n\to+\infty} P_{ij}^{(n)}(t)=P_{ij}(t)$。

证 若能证 $x(t,\omega)$，$t\geqslant 0$ 是齐次马氏过程，则由(6.29)及(2.10)(2.11)立知它是 Q 过程。显然它的相空间是 $\overline{E}$。由上述 $P(x(t,\omega)=+\infty)=0$ 及定理 6.2，对任一组 $0\leqslant t_1<t_2<\cdots<t_k$，随机向量 $(x_n(t_1),x_n(t_2),\cdots,x_n(t_k))$，当 $n\to+\infty$ 时以概率 1 收敛于 $(x(t_1),x(t_2),\cdots,x(t_k))$，故也依分布收敛，即
$$\lim_{n\to+\infty} P(x_n(t_1)=i_1,\ x_n(t_2)=i_2,\ \cdots,\ x_n(t_k)=i_k)$$
$$=P(x(t_1)=i_1,\ x(t_2)=i_2,\ \cdots,\ x(t_k)=i_k), \tag{6.35}$$

由此式并利用 $x_n(t,\omega)$，$t\geqslant 0$ 是齐次马氏过程，即知 $x(t,\omega)$，$t\geqslant 0$ 也是齐次马氏过程，而且 $\lim\limits_{n\to+\infty} P_{ij}^{(n)}(t)=P_{ij}(t)$。∎

过程 $x(t,\omega)$，$t\geqslant 0$ 自然地称为 (Q,V) 过程，其中 $V=(v_n)$ 是满足(6.1)的任一序列。回忆(5.6)，有

定理 6.4 (Q,V) 过程 $x(t,\omega)$，$t\geqslant 0$ 是满足
$$P(x(\beta_1^{(n)})=n)=v_n,\ n\geqslant 0 \tag{6.36}$$
的唯一 Q 过程。

证 先证 (Q,V) 过程满足(6.36)。设 $(v_0^{(n)},v_1^{(n)},\cdots,v_n^{(n)})$，$n\geqslant 1$ 是(3.14)的任一非负解，利用 $c_{l+1,l}c_{l,l-1}=c_{l+1,l-1}$,

用归纳法易见

$$v_i^{(n)} = \frac{v_i^{(n+k)}}{\sum_{j=0}^{n+k} v_j^{(n+k)} c_{jn}}, \quad v_n^{(n)} = \frac{\sum_{j=n}^{n+k} v_j^{(n+k)} c_{jn}}{\sum_{j=0}^{n+k} v_j^{(n+k)} c_{jn}}, \quad (6.37)$$

$k, n \in \mathbf{N}^*, i = 0, 1, 2, \cdots, n-1$. 今考虑 $(Q, V^{(k)})$ 过程 $x_k(t, \omega), t \geqslant 0$，对此过程用(6.5)定义 $\beta_1^{(k,n)}$. 由(6.3) $\beta_1^{(k,n)}(\omega) \leqslant \beta_1^{(k+1,n)}(\omega)$ $(k > n)$，容易看出，以概率 1 有

$$\lim_{k \to +\infty} \beta_1^{(k,n)}(\omega) = \beta_1^{(n)}(\omega); \quad \lim_{k \to +\infty} x_k(\beta_1^{(k,n)}) = x(\beta_1^{(n)}).$$

根据(6.37)得

$$P(x(\beta_1^{(n)}) = n) = \lim_{k \to +\infty} P(x_k(\beta_1^{(k,n)}) = n)$$

$$= \lim_{k \to +\infty} P(x_{n+k}(\beta_1^{(n+k,n)}) = n)$$

$$= \lim_{k \to +\infty} \frac{\sum_{j=n}^{n+k} v_j^{(n+k)} c_{jn}}{\sum_{j=0}^{n+k} v_j^{(n+k)} c_{jn}} = v_n^{(n)} = v_n, n \geqslant 0.$$

次证若 $\tilde{x}(t, \omega), t \geqslant 0$ 为某满足(6.36)的可分、Borel 可测的 Q 过程，则它与 (Q, V) 过程有相同的转移概率. 实际上，利用(5.5)~(5.8)对 $\tilde{x}(t, \omega)$ 定义 $(\tau_m, \beta_m^{(n)})$, $m \geqslant 1$，并对它进行 $c(\tau_m, \beta_m^{(n)})$ 变换，由定理 5.2，所得过程 $x_n(t, \omega), t \geqslant 0$ 是 $(Q, V^{(n)})$ 过程. 由(6.36), $v_n^{(n)} = v_n$. 根据定理 6.3，它们的转移概率 $p_{ij}^{(n)}(t)$ 收敛于 (Q, V) 过程的转移函数 $P_{ij}(t)$.

另一方面，可证对任意固定的 $t \geqslant 0$, $P(\lim_{n \to +\infty} x_n(t, \omega) = \tilde{x}(t, \omega)) = 1$，从而 $p_{ij}^{(n)}(t) \to \tilde{p}_{ij}(t)$ ($\tilde{p}_{ij}(t)$ 是 $\tilde{x}(t, \omega)$ 的转移函数)，于是 $\tilde{p}_{ij}(t) = p_{ij}(t)$. 为证此令

$$S_{+\infty}^{(b)}(\omega) = (t: t \in [0, b], t \text{ 是 } \tilde{x}(t, \omega) \text{ 的飞跃点}),$$

$$S_i^{(b)}(\omega) = (t: t \in [0, b], x(t, \omega) = i),$$

则由定理 4.1 及 [1, P160 系] 以概率 1, $S_{+\infty}^{(b)}(\omega)$ 是 L 测度为零

的闭集，既然$[0,b]=S^{(b)}_{+\infty}(\omega)\bigcup \bigcup\limits_{i\neq+\infty} S^{(b)}_i(\omega)$，故以概率1

$$\sum_{i\neq+\infty} L(S^{(b)}_i(\omega)) = b. \qquad (6.38)$$

根据$x_n(t,\omega)$的定义，在$[0,b]$中，它至少包含$\tilde{x}(t,\omega)$的对应于$S^{(b)}_i(\omega)$，$i\leqslant n$的段．由于

$$x_n(t,\omega)=\tilde{x}(t+\tau^{(n)}_t,\omega), \quad t\in[0,b], \qquad (6.39)$$

这里$\tau^{(n)}_t(\omega)$是自$\tilde{x}(t,\omega)$变到$x_n(t,\omega)$时，自$[0,b]$中所抛去的部分段的总长，故

$$\tau^{(n)}_t(\omega)\leqslant \sum_{\substack{i\geqslant n+1 \\ i\neq+\infty}} L(S^{(b)}_i(\omega)), \quad t\in[0,b].$$

由(6.38)得$\lim\limits_{n\to+\infty}\tau^{(n)}_t(\omega)\leqslant \lim\limits_{n\to+\infty}\sum\limits_{\substack{i\geqslant n+1 \\ i\neq+\infty}}L(S^{(b)}_i(\omega))=0$．由此及(6.39)得知，$P(\lim\limits_{n\to+\infty}x_n(t,\omega)=\tilde{x}(t,\omega))=1$对任一固定点$t\in[0,b]$成立，因为$t$以概率1是$\tilde{x}(s,\omega)$，$s\geqslant 0$的连续点，由$b>0$的任意性即得所欲证．∎

总结以上主要结果，得

基本定理1 (i) 设已给可分、Borel可测Q过程$x(t,\omega)$，$t\geqslant 0$，使$R<+\infty$，$S<+\infty$，则序列(v_n)

$$v_n=P(x(\beta^{(n)}_1)=n), \quad n\geqslant 1 \qquad (6.40)$$

满足关系(6.1)．

(ii) 反之，设已给Q使$R<+\infty$，$S<+\infty$，则对满足(6.1)的序列(v_n)，存在唯一Q过程$x(t,\omega)$，$t\geqslant 0$，使(6.40)成立．它的转移概率$p_{ij}(t)=\lim\limits_{n\to+\infty}p^{(n)}_{ij}(t)$，这里$p^{(n)}_{ij}(t)$是$(Q,V^{(n)})$过程的转移概率，而$V^{(n)}=(v^{(n)}_0,v^{(n)}_1,\cdots,v^{(n)}_n)$是(3.14)在条件$v^{(n)}_n=v_n$下的解，$n\geqslant 1$．

实际上，由引理6.1得(i)；(ii)则自定理6.3及6.4推出．

§7. 方程组的非负解与结果的深化

§6中结果的深化有待于对方程组(3.14)的非负解的研究，本节中首先求出(3.14)的全部非负解，然后应用此结果来进一步刻画全体Q过程. 以下恒设$R<+\infty$，令

$$R_n = \sum_{i=0}^{n-1} v_i^{(n)} c_{i0}, \quad S_n = v_n^{(n)} c_{n0}, \quad \Delta_n = R_n + S_n.$$

引理 7.1 设$V^{(n)} = (v_0^{(n)}, v_1^{(n)}, \cdots, v_n^{(n)})$，$n \geqslant 1$是(3.14)的非负解，则$\lim\limits_{n\to+\infty} \dfrac{R_n}{\Delta_n} = p(\geqslant 0)$，$\lim\limits_{n\to+\infty} \dfrac{S_n}{\Delta_n} = q(\geqslant 0)$；$p+q=1$，当且仅当$v_n^{(n)}=1$ $(n\geqslant 1)$时，$q=1$，$p=0$.

证 改写(3.14)为

$$v_j^{(n+1)} = v_j^{(n)}\left(1 - v_{n+1}^{(n+1)} \frac{z_{n+1} - z_n}{z - z_n}\right), j = 0, 1, 2, \cdots, n-1,$$

$$v_n^{(n+1)} = v_n^{(n)}\left(1 - v_{n+1}^{(n+1)} \frac{z_{n+1} - z_n}{z - z_n}\right) - v_{n+1}^{(n+1)} \frac{z - z_{n+1}}{z - z_n}, \quad (7.1)$$

$$\sum_{i=0}^n v_i^{(n)} = \sum_{i=0}^{n+1} v_i^{(n+1)} = 1,$$

并令$\delta_{n+1} = \sum\limits_{i=0}^n v_i^{(n+1)} + v_{n+1}^{(n+1)} c_{n+1,n} = 1 - v_{n+1}^{(n+1)} \dfrac{z_{n+1} - z_n}{z - z_n} > 0$. 由(3.14)及(6.37)经简单计算后得

$$\frac{R_{n+1}}{\Delta_{n+1}} = \frac{R_n}{\Delta_n} + \frac{v_n^{(n+1)} c_{n0}}{\Delta_n \delta_{n+1}},$$

故$\dfrac{R_n}{\Delta_n} \uparrow p$，$0 \leqslant p \leqslant 1$. 由$\dfrac{R_n}{\Delta_n} + \dfrac{S_n}{\Delta_n} = 1$，得$\dfrac{S_n}{\Delta_n} \downarrow q$，$0 \leqslant q \leqslant 1$，$p+q=1$. 若$v_n^{(n)}=1(n\geqslant 1)$，则$R_n=0$，$q=1$；反之，若$q=1$，则$\dfrac{S_n}{\Delta_n}=1$，$n\geqslant 1$，$R_n=0$. 因$c_{i0}>0$，故$v_i^{(n)}=0$，$i<n$，从而$v_n^{(n)}=$

1 $(n \geqslant 1)$. ∎

引理 7.2 设 $V^{(n)}$, $n \geqslant 1$ 为 (3.14) 非负解,若存在 k 使 $v_i^{(i)} = 1$, $i \leqslant k$, 但 $v_{k+1}^{(k+1)} < 1$, 则

(i) $v_k^{(n)} > 0$, $n \geqslant k$;

(ii) $\dfrac{v_j^{(n)}}{v_k^{(n)}}$ 不依赖于 n $(> \max(j, k))$;

(iii) 任取一数 $r_k > 0$,并令 $r_j = \dfrac{v_j^{(n)}}{v_k^{(n)}} r_k$ $(n > \max(j, k))$,则

$$0 < \sum_{m=0}^{+\infty} r_m c_{m0} < +\infty. \quad (7.2)$$

证 因 $(v_0^{(i)}, \cdots, v_{i-1}^{(i)}, v_i^{(i)}) = (0, \cdots, 0, 1)$ $(i \leqslant k)$ 由 (7.1) 得

$$v_j^{(n)} = 0, \text{ 一切 } n > j, j = 0, 1, 2, \cdots, k-1, \quad (7.3)$$

特别,$v_j^{(k+1)} = 0$, $j \leqslant k-1$. 由假定 $v_{k+1}^{(k+1)} < 1$, $\sum_{j=0}^{k+1} v_j^{(k+1)} = 1$, 故 $v_k^{(k+1)} > 0$, 由 (7.1) 得证 (i). 任取 $n > m > \max(j, k)$, 由 (6.37) 得

$$\frac{v_j^{(m)}}{v_k^{(m)}} = \frac{v_j^{(n)} \div \sum_{l=0}^{n} v_l^{(n)} c_{lm}}{v_k^{(n)} \div \sum_{l=0}^{n} v_l^{(n)} c_{lm}} = \frac{v_j^{(n)}}{v_k^{(n)}},$$

此即 (ii). 因 $r_k > 0$ 而一切 $r_m \geqslant 0$, 故得 (7.2) 中前一不等式. 取 $n > k$, 由 (7.3) $v_1^{(n+1)} = v_2^{(n+1)} = \cdots = v_{k-1}^{(n+1)} = 0$. 由 p 的定义及 (ii) 与 (6.37) 得

$$p = \lim_{n \to +\infty} \frac{\sum_{l=k}^{n} v_l^{(n+1)} c_{l0}}{v_k^{(n+1)}} \cdot \frac{v_k^{(n+1)}}{v_k^{(n+1)} c_{k0} + \left(\sum_{l=k+1}^{n+1} v_l^{(n+1)} c_{l,k+1} \right) c_{k+1,0}}$$

$$= \lim_{n \to +\infty} \frac{1}{r_k} \left(\sum_{l=k}^{n} r_l c_{l0} \right) \frac{v_k^{(k+1)}}{v_k^{(k+1)} c_{k0} + v_{k+1}^{(k+1)} c_{k+1,0}}.$$

因由(7.3)，$r_l=0$，$l<k$，故得

$$\sum_{l=0}^{+\infty} r_l c_{l0} = \frac{pr_k(v_k^{(k+1)}c_{k0}+v_{k+1}^{(k+1)}c_{k+1,0})c_{k+1,0}}{v_k^{(k+1)}} < +\infty. \blacksquare$$

注 p, q 由解 $v^{(n)}$, $n \geqslant 1$ 唯一决定，(r_j) 则除一常数因子外唯一决定.

定理 7.1[①] 为使 $(v_0^{(n)}, v_1^{(n)}, \cdots, v_n^{(n)})$, $n \geqslant 1$ 是 (3.14) 的一非负解，充分必要条件是存在非负数 p, q, r_n, $n \geqslant 0$, 满足关系式

$$p+q=1, \tag{7.4}$$

$$0 < \sum_{n=0}^{+\infty} r_n c_{n0} < +\infty, \quad p > 0, \tag{7.5}$$

$$r_n = 0 \quad (n \geqslant 0), \quad p = 0, \tag{7.6}$$

解 $(v_0^{(n)}, v_1^{(n)}, \cdots, v_n^{(n)})$ 可表为

$$v_j^{(n)} = X_n \frac{r_j}{A_n}, \quad j=0, 1, 2, \cdots, n-1, \tag{7.7}$$

$$v_n^{(n)} = Y_n + X_n \frac{\sum_{l=n}^{+\infty} r_l c_{ln}}{A_n}, \tag{7.8}$$

其中[②] $0 \leqslant A_n = \sum_{l=0}^{+\infty} r_l c_{ln} < +\infty$，而

$$X_n = \frac{pA_n(z-z_n)}{pA_n(z-z_n)+qA_0 z}, \quad Y_n = \frac{qA_0 z}{pA_n(z-z_n)+qA_0 z}. \tag{7.9}$$

此时 p, q 唯一决定，而 r_n, $n \geqslant 0$ 则除一常数因子外唯一决定.

证 充分性 设已给满足 (7.4)～(7.6) 的 p, q, r_n. 如

---

① 若理解 q 为到达 $+\infty$ 后，自 $+\infty$ "连续地" 回到有穷状态的概率，p 为自有穷状态跳出的概率，则不难理解本定理的概率意义.

② 认为 $\frac{0}{0}=0$.

$p=0$，则由(7.7)(7.8)得$(v_0^{(n)}, \cdots, v_{n-1}^{(n)}, v_n^{(n)}) = (0, \cdots, 0, 1)$，$n \geqslant 1$，它显然是(3.14)的一非负解．如$p>0$，由(7.5)及

$$A_n - A_{n+1} c_{n+1,n} = \big(\sum_{i=0}^{n} r_i\big)(1 - c_{n+1,n}), \qquad (7.10)$$

$$A_n \geqslant A_0, \qquad (7.11)$$

可见$0 < A_n < +\infty$，$0 < X_n < +\infty$．由(7.7)(7.8)及(2.16)，有

$$\sum_{i=0}^{n+1} v_i^{(n+1)} c_{in} = \frac{X_{n+1}}{A_{n+1}} A_n + Y_{n+1} c_{n+1,n}, \qquad (7.12)$$

由(7.9)得$\dfrac{Y_{n+1}}{X_{n+1}} \dfrac{z-z_{n+1}}{z-z_n} A_{n+1} = A_n \dfrac{Y_n}{X_n}$，利用$X_n + Y_n = 1$及

(2.16)有$\dfrac{X_n}{A_n} = \dfrac{X_{n+1}}{A_{n+1}} \div \Big(\dfrac{X_{n+1}}{A_{n+1}} A_n + Y_{n+1} c_{n+1,n}\Big)$．以(7.12)代入得

$$\frac{X_n}{A_n} = \frac{\dfrac{X_{n+1}}{A_{n+1}}}{\displaystyle\sum_{i=0}^{n+1} v_i^{(n+1)} c_{in}}.$$

上式两边乘$r_j (j=0, 1, 2, \cdots, n-1)$并利用(7.7)(7.8)得

$$v_j^{(n)} = \frac{v_j^{(n+1)}}{\displaystyle\sum_{i=0}^{n+1} v_i^{(n+1)} c_{in}} \quad (j=0,1,2,\cdots,n-1). \qquad (7.13)$$

由(7.7)(7.8)有$\displaystyle\sum_{j=0}^{n} v_j^{(n)} = \sum_{j=0}^{n+1} v_j^{(n+1)} = 1$，按(7.13)及$c_{jn}=1 (j \leqslant n)$，

$$v_n^{(n)} = 1 - \sum_{j=0}^{n-1} v_j^{(n)} = \frac{v_n^{(n+1)} + v_{n+1}^{(n+1)} c_{n+1,n}}{\displaystyle\sum_{i=0}^{n+1} v_i^{(n+1)} c_{in}}, \qquad (7.14)$$

因而充分性得证．

必要性 设已给(3.14)的一非负解$(v_0^{(n)}, \cdots, v_{n-1}^{(n)}, v_n^{(n)})$．取引理7.1中的$p, q$，此时若$p=0$，则取$r_n = 0 (n \geqslant 1)$而结论显然正确；否则由引理7.1必存在$k (\geqslant 0)$使满足引理7.2的条件，于是如引理7.2定义$r_n (n \geqslant 1)$，由此两引理知(7.4)～

(7.6)满足. 下证(7.7)(7.8)成立. 为此定义

$$u_j^{(n)} = X_n \frac{r_j}{A_n}, \; j=0,1,2,\cdots,n-1; u_n^{(n)} = Y_n + X_n \frac{\sum_{l=n}^{+\infty} r_l c_{ln}}{A_n},$$
(7.15)

其中 $A_n = \sum_{l=n}^{+\infty} r_l c_{ln} > 0$, 而 X_n, Y_n 由(7.9)给出. 由充分性之证知 $(u_0^{(n)}, u_1^{(n)}, \cdots, u_{n-1}^{(n)}, u_n^{(n)})(n \geqslant 1)$ 是(3.14)的非负解, 故为了证 $(v_0^{(n)}, v_1^{(n)}, \cdots, v_n^{(n)}) = (u_0^{(n)}, u_1^{(n)}, \cdots, u_n^{(n)})(n \geqslant 1)$, 只要证

$$v_n^{(n)} = u_n^{(n)}, \; n \geqslant 1. \qquad (7.16)$$

注意, 若 $n \leqslant k$, 则由(7.3)得 $u_n^{(n)} = 1 = v_n^{(n)}$, 而此时(7.16)成立, 故只要对 $n \geqslant k+1$ 证明(7.16). 对 n 用归纳法, 经过一些计算后, 可证

$$u_k^{(n)} = v_k^{(n)}, \; n \geqslant k+1, \qquad (7.17)$$

由此式并注意 $\dfrac{u_{k+i}^{(n)}}{u_k^{(n)}} = \dfrac{r_{k+i}}{r_k} = \dfrac{v_{k+i}^{(n)}}{v_k^{(n)}}$ $(k+i<n)$

立得

$$\frac{1-\sum_{i=k}^{n-1} u_i^{(n)}}{u_k^{(n)}} = \frac{1-\sum_{i=k}^{n-1} v_i^{(n)}}{v_k^{(n)}},$$

即

$$1-\sum_{i=k}^{n-1} u_i^{(n)} = 1-\sum_{i=k}^{n-1} v_i^{(n)}.$$

今因 $\sum_{i=0}^{n} u_i^{(n)} = \sum_{i=0}^{n} v_i^{(n)} = 1$ 而且 $u_i^{(n)} = v_i^{(n)} = 0$ 对一切 $i<k$ 成立, 故得证 $u_n^{(n)} = v_n^{(n)}, \; n \geqslant k+1$.

唯一性 今设有两组满足(7.4)~(7.6)的非负数 p, q, r_n, $n \geqslant 0$, 及 $\bar{p}$, $\bar{q}$, $\bar{r}_n$, $n \geqslant 0$, 均使已给非负解 $(v_0^{(n)}, v_1^{(n)}, \cdots, v_n^{(n)})$ 能表成(7.7)~(7.9)的形式. 若 $v_n^{(n)} = 1 (n \geqslant 1)$, 即 $v_j^{(n)} = 0 (n \geqslant 1, j<n)$, 则由(7.7)及 $\dfrac{0}{0} = 0$ 的协定知 $p = \bar{p} = 0$, $r_n = \bar{r}_n = 0$,

$q = \bar{q} = 1$. 若存在 k 使 $v_i^{(k)} = 1$, $i \leqslant k$, $v_{k+1}^{(k+1)} < 1$, 则由 (7.7)(7.9)(7.5) 知 $p > 0$, $X_n > 0$, $A_0 > 0$, 于是由 (7.7)(7.8)

$$\frac{\sum_{k=0}^{n-1} v_k^{(n)} c_{k0}}{\sum_{k=0}^{n} v_k^{(n)} c_{k0}} = \frac{\sum_{k=0}^{n-1} r_k c_{k0}}{\sum_{k=0}^{+\infty} r_k c_{k0} + \frac{Y_n}{X_n} A_n c_{n0}} = \frac{\sum_{k=0}^{n-1} r_k c_{k0}}{A_0 + \frac{q}{p} A_0} \uparrow p, \; n \to +\infty.$$

同样，上式右方也应收敛于 $\bar{p}$, 故 $p = \bar{p}$, 由 (7.4) 得 $q = \bar{q}$; 其次，当 $j < k$ 时，由于 $v_j^{(k)}$ 等于 0 而由 (7.7) 得 $r_j = 0 = \bar{r}_j$. 既然 $v_k^{(k+1)} > 0$, 故 $r_k > 0$, $r'_k > 0$, 于是由 (7.7) 得知对 $j > k$, 有

$$\frac{r_j}{r_k} = \frac{v_j^{(n)}}{v_k^{(n)}} = \frac{\bar{r}_j}{\bar{r}_k}, \; n > \max(j, k). \; \blacksquare$$

考虑任一可分 Borel 可测 Q 过程 $x(t, \omega)$, $t \geqslant 0$, 如前所述，由 (6.40) 定义的 (v_n) 决定 (3.14) 一非负解 $(v_0^{(n)}, v_1^{(n)}, \cdots, v_n^{(n)})$, $n \geqslant 1$ (更明确些，$v_j^{(n)} = P(x(\beta_1^{(n)}) = j)$, $j = 0, 1, 2, \cdots, n$, $n \geqslant 1$), 通过定理 7.1, 它给出一列非负数 $p, q, r_n, n \geqslant 1$.

定义 7.1 称 $p, q, r_n, n \geqslant 1$ 为 Q 过程 $x(t, \omega)$, $t \geqslant 0$ 的特征数列.

综合定理 7.1 与基本定理 1, 立得

基本定理 2 (i) 设已给可分 Borel 可测 Q 过程 $x(t, \omega)$, $t \geqslant 0$, 使 $R < +\infty$, $S < +\infty$, 则它的特征数列满足条件 (7.4) ~ (7.6);

(ii) 反之, 设已给一列非负数 $p, q, r_n, n \geqslant 1$, 满足 (7.4) ~ (7.6), 则存在唯一 Q 过程 $x(t, \omega)$, $t \geqslant 0$, 其特征数列重合于此已给数列, 而且此过程的转移概率 $P_{ij}(t) = \lim_{n \to +\infty} P_{ij}^{(n)}(t)$, 这里 $P_{ij}^{(n)}(t)$ 是 $(Q, V^{(n)})$ 过程的转移概率, 而 $V^{(n)} = (v_0^{(n)}, v_1^{(n)}, \cdots, v_n^{(n)})$ 由 (7.7) ~ (7.9) 决定.

今研究特征数列为 $p = 0$, $q = 1$, $r_n = 0 (n \geqslant 0)$ 的 Q 过程, 记为 $(Q, 1)$ 过程, 它是 §6 中 $(Q, \pi^{(n)})$ 过程列 $x_n(t, \omega)$, $t \geqslant 0$

($n \geqslant 1$)的极限，$\pi^{(n)} = (0, 0, \cdots, 0, 1)$. 其特殊性已见于§6，即它是"最难"收敛的过程列的极限. 换言之，由$(Q, \pi^{(n)})$过程列的收敛，可推出其他$(Q, V^{(n)})$过程列的收敛. 它的特殊性还在于下列

定理 7.2 $(Q, 1)$过程 $x(t, \omega)$, $t \geqslant 0$ 是唯一的既满足向后方程组(1.8)又满足向前方程组(1.9)的Q过程.

证 因每Q过程都满足(1.8), 故只要证它满足(1.9). 为此只要证在任一固定的$t_0 > 0$前，样本函数以概率1有最后一断点[①]，而且是跳跃点[1,P227]. 令$\Omega_0 = (x(t_0, \omega) \in E)$，则$P(\Omega_0) = 1$. 固定$\omega \in \Omega_0$, 设 $x(t_0, \omega) = k$. 由于 $x_n(t_0, \omega) \to x(t_0, \omega)$ 及 E 的离散性，存在 $N(= N(\omega))$，使 $n \geqslant N$ 时, $x_n(t_0, \omega) = k$. 取 $M > \max(N, k)$，由于在飞跃点 τ 上, $x_n(\tau, \omega) = n$, 再注意(1.7), 可见在 t_0 以前, $x_M(t, \omega)$必有最后一断点，而且是跳跃点，这个点是某k区间的闭包的左端点. 由于(6.3)，此k区间保留在一切$x_n(t, \omega)$, $n \geqslant M$之中，故也保留在$x(t, \omega)$之中，从而得证$x(t, \omega)$在t_0以前有最后一断点为跳跃点($\omega \in \Omega_0$)，故$(Q, 1)$过程满足(1.8)与(1.9). 至于这种过程的唯一性则已有[11, 定理11]中证明. ∎

系 7.1 设 $\tau(\omega)$为$(Q, 1)$过程的第一个飞跃点，则
$$P(\lim_{t \uparrow \tau} x(t, \omega) = +\infty) = 1.$$

证 此由定理7.2及[1, P227 定理5]推出. ∎

---

① $t = 0$ 也看作一跳跃点.

§8. $S=+\infty$ 时 Q 过程的构造

本节的主要任务是当 $R<+\infty$，$S=+\infty$ 时，求出一切 Q 过程，所用方法与 §6 相同，故证明扼要. 以下固定一如此的 Q，考虑某可分 Borel 可测 Q 过程 $x(t,\omega)$，$t\geq 0$ 以及 (5.11) 中的 $a_1^{(n)}(\omega)$，由定理 5.3，存在非负整数 l，当 $n\geq l$ 时，$P(a_1^{(n)}<+\infty)=1$. 令 $\mathscr{S}_i^{(n)}=P(x(a_1^{(n)})=i)$，则 $\sum_{i=0}^{n}\mathscr{S}_i^{(n)}=1(n\geq l)$，故至少存在一个 $\mathscr{S}_k^{(n)}>0$. 再由定理 5.3 及 (3.8)，有

$$\mathscr{S}_j^{(n)}=\frac{\mathscr{S}_j^{(n+1)}}{\sum_{i=0}^{n}\mathscr{S}_i^{(n+1)}} \quad (j\leq n), \tag{8.1}$$

故 $\mathscr{S}_k^{(m)}>0(n\geq k)$ 而且 $\frac{\mathscr{S}_j^{(m)}}{\mathscr{S}_k^{(m)}}$ 不依赖于 $m(\geq\max(j,k))$. 今任意取定 $\mathscr{S}_k>0$，而定义

$$\mathscr{S}_i=\mathscr{S}_k\frac{\mathscr{S}_j^{(m)}}{\mathscr{S}_k^{(m)}} \quad (\geq 0), \tag{8.2}$$

显然，除差一常数因子外，$(\mathscr{S}_i)j\geq 0$ 由过程唯一决定.

定义 8.1 称 $(\mathscr{S}_i)j\geq 0$ 为 Q 过程 $x(t,\omega)$，$t\geq 0$ 的特征数列.

以 $\tau_1(\omega)$ 表此过程的第一个飞跃点，令

$$n_j=E_j(\tau_1(\omega))=\sum_{i=j}^{+\infty}m_i \tag{8.3}$$

[见(2.1)]，以后会证明

$$0<\sum_{j=0}^{+\infty}\mathscr{S}_j n_j<+\infty. \tag{8.4}$$

现在考虑反面的问题：设已给满足 (8.4) 的非负数列 $(\mathscr{S}_j)$，$j\geq$

0，由(8.4)至少有一 $\mathscr{S}_k > 0$，不失以下讨论的一般性，设 $\mathscr{S}_0 > 0$，按(3.20)作分布 $S^{(n)} = (\mathscr{S}_0^{(n)}, \mathscr{S}_1^{(n)}, \cdots, \mathscr{S}_n^{(n)})$。由定理 3.3，存在 $(\Omega, \mathscr{F}, P)$，在其上定义 $(Q, S^{(n)})$ 过程列 $x_n(t, \omega)$，$t \geq 0 (n \geq 0)$，满足(3.7)。以 $\tau_i^{(n)}(\omega)$ 表 $x_n(t, \omega)$ 的第 i 个飞跃点。$\beta_0^{(n)}(\omega) = \min(\tau_i^{(n)}(\omega), x_n(\tau_i^{(n)}) = 0)$，由于 $\mathscr{S}_0 > 0$，$P(\beta_0^{(n)} < +\infty) = 1$。$\xi^{(n,k)}(\omega)$ 为满足 $\tau_i^{(n)}(\omega) \leq \beta_0^{(n)}(\omega)$ 及 $x_n(\tau_i^{(n)}) = k$ 的 i 的个数。

引理 8.1 $E\xi^{(n,k)}(\omega) = \sum_{i=0}^{+\infty} P(\tau_i^{(n)} \leq \beta_0^{(n)}, x_n(\tau_i^{(n)}) = k) = \frac{\mathscr{S}_k^{(n)}}{\mathscr{S}_0^{(n)}} = \frac{\mathscr{S}_k}{\mathscr{S}_0}$。

证 令 $\eta_i(\omega) = \begin{cases} 1, & \tau_i^{(n)}(\omega) \leq \beta_0^{(n)}(\omega), x_n(\tau_i^{(n)}(\omega)) = k, \\ 0, & \tau_i^{(n)}(\omega) > \beta_0^{(n)}(\omega) \text{ 或 } x_n(\tau_i^{(n)}(\omega)) \neq k. \end{cases}$

$$E\xi^{(n,k)}(\omega) = \sum_{i=1}^{+\infty} E\eta_i = \sum_{i=1}^{+\infty} P(\tau_i^{(n)} \leq \beta_0^{(n)}, x_n(\tau_i^{(n)}) = k)$$
$$= \mathscr{S}_k^{(n)} \sum_{l=0}^{+\infty} (1 - \mathscr{S}_0^{(n)})^l = \frac{\mathscr{S}_k^{(n)}}{\mathscr{S}_0^{(n)}} = \frac{\mathscr{S}_k}{\mathscr{S}_0}. \blacksquare$$

今定义

$$L_{nm}^{(i)} = \begin{cases} \tau_{i+1}^{(n)}(\omega) - \tau_i^{(n)}(\omega), & \tau_i^{(n)}(\omega) \leq \beta_0^{(n)}(\omega), n \geq x_n(\tau_i^{(n)}) > m, \\ 0, & \tau_i^{(n)}(\omega) > \beta_0^{(n)}(\omega) \text{ 或 } x_n(\tau_i^{(n)}) \leq m, \end{cases}$$

(8.5)

因而 $\sum_{i=1}^{+\infty} L_{nm}^{(i)}(\omega)$ 是在变换(3.7)下在 $[0, \beta_0^{(n)}(\omega))$ 中自 $x_n(t, \omega)$ 变到 $x_m(t, \omega)$ 时所抛去区间之总长，记 $\sum_{i=1}^{+\infty} L_{nm}^{(i)}(\omega) \uparrow L_n(\omega)$。

引理 8.2 $\lim_{m \to +\infty} EL_m(\omega) = 0$，$P(\lim_{m \to +\infty} L_m(\omega) = 0) = 1$。

证 由(8.4)只要证

$$EL_m(\omega) \leq \frac{1}{\mathscr{S}_0} \sum_{k=m+1}^{+\infty} \mathscr{S}_k n_k.$$

令 $C = (x_n(\tau_i^{(n)}) > m, \tau_i^{(n)} \leqslant \beta_0^{(n)})$, $\Delta = \sum_{j=m+1}^{n} \mathscr{S}_j^{(n)}$, 由(8.3)及引理 8.1 得

$$E\left(\sum_{i=1}^{+\infty} L_{mn}^{(i)}\right) = \sum_{i=1}^{+\infty} E(\tau_{i+1}^{(n)} - \tau_i^{(n)} \mid C) \cdot P(C)$$

$$= \sum_{i=1}^{+\infty} E(\tau_{i+1}^{(n)} - \tau_i^{(n)} \mid x_n(\tau_i^{(n)}) > m) \cdot P(C)$$

$$= \sum_{i=1}^{+\infty} \Big[\sum_{k=m+1}^{n} E(\tau_{i+1}^{(n)} - \tau_i^{(n)} \Big| x_n(\tau_i^{(n)}) = k) \cdot P[x_n(\tau_i^{(n)}) = k \mid x(\tau_i^{(n)} > m)]\Big] \cdot P(C)$$

$$= \sum_{i=1}^{+\infty} \Big[\sum_{k=m+1}^{n} n_k \frac{\mathscr{S}_k^{(n)}}{\Delta}\Big] \cdot P(C)$$

$$= \Big[\sum_{k=m+1}^{n} n_k \frac{\mathscr{S}_k^{(n)}}{\Delta}\Big] \cdot \sum_{l=m+1}^{n} \sum_{i=1}^{+\infty} P(x_n(\tau_i^{(n)}) = l, \tau_i^{(n)} \leqslant \beta_0^{(n)})$$

$$= \Big[\sum_{k=m+1}^{n} n_k \frac{\mathscr{S}_k^{(n)}}{\Delta}\Big] \cdot \frac{\Delta}{\mathscr{S}_0^{(n)}} = \frac{1}{\mathscr{S}_0} \sum_{k=m+1}^{n} n_k \mathscr{S}_k. \blacksquare \quad (8.6)$$

今令 $\tau_{ij}^{(n)}(\omega)$ 为 $\tau_i^{(n)}(\omega)$ 后第 j 个跳跃点,利用(6.16),定义

$$L_{n\varepsilon}^{(i)}(\omega) = \sum_{j=0}^{+\infty} f_\varepsilon(\tau_{i,j+1}^{(n)}(\omega) - \tau_{ij}^{(n)}(\omega)) \text{ 或 } 0, \text{ 视 } \tau_i^{(n)}(\omega) \leqslant \beta_0^{(n)}(\omega)$$

与否而定,再令

$$T_\varepsilon^{(n)}(\omega) = \sum_{i=1}^{+\infty} L_{n\varepsilon}^{(i)}(\omega) = \sum_{\tau_1^{(n)} \leqslant \tau_{ij}^{(n)} < \beta_0^{(n)}} f_\varepsilon(\tau_{i,j+1}^{(n)}(\omega) - \tau_{ij}^{(n)}(\omega)).$$

(8.7)

又 $T_\varepsilon^{(n)}(\omega) \uparrow T_\varepsilon(\omega)$, $n \to +\infty$.

若令

$$F_\varepsilon^{(i)} = \sum_{j=0}^{+\infty} f_\varepsilon(\tau_{i,j+1}^{(n)} - \tau_{ij}^{(n)}),$$

$$n_k^{(\varepsilon)} = E_k F_\varepsilon^{(i)}$$

$$= \sum_{l=k}^{+\infty} \Big[\frac{1}{b_l}(1 - e^{-c_l \varepsilon}) + \sum_{k=0}^{l-1} \frac{a_l a_{l-1} \cdots a_{l-k}(1 - e^{-c_{l-k}\varepsilon})}{b_l b_{l-1} \cdots b_{l-k} b_{l-k-1}}\Big]$$

（因而 $n_k^{(+\infty)}=n_k$），则与上面的证明同样得

$$ET_\varepsilon^{(n)}(\omega)=\frac{1}{\mathscr{S}_0}\sum_{k=1}^{n}n_k^{(\varepsilon)}\mathscr{S}_k\leqslant\frac{1}{\mathscr{S}_0}\sum_{k=1}^{n}n_k\mathscr{S}_k$$

（为此只要以 $F_\varepsilon^{(i)}$ 换(8.6)中的 $\tau_{i+1}^{(n)}-\tau_i^{(n)}$）. 故得证

引理 8.3 $\lim_{\varepsilon\to 0}ET_\varepsilon(\omega)=0, \quad P(\lim_{\varepsilon\to 0}T_\varepsilon(\omega)=0)=1.$

定理 8.1 若 $0<\sum_{i=0}^{+\infty}\mathscr{S}_in_i<+\infty$，则$(Q,S^{(n)})$过程的样本函数 $x_n(t,\omega)$，当 $n\to +\infty$ 时几乎对一切 t 收敛.

证 只需利用引理 8.2 及 8.3，重复定理 6.1 的证即可. ∎

仿§6补定义极限函数后，同样证明所得为 Q 过程 $x(t,\omega)$, $t\geqslant 0$，称为(Q,S)过程，其转移概率 $P_{ij}(t)=\lim_{n\to +\infty}P_{ij}^{(n)}(t)$，而 $P_{ij}^{(n)}(t)$ 是 $(Q,S^{(n)})$ 过程的转移概率.

定理 8.2 (Q,S)过程 $x(t,\omega)$，$t\geqslant 0$ 是唯一以已给的 $(\mathscr{S}_j)j\geqslant 0$ 为特征数列的 Q 过程.

证 用(5.11)于(Q,S)过程及$(Q,S^{(l)})$过程而得 $\alpha_1^{(n)}$ 及 $\alpha_1^{(l,n)}$，由(8.1)得

$$P(x(\alpha_1^{(n)})=i)=\lim_{l\to +\infty}P(x_l(\alpha_1^{(l,n)})=i)=\lim_{l\to +\infty}P(x_{n+l}(\alpha_1^{(n+l,n)})=i)$$
$$=\lim_{l\to +\infty}\frac{\mathscr{S}_j^{(n+l)}}{\sum_{j=0}^{n}\mathscr{S}_j^{(n+l)}}=\mathscr{S}_i^{(n)}, i=0,1,2,\cdots,n.$$

由此即知 $x(t,\omega)$，$t\geqslant 0$ 的特征数列为$(\mathscr{S}_j)$.

唯一性的证仍仿定理 6.4，但要作下列改变：设 $\tilde{x}(t,\omega)$，$t\geqslant 0$ 是以$(\mathscr{S}_j)$为特征数列的 Q 过程，它可分、Borel 可测. 对它用(5.10)~(5.13)定义 τ_m，$\alpha_m^{(n)}$，$m\geqslant 1$，并施行 $c(\tau_m,\alpha_m^{(n)})$ 变换而得$(Q,S^{(n)})$过程 $x_n(t,\omega)$，$t\geqslant 0$. 由特征数列的定义 $\mathscr{S}_i^{(n)}=\dfrac{\mathscr{S}_i}{\sum_{j=0}^{n}\mathscr{S}_j}$ （$i=0,1,2,\cdots,n$）. 故 $x_n(t,\omega)$ 的转移概率 $P_{ij}^{(n)}(t)$

如上述应收敛于(Q,S)过程的转移概率$P_{ij}(t)$.

今证$P_{ij}^{(n)}(t)$也收敛于$\widetilde{x}(t,\omega)$的转移概率$\widetilde{P}_{ij}(t)$. 固定$(0,b)$而令

$$S_{+\infty}^{(b)}(\omega)=(t:t\in(0,b),t \text{ 是 }\widetilde{x}(t,\omega)\text{ 的飞跃点}),$$

因以概率1, $S_{+\infty}^{(b)}(\omega)$是闭集, 故$\dfrac{(0,b)}{S_{+\infty}^{(b)}(\omega)}$是可列多个不相交的开区间$T_j(\omega)=(\eta_j(\omega),\gamma_j(\omega))$的和, 在每$T_j(\omega)$中, $\widetilde{x}(t,\omega)\neq+\infty$. 定义$\widetilde{x}(\eta_j)=\lim\limits_{s\downarrow\eta_j}\widetilde{x}(s)$, 试证

$$P(\widetilde{x}(\eta_j,\omega)=+\infty)=0. \tag{8.8}$$

否则, 若说$P(\widetilde{x}(\eta_j)=+\infty)>0$, 则必存在$k\in E$, 使$P(\widetilde{x}(\eta_j))=+\infty$; 对某$t\in(\eta_j(\omega),\gamma_j(\omega)),\widetilde{x}(t,\omega)=k)>0$, 于是, 由上式得$P(\xi_k(\omega)<+\infty)>0$(这里$\xi_k(\omega)$由(4.5)对$\widetilde{x}(t,\omega)$定义), 此与(4.7)矛盾. 今令$t$-集

$$S_i^{(b)}(\omega)=(\text{使 }\widetilde{x}(\eta_j)=i\text{ 的区间}(\eta_j(\omega),\gamma_j(\omega))\text{ 之和}),$$

由(8.8)得$\bigcup\limits_{i\neq\infty}S_i^{(b)}(\omega)=\bigcup\limits_{j}(\eta_j(\omega),\gamma_j(\omega))=\dfrac{(0,b)}{S_{+\infty}^{(b)}(\omega)}$. 既然以概率1, $L(S_{+\infty}^{(b)}(\omega))=0$, 故$\sum\limits_{i\neq+\infty}L(S_i^{(b)}(\omega))=b$的概率为1. 然后自(6.38)起逐句重复定理6.4的证并注意$x(0,\omega)=x_n(0,\omega)$即可. ∎

基本定理3 (i) 设已给可分、Borel可测Q过程$x(t,\omega)$, $t\geqslant 0$, 使$R<+\infty$, $S=+\infty$, 则它的特征数列$(\mathscr{S}_i)$满足条件(8.4);

(ii) 反之, 设已给一列非负数$(\mathscr{S}_i)$满足(8.4), 则存在唯一Q过程$x(t,\omega)$, $t\geqslant 0$, 其特征数列重合于此已给数列, 而且此过程的转移概率$P_{ij}(t)=\lim\limits_{n\to+\infty}P_{ij}^{(n)}(t)$, 这里$P_{ij}^{(n)}(t)$是$(Q,S^{(n)})$过程的转移概率, 而$S^{(n)}=(\mathscr{S}_0^{(n)},\mathscr{S}_1^{(n)},\cdots,\mathscr{S}_n^{(n)})$由(3.20)决定.

证 (ii) 由定理 8.2 及其上一段论述推出. 今证(i). 由定理 5.3, 存在 $l \geqslant 0$, 当 $n \geqslant l$ 时, $P(\alpha_1^{(n)} < +\infty) = 1$. 因 $\sum_{k=0}^{n} \mathscr{S}_k^{(n)} = \sum_{k=0}^{n} P(x(\alpha_1^{(n)}) = k) = 1$, 故至少有一 $\mathscr{S}_k^{(n)} > 0$. 不失以下讨论的一般性, 可设 $\mathscr{S}_0^{(n)} > 0$. 定义

$$\alpha(\omega) = \inf(u: u \text{ 是 } x(t, \omega) \text{ 的飞跃点}; x(u, \omega) = 0), \tag{8.9}$$

则 $P(\alpha(\omega) < +\infty) \geqslant \mathscr{S}_0^{(n)} > 0$. 故存在 $T > 0$, $\beta > 0$ 使 $P_0(\alpha(\omega) < T) = \beta > 0$. 由引理 4.1 即得

$$E\alpha(\omega) < +\infty. \tag{8.10}$$

今由 $x(t, \omega)$ 作 $(Q, S^{(n)})$ 过程 $x_n(t, \omega)$ (参看定理 5.3), 令

$$\beta_0^{(n)}(\omega) = \inf(u: u \text{ 是 } x_n(t, \omega) \text{ 的飞跃点}; x_n(u, \omega) = 0).$$

显然 $\beta_0^{(n)}(\omega) \leqslant \alpha(\omega)$, 故由(8.6)得

$$E\alpha \geqslant E\beta_0^{(n)} = E\left(\tau_1^{(n)} + \sum_{i=1}^{+\infty} L_{n0}^{(i)}\right) = E\tau_1^{(n)} + \frac{1}{\mathscr{S}_0} - \sum_{k=1}^{n} n_k \mathscr{S}_k,$$

注意(8.10)即得 $0 < n_0 \mathscr{S}_0 \leqslant \sum_{k=0}^{+\infty} n_k \mathscr{S}_k < +\infty$. ∎

§9. 进一步的问题

(i) 研究一般的满足(1.6)的 Q 过程的构造. 首先值得注意的是所谓双边生灭过程，此时相空间 $E=\mathbf{Z}$，(1.7)仍成立，但其中 i 可为一切整数. 增添 $+\infty$ 与 $-\infty$ 于 E 中而使之紧化. 对已给的 Q，Q 可能是正则的，否则，质点可能以正的概率，在有穷时间内到达 $+\infty$，也可能自 $+\infty$ 连续流入；对 $-\infty$ 也如此，因而情况比较复杂. 双边生灭过程，对一般 Q 过程的构造问题的研究，是有很大帮助的. 当相空间为 (a, b) 时，对扩散过程，类似的问题近年来由 Feller 研究解决.

(ii) 本文中研究的生灭过程以 0 为反射壁，类似地可研究对 0 为吸引壁、或以概率 u 为反射以概率 v 为吸引的情形 $(u+v=1)$.

(iii) 为了叙述 §1 中所说的第三种方法的逻辑基础，我们用两种不同的 Doob 过程的变换来分别处理 $S<+\infty$ 与 $S=+\infty$ 的情形. 自然地提出下列问题：能否用处理 $S<+\infty$ 时所用的变换(即变换(6.3))，来研究 $S=+\infty$ 的情形？看来也许是可能的[①].

(iv) 我们证明了：任一 Q 过程的转移函数 $P_{ij}(t)$ 可用某一列 Doob 过程列的转移函数 $P_{ij}^{(n)}(t)$ 所逼近. 事实上证明了更强的收敛性成立(见定理 6.2). 由于 Doob 过程的结构较简单，因此，为了研究 Q 过程的某一性质，自然想到先对 Doob 过程研究此性质，然后过渡到极限. 这方面的工作还待深入考虑.

---

① 此问题今已解决.

附 录

本文中所用到的关于生灭过程的性质,在此补加证明[①],以便于阅读正文. 以下无特别声明时,恒设 Q 为生灭过程的密度矩阵,即满足(1.7).

定理 1(Добрушин[2]) 对已给 Q,Q 过程唯一的充分必要条件是 $R=+\infty$.

证 考虑任一可分 Q 过程 $x(t,\omega)$,$t\geqslant 0$,它在跳跃点上右连续. 利用(2.6)定义 $\tau(\omega)$,则当且仅当 $P(\tau(\omega)=+\infty)=1$ 时,Q 过程唯一. 故只要证明:当且仅当 $R=+\infty$ 时,$P(\tau(\omega)=+\infty)=1$.

用(2.5)定义 ξ_{i+1},令 $m_i = E_i \xi_{i+1}$. 由强马尔可夫性

$$m_i = \frac{b_i}{a_i+b_i} \cdot \frac{1}{a_i+b_i} + \frac{a_i}{a_i+b_i}\left(\frac{1}{a_i+b_i}+m_{i-1}+m_i\right) \quad (i>0),$$

$$m_0 = \frac{1}{b_0}.$$

解此即得 $m_i = \dfrac{1}{b_i} + \sum\limits_{k=0}^{i-1} \dfrac{a_i a_{i-1}\cdots a_{i-k}}{b_i b_{i-1}\cdots b_{i-k}b_{i-k-1}}$,$i \geqslant 0$,此即(2.1).

因 $\tau(\omega) = \lim\limits_{n\to+\infty} \xi_n(\omega)$,由积分单调收敛定理,$E_0 \tau(\omega) = \lim\limits_{n\to+\infty} E_0 \xi_n(\omega)$. 对 $\xi_i(\omega)$ 用强马尔可夫性,令 $\xi_0 \equiv 0$,得

$$E_0 \xi_n = E_0 \Big[\sum_{i=0}^{n-1}(\xi_{i+1}-\xi_i)\Big] = \sum_{i=0}^{n-1} E_0(\xi_{i+1}-\xi_i)$$
$$= \sum_{i=0}^{n-1} E_0[E_0(\xi_{i+1}-\xi_i \mid \mathscr{B}_{[0,\xi_j]})] = \sum_{i=0}^{n-1} E_i \xi_{i+1}$$

---

① 请参看 §1 最后一段.

$$= \sum_{i=0}^{n-1} m_i,$$

因此

$$E_0 \tau(\omega) = \sum_{i=0}^{+\infty} m_i = R.$$

类似地有

$$E_j \tau(\omega) = \sum_{i=j}^{+\infty} m_i \leqslant R \quad (j \in E).$$

若 $P(\tau = +\infty) = 1$，由于 $x(t, \omega)$，$t \geqslant 0$ 是任意的 Q 过程，则 $P_0(\tau = +\infty) = 1$，于是 $R = E_0 \tau = +\infty$.

反之，若 $P(\tau = +\infty) < 1$，则必 $R < +\infty$. 事实上，由于 $P(\tau < +\infty) > 0$ 及 (1.7)，得 $P_0(\tau < +\infty) > 0$. 存在 $T > 0$ 及 $\alpha > 0$，使 $P_0(\tau \leqslant T) \geqslant \alpha$，从而对一切 $k \in E$，$P_k(\tau > T) \leqslant P_0(\tau > T) \leqslant 1 - \alpha$. 由引理 4.1（那里的其他条件均满足）即得 $R = E_0 \tau < +\infty$. ∎

定理 2(Harris) 设齐次马氏链的一步转移概率为

$$p_{01} = 1, \quad p_{i,i+1} = \frac{b_i}{a_i + b_i}, \quad p_{i,i-1} = \frac{a_i}{a_i + b_i}, \quad a_i > 0, \ b_i > 0, \ i > 0, \tag{1}$$

则对 $n > k > j$，自 k 出发，在到达 n 以前先到达 j 的概率为 $\dfrac{z_n - z_k}{z_n - z_j}$，这里 z_n 由 (2.4) 定义.

证 在直线上取点集 $A = (z_0, z_1, \cdots)$，设 $y(t, \omega)$，$t \geqslant 0$ 是以 0 为反射壁的布朗运动(Wiener 过程)，$P(y(0) = z_k) = 1$. 考虑此过程的轨道：对任一固定的 t，以 x 表在 t 以前（包括 t 在内）最后一次落在 A 时所处的点的下标，x 是随机变量. 当 t 自 0 变向 $+\infty$ 时得一列随机变量 x_j，$j \geqslant 0$，$P(x_0 = k) = 1$. 由于 $y(t, \omega)$，$t \geqslant 0$ 是齐次马氏过程，知 x_j，$j \geqslant 0$ 是齐次马氏链. 利用布朗运动的一性质：设布朗质点开始时位在点 B，$A < B <$

C,则它在到达 C 以前先到达 A 的概率为 $\dfrac{C-B}{C-A}$. 由此得

$$P(x_1=k-1 \mid x_0=k)=\frac{z_{k+1}-z_k}{z_{k+1}-z_{k-1}}=\frac{a_k}{a_k+b_k},$$

$$P(x_1=k+1 \mid x_0=k)=\frac{z_k-z_{k-1}}{z_{k+1}-z_{k-1}}=\frac{b_k}{a_k+b_k}.$$

换言之,x_j,$j\geqslant 0$ 的转移概率由(1)决定.

今考虑 $n>k>j$. 对 x_j,$j\geqslant 0$,质点自 k 出发,在到达 n 以前先到达 j 的概率,等于对 $y(t,\omega)$,$t\geqslant 0$,质点自 z_k 出发,在到达 z_n 以前先到达 z_j 的概率. 再次利用上述布朗运动的性质,即得证此概率为 $\dfrac{z_n-z_k}{z_n-z_j}$. ∎

由于 Doob 过程在本文中是构造全体 Q 过程的基石,故在此回忆一下它的定义. 设已给任一满足(1.6)的密度矩阵 Q,使 $P(\tau<+\infty)=1$,τ 由(2.6)定义. 在 $(\Omega,\mathscr{F},P)$ 上考虑一列相互独立的 Q 过程 $x^{(n)}(t,\omega)$,$t\geqslant 0(n\in \mathbf{N}^*)$,它们可分,在跳跃点上右连续,又 $x^{(n)}(t,\omega)(n>1)$ 有共同的开始分布为 π. 用 (2.6) 对 $x^{(n)}(t,\omega)$ 定义得 $\tau^{(n)}(\omega)(n\geqslant 1)$. 令 $\tau_0=0$,$\tau_n=\sum_{v=1}^{n}\tau^{(v)}$,定义

$$x(t,\omega)=x^{(n)}(t-\tau_{n-1}(\omega),\omega),\ \text{如}\ \tau_{n-1}(\omega)\leqslant t<\tau_n(\omega),$$

则称 $x(t,\omega)$,$t\geqslant 0$ 为 Doob 过程. 由于 $x(t,\omega)$ 的转移概率完全由 Q 及 π 决定,故也称它为 (Q,π) 过程.

参考文献

[1] Chung K L. Markov Chains with Stationary Transition Probabilities. 1960.

[2] Добрушин Р Л. Об условиях регулярности однородных по времени марковских процессов. Со счетным числом

возможных состояний. Успехи Матем. Наук, 1952, 7(6): 186-191.

[3] Добрушин Р Л. Некоторые классы однородных счетных марковских процессов. Теория Вероят. и ее Прим., Ⅱ, 1957, 3: 377-380.

[4] Doob J L. Stochastic Processes. 1953.

[5] Дынкин Е Б. Основания Теории Марковских Процессов. 1959(马尔可夫过程论基础，王梓坤，译. 北京：科学出版社，1962).

[6] Дынкин Е Б. Скачкообразные Марковские процессы. Теория Вероят. и ее Прим., Ⅲ, 1958, 1: 41-60.

[7] Feller W. On the integro-differential equations of purely discontinuous Markov process. Trans. Am. Math. Soc., 1940, 48: 488-515; Trans. Am. Math. Soc., 1945, 58: 474.

[8] Feller W. On boundaries and lateral conditions for the Kolmogorov differential equations. Ann. of Math., 1957, 65: 527-570.

[9] Harris T E. First passage and recurrence distribution. Trans. Am. Math. Soc., 1952, 73(3): 471-486.

[10] Karlin S. Mcgregor J. The differential equations of birth and death processes and the Sticltjies moment problem. Trans. Am. Math. Soc., 1957, 85(2): 489-546.

[11] Reuter G E H. Denumerable Markov processes and the associated semigroup on I. Acta Math., 1957, 97: 1-46.

[12] Хинчин А Я. Математические Методы Теории Массового Обслуживания. 1955(有中译本).

[13] 孙振祖. 一类马氏过程的表达式(毕业论文).

[14] 王梓坤. Класснфикация всех процессов размножения н гибели. Научные Доклады Высшей Школы, физ. - Матем. Науки, 1958, 4: 19-25.

[15] 王梓坤. 一个生灭过程. 科学记录新辑, 1959, 3(8): 266-268.

[16] Wang Tzu-Kwen(王梓坤). On distributions of functionals of birth and death processes and their applications in the theory of queues. Scientia Sinica, 1961, 10(2): 160-170.

Construction Theory of Birth and Death Processes

Abstract Let $Z = \{x_t, t \geqslant 0\}$ be a separable birth and death process (BDP) with transition probabilities $(P_{ij}(t))$ and density matrix Q

$$Q = \begin{pmatrix} -b_0 & b_0 & & & & 0 \\ a_1 & -(a_1+b_1) & b_1 & & & \\ & \vdots & \vdots & & \vdots & \\ & & & a_n & -(a_n+b_n) & b_n \\ 0 & & & & \vdots & \vdots \end{pmatrix} \quad (B)$$

where $a_i > 0 (i > 0)$; $b_i > 0 (i \geqslant 0)$. Every process has a unique Q, but there are infinitely many BDP with a same Q if $R < +\infty$, where

$$R = \sum_{i=0}^{+\infty} m_i, \quad m_i = \frac{1}{b_i} + \sum_{l=1}^{i-1} \frac{a_i a_{i-1} \cdots a_{i-k}}{b_i b_{i-1} \cdots b_{i-k} b_{i-k-1}}, \quad i \geqslant 0,$$

Moreover, introduce

$$S = \sum_{i=1}^{+\infty} e_i, \quad e_i = \frac{1}{a_i} + \sum_{i=0}^{+\infty} \frac{b_i b_{i+1} \cdots b_{i+k}}{a_i a_{i+1} \cdots a_{i+k} a_{i+k+1}}, \quad i > 0,$$

$$z_0 = 0, \quad z_n = 1 + \sum_{k=1}^{n-1} \frac{a_1 a_2 \cdots a_k}{b_1 b_2 \cdots b_k}, \quad z = \lim_{n \to +\infty} z_n,$$

$$\tau = \inf(t : \lim_{s \uparrow t} x(s) = +\infty),$$

We call τ the first infinite. It is shown that $P_0(\tau < +\infty) = 1$ iff $R < +\infty$; $E_0 \tau = R$. Hence R is the mean time from 0 to $+\infty$, and S in some sense is the mean time from ∞ to 0.

Now suppose a matrix Q of type (B) is given with $R < +\infty$. We are going to construct all BDP whose density matrixes coincide with the given Q. We call such process Q-process. In 1945 Doob found a class of Q-processes, but not all. He took a probability distribution $V = (v_i)$ and put $P(x(\tau) = i) = v_i$, then Q and V define a Q-process, Known as Doob (Q, V)-process.

We introduce a sequence of characteristic numbers which describe how the process Z returns from $+\infty$ into finite states after the first infinite.

Put $\beta^{(n)} = \inf(t : t \geq \tau, x(t) \leq n)$, $v_j^{(n)} = P(x(\beta^{(n)}) = j)$.

$$c_{ij} = \frac{z - z_i}{z - z_j} \text{ (if } i > j); \quad = 1 \quad \text{(if } i \leq j).$$

$$R_n = \sum_{i=0}^{n-1} v_i^{(n)} c_{i0}, \quad S_n = v_n^{(n)} c_{n0}, \quad \Delta_n = R_n + S_n.$$

$$\lim_{n \to +\infty} \frac{R_n}{\Delta_n} = p(\geq 0), \quad \lim_{n \to +\infty} \frac{S_n}{\Delta_n} = q(\geq 0).$$

$$k = \max(i, v_i^{(i)} = 1),$$

$$r_j = \begin{cases} \dfrac{v_j^{(n)}}{v_k^{(n)}}, & \text{independent on } n > \max(j, k), \quad \text{if} \quad k < +\infty, \\ 0, & \text{if} \quad k = +\infty. \end{cases}$$

Then it is proved that

$$p+q=1, \tag{7.4}$$

$$0 < \sum_{n=0}^{+\infty} r_n c_{n0} < +\infty, \quad \text{if} \quad p>0, \tag{7.5}$$

$$r_n = 0, \quad \text{if} \quad p=0. \tag{7.6}$$

We call $\{p, q, r_n, n \geqslant 0\}$ the characteristic numbers of the process Z.

Theorem Let a matrix Q of type (B) be given, satisfying the conditions $R < +\infty$, $S < +\infty$. Then the following conclusions hold:

(i) The sequence of characteristic numbers $\{p, q, r_n, n \geqslant 0\}$ of any Q-process Z must satisfy $(7.4) \sim (7.6)$.

(ii) Conversely, if we have given a sequence of nonnegative numbers $p, q, r_n, n \geqslant 0$, satisfying $(7.4) \sim (7.6)$, then there exists a unique Q-process Z, whose sequence of characteristic numbers coincides with the given sequence of numbers and whose transition probabilities $P_{ij}(t)$ satisfy

$$P_{ij}(t) = \lim_{n \to +\infty} P_{ij}^{(n)}(t),$$

where $P_{ij}^{(n)}(t)$ are the transition probabilities of Doob $(Q, v^{(n)})$-process, $v^{(n)} = (v_0^{(n)}, v_1^{(n)}, \cdots, v_n^{(n)})$ is given by

$$v_j^{(n)} = Z_n \frac{r_j}{A_n}, \quad j=0, 1, 2, \cdots, n-1,$$

$$v_n^{(n)} = Y_n + Z_n \frac{\sum_{l=n}^{+\infty} r_l c_{ln}}{A_n},$$

and $0 \leqslant A_n = \sum_{l=0}^{+\infty} r_l c_{ln} < +\infty,$

$$Z_n = \frac{pA_n(z-z_n)}{pA_n(z-z_n)+qA_0 z}, \quad Y_n = \frac{qA_0 z}{pA_n(z-z_n)+qA_0 z}$$

(with convention $\frac{0}{0}=0$).

In particular, $\{p=0, q=1, r_n \equiv 0\}$ gives the unique Q-process, satisfying both Kolmogorov forward and backward equations. The process returns from $+\infty$ to finite states "continuously".

In the case $R<+\infty$, $S=+\infty$, we can construct all Q-processes similarly. Now we have fewer Q-processes because "continuous" return is impossible by $S=+\infty$.

Complement

The construction theory was improved later by the following theorem, which deals with both $S<+\infty$, and $S=+\infty$ simultaneously. Let

$$R^{(i)} = \sum_{j=i}^{+\infty} m_j.$$

Fundamental theorem Let a matrix Q of type (B) be given, satisfying the condition $R<+\infty$. Then the following conclusions hold:

(i) The sequence of characteristic numbers p, q, r_n, $n \geqslant 0$ of any Q-process Z must satisfy the relations

$$p+q=1, \tag{C.1}$$

$$0 < \sum_{i=0}^{+\infty} r_i R^{(i)} < +\infty, \text{ if } p>0, \tag{C.2}$$

$$r_n=0, \text{ if } p=0, \tag{C.3}$$

$$q=0, \text{ if } S=+\infty. \tag{C.4}$$

(ii) Conversely, if we have given a sequence of nonnegative numbers p, q, r_n, $n \geq 0$, satisfying (C.1)~(C.4), then there exists a unique Q-process Z, whose sequence of characteristic numbers must coincide with the given sequence of numbers and whose transition probabilities $P_{ij}(t)$ satisfy

$$P_{ij}(t) = \lim_{n \to +\infty} P_{ij}^{(n)}(t).$$

Here $P_{ij}^{(n)}(t)$ are the transition probabilities of the Doob $(Q, v^{(n)})$-process, while the distribution

$v^{(n)} = (v_0^{(n)}, v_1^{(n)}, \cdots, v_n^{(n)})$ is given as follows:

if $p = 0$, then set $v_j^{(n)} = 0$ $(0 \leq j < n)$, $v_n^{(n)} = 1$;

if $p > 0$, then set

$$v_j^{(n)} = Z_n \frac{r_j}{A_s} \quad (0 \leq j < n),$$

$$v_n^{(n)} = Y_n + Z_n \frac{\sum_{l=n}^{+\infty} r_l c_{ln}}{A_n};$$

where

$$0 < A_n = \sum_{l=0}^{+\infty} r_l c_{ln} < +\infty,$$

$$Z_n = \frac{pA_n(z - z_n)}{pA_n(z - z_n) + qA_0 z},$$

$$Y_n = \frac{qA_0 z}{pA_n(z - z_n) + qA_0 z}.$$

the proof can be found in monographs:

[1] Wang Zikun, Yang Xiangqun. Birth and Death Processes and Markov Chains. Springer-Verlag, Science Press, 1992, §6.6: 234.

[2] 王梓坤. 生灭过程与马尔可夫链. 北京: 科学出版社, 1979, 第 6 章.

王梓坤，译. 数学进展，1963，6(1)

波兰应用数学中若干结果的概述
J. Łukaszewicz（波兰科学院数学研究所）

§1. 前　言

本文是 1958 年秋我在北京中国科学院数学研究所所作的一些报告所组成．在这些报告中，我给自己提出的任务是：指出伏洛茨瓦夫应用数学学派的特点．这学派由波兰著名数学家 Hugo Steinhaus 所创立，它是波兰应用数学最大的中心．我深以自己是 H. Steinhaus 教授的学生与合作者为荣．

H. Steinhaus 教授认为数学是应用科学，它的目的是各种可能的应用．数学应用的本质不是作为武器的数学理论，而是数学所特有的、固有的观察实际的方式．应用数学中最好的研究方法是数学家与其他专家的集体合作．但在应用数学方面工作的数学家不应该坐待工程师、经济学家或生物学家带着他们不会解的微分方程走上门来，而应该走向实际中去迎接他们，在问题一开始发现时就和他们合作．必须讨论工作的目的、问题的提法和实验的组织，然后还要分析观察的结果．在这种工作风格下，数学家常在实际工作者不注意的地方，发现数学问题. 在数学家与他们多次讨论中，双方找到了共同的语言，以使彼此能相互了解，当问题有了严格的数学提法时，就可认为它已解决一半.

本文中收集了 14 个问题. 我认为，这已足够说明我们在伏洛茨瓦夫所研究的问题和所用的数学方法的多样性. 除了直接由 H. Steinhaus 所领导的波兰科学院数学研究所生物与农业中的应用讨论班的某些结果外，在本文中还叙述了研究所其他讨论班的两个结果（§4 问题 3 与 §6 问题 5），这两结果与我们的讨论班的工作有紧密关系.

本文只是概述性的，其中某些问题讲得较仔细，而其他的则只是提出了问题. 如果读者能寄给我们有关本文中内容或数学应用中其他问题的意见或问题，我和我的波兰同事会感到非常的高兴. 如能在我的报告和本文的基础上，开始我们的科学研究合作那就更好了. 请读者把信寄往

J. Łukaszewicz, Mathematical Institute of the Polish Academy of Sciences，ul，Kopernika 186，Wroclaw 9 Poland.

最后我愉快地感谢在北京的中国科学院和数学研究所的领导，使我能荣幸地在这里作报告和发表这篇文章. 衷心感谢南开大学王梓坤先生在翻译这些报告和准备本文出版中所付出的劳动.

J. Łukaszewicz
1958 年 12 月 17 日于北京

§2. 统计估值

在接收一批商品时,古典的统计检查方法在于制订验收方案,这方案由两个整数 n, m 所完全决定,其中 n 是为了检查而随机地选出的商品件数(或样本个数),而 $m(m<n)$ 是样本中坏商品件数所能达到的最大值;对此值来说,全批商品可以接收(如果样本中坏商品个数大于 m,那么全批拒受).用通常的符号 $m/\!/n$ 记此方案,数 n, m 应选得使这方案满足下列两条件:

(i) 质量为 k_1 的一批商品(一批商品中好商品的频率称为其质量)以概率 β_1 被接受.

(ii) 质量为 $k_2(k_2<k_1)$ 的一批商品以概率 β_2 被拒收.

通常都取 β_1, β_2 近于 1(例如:$\beta_1=\beta_2=0.95$),因而条件(i)保证卖方的质量为 $k\geqslant k_1$ 的商品几乎常被接受.而条件(ii)则保证买方几乎总可拒收质量为 $k\leqslant k_2$ 的商品.例如方案 $3/\!/60$ 就是根据 $\beta_1=\beta_2=0.95$, $k_1=0.9772$, $k_2=0.8708$ 制订的.

上述方法有许多缺点,因而在实际中常不能应用.首先,古典方法在实际中难以叙述,买方常要求数学家给他一个方案,使能保证在他所接收的一批商品中(譬如说,其质量为 $k\leqslant K$)坏商品的频率不大于他所提出的定数(譬如 B);他不了解,如事先不知道卖方所提出的那批商品质量的先验分布,这是做不到的.其次,在实际中数 k_1, k_2, β_1, β_2 很难决定.第三,这方案不依赖于该批商品的件数,也不依赖一件商品的价值及检查费用.

这些缺点使得 Steinhaus 教授创造统计估值法.这方法的原

则在于根据件数为 n 的样本来计算个体的平均价值

$$\bar{x}_n = \frac{1}{n}\sum_{i=1}^{n} x_i.$$

总体的价值 z 则由下式

$$z = N \cdot \bar{x}_n$$

决定，其中 N 为总体的件数.

不仅当检查一件商品后，检查结果只能取两个定性的值：好或坏时，这方法可以应用；而且可应用于个体可取任意数值时，这些值甚至可以是负的；例如，一件坏零件，使用它后可损坏贵重的机器，就是如此. 但这里需要假定总体的价值是其中全部个体价值的和. 再者还要假定统计估计法是买方与卖方协商中的主要部分. 因而卖方必须按价格 z 把该批商品卖给买方，而 z 是由统计估值法所求得的. 这一原则合理，因为样本是随机选出的，故

$$E(z) = w.$$

这里 $E(z)$ 表 z 的数学期望（z 显然是随机变量），w 表总体的精确价值.

要完全订出验收计划还要定出样本的个数 n，它可由经济损失最小原则求出. 由估计总体的真值而引起误差所造成的经济损失为绝对值 $|z-w|$. 须知经济损失既可由卖方所得多于总体的真值（$z>w$）而产生，也可由他所得少于真值（$z<w$）而造成. 在两种情形下人们劳动的成果：商品或金钱的来往还没有计算在内. 此外，关于经济损失还要估计到由于检查而付出的费用；因为检查绝不会增加价值. 设后一损失可表为

$$U + nu,$$

其中 U 为不依赖于 n 的固定开支，u 为一件商品的检查费用. 经济损失最小原则是要选定使经济损失达到最小的 n，但因损失中的第一部分 $|z-w|$ 是随机变量，故将求出下式之最小值

$$S = E|z-w| + U + nu.$$

为了要求出

$$E|z-w| = \frac{N}{n} E|(x_1-\bar{x})+(x_2-\bar{x})+\cdots+(x_n-\bar{x})|,$$

(2.1)

其中 $\bar{x}$ 为总体中一件商品的平均值，必须要知道总体中个别商品价值的分布，而这等于要了解整批商品，因而统计检查失去了意义．但实际上常常发生下述情况，即预知整体中任一商品的价值都位于某一区间 $[a, b]$ 中，这时可由求极大中的极小方法来解决统计估值问题．

(2.1)右方为独立随机变量 $(x_i-\bar{x})(i=1, 2, \cdots, n)$ 的和的期望，且 $E(x_i-\bar{x})=0$，$\sigma(x_i-\bar{x}) \leqslant \frac{1}{2}(b-a)$，这里 $\sigma(x_i-\bar{x})$ 为 $(x_i-\bar{x})$ 的标准差．若 n 不很小，则按中心极限定理可认为 $\sum_{i=1}^{n}(x_i-\bar{x})$ 具有正态分布，它的平均值为 0，标准差不超过 $\frac{\sqrt{n}}{2}(b-a)$．最后注意若 ξ 为遵守正态分布之变量，其平均值为 0，标准差为 σ，则

$$E|\xi| = \sqrt{\frac{2}{\pi}}\sigma,$$

因此得

$$E|z-w| \leqslant \frac{N}{n}\sqrt{\frac{2}{\pi}}\sqrt{\frac{n}{2}}(b-a) = \frac{Nd}{\sqrt{n}\sqrt{2\pi}} \approx 0.4 \frac{Nd}{\sqrt{n}},$$

(2.2)

其中 $\bar{d}=b-a$ 为一件商品的最大价值与最小价值的差．

若总体中一件商品取最低价值 a，一件取最高价值 b，则 (2.2)取等号，因之

$$\max S = 0.4 \frac{Nd}{\sqrt{n}} + U + un. \qquad (2.3)$$

这里 max 对一切集中在区间 $[a, b]$（其长为 $d=b-a$）的商品价值分布来取，现在容易求得

$$\min_n \max S = 3\left(\frac{Nd}{5}\right)^{\frac{2}{3}} u^{\frac{1}{3}} + U \qquad (2.4)$$

且达到极小值的 n 为

$$n = \left(\frac{Nd}{5u}\right)^{\frac{2}{3}}, \qquad (2.5)$$

因而 Steinhaus 向实际工作者推荐公式 (2.5) 来决定样本的件数. 根据样本平均值 $\bar{x}_n$ 可估计总体的价值为

$$Z = N \cdot \bar{x}_n. \qquad (2.5')$$

容易验证，常数 U 显然不影响样本件数，并且若不考虑 U，则由估价错误而引起的损失为 (2.4) 右边第一项的三分之二. 因而由估价错误的损失的最大期望值为样本检查费用的两倍.

注意，我们用了中心极限定理，才得到公式 (2.5)；由于对有限和 (2.1) 用了渐近式. 如 n 相当小则结论可能不正确，因之式 (2.5) 不能应用，而用此方法还要其他的更复杂的计算. 若 d 为大于 0 的很小数，则由 (2.5) 有时可得 $n > N$，初看起来，这结果没有意义，其实抽样时，抽出的个体可放回到总体中去（这里需要假定检查不损失个体的价值），故同一个体可能被检查几次. 采用此方案可减少理论上的计算，但如 (2.5) 所定出的 n 近于或大于 N 时，最好采用其他的抽出后不放回的检查方案.

根据经济损失最小原则，对总体中个体价值分布作不同的假定，可得到其他的方法与公式. 例如：设此分布为正态的，且 $\alpha = E|x - Ex|$，Steinhaus 得到类似于 (2.5) 的另一公式

$$n = \left(\frac{N\alpha}{2u}\right)^{\frac{2}{3}}. \qquad (2.6)$$

如果此分布非正态，仍可设法化为正态情况。为此不要考虑个别的商品，而以十个或一打（所谓分打法 Метод дюжин）商品为单位，并且相应地应用(2.6)于以打为单位的商品（作为独立随机变数的和的一打的价值的分布已接近于正态）。

比较公式(2.5)(2.6)，由于求得(2.5)时，首先不知价值的分布，故要求出极大中的极小，以预防最坏的分布。而在(2.6)中则假定了分布的正态性，故(2.6)的优点是可大大减少样本件数 n。

如已知个体的价值具有正态分布，但 $\alpha = E \mid x - Ex \mid$ 未知而应用(2.6)时，Steinhaus 推荐下列序贯方法：先检查 n_0 个商品（n_0 任意），根据经验平均绝对差的公式

$$\alpha_1 = \frac{1}{n_0} \sum_{i=1}^{n_0} \mid x_i - \bar{x}_{n_0} \mid,$$

以 α_1 代入(2.6)中的 α，得数 n_1，若 $n_0 < n_1$，则要再检查 $n_1 - n_0$ 个商品。这步骤可重复几次。

§3. 量水计的最佳验收方法

度量仪器，例如量水计，所指示的数字常带有两类误差：系统的和随机的．验收方法是要在所提交的一批量水计中，挑选相对系统误差不太大（按绝对值计）的个体来．在这篇报告中，我想谈谈波兰关于验收量水计方法的讨论、这方法是根据波兰度量总局的规则拟定的．根据这个方法，需将每个要检验的量水计测量后所得的值，与一精确的钢制测量仪器所测得的值加以比较，并以

$$m_1 = \frac{\text{量水计所测得值} - \text{精确仪器所测得值}}{\text{精确仪器所测得值}}$$

表示比较的结果，于是有三种可能情况．

如 $|m_1| \leqslant 0.9q$，量水计算是上品．

如 $|m_1| > 1.1q$，量水计算是次品而拒收．

如 $0.9q < |m_1| \leqslant 1.1q$，再如上检查一次，得 m_2，于是

如 $\dfrac{|m_1 + m_2|}{2} \leqslant q$，量水计算是上品．

如 $\dfrac{|m_1 + m_2|}{2} > q$，量水计算是次品而拒收．

这里 q 是经协商后的定数，如果量水计的相对系统误差的绝对值不超过 q，就公认它是上品．

现在来对量水计作一些统计假定．对于某一确定的量水计，比较后的结果 m 是一正态的随机变量，它的标准差为 σ_1，这分布的数学期望 x 就是量水计的相对系统误差．各次检查测量假定是相互独立的．其次，还要假设受验的总体中，或量水计工厂的全部产品中，个体的相对系统误差也遵守正态分布，它的

期望为 a，标准差为 σ_0．这些假设都在伏洛茨瓦夫的量水计工厂检验过，此外还要设参数 a，σ_0，σ_1 已知.

由上述可提出两个理论问题：

(i) 利用度量总局的检查方法，全部接受的量水计是上品的概率为若干？量水计的 x 如满足 $|x|<q$，就算是上品.

(ii) 该工厂出产的某一量水计被接受的概率为若干？

第一个问题是买方提出的，他希望知道用上述验收方案所选定的量水计的质量如何．第二个问题则由卖方提出，他熟悉自己的技术条件，并想知道他的产品经验收检查后的结果．当他知道量水计是上品的概率大小、并知道这概率如何依赖于他的技术参数 a，σ_0，σ_1 后，他就会受到启发，知道应朝什么方向来改进产品的质量.

我们不准备在这里来研究这些问题（它们已在 J. Obalski[1]，Oderfeld 及 Zubrzycki[2] 中得到讨论和解决），而立即转来研究一个更有趣味的问题：试求验收量水计的最佳方法.

为了用数学形式来叙述这个问题，先注意下列定义：如果量水计的相对系统误差的绝对值不超过 q（$|x|\leqslant q$），则说它是上品，我们商定量水计的质量是连续的，且其自然的数值度量就是 x. 检验方法的质量的度量是被接收的量水计的质量. 后者可用，例如被接收的全体量水计的相对系统误差平方的数学期望，来度量. 因此，为了要比较两个检验方法时，譬如说这一方法较好，就是说，这方法保证被验收的量水计的 x^2 的数学期望较小. 容易证明，可以拟定一验收方法，它保证 x^2 的数学期望任意地小. 但这方法会使得测量的次数迅速增加，同时会使大多数受验的量水计遭到拒收. 从经济观点出发，这是毫无意义的，故 Zubrzycki[3] 在保持下列三条件下，解决了 Steinhaus 提出的寻求最佳验收方法的问题：

(i) 检验一个量水计所需的试验次数的数学期望应与度量总局的方法相同.

(ii) 量水计被接收的概率也相同.

(iii) 对一个量水计只能度量一次和两次. 由推广度量总局的方法, Zubrzycki 找到了最佳的验收方法. 他在第一次检查之后, 采用了下列方案:

如 $|m_1| \leqslant \alpha$, 接收该量水计.

如 $|m_1| > \beta$, 拒收该量水计.

如 $\alpha < |m_1| \leqslant \beta$, 进行第二次检验, 于是

如 $\left|\dfrac{m_1+m_2}{2}\right| \leqslant \gamma$, 接受该量水计.

如 $\left|\dfrac{m_1+m_2}{2}\right| > \gamma$, 拒收该量水计.

已知 a, σ_0, σ_1, q 时, 只需求出参数 α, β, γ 的最佳值, 使能保持两个理论问题的条件(i)及(ii), 并且使接受的量水计的相对系统误差平方的数字期望最小.

按照我们的假设可以写出

$$f(x) = \frac{1}{\sigma_0 \sqrt{2\pi}} l^{-\frac{(x-a)^2}{2\sigma_0^2}}, \tag{3.1}$$

$$f_x(m_1) = \frac{1}{\sigma_1 \sqrt{2\pi}} l^{-\frac{1}{2\sigma_1^2}(m_1-x)^2}, \tag{3.2}$$

$$f_x(m_2) = \frac{1}{\sigma_1 \sqrt{2\pi}} l^{-\frac{1}{2\sigma_1^2}(m_2-x)^2}, \tag{3.3}$$

$$f_x(m_1, m_2) = f_x(m_1) f_x(m_2), \tag{3.3}$$

$$f(m_1) = \frac{1}{\sqrt{\sigma_0^2+\sigma_1^2} \sqrt{2\pi}} l^{-\frac{(m_1-a)^2}{2(\sigma_0^2+\sigma_1^2)}}, \tag{3.4}$$

利用恒等式

$$f(x) f_x(m_1) = f(m_1) f_{m_1}(x) \tag{3.5}$$

求得

$$f_{m_1}(x) = \frac{1}{A\sqrt{2\pi}} l^{-\frac{(x-A)^2}{2A^2}}, \quad (3.6)$$

其中

$$A = \frac{\sigma_0 \sigma_1}{\sqrt{\sigma_0^2 + \sigma_1^2}}, \quad B = \frac{m_1 \sigma_0^2 + a\sigma_1^2}{\sigma_0^2 + \sigma_1^2}.$$

再估计到 $m_2 = x + (m_2 - x)$,但 x 与 $(m_2 - x)$ 独立,得

$$f_{m_1}(m_2) = \frac{1}{A_1 \sqrt{2\pi}} l^{-\frac{(m_2-\beta)^2}{2A_1^2}}, \quad (3.7)$$

其中

$$A_1 = \sigma_1 \sqrt{\frac{2\sigma_0^2 + \sigma_1^2}{\sigma_0^2 + \sigma_1^2}}$$

再由计算得

$$f_{m_1 m_2}(x) = \frac{f(x) f_x(m_1, m_2)}{f(m_1, m_2)} = \frac{1}{A\sqrt{2\pi}} l^{-\frac{(x-\beta_2)^2}{2A_2^2}}, \quad (3.8)$$

其中

$$A_2 = \frac{\sigma_0 \sigma_1}{\sqrt{2\sigma_0^2 + \sigma_1^2}}, \quad B_2 = \frac{(m_1 + m_2)\sigma_0^2 + a\sigma_1^2}{2\sigma_0^2 + \sigma_1^2}.$$

由公式(3.6)可见

$$E_{m_1}(x^2) = A^2 + B^2. \quad (3.9)$$

而从(3.8)得

$$E_{m_1 m_2}(x^2) = A_2^2 + B_2^2. \quad (3.10)$$

量水计被接受的概率由下式可得

$$P = \iint_L f(m_1, m_2) \mathrm{d}m_1 \mathrm{d}m_2, \quad (3.11)$$

其中 L 为图 3-1 中的二维积分

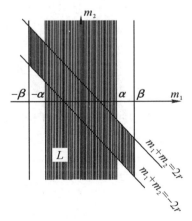

图 3-1

区域.

量水计要检验两次的概率 P_1 为

$$P_1 = \int_{-\beta}^{-\alpha} f(m_1) dm_1 + \int_{\alpha}^{\beta} f(m_1) dm_1. \qquad (3.12)$$

最后,被接受的量水计的相对系统误差的平方为

$$E^*(x^2) = \frac{1}{\rho} \iint_L (A_2^2 + B_2^2) f(m_1, m_2) dm_1 dm_2. \qquad (3.13)$$

由问题的提法容易看到,新验收方法中三个参数 α, β, γ 中只有一个,例如 α,可以任意. 其余两个由寻求最佳验收方法的条件(i)及(ii)完全决定. 因而,根据已知的 a, σ_0, σ_1, q, 利用上述诸公式可近似地算出 β, γ 及 $E^*(x^2)$, 它们都是未知参数 α 的函数. Zubrzycki[3] 算出当 $a=0$, $\sigma_0 = 1\%$, $q = 1.5\%$ (这些数字取自伏罗茨拉夫量水计工厂)时, 得下列结果:

$\alpha(\%)$	$\beta(\%)$	γ	$E^*(x^2)$	$\sqrt{E^*(x^2)}(\%)$
1.280	1.563	$+\infty$	0.6723×10^{-1}	0.8200
1.350	1.650	1.500	0.6443×10^{-4}	0.8027
1.400	1.713	1.242	0.6422×10^{-4}	0.8014
1.430	1.752	1.101	0.6411×10^{-4}	0.8007
1.563	1.929	0	0.6723×10^{-4}	0.8200

这里 $\alpha = 1.350\%$ 是度量总局的方法, $\alpha = 1.430\%$ 给出 $E^*(x^2)$ 的极小, 但相差甚小, 不值得在这方面改变验收方法. 但还值得注意的是, 上表中两端的数 $\alpha = 1.280\%$ 及 1.563% 给出几乎相同的结果. 这两种情形都使方法简化而只需对每一量水计检验一次. 这一简化使结果更经济, 因为它减少了观察次数并缩短了检验时间.

参考文献

[1] Obalski J. O pewności wyników sprawdzania narzedzi mierniczych. Zastosowania Matematyki, 1953, 1: 105-124.

[2] Oderfeld J, Zubrzycki S. O sprawdzanin wodomierzy. Zastosowania Matematyki, 1953, 1: 125-137.

[3] Zubrzycki S. O optymalnej metodzie odbioru wodomierzy. Zastosowania Matematyki, 1955, 2: 199-209.

§4. 估计到观察错误的产品质量的验收检查

一批产品的验收检查理论只在最简单的情况下已分析清楚，即当个体的类别只有两种时（个体是上品或次品），但各种验收检查方法都不估计到检查员所可能犯的两种错误（误认上品为次品或次品为上品）. 有时，这些错误相当严重，这可从 Kennedy 所描述的试验或从华沙统计与设计高等学校的女大学生 H. Mierzejewska 在量水计工厂所作的试验得到证明. Kennedy 的试验是：在一批全为上品的总体（其个数未知）添进 100 个次品，然后把这批总体拿去检查，结果检查者只找出了 68 个次品，剩下来的总体请他再检查一次，而不告诉他，这就是他先已检查过的且经他认为全是上品的总体. 在第二次分类中，他又找到了 18 个次品. 再让他作第三次检查，又发现了 8 个. 第四次特别组织了人去检查又找到了 4 个次品，但还有 2 个次品未找出来. Mierzejewska 所做的一次试验是：有 1 873 个小齿轮，检查者在分类时挑出了 169 个次品后，她又去检查了几次. 第一次她找出了 70 个次品，第二次 24 个，第三次 4 个，第四次 1 个次品. 然后她再一次检查了那些认为是次品的小齿轮，结果发现在第一次所挑出的 70 个中有 2 个是上品，第二次的 24 个中有 1 个也是上品. 这些例子表示，在某些技术条件下，检查者犯错误（第一类及第二类）的概率不会太小，如不注意这点就可能得到本质上错误的结论. 同样的情形不仅在品质检查中，而且也在排版校对，X 射线摄片的阅读以及其他场合碰到. 这里我想谈谈在估计到检查员的错误时的质量控制理论中的一些结果．它们是K. Wiśniewski所得到的，我认为很有趣味而且有不少应用．

我们假定，在接受某一个总体时，检查员以概率$1-p$认上品为次品，以概率$1-q$认次品为上品（因而p及q为得到正确结论的概率：认上品为上品及认次品为次品）．注意，概率p及q依赖于许多难以细致研究的检查条件（检查员的工作速度，受检查的样本或总体的个数，检查员的疲劳度，总体的次品率），因此，它们其至在检查一批总体时也会显著地变化．

若已知总体的次品率w，则容易算出$P(c \mid w, p, q)$，这是从总体中随机地挑出的个体被认为是上品的概率，

$$P(c \mid w, p, q) = (1-w)p + w(1-q).$$

从总体中随机地挑出的个体被认为是次品的概率$P(\bar{c} \mid w, p, q)$为$1 - P(c \mid w, p, q)$，即

$$P(\bar{c} \mid w, p, q) = 1 - (1-w)p - w(1-q)$$
$$= wq + (1-w)(1-p) = w_1,$$

后一概率w_1为总体的伪次品率，因为运用全面检查时，被认为是次品而拒收的个体的频率就是w_1．

由全面的验收检查所制订的验收计划常是这样：检查全部总体，坏的个体拒收，好的则接受．若检查员不犯错误，且若不顾及检查费用，则这种方法是最好的，因为它保证了一切次品的拒收，所以经过挑选而接受的总体中，次品为零．

但若检查员可能犯错误，则虽然全部检查，仍不能保证把次品全部挑出，且此外还可能把某些上品错误地抛掉．本文的目的就是仿Wiśniewski而证明：在某些条件下，当检查员不常正确时，抽样检查给出比全面检查更好的结果．

设有个体数为N，次品率为w的总体，进行全面检查后，恰有k个个体被认为是次品的概率$P_k(N, w, p, q)$为

$$P_k(N,w,p,q)$$
$$= \sum_{i=0}^{k} C_{N(1-w)}^{i} C_{Nw}^{k-i} p^{N(1-w)-i}(1-p)^i q^{k-i}(1-q)^{Nw-(k-i)}, \quad (4.1)$$

其中因子
$$C_{N(1-w)}^{i} p^{N(1-w)-i}(1-p)^i$$

为总体中 $N(1-w)$ 个好的个体里，恰有 i 个被认为是次品的概率，而另一因子

$$C_{Nw}^{k-i} q^{k-i}(1-q)^{Nw-(k-i)}$$

为 Nw 个次品中，有 $k-i$ 个被认为是次品的概率. 公式中的求和取自一切可能的 i，即自 0 到 k.

今考虑按验收方案 $m /\!/ n$ 的抽样检查. 若检查员不犯错误，则方案 $m /\!/ n$ 的特征数由下式计算，

$$f(w) = P(z \leqslant m) = \sum_{i=0}^{m} \frac{C_{Nw}^{i} C_{N(1-w)}^{n-i}}{C_{N}^{n}}.$$

若检查员可能犯错误，则方案 $m /\!/ n$ 的特征数为

$$\varphi(w) = P(k \leqslant m)$$
$$= \sum_{j=0}^{m} \sum_{i=0}^{n} \frac{C_{Nw}^{i} C_{N(1-w)}^{n-i} P_i\left(n, \frac{i}{n}, p, q\right)}{C_{N}^{n}}, \quad (4.2)$$

其中 $\dfrac{C_{Nw}^{i} C_{N(1-w)}^{n-i}}{C_{N}^{n}}$ 为下事件的概率：自个数为 N，次品率为 w 的总体中，抽选个数为 n 的样本，此样本中恰含 i 个次品. 而 $P_i\left(n, \dfrac{i}{n}, p, q\right)$ 是检查员检查此样本后，其中恰有 j 件被认为是次品的概率，它由(4.1)表达.

若 $k \leqslant m$，则应用抽样检查法时，总体全被接受. 这等于把未列入样本中的一切个体划为上品. 若 $k > m$，则总体被拒收，这等于承认全体未经检查的个体都是次品. 我们现在来比较全面检查法和抽样检查法中，被错误地分类的个体件数的数学

期望.

在全面检查法中,被错误地分类的个体件数的数学期望 R_1 显然为

$$R_1 = N(1-(1-w)p-wq). \tag{4.3}$$

若采用验收方案,则在样本中被错误地分类的个体件数的数学期望为 τ_1,

$$\tau_1 = n(1-(1-w)p-wq). \tag{4.4}$$

如果根据抽出来的样本,总体被拒收,那么其他的一切个体都算是次品.因为总体被拒收的概率为 $1-\varphi(w)$,所以未经检查的个体中,原是上品而被认作次品的件数的数学期望为

$$\tau_2 = (N-n)(1-w)(1-\varphi(w)). \tag{4.5}$$

同样,根据验收方案 $m/\!/n$,当总体被接受时,未经检查的个体中,原为次品而被认为是上品的件数的数学期望为

$$\tau_3 = (N-n)w\varphi(w). \tag{4.6}$$

把 τ_1,τ_2,τ_3 加起来,就得抽样方法中被错误地分类的个体件数的数学期望 R_2,

$$\begin{aligned}R_2 &= n[1-(1-w)p-wq]+(N-n)(1-w)(1-\varphi(w))+\\&\quad (N-n)w\varphi(w)\\&= n[1-(1-w)p-wq]+(N-n)[1-w-(1-2w)\varphi(w)].\end{aligned} \tag{4.7}$$

比较公式(4.3)与(4.7),可见当下不等式满足时,抽样方法就优于全面检查($R_2 \leqslant R_1$),

$$(1-2w)\varphi(w) \geqslant p(1-w)-w(1-q). \tag{4.8}$$

因为 $\varphi(w) \geqslant 0$,故如上面的不等式的右方取负值时,(4.8)成立.为此只要当 $1-2w>0$ 时,

$$\frac{p}{1+p-q} < w < \frac{1}{2}, \quad 如 \frac{p}{1+p-q} < \frac{1}{2}.$$

如 $1-2w<0$ 时,由于 $\varphi(w) \leqslant 1$,得

$$\frac{1}{2}<w<\frac{1-p}{1+q-p}, \text{如} \frac{1-p}{1+q-p}>\frac{1}{2}. \qquad (4.9)$$

(4.9)中两不等式表示，在某些情况下，抽样方法优于对总体的全面检查．这结果有很大的实际意义．

第一，运用抽样检查，可以节省许多经费，因为抽样方法几乎总是比全面检查便宜．

第二，不检查总体而只检查样本，在许多情况下可以做得更好，因而缩小错误概率 $1-p$ 及 $1-q$．

Wiśniewski还考虑了其他的验收方案，如采用方案 $m/\!/n$，但此时条件为，当在样本中找出多于 m 个次品时，不是拒收总体，而是再进行全面检查．当总体的次品率满足不等

$$w \leqslant \frac{1-p}{1-p+q}$$

式时，新方案优于全面检查．

参考文献

[1] Oderfeld J, Wiśniewski K. Odbiór statystycrny z uwrględnieniem błędów kontrolera. Zastosowanta Matematyki, 1955, 2: 312-327.

[2] Kennedy C W. An unconventional approach to statistical quality control. The Machinist, 1950, 20.

§5. 两个生产过程的比较与对偶性原则

这篇报告主要的根据是 H. Steinhaus 及 S. Zubrzycki 的文章. 报告的出发点是关于两个连续生产过程损耗度的比较. 连续生产过程的例子可举自动机生产人造丝带. 为了要得到比较两个生产过程的例子，可设想两根丝带并排地以同一速度移动. 在观察它们时，需要检查下列假定：带 II 的损耗度超过带 I 的损耗度 2 倍，如果在带 II 及带 I 上分别发现了 m 及 n 个缺陷. 这里所谓缺陷是指斑点或小洞，此外还要假设每根丝带上缺陷的位置构成泊松(Poisson)流即最简单流[①]，其强度分别为 c_I 及 c_{II}，且两丝带上的缺陷流是彼此独立的.

我们要研究两种不同的概型. 第一种是，继续观察丝带，直到缺陷的总数 $m+n$ 达到预定数 N 时为止，我们称这种概型为古典的. 第二种概型称为序贯的，它不同于前者的是，观察继续到带 I 上的缺陷数达到预定数 n 时为止.

分别以随机变量 μ 及 ν 表带 II 及 I 上被观察到的缺陷个数，问题化为要计算概率

$$P(c_{II} > \alpha c_I \mid \mu = m, \nu = n), \tag{5.1}$$

如果观察是序贯式的；或

$$P(c_{II} > \alpha c_I \mid \mu = m, \mu + \nu = N), \tag{5.2}$$

如果观察是古典式的.

众所周知，概率 (5.1) 及 (5.2) 不是唯一决定的. Steinhaus[4,5,6] 预先指出了这些困难而试图解释似然观念（Понятия

[①] 参阅 Хинчин. 张里千，等译. 公用事业中的数学方法. 北京：科学出版社，1958.

правдоподобия)或所谓可信论(Фидуциалъный аргумент)的作用. 在这方面有 Sarkadi[3] 及 Oderfeld[1] 的文章, 后者最先发现了一些法则, 他称为对偶法则. 我们先研究一个泊松流, 借以说明整个问题. 这例子取自 Sarkadi[3].

观察一个泊松流, 试求此流的强度 c 小于 α 的概率, 如果在单位时间内找到了 k 个缺陷. 若不知 c 的先验分布, 这概率是不能算出的. 但通常运用可信论而解决另一问题, 即求出可以计算的概率 $P(k>K \mid c=\alpha)$——已知流的强度 $c=\alpha$, 在单位时间内, 观察到的缺陷个数 k 大于数 K 的概率. 这一概率称为在已知 $k=K$ 时, $c<\alpha$ 的似然 $W(c<\alpha \mid k=K)$. 因而这里似然的定义是

$$W(c<\alpha \mid k=K)=P(k>K \mid c=\alpha). \tag{5.3}$$

但若假定存在某一先验分布, 譬如采用贝叶斯假设, 即强度的一切可能值都等概率, 则仍可计算不等式 $c<\alpha$ 的概率. 上述假定的意义将在下面的计算中仔细说明. 按贝叶斯假设所计算得的概率以

$$P_{HB}(c<\alpha \mid k=K)$$

表示. 为算此概率先假定 c 的先验分布为在区间 $(0, L)$ 中均匀的. 则按贝叶斯公式得

$$\begin{aligned}
&P_{HB}^{(L)}(c<\alpha \mid k=K)\\
&=\frac{\int_0^\alpha P_L(\alpha)P(k=K \mid c=\alpha)\mathrm{d}\alpha}{\int_0^L P_L(\alpha)P(k=K \mid c=\alpha)\mathrm{d}\alpha}=\frac{\int_0^\alpha \frac{1}{L}\mathrm{e}^{-\alpha}\frac{\alpha^K}{K!}\mathrm{d}\alpha}{\int_0^L \frac{1}{L}\mathrm{e}^{-\alpha}\frac{\alpha^K}{K!}\mathrm{d}\alpha}\\
&=\frac{1-\mathrm{e}^{-\alpha}\left(1+\alpha+\cdots+\frac{\alpha^K}{K!}\right)}{1-\mathrm{e}^{-L}\left(1+L+\cdots+\frac{L^K}{K!}\right)},
\end{aligned}$$

然后定义

$$P_{HB}(c<\alpha \mid k=K) = \lim_{L \to +\infty} P_{HB}^{(L)}(c<\alpha \mid k=K),$$

上式可作为贝叶斯假设的准确意义,于是得

$$P_{HB}(c<\alpha \mid k=K) = 1 - e^{-\alpha}\left(1 + \alpha + \cdots + \frac{\alpha^K}{K!}\right), \quad (5.4)$$

但(5.4)的右方也是似然(5.3)的值,因而得等式

$$W(c<\alpha \mid k=K) = P_{HB}(c<\alpha \mid k=K). \quad (5.5)$$

此式表达了关于泊松流的对偶法则.

现在转来研究如何比较两个生产过程. 这问题易化为抽球问题,若带 I 单位长度内的缺陷数期望值为 c_I,带 II 为 c_{II},则某一确定的缺陷(譬如在时间 t_0 后第一个被察觉的缺陷)为缺陷 I (即属于带 I)的概率为 $\frac{c_I}{c_I + c_{II}}$,为缺陷 II (即属于带 II)的概率是 $\frac{c_{II}}{c_I + c_{II}}$. 于是登记两条并列丝带上缺陷的程序可化以按序地从某一箱中抽球的程序,抽得白球(缺陷 I)的概率为 $p = \frac{c_I}{c_I + c_{II}}$,得黑球(缺陷 II)的概率为 $q = \frac{c_{II}}{c_I + c_{II}}$.

但抽球问题与质量统计检查中的主要问题,即估计总体中的次品率是属于同一类型的. 如果认黑球为次品,那么问题化为要估计概率 q. 又我们原来要研究的是关系式 $\frac{c_{II}}{c_I} = \frac{q}{1-q} = \tau$ (即要检查假定 $\tau > \alpha$),但因 τ 及 q 间有关系 $\tau = \frac{q}{1-q}$,故可把此问题化为上问题,反之亦然.

现在采取假设 $HB1$ 来计算概率(5.1),即假定两强度的先验分布满足下列条件:$\frac{c_{II}}{c_I}$ 之一切可能值等概率. 因为整个模型不依赖于丝带长度量度单位的选择,所以不失普遍性可令 $c_I = 1$(1 单位长度等于带 I_1 上相邻两缺陷间距离的平均值). 因而

假设 $HB1$ 保证强度 c_{II} 的一切可能值等概率(准确意义见公式 (5.4)的证明),且条件 $c_{\mathrm{II}}>ac_{\mathrm{I}}$ 化为 $c_{\mathrm{II}}>\alpha$. 先注意

$$P(\mu=m\mid c_{\mathrm{II}}=\alpha,\nu=n)=C_{n+m-1}^{m}\left(\frac{\alpha}{1+\alpha}\right)^{m}\left(\frac{1}{1+\alpha}\right)^{n}, \quad (5.6)$$

这式可如下得到:在抽球问题中 $P(\mu=m\mid c=\alpha,\nu=n)$ 是在前 $n+m-1$ 个抽出的球中,有 $n-1$ 白球及 m 黑球而第 $n+m$ 个球是白球的概率(因在序贯概型中,观察是在发现带 I 上第 n 个缺陷时结束的). 抽得白球与黑球的概率分别为 $p=\dfrac{1}{1+\alpha}$, $q=\dfrac{\alpha}{1+\alpha}$ (因 $c_{\mathrm{I}}=1$, $c_{\mathrm{II}}=\alpha$). 按贝叶斯公式得

$$P_{HB1}^{(L)}(c_{\mathrm{II}}>\alpha\mid\mu=m,\nu=n)=\frac{\int_{a}^{L}P_{L}(\alpha)P(\mu=m\mid c=\alpha,\nu=n)\mathrm{d}\alpha}{\int_{0}^{L}P_{L}(\alpha)P(\mu=m\mid c=\alpha,\nu=n)\mathrm{d}\alpha}$$

$$=\frac{\int_{a}^{L}\dfrac{1}{L}C_{n+m-1}^{m}\left(\dfrac{\alpha}{1+\alpha}\right)^{m}\left(\dfrac{1}{1+\alpha}\right)^{n}\mathrm{d}\alpha}{\int_{0}^{L}\dfrac{1}{L}C_{n+m-1}^{m}\left(\dfrac{\alpha}{1+\alpha}\right)^{m}\left(\dfrac{1}{1+\alpha}\right)^{n}\mathrm{d}\alpha}$$

$$=\frac{\int_{a}^{L}\left(\dfrac{\alpha}{1+\alpha}\right)^{m}\left(\dfrac{1}{1+\alpha}\right)^{n}\mathrm{d}\alpha}{\int_{0}^{L}\left(\dfrac{\alpha}{1+\alpha}\right)^{m}\left(\dfrac{1}{1+\alpha}\right)^{n}\mathrm{d}\alpha}.$$

令 $L\to+\infty$ 得

$$P_{HB1}(c_{\mathrm{II}}>\alpha\mid\mu=m,\nu=n)=\frac{\int_{a}^{+\infty}\left(\dfrac{\alpha}{1+\alpha}\right)^{m}\left(\dfrac{1}{1+\alpha}\right)^{n}\mathrm{d}\alpha}{\int_{0}^{+\infty}\left(\dfrac{\alpha}{1+\alpha}\right)^{m}\left(\dfrac{1}{1+\alpha}\right)^{n}\mathrm{d}\alpha}.$$

$$(5.7)$$

这里需要指出,只是当 $n\geqslant 2$ 时,上面取的极限才是合法的,否则积分不收敛而对一切 α,概率 $P_{HB1}^{(L)}$ 都趋近于 1. (5.7)积分后得

$$P_{HB1}(c_{\text{II}} > \alpha c_{\text{I}} \mid \mu = m, \nu = n)$$
$$= \frac{1}{(1+\alpha)^{n-1}} \sum_{k=0}^{m} C_{n+k-2}^{k} \left(\frac{\alpha}{1+\alpha}\right)^{k}. \tag{5.8}$$

若仿 Pearson 引进记号

$$I_x(a,b) = \frac{\int_0^x \beta^{a-1}(1-\beta)^{b-1}\,d\beta}{\int_0^1 \beta^{a-1}(1-\beta)^{b-1}\,d\beta} = \frac{1}{\beta(a,b)}\int_0^x \beta^{a-1}(1-\beta)^{b-1}\,d\beta,$$

则化为更便于实际计算的形式

$$P_{HB1}(c_{\text{II}} > \alpha c_{\text{I}} \mid \mu = m, \nu = n) = 1 - I_{\frac{\alpha}{1+\alpha}}(m+1, n-1). \tag{5.8'}$$

因此要求 P_{HB1}，只要查不完全 Beta 函数表．在公式(5.8)及 (5.8′)中已不用假定 $c_{\text{I}} = 1$．但在实际的实量统计检查中，到处都采用可信论，公式(5.8)或(5.8′)不被实际运用．但若能像 Sarkadi 对一个流所做的那样，也能证明对偶法则的存在，则情况就两样．为此，首先要说明如何理解不等式 $c_{\text{II}} > \alpha c_{\text{I}}$ 在条件 $\mu = m$，$\nu = n$ 下的似然．利用公式(5.6)得

$$P(\mu \leqslant m \mid c_{\text{II}} = \alpha c_{\text{I}}, \nu = n-1)$$
$$= \frac{1}{(1+\alpha)^{n-1}} \sum_{k=0}^{m} C_{n+k-2}^{k} \left(\frac{\alpha}{1+\alpha}\right)^{k}.$$

将此与(5.8)比较求得对偶性

$$W(c_{\text{II}} > \alpha c_{\text{I}} \mid \mu = m, \nu = n)$$
$$= P_{HB1}(c_{\text{II}} > \alpha c_{\text{I}}, \mu = m, \nu = n), \tag{5.9}$$

如果取似然的定义为

$$W(c_{\text{II}} > \alpha c_{\text{I}} \mid \mu = m, \nu = n)$$
$$= P(\mu \leqslant m \mid c_{\text{II}} = \alpha c_{\text{I}}, \nu = n-1). \tag{5.10}$$

上面已指出两个泊松流的比较可化为抽球问题，其中 $p = q$ （白球概率：黑球概率）等于 $c_{\text{I}} : c_{\text{II}}$．因此(5.8)与(5.9)也可用来研究总体的次品率．只要引入 $W = q = \dfrac{c_{\text{II}}}{c_{\text{I}} + c_{\text{II}}}$ 及 $c_{\text{II}} : c_{\text{I}} =$

$q: p = \dfrac{w}{1-w}$ 并把缺陷 I 看作上品，缺陷 II 看作次品. 于是得下定理:

如果对总体的检查进行到发现第 n 个上品时为止，这里 n 是一预先确定的数，且若 m 是找出的次品件数，则不等式 $W > \beta$ 的 $HB1$ 概率由下式计算，

$$P_{HB1}(W > \beta \mid \mu = m, \nu = n) = W(w > \beta \mid \mu = n, \nu = n)$$

$$= (1-\beta)^{n-1} \sum_{k=0}^{m} C_{n+k-2}^{k} \beta^{k}. \tag{5.11}$$

要注意的是 (5.11) 中的概率 P_{HB1} 对应于 $\tau = \dfrac{w}{1-w}$ 的均匀先验分布. 但容易证明不存在次品率 w 的对应于假定 $HB1$ 的分布. 因为假定 $HB1$ 是假定 $HB1^{(L)}$ 的极限形式，故先求出对应于假定 $HB1^{(L)}$ 的分布

$$P_L(w < \beta) = \begin{cases} 0, & \beta \leqslant 0, \\ \dfrac{1}{L} \dfrac{\beta}{1-\beta}, & 0 \leqslant \beta < \dfrac{L}{1+L}, \\ 1, & \beta \geqslant \dfrac{L}{1+L}, \end{cases} \tag{5.12}$$

但当 $L \to +\infty$ 时，分布 (5.12) 退化，即 $w=1$ 的概率为 1.

我们已解决的总体次品率的估计问题可重新在古典型（不用序贯型）的前提下重新考虑，此时样本的件数是预定的. 这问题的对偶法则由 Oderfeld[1] 找到. 设在件数为 N 的样本中有 m 件次品，定义在此条件下次品率 w 大于 β 的似然为下事件的概率：从次品率为 β 的总体中，抽出件数为 $N+1$ 的样本，其中次品件数不多于 m. 换言之，

$$W(w > \beta \mid \mu = m, \mu + \nu = N) = P(\mu \leqslant m \mid w = \beta, \mu + \nu = N+1). \tag{5.13}$$

Oderfeld 求得

$$P_{HB2}(w > \beta \mid \mu = m, \mu + \nu = N)$$
$$= \sum_{k=0}^{m} C_{N+1}^{k} \beta^{k} (1-\beta)^{N-k+1}. \tag{5.14}$$

由定义(5.13)有

$$P_{HB2}(w > \beta \mid \mu = m, \mu + \nu = N)$$
$$= W(w > \beta \mid \mu = m, \mu + \nu = N). \tag{5.15}$$

这里 $HB2$ 表示次品率 w 在区间 $(0, 1)$ 中有均匀先验分布，这是 Oderfeld 在其对偶法则中所假设的.

引进不完全 Beta 函数，则(5.14)可化为更便于计算的形式

$$P_{HB2}(w > \beta \mid \mu = m, \mu + \nu = N)$$
$$= 1 - I_{\beta}(m+1, N-m+1). \tag{5.16}$$

现在利用 Oderfeld 的结果来比较两生产过程，这时观察是古典型的，即观察继续到发现两根丝带上的缺陷总数达到一预定数 N 时为止，为此在(5.13)~(5.16)中要以 $\dfrac{c_{\text{II}}}{c_{\text{I}} + c_{\text{II}}}$ 代 w，以 $\dfrac{\alpha}{1+\alpha}$ 代 β，以 $c_{\text{II}} > \alpha c_{\text{I}}$ 代 $w > \beta$，并把 μ 及 ν 解释为带 I 及带 II 的缺陷数，于是得到比较两个泊松流的公式. 作为似然的定义我们令

$$W(c_{\text{II}} > \alpha c_{\text{I}} \mid \mu = m, \mu + \nu = N)$$
$$= P(\mu \leqslant m \mid c_{\text{II}} = \alpha c_{\text{I}}, \mu + \nu = N+1), \tag{5.17}$$

则对偶法则取形式

$$W(c_{\text{II}} > \alpha c_{\text{I}} \mid \mu = m, \mu + \nu = N)$$
$$= P_{HB2}(c_{\text{II}} > \alpha c_{\text{I}} \mid \mu = m, \mu + \nu = N), \tag{5.18}$$

其中

$$P_{HB2}(c_{\text{II}} > \alpha c_{\text{I}} \mid \mu = m, \mu + \nu = N)$$
$$= \sum_{k=0}^{m} C_{n+1}^{k} \left(\frac{\alpha}{1+\alpha}\right)^{k} \left(\frac{1}{1+\alpha}\right)^{n-k+1}$$

$$= 1 - I_{\frac{a}{1+a}}(m+1, N-m+1). \quad (5.19)$$

这里 $HB2$ 表示 $\dfrac{c_{\text{II}}}{c_{\text{I}}+c_{\text{II}}}$ 的先验分布在区间 $(0,1)$ 中是均匀的,亦即 $\dfrac{c_{\text{II}}}{c_{\text{I}}}=c$ 的先验分布为

$$P(c<a) = \frac{a}{1+a}, \quad 0<a<+\infty. \quad (5.20)$$

注意,倒数 c^{-1} 即分数 $\dfrac{c_{\text{I}}}{c_{\text{II}}}$ 有同样的先验分布,这是下列命题的推论:若次品率 w 有均匀分布,则 $1-w$ 的分布也是均匀的.

至今我们讲了两种不同的对偶法则. 其一表达为公式 (5.11),它对应于序贯型. 在此式中,概率 P_{HB1} 及似然 W 均对应于序贯型,即抽样进行到发现第 n 个上品为止. 另一由公式 (5.18) 表达,它是古典(或古典—古典的)的法则,因为这时不论概率 P_{HB2} 或似然都是对古典型的观察计算的,即观察的总次数为定数. 但还可以找到其他一些混合型的对偶法则:序贯—古典的及古典—序贯的. 例如由公式 (5.7) 出发,积分后得

$$P_{HB1}(w>\beta \mid \mu=m, \nu=n) = \frac{\int_{\beta}^{1} \beta^m (1-\beta)^{n-2} d\beta}{\int_{0}^{1} \beta^m (1-\beta)^{n-2} d\beta}$$

$$= \sum_{k=0}^{m} C_{m+n-1}^{k} \beta^k (1-\beta)^{n+m-1-k}. \quad (5.21)$$

但后一数可读为下列事件的概率:从次品率为 β 的总体中抽取个数为 $n+m-1$ 的样本,后者所含次品数不多于 m. 因此,可把公式 (5.21) 视为序贯—古典对偶法则的形式:

$$P_{HB2}(w>\beta \mid \mu=m, \nu=n)$$
$$= P(\mu \leqslant m \mid w=\beta, \mu+\nu=m+n-1). \quad (5.22)$$

这名称的来源是由于 P_{HB2} 是按序贯型的观察所算得的概率,而似然((5.22)式的右方需了解为不等式 $w>\beta$ 的似然)是

按古典型的观察所定义的. 像以前一样，这是 $HB1$ 仍然表示 $\frac{w}{1-w}$ 有均匀的先验分布. 利用这假定，还可以得到两条对偶法则. 为此，先注意下列事实: 当 $\frac{w}{1-w}$ 有均匀先验分布时，在按序贯型观察 $(\nu=n)$ 及 $\mu=m$ 的条件下，次品率 w 的后验分布与在按古典型观察 $\mu+\nu=N$，观察结果为 $\mu=m$, $\nu=n$ ($m+n=N$) 的条件下，次品率 w 的后验分布一样. 这事实可写成下式：

$$P_{HB1}(w>\beta \mid \mu=m, \nu=n)$$
$$=P_{HB1}(w>\beta \mid \mu=m, \mu+\nu=m+n). \qquad (5.23)$$

因式中第一个概率由(5.21)定出，故要证明此式，只要证第二个概率也由同一式决定. 为此先取假定 $HB1^{(L)}$: w 的先验分布为(5.12). 由贝叶斯公式有

$$P_{HB1}^{(L)}(w>\beta \mid \mu=m, \mu+\nu=m+n)$$

$$=\frac{\int_{\beta}^{\frac{L}{1+L}} P_L(\beta) P(\mu=m \mid w=\beta, \mu+\nu=m+n) \mathrm{d}\beta}{\int_{0}^{\frac{L}{1+L}} P_L(\beta) P(\mu=m \mid w=\beta, \mu+\nu=m+n) \mathrm{d}\beta}$$

$$=\frac{\int_{\beta}^{\frac{L}{1+L}} \frac{1}{L} \frac{1}{(1-\beta)^2} C_{m+n}^{m} \beta^m (1-\beta)^n \mathrm{d}\beta}{\int_{0}^{\frac{L}{1+L}} \frac{1}{L} \frac{1}{(1-\beta)^2} C_{m+n}^{m} \beta^m (1-\beta)^n \mathrm{d}\beta}$$

$$=\frac{\int_{\beta}^{\frac{L}{1+L}} \beta^m (1-\beta)^{n-2} \mathrm{d}\beta}{\int_{0}^{\frac{L}{1+L}} \beta^m (1-\beta)^{n-2} \mathrm{d}\beta},$$

令 $L \to +\infty$ (当 $n \geq 2$ 时) 得

$$P_{HB1}(w>\beta \mid \mu=m, \mu+\nu=m+n) = \frac{\int_{\beta}^{1} \beta^m (1-\beta)^{n-2} \mathrm{d}\beta}{\int_{0}^{1} \beta^m (1-\beta)^{n-2} \mathrm{d}\beta},$$

由(5.21)得证(5.23).

由(5.23)(5.11)及(5.22)得两新对偶法则,即古典-序贯的

$$P_{HB1}(w>\beta\mid\mu=m,\mu+\nu=m+n)$$
$$=P(\mu\leqslant m\mid w=\beta,\nu=n-1) \quad (5.24)$$

及古典-古典的

$$P_{HB2}(w>\beta\mid\mu=m,\mu+\nu=m+n)$$
$$=P(\mu\leqslant m\mid w=\beta,\mu+\nu=m+n-1).$$
$$(5.25)$$

现在来研究对偶法则(5.15),它是在假定 $HB2$ 下,即假定次品率 w 在 $(0,1)$ 中有均匀先验分布时得到的,我们不进行全部计算只是指出,在这种情况下,Steinhaus 及 Zubrzycki 也得到了其他一些对偶法则即古典-序贯的(法则(5.15)为古典-古典的)

$$P_{HB2}(w>\beta\mid\mu=m,\mu+\nu=N)$$
$$=P(\mu\leqslant m\mid w=\beta,\nu=N-m+1), \quad (5.26)$$

序贯-古典的

$$P_{HB2}(w>\beta\mid\mu=m,\nu=n)$$
$$=P(\mu\leqslant m\mid w=\beta,\mu+\nu=m+n+1), \quad (5.27)$$

及序贯-序贯的

$$P_{HB2}(w>\beta\mid\mu=m,\nu=n)$$
$$=P(\mu\leqslant m\mid w=\beta,\nu=n+1). \quad (5.28)$$

值得注意的是,由于采取了假定 $HB1$,即假定 $\dfrac{w}{1-w}$ 有均匀先验分布,我们得到对偶法则(5.11)(5.22)(5.24)及(5.25),其右方是 $\mu\leqslant m$ 的条件概率,但条件里总是样本中上品的个数 ν 比观察次数 n 小 1. 采取另一假定 $HB2$,即假定次品率 w 在 $(0,1)$ 中有均匀先验分布,则得对偶法则(5.15)(5.26)~(5.28),其右方是 $\mu\leqslant m$ 的条件概率,但条件里总是样本中上品的个数 ν 比观察次数 n 大 1. 但实际上作者及一切可信论者会

再一次辩证说,这里所说的对偶法则与可信论的方法无关,因为实际中常取不等式 $w>\beta$ 的似然为 $\mu\leqslant m$ 在条件 $w=\beta$ 与样本个数(古典型)或上品个数(序贯型)符合于观察结果下的概率. 若 n 相当大,则因只相差 1,而可忽略此差别. 但可能 n 相当小,此对 Steinhaus 及 Zubrzycki 提出了下列问题: 试求假定 $HB3$,使当此条件成立时,既满足等式

$$P_{HB3}(w>\beta \mid \mu=m, \mu+\nu=N)$$
$$=P(\mu\leqslant m \mid w=\beta, \mu+\nu=N) \qquad (5.29)$$

(它表达新的古典-古典对偶法则),也使其他的法则: 古典-序贯的

$$P_{HB3}(w>\beta \mid \mu=m, \mu+\nu=N)$$
$$=P(\mu\leqslant m \mid w=\beta, \nu=n), \qquad (5.30)$$

序贯-古典的

$$P_{HB3}(w>\beta \mid \mu=m, \mu+\nu=N)$$
$$=P(\mu\leqslant m \mid w=\beta, \mu+\nu=N), \qquad (5.31)$$

及序贯-序贯的

$$P_{HB3}(w>\beta \mid \mu=m, \nu=n)$$
$$=P(\mu\leqslant m \mid w=\beta, \nu=n), \qquad (5.32)$$

法则都成立.

解决此问题时,Steinhaus 及 Zubrzycki 证明了,若把 $\lg\dfrac{1}{1-w}$ 有均匀先验分布(这不对应于次品率的任何分布)当作假定 $HB3$,则等式(5.29)~(5.32)都成立. 这里只证明公式(5.29). 以 $HB3^{(L)}$ 表 $\lg\dfrac{1}{1-w}$ 在 $(0,1)$ 中有均匀先验分布的假定,即令

$$P\left(\lg\dfrac{1}{1-w}<\alpha\right)=\dfrac{\alpha}{L}, \quad 0\leqslant\alpha\leqslant L.$$

这等于假定次品率 w 有下列先验分布

$$P(w<\beta)=\frac{1}{L}\lg\frac{1}{1-\beta}, \ 0<\beta<1-e^{-L}.$$

其密度为

$$P_L(\beta)=\frac{1}{L}\frac{1}{1-\beta}, \ 0<\beta<1-e^{-L}.$$

运用贝叶斯公式得

$$P_{HB3}^{(L)}(w>\beta) \mid \mu=m, \mu+\nu=N) =$$

$$= \frac{\int_{\beta}^{1-e^{-L}} P_L(\beta) P(\mu=m \mid w=\beta, \mu+\nu=N) d\beta}{\int_{0}^{1-e^{-L}} P_L(\beta) P(\mu=m \mid w=\beta, \mu+\nu=N) d\beta}$$

$$= \frac{\int_{\beta}^{1-e^{-L}} \beta^{-m}(1-\beta)^{N-m-1} d\beta}{\int_{0}^{1-e^{-L}} \beta^{m}(1-\beta)^{N-m-1} d\beta},$$

令 $L \to +\infty$（为使极限存在，须设 $n=N-m \geqslant 1$），得

$$P_{HB3}(w>\beta \mid \mu=m, \mu+\nu=N) = \frac{\int_{\beta}^{1}\beta^{m}(1-\beta)^{N-m-1}d\beta}{\int_{0}^{1}\beta^{m}(1-\beta)^{N-m-1}d\beta}.$$

计算此积分，最后得

$$P_{HB3}(w>\beta \mid \mu=m, \mu+\nu=N) = \sum_{k=0}^{m} C_N^k \beta^k (1-\beta)^{N-k}$$
$$= P(\mu \leqslant m \mid w=\beta, \mu+\nu=N).$$

Steinhaus[4,5,6] 及 Oderfeld[1] 早就指出，统计中可信论的应用并不解决下述困难，如不采用一些协议作为前提，就不可能回答实际工作者的问题. 上面各结果解释了这些前提在可信论方法中的作用. 从实用观点来看，要求 $-\lg(1-w)$ 有均匀分布的假定 HB3 更复杂，故不如贝叶斯假定容易接受. 但从上诸结果，可见总体次品率的各种不同的先验分布，给出甚为接近的后验概率.

参考文献

[1] Oderfeld J. On the dual aspect of sampling plans. Colloquium Mathematicum, 1951, 2: 89-97.

[2] Romejko A. Porównywanie dwóch partii towaru. Zastosowania Matematyhi, 1957, 3: 217-228.

[3] Sarkadi K. A Selejtaráng Bayes-féle Valószinūségi határaita Vonatkoxó dualitási elvröl (On the rule of dualism Concerning the Bayes' probability limits of the fraction defective). Alkalmazott Matematikai Intézetének Közleményei 11, Budapest, 1953: 275-285.

[4] Steinhaus H. Quality control by sampling (A plea for Bayés rule). Colloquium Mathematicum, 1951, 2: 98-108.

[5] Steinhaus H. Podstawy kontroli statystycznej. Zastosowania Matematyki. 1953, 1: 4-25.

[6] Steinhaus H. Prawdopodobieústwo, Wiarogodnosć, moźliwosć. Zastosowania Matematyki, 1954, 1: 149-171.

[7] Steinhaus H, Zubrzycki S. O poróonywaniu dwóch procesów produkcyinych i Zasadzie dualizmu. Zastosowania Matematyki. 1958, 3: 229-257.

§6. 度量分析的新代数理论及其在产品抽样实验中的应用

量度分析是力学物理及其他自然科学中常用的方法. 这方法研究各种具体问题中的度量, 有时只需要初等的理论和计算, 就可得到有趣的结果. 虽然量度分析应用已久甚至不知谁是创始者, 但至今这方法还不简明, 而且从数学观点看来也不准确. 在这方面的各种努力都未得到令人满意的结果. 直到几年前, 伏洛茨瓦夫的数学家 S. Drobot 才建立了量度分析的代数方法, 据我看来, 它能满足简明性与准确性的一切要求. 这作者还与伏洛茨瓦夫的另一数学家 M. Warmus 运用此新理论于产品的抽样试验问题上. 由于这两结果相当有趣, 我希望能在这里报告.

假定量度是空间 Π 的元, Π 满足下列公理:

(i) 对空间 Π 的任两元 A 及 B, 其乘积 AB 也属于 Π, 且使交换律

$$AB = BA \qquad (6.1)$$

及结合律

$$(AB)C = A(BC) \qquad (6.2)$$

成立, 并且对每 A 及 B, 存在 Π 中的元 x 使 $Ax = B$.

(ii) 空间 Π 的元 A 可配以实指数 a, 且 A^a 属于 Π, 并满足下列条件

$$A^{a+b} = A^a \cdot A^b, \qquad (6.3)$$

$$(AB)^a = A^a B^a, \qquad (6.4)$$

$$(A^a)^b = A^{ab}, \qquad (6.5)$$

$$A^1 = A. \qquad (6.6)$$

(iii) 实数 α, β, γ, $\cdots$ 也是空间 Π 的元.

空间中非正数的元称为量纲,而正数则称为非量纲.

空间 Π 中的元 A_1, A_2, $\cdots$, A_m 称为量度独立,如从等式
$$A_1^{a_1} A_2^{a_2} \cdots A_m^{a_m} = \alpha,$$
其中 α 为正数,a_1, a_2, $\cdots$, a_m 为实数,即有
$$a_1 = a_2 = \cdots = a_m = 0$$
及
$$\alpha = 1.$$

其次,若在空间 Π 中存在 n 个量度独立的元但不存在 $n+1$ 个量度独立的元,则说空间 Π 有 n 个单元. 每 n 个独立的元所成的组称为单元组.

容易证明,若 x_1, x_2, $\cdots$, x_n 是单元组,则空间 Π 中每一元可唯一地表为
$$A = \alpha x_1^{a_1} x_2^{a_2} \cdots x_n^{a_n}, \tag{6.7}$$
其中 α 为非量纲,而 a_1, a_2, $\cdots$, a_n 为实数.

若 A_1, A_2, $\cdots$, A_m 由单元组 x_1, x_2, $\cdots$, x_n 表为
$$A_i = a_i x_1^{a_{1i}} x_2^{a_{2i}} \cdots x_n^{a_{ni}}, \quad i = 1, 2, \cdots, m, \tag{6.8}$$
则 A_1, A_2, $\cdots$, A_m 为量度独立的充分必要条件是指数矩阵
$$\begin{pmatrix} a_{11} & a_{12} & \cdots & a_{1m} \\ \vdots & \vdots & & \vdots \\ a_{n1} & a_{n2} & \cdots & a_{nm} \end{pmatrix} \tag{6.9}$$
的秩为 m.

若 x_1, x_2, $\cdots$, x_n 及 y_1, y_2, $\cdots$, y_n 为两个单元组,则单元组 x_1, x_2, $\cdots$, x_n 中每一元可用 y_1, y_2, $\cdots$, y_n 表为
$$x_i = \xi_i y_1^{t_{i1}} y_2^{t_{i2}} \cdots y_n^{t_{in}}, \quad i = 1, 2, \cdots, n, \tag{6.10}$$
其中 ξ_1, ξ_2, $\cdots$, ξ_n 为非量纲而 t_{ij} (i, $j = 1, 2, \cdots, n$) 是使行列式 $|t_{ij}| \neq 0$ 的实数.

设元 A 在单元组 x_1, x_2, $\cdots$, x_n 中由公式(6.7)表示,于其中代以新单元(6.10)后,得 A 在单元组 y_1, y_2, $\cdots$, y_n 中的表式

$$A = a\xi_1^{a_1}\xi_2^{a_2}\cdots\xi_n^{a_n} y_1^{t_{11}a_1+t_{21}a_2+\cdots+t_{n1}a_n}\cdots y_n^{t_{1n}a_1+t_{2n}a_2+\cdots t_{nn}a_n},$$

(6.11)

公式(6.11)是由一单元组向另一单元组的变换式. 但若在(6.11)中以 x_1, x_2, $\cdots$, x_n 代 y_1, y_2, $\cdots$, y_n, 则得空间 Π 的一新元

$$\theta_x A = a\xi_1^{a_1}\xi_2^{a_2}\cdots\xi_n^{a_n} x_1^{t_{11}a_1+t_{21}a_2+\cdots+t_{n1}a_n}\cdots x_n^{t_{1n}a_1+t_{2n}a_2+\cdots t_{nn}a_n}.$$

(6.12)

公式(6.12)是空间 Π 到自身的变换,这变换称为量度变换. 它有下列性质:

(i) 变换 θ 是互为单值的;

(ii) 对每两元 A 及 B, 有 $\theta(AB) = (\theta A)(\theta B)$;

(iii) 对每实数 a 有 $\theta(A^a) = (\theta A)^a$;

(iv) 对每非量纲 α 有 $\theta\alpha = \alpha$.

可以证明,这些条件完全决定量度变换. 对每一单元组 x_1, x_2, $\cdots$, x_n, 任一满足条件(i)～(iv)的变换可写为公式(6.12).

再引进函数 $\Phi(Z_1, Z_2, \cdots, Z_s)$ 的概念,它给空间 Π 的元所成的每一集配以此空间某一元. 我们只研究满足下列条件的函数:

(i) 量度不变性条件:对每一量度变换 θ 有恒等式

$$\Phi(\theta Z_1, \theta Z_2, \cdots, \theta Z_s) = \theta\Phi(Z_1, Z_2, \cdots, Z_s); \quad (6.13)$$

(ii) 量度齐次性条件:对每一组非量纲 ζ_1, ζ_2, $\cdots$, ζ_s, 存在非量纲 ζ 使

$$\Phi(\zeta_1 Z_1, \zeta_2 Z_2, \cdots, \zeta_s Z_s) = \zeta\Phi(Z_1, Z_2, \cdots, Z_s). \quad (6.14)$$

现在来叙述两个基本定理,它们对量度分析的应用有头等

意义：

定理 1 如果量变不变函数 $\Phi(Z_1, Z_2, \cdots, Z_m)$ 中的 Z_1, $Z_2, \cdots, Z_m$ 是量度独立的，那么函数的形状为

$$\Phi(Z_1, Z_2, \cdots, Z_m) = \varphi Z_1^{t_1} Z_2^{t_2} \cdots Z_m^{t_m}, \tag{6.15}$$

其中非量纲系数 φ 及实指数均不依赖于 $Z_1, Z_2, \cdots, Z_m$. 反之由公式(6.15)所定义的每一函数满足量度不变性条件.

定理 2 如果量度不变及齐次函数 $\Phi(Z_1, Z_2, \cdots, Z_m, P_1, P_2, \cdots, P_q)$ 中的 $Z_1, Z_2, \cdots, Z_m$ 量度独立，但 $P_1, P_2, \cdots, P_q$ 量度依赖于 $Z_1, Z_2, \cdots, Z_m$

$$P_k = \pi_k Z_1^{r_{1k}} Z_2^{r_{2k}} \cdots Z_m^{r_{mk}}, \quad k = 1, 2, \cdots, q, \tag{6.16}$$

其中 $\pi_k (k = 1, 2, \cdots, q)$ 为非量纲（正数），但指数 $r_{ik} (i = 1, 2, \cdots, m; k = 1, 2, \cdots, q)$ 是实数，那么函数 Φ 可表为

$$\Phi(Z_1, Z_2, \cdots, Z_m, P_1, P_2, \cdots, P_q)$$
$$= \varphi(\pi_1, \pi_2, \cdots, \pi_q) Z_1^{t_1} Z_2^{t_2} \cdots Z_m^{t_m}, \tag{6.17}$$

其中 $\varphi(\pi_1, \pi_2, \cdots, \pi_q)$ 为不依赖于 $Z_1, Z_2, \cdots, Z_m$ 的数值变数 $\pi_1, \pi_2, \cdots, \pi_q$ 的普通数值函数，而实指数 $t_1, t_2, \cdots, t_m$ 不依赖于 $\pi_1, \pi_2, \cdots, \pi_q$ 及 $Z_1, Z_2, \cdots, Z_m$.

反之，每一由公式(6.17)所表达的函数是量度不变和齐次的.

定理 2 由 Buckingham 证明，在量度分析中称为定理 π. 注意在定理 2 中令 $q = 0$ 易得定理 1 的结果，但这只对量度不变及齐次函数而言，而定理 1 的证明则只需假定函数的量度不变性. 但由于上述诸定理，每一量度无关变元的量度不变函数是量度齐次的.

再引进一非常直观且为实际工作者所常用的量度观念，虽然从理论上看来这并非必需. 作为等量度关系的抽象族，我们来引进这观念. 说两个量有相同的量度，如果它们的关系式 AB^{-1} 是非量纲. 记此事实为

$$[A] = [B].$$

易证等量度关系是等价关系,因为它满足条件
$$[A]=[A].$$
若$[A]=[B]$,则$[B]=[A]$.
若$[A]=[B]$及$[B]=[C]$,则$[A]=[C]$.

因此空间 Π 分裂为不相交的等量度的元所成的族,后者称为量度. 实践中常采用不同的记号来记量度. 例如 5 cm 的量度将记为
$$[5 \text{ cm}] \quad 或 \quad 厘米 \quad 或 \quad 长度.$$
还可以引进量度的乘法与乘方. 这运算的定义是
$$[A][B]=[AB] 及 [A]^a=[A^a].$$
因对每一正数又有
$$[\alpha]=[1]=1,$$
故由量度的乘法定义易证
$$[\alpha][B]=[\alpha B]=[B].$$
容易验证,量度及由上定义的对它们的运算满足构成空间 Π 的一切公理,因而它们构成一新空间 Π.

运用量度的记号,现在可以化量度齐次性的要求(6.15)为更直观的形式
$$[\Phi([Z_1], [Z_2], \cdots, [Z_s])]=[\Phi(Z_1, Z_2, \cdots, Z_s)].$$
再讲量度分析中的一定理,在叙速它时要用到量度观念.

定理 3 设在单元组 $x_1, x_2, \cdots, x_n$ 中,量度无关的量 $Z_1, Z_2, \cdots, Z_m$ 有量度
$$[Z_k]=[x_1^{a_{1k}} x_2^{a_{2k}} \cdots x_n^{a_{nk}}], \quad k=1, 2, \cdots, m.$$
则存在函数 Φ,它给 $Z_1, Z_2, \cdots, Z_m$ 配以某已给量 $F([F]=[x_1^{a_1} x_2^{a_2} \cdots x_n^{a_n}])$ 的充分必要条件是下列含未知元 $f_1, f_2, \cdots, f_m$ 的线性方程组有解且解唯一.
$$a_{11}f_1+a_{12}f_2+\cdots+a_{1m}f_m=a_1,$$
$$a_{21}f_1+a_{22}f_2+\cdots+a_{2m}f_m=a_2,$$
$$\cdots$$

$$a_{n1}f_1+a_{n2}f_2+\cdots+a_{nm}f_m=a_n.$$

在理论部分讲完之后，现在来讲量度分析在产品抽样试验问题中的应用．先讲几点方法论上的注意．

产品抽样试验理论是建立在对产品总体，抽样方法及其他等所加的一些前提上的统计理论，其中要用概率方法．虽然概率方法回答了抽样试验中的许多实际问题，但它有时仍有缺点：冗长，繁重，需要把现象概型化而得到的结果常不能吻合实际．并且在这些方法中经验的作用也不十分明显．

因为概率判断表达了研究对象的客观性质，故可以用来陈述抽样试验的现象理论．类似地，例如现象热力学和统计热力学描述同一类现象．统计理论描述热力现象较细致，而在许多实际问题中，现象理论所得的结果却很精确．可以断定，在各种统计检查问题中，常有一些问题相当粗糙，以致不必应用统计方案，何况后者常需要应用复杂的数学工具．

产品抽样试验理论的基本观念是总体．产品的总体是对象的集合 Ω，它根据假定满足下列条件：

若 Ω_1, Ω_2 是总体 Ω 的子集（部分），则可将其相加而得总体 Ω 的一新子集 $\Omega_1 \cup \Omega_2$．

对总体的一切部分定义如下的两测度 N 及 W：

(i) 其中每个都是量纲．

(ii) 对总体 Ω 的一切子集，测度 N 及 W 分别各有一定的量度 $[N]$ 及 $[W]$．

(iii) $N(\Omega_i)$ 及 $W(\Omega_i)$ 对总体的每一子集是量度无关的．

(iv) 测度 N 及 W 都有可加性；若 Ω_1 及 Ω_2 没有公共部分，则满足等式

$$N(\Omega_1 \cup \Omega_2)=N(\Omega_1)+N(\Omega_2),$$
$$W(\Omega_1 \cup \Omega_2)=W(\Omega_1)+W(\Omega_2).$$

$N(\Omega)$ 称为总体的体积而 $W(\Omega)$ 称为总体 Ω 的价值. 总体子集的两测度分别称为该子集的体积与价值. 虽然在实际中总体的体积不常是商品的个数,我们仍约定取体积的量度为"个",而总体或其部分的价值量度则取为"圆".

同样地对样本引进对应的观念. 产品总体 Ω 的子集 ω 称为样本,它有下列性质:

若 ω_1, ω_2, … 为样本 ω 的子集,则定义了它们的加法,其结果仍是样本 ω 的子集. 因为样本的部分的加法未必与总体的部分的加法重合,故我们将用另一记号"$\curlyvee$". 对样本的一切部分定义两测度 n 及 w,分别称为整个样本或其部分的体积与价值. 按假定 n 及 w 满足下列条件:

(v) 测度 n 及 w 为量纲.

(vi) 测度 n 及 w 对样本 ω 的一切子集有一定的量度 $[n]$ 及 $[w]$.

(vii) $N(\Omega_i)$,$W(\Omega_i)$,$n(\omega_j)$,$w(\omega_j)$ 为量度无关.

(viii) 对样本部分的加法"$\curlyvee$",测度 n 及 w 可加;若 ω_1 及 ω_2 没有公共部分,则满足等式

$$n(\omega_1 \curlyvee \omega_2) = n(\omega_1) + n(\omega_2),$$
$$w(\omega_1 \curlyvee \omega_2) = w(\omega_1) + w(\omega_2).$$

按照假定,样本及总体的体积是量度无关的,为了强调它们间的区别,我们约定称样本的体积是"个". 同样地,样本的价值称为"元".

上述的假定需要一些证明. 首先在条件(iv)及(viii)中,我们曾将量纲相加,而等式的右方为和 $N(\Omega_1) + N(\Omega_2)$,$W(\Omega_1) + W(\Omega_2)$,$n(\omega_1) + n(\omega_2)$,$w(\omega_1) + w(\omega_2)$. 这需要特别解释,因为在造空间 Π 时我们未提到量纲相加,且一般情况下量纲的和不属于空间 Π. 但因这里被加量有相同的量度,我们可设,若 α 及 β 为两实数,又 A 为任一量纲,则令

$$\alpha A + \beta A = (\alpha + \beta) A.$$

其次还要解释假定(vii),因为实际工作者可能不同意下列假定:总体与样本的体积或总体与样本的价值有不同的量度而且它们量度无关.但不难举例说明这些假定完全成立.例如运输煤时,总体的体积单位自然取为吨,而其价值单位则为圆.但因煤的检查是用某种方法抽样本,把这样本磨碎调匀,再进行第二次抽样,然后对新样本作化学分析或把它作煤球后燃烧以计算热量,故作为样本的体积单位自然可取立方厘米(因这里样本的体积是煤球的体积)而样本的价值单位取为卡.甚至总体与样本的体积(或价值)有相同的量度时(例如被检查的煤球体积用重量单位克表达时),我们以后也不利用这点而假定这些量度不同且满足条件(vii)(当"个"=吨及"个"=克时,我们也不利用等式 1 t=10^6 g).

根据上面引进的概念及公理,我们来叙述产品的抽样检查理论.

总体(或其部分)的价值与体积之比称为总体的价格或出售价格 C

$$\frac{W}{N} = C, \tag{6.18}$$

平行地引进样本的价格或实验价格 c,

$$\frac{w}{n} = c. \tag{6.19}$$

这些价格的量度显然是

$$[C] = 圆 \cdot 个^{-1}, \quad [c] = 元 \cdot 个^{-1}. \tag{6.20}$$

这里,总体与样本价格的观念比普通的价格更广泛.例如样本的价格可以是从 1 cm^3 的煤中取出的卡数,也可能是总体中上品的频数,只要我们认为总体的体积是其个数而价值是总体中的上品数.

我们再假定,如果知道了样本的实验价值,就有计算总体

出售价格的法则. 我们这里只考虑出售价格与实验价格线性相关的情况, 即
$$C = qc + C_0, \qquad (6.21)$$
其中 q 称为会计系数, C_0 不依赖于实验价格而可认为是不变的发货费. 还假定参数 q, C_0 已知, 且对已确定那批产品的统计质量检查问题是不变的.

为使等式 (6.21) 成立, 且使此式右方的加法合法, 必须要会计系数 q 有量度
$$[q] = 圆 \cdot 个^{-1} \cdot 元^{-1} \cdot 个. \qquad (6.22)$$
而 C_0 的量度应与总体出售价格的一样, 即
$$[C_0] = [c] = 圆 \cdot 个^{-1}.$$

今令 Ω_1 为总体 Ω 的某一确定部分, 又 $N_1 = N(\Omega_1)$, 则
$$\overline{W}_1 = W \frac{N_1}{N} = CN_1 \qquad (6.23)$$
称为总体部分 Ω_1 的平均(或期望)价值.

类似地定义样本部分 w_1 的平均价值 $\overline{w}_1$ 为
$$\overline{w}_1 = w \frac{n_1}{n} = cn_1, \qquad (6.24)$$
其中 $n_1 = n(\omega_1)$ 为样本部分的体积 $n_1 = n(\omega_1)$. 总体部分和样本部分的平均价值分别与总体和样本的量度相同, 即
$$[\overline{W}] = [W] = 圆, \quad [\overline{w}] = [w] = 元.$$

今设总体 Ω 由 v 个互不相交的子集 $\Omega_1, \Omega_2, \cdots, \Omega_v$ 构成, 各有相同的体积 $N(\Omega_i) = N_i (i = 1, 2, \cdots, v)$. 则这些部分有相同的平均价值 $\overline{W} = N_1 C$. 今令
$$D = \sqrt{\frac{N_1}{N} \sum_{t=1}^{v} (W_i - \overline{W}_1)^2}, \qquad (6.25)$$
其中 $W_i = W(\Omega_i)$, $i = 1, 2, \cdots, v_0$, 这里需补充 $(W_i - \overline{W}_1)^2$ 的定义, 因它在空间 Π 中无定义. 但令 $W_i = \alpha_i \cdot 圆$ 及 $\overline{W}_1 = \bar{\alpha} \cdot$

元. $\left(\bar{\alpha} = \dfrac{1}{v}\sum\limits_{i=1}^{v}\alpha_i\right)$ 则自然地取 $(W_i - \overline{W}_1)^2$ 的定义为

$$(W_i - \overline{W}_1)^2 = (\alpha_i - \bar{\alpha})^2 \cdot 圆^2.$$

因此，全部 $(W_i - \overline{W}_1)^2 (i=1, 2, \cdots, v)$ 有相同的量度圆^2，而 (6.25) 中根号下的和数合法，且除 $\Omega_i (i=1, 2, \cdots, v)$ 有相同的价值 $W_i = \overline{W}$ 的特殊情形外，D 有定义. 不过这特殊情形实际上很少碰到，对它运用统计检查没有实际意义.

总体价值的平均二次离差 S 定义为

$$S = \dfrac{D}{\sqrt{N}}, \tag{6.26}$$

它的量度为 $[S] = 圆 \cdot 个^{-\frac{1}{2}}$.

同样定义样本价值的平均二次离差 s 为

$$s = \dfrac{d}{\sqrt{n}}, \tag{6.27}$$

其中

$$d = \sqrt{\dfrac{n_1}{n}\sum_{i=1}^{\mu}(w_i - \overline{w}_1)^2}, \tag{6.28}$$

$w_i (i=1, 2, \cdots, \mu)$ 是样本的子集 $\omega_1, \omega_2, \cdots, \omega_\mu$ 的价值，这些子集互不相交，有相同的体积 $n_1 = n(\omega_i)$ 及平均价值 $\overline{w}_1 = cn_1$，且 $\omega = \omega_1 \curlyvee \omega_2 \curlyvee \cdots \curlyvee \omega_\mu$.

s 的量度为 $[s] = 元 \cdot 个^{-\frac{1}{2}}$.

当总体部分 $\Omega_i (i=1, 2, \cdots, v)$ 有体积 $N_1 = 1$ 个，且总体的价值由上品的个数所标志时，1 圆 = 1 好个，公式 (6.27) 的形式特别简单. 此时得

$$s = \sqrt{\Gamma(1-\Gamma)} \cdot \dfrac{1\ 好个}{\sqrt{N}},$$

这里 Γ 为非量纲，它是价值为 1 好个的总体子集的平均体积除以总体体积 N 的商，或即普通所谓的总体中上品的频率.

最后引进检查价值 B 的观念，假定此价值与总体价值有相同的量度，即 $[B]=[W]=$ 圆. 检查价值 B 与待检查的样本的体积 n 之比

$$k=\frac{B}{n} \qquad (6.29)$$

称为样本检查价格，它的量度为

$$[k]=圆\cdot 个^{-1}. \qquad (6.30)$$

作过产品抽样检查理论的主要参数与指标的概述后，现在来运用量度分析以解决统计检查主要问题的各种方案. 首先分这问题的各种解为两类，一类为统计型解，其中假定样本的体积必须依赖于总体价值的平均二次离差 S 或样本价值的平均二次离差 s. 另一类为非统计型解，其中假定样本的体积不依赖于 S 及 s.

先看一些统计型问题提法的例子.

例 1 假定样本体积只依赖于总体体积 N，样本检查价格 k，样本价值平均二次离差 s 及会计系数 q，

$$n=\Phi(N, k, s, q),$$

其中 Φ 为某量度函数. 试问函数 Φ 的形状如何？

在单元组

$$个，个，圆，元$$

中，N, k, s, q 的量度为

$[N]=$ 个

$[k]\quad\ 个^{-1}$ 圆

$[s]\quad\ 个^{-\frac{1}{2}}\ $ 元

$[q]=$ 个　个　圆　元$^{-1}$

易证指数行列式不为 0，

$$\begin{vmatrix} 1 & 0 & 0 & 0 \\ 0 & -1 & 1 & 0 \\ 0 & -\dfrac{1}{2} & 0 & 1 \\ -1 & 1 & 1 & -1 \end{vmatrix} = \frac{3}{2} \neq 0,$$

这等价于 N, k, s, q 的量度无关性. 按定理 9, 这些量的每一量度不变函数之形状为

$$\Phi(N, k, s, q) = a N^{a_1} k^{a_2} s^{a_3} q^{a_4}. \tag{6.31}$$

由于我们要找关系式 $n = \Phi(N, k, s, q)$, 需求出 (6.31) 的各指数, 它们可由线性方程

$$a_1 - a_4 = 0, \; -a_2 - \frac{1}{2} a_3 + a_4 = 1, \; a_2 + a_4 = 0, \; a_3 - a_4 = 0$$

求出. 后者有唯一解

$$a_1 = a_3 = a_4 = \frac{2}{3}, \; a_2 = -\frac{2}{3}.$$

故所求的关系式为

$$n = 2\left(\frac{Nqs}{k}\right)^{\frac{2}{3}}, \tag{6.32}$$

其中正的常系数不能用量度分析方法求出. 有趣的是, 当 $q = 1$ 圆个$^{-1}$元$^{-1}$个时, 公式 (6.32) 重合于 Steinhaus 用统计估值法所得的公式.

例 2 今设样本的体积除依赖于例 1 中的四个量 N, k, s, q 外, 还依赖于总体的价格 c, 即

$$n = \Phi(N, k, s, q, c). \tag{6.33}$$

因 N, k, s, q 量度独立且量度空间 Π 这时又只有四个单元, 故这空间的任一其他量纲, 例如 c, 可由 (6.31) 表达. 因 c 的量度 $[c] = $ 圆个$^{-1}$, 又 N, k, s, q 的量度已在上面写出, 故比较对应的指数后易得

$$c=\beta_1\left(\frac{kq^2s^2}{N}\right)^{\frac{1}{3}}. \tag{6.34}$$

今设函数(6.33)为量度不变且为齐次的，则可对它运用定理Ⅱ得函数之形状只可能是

$$\gamma=\varphi(\beta_1)\left(\frac{Nqs}{k}\right)^{\frac{2}{3}}, \tag{6.35}$$

其中 $\varphi(\beta_1)$ 是非量纲参数

$$\beta_1=cN^{\frac{1}{3}}k^{-\frac{1}{3}}q^{-\frac{2}{3}}s^{-\frac{2}{3}}$$

的任意函数. 对总体的一定的检查法，这参数称为产品总体相似判定，因当两个不同的总体有相同的参数 β_1 时，则(6.35)中的系数 $\varphi(\beta_1)$ 对两个总体一样. 量度分析理论没有给出任何便于找到函数 $\varphi(\beta_1)$ 的启示. 在这报告末尾还会谈到如何根据实验来研究 $\varphi(\beta_1)$，现在暂时结束例 2 的研究而假定 $\varphi(\beta_1)$ 可用它的麦克劳林展开式的前两项来代替，这对实际需要是的确精确的，于是令

$$\varphi(\beta_1)=\alpha_0+\alpha_1\beta.$$

这时一般公式(6.35)化为

$$n=\alpha_0\left(\frac{Nqs}{k}\right)^{\frac{2}{3}}+\alpha_1\frac{Nc}{k}, \tag{6.36}$$

若 $\alpha_1=0$，则这公式化为(6.32)，在一般情况下，此式第一项除一常系数外重合于(6.32)，(6.36)中第二项与总体价值 $w=Nc$ 成正比，与检查价格 k 成反比.

例 3 今设样本体积 n 依赖于总体体积 N，检查价格 k 及总体价值的平均二次离差 S.

在这问题中变量个数小于量度空间的单元个数，故事先不知道是否存在所求的函数

$$n=\Phi(N, k, s). \tag{6.37}$$

这问题由定理Ⅲ回答. 因变量 N, k, s 的量度为

$$[N]=个，$$

$[k] =$ 个$^{-1}$圆,

$[s] =$ 个$^{-\frac{1}{2}}$ 圆,

且易验证它们为量度独立的,又 n 的量度为

$[n] =$ 个,

故按定理 3 为使函数(6.37)存在,充分必要条件为下线性方程组有唯一解

$$\begin{cases} 1f_1 + 0f_2 + \frac{1}{2}f_3 = 0, \\ 0f_1 - 1f_2 + 0f_3 = 1, \\ 0f_1 + 1f_2 + 1f_3 = 0, \\ 0f_1 + 0f_2 + 0f_3 = 0. \end{cases} \quad (6.38)$$

但因最后一方程常满足,而前三方程含三未知数,且系数行列式不为 0,故(6.38)有唯一解

$$f_1 = \frac{1}{2}, \ f_2 = 1, \ f_3 = -1.$$

这就是最后所要求的公式 f 的系数而得

$$n = \delta \frac{sN^{\frac{1}{2}}}{k}, \quad (6.39)$$

其中 δ 是正的常系数.

公式(6.39)为实际工作所熟知,在抽样验收方案中常用. 但至今以前它只是根据经验而不是建立在统计理论的基础上.

例 4 试推广公式(6.39)而假定关系式中还含其他的量纲. 例如 n 除含例 3 中的 N, k, s 外还含会计系数 q,

$$n = \Phi(N, k, s, q).$$

这类似于例 1,只是用总体价值平均二次离差 s 代替了样本价值平均二次离差 s. 根据上面的结果容易预见这问题的结果. 因为 N, k, s, n 是空间 π 的某子空间 π_1 的元,π_1 也是量度空间,但它只有三个单元(例如:1 个,1 个,1 圆)而 q 不属于此子空

间，所以在公式
$$n = aN^{a_1}k^{a_2}s^{a_3}q^{a_4}$$
中，指数 a_4 必须等于 0，而其他的指数必须与(6.39)中对应的指数重合．容易验明，计算后的结果完全肯定这些预见．因而(6.39)是我们的问题的一般解．将此式与(6.32)比较，有趣的是，当令 $n \to +\infty$ 而固定其余的参数时，按(6.32)所求得的样本体积 n（与 $N^{\frac{2}{3}}$ 成正比）比按(6.39)所求得的 n（与 $N^{\frac{1}{2}}$ 成正比）增长得更快．这说明了实际工作者周知的事实：知道总体价值的离散度比知道样本的离散度更可贵．

例 5 今推广公式(6.39)而假定样本体积还可依赖于总体的价格 c：
$$n = \Phi(N, k, s, q, c),$$
这里样本体积 n 不依赖于会计系数，又因 c 可表为
$$c = \beta \frac{s}{N^{\frac{1}{2}}},$$
得最后结果为公式
$$n = \psi(\beta) \frac{sN^{\frac{1}{2}}}{k}, \tag{6.40}$$
这里 $\beta = cN^{\frac{1}{2}}s^{-1}$ 而非量纲，而 $\psi(\beta)$ 是任一数值函数．

例 6 再举一非统计型的问题为例，假定样本体积既不依赖于总体价值的平均二次离差，也不依赖于样本价值的平均二次离差．设样本体积只决定于总体体积 N，总体价格 c 及检查价格 k．因而问题在于决定函数
$$n = \Phi(N, c, k) \tag{6.41}$$
的可能形状．因这时 N, c, k 量度无关，按定理 1 函数(6.41)除一非量纲常系数外完全决定，且得公式
$$n = \alpha_1 \frac{Nc}{k}. \tag{6.42}$$

此式为实际工作者所知并在某些情况下得以运用.

在上述诸例中,我们只假定了样本个数 n 依赖于某些而不依赖于其他量纲. 这些假定由实际情况所提示. 但此外实际工作者还应决定抽样检查的目的性. 在解上述诸例时关于这点毫未提到. 甚至读者也会感到奇怪,因为没有明确的目的性的研究只是为了研究而研究. 知道抽样检查的目的,就等于在我们已得到的公式和实际间搭上了桥梁,因为它可决定未知系数或未知数值函数的形状.

在实际中常碰到的且也最合理的原则是,统计检查的目的是使某损失最小,在特殊的提法下,这一原则使 Steinhaus 得到了统计估值(见问题 1)的各公式,而在一般情况下则把问题化为博弈论中的一些问题. 这里我们提出如何叙述这一原则及如何把它运用到由量度分析所建立的抽样检查理论中去.

设在每次产品总体的抽样检查时,定义了某一量度函数
$$R=\Phi(N, n, C, c, q, k, S, s). \tag{6.43}$$
假定 R 的量度与总体价值的相同.
$$[R]=[W]=圆$$
R 称为试验的风险. 由(6.43)风险可依赖于上面所引进的全部抽样检查的参数(这里未明显指出 R 依赖于总体价值 W 及样本价值 w,因它们完全由价格及体积所决定: $W=CN$, $w=cn$).

函数(6.43)中有 8 个参数,但因这里的量度空间只有 4 个单元,故其中只可能有 4 个是量度无关的. 取 N, n, S, s 为独立变量,则其余可由这单元组表出,此时得到风险 R 的 4 个特征数值参数
$$\xi_1=CN^{\frac{1}{2}}S^{-1}, \qquad \xi_2=Cn^{\frac{1}{2}}S^{-1},$$
$$\xi_3=2N^{\frac{1}{2}}sn^{-\frac{1}{2}}S^{-1}, \quad \xi_4=knN^{-\frac{1}{2}}S^{-1}.$$
然后运用定理 2 得试验风险 R 为

$$R=\rho(\xi_1, \xi_2, \xi_3, \xi_4)N^{\frac{1}{2}}S, \qquad (6.44)$$

其中 ρ 是数值变量 $\xi_1, \xi_2, \xi_3, \xi_4$ 的数值函数.

函数 $\rho(\xi_1, \xi_2, \xi_3, \xi_4)$ 应由经济工作者决定，但我们再把它简化一下而设它是一线性函数，这在实际中已足够精确，即

$$\rho=\rho_0+\rho_1\xi_1+\rho_2\xi_2+\rho_3\xi_3+\rho_4\xi_4, \qquad (6.45)$$

其中 $\rho_0, \rho_1, \rho_2, \rho_3, \rho_4$ 为数值参数.

以 (6.45) 代入 (6.44) 并利用上面非量纲 $\xi_1, \xi_2, \xi_3, \xi_4$ 的表达式，得风险的简化式为

$$R=\rho_0 SN^{\frac{1}{2}}+\rho_1 CN+\rho_2 Cn^{\frac{1}{2}} s^{-1} N^{\frac{1}{2}} S+\rho_3 Nqsn^{-\frac{1}{2}}+\rho_4 kn. \qquad (6.46)$$

再来研究例 1，那里假定样本体积 n 只依赖于参数 N, s, k, q. 我们可假定试验风险也只依赖于这些量及样本体积 n. 由于这假定，在 (6.46) 中需设 $\rho_0=\rho_1=\rho_2=0$，因而得

$$R=\rho_3 Nqsn^{-\frac{1}{2}}+\rho_4 kn. \qquad (6.47)$$

但因在例 1 中得到了（见公式 (6.32)）

$$n=\alpha\left(\frac{Nqs}{k}\right)^{\frac{2}{3}}, \qquad (6.48)$$

以之代入 (6.47) 后得

$$R=rkn,$$

这表明在所设前提下，试验风险与检查价格成正比. 同样的结果也由 Steinhaus 在其统计估值论中得到. 那里若不顾及常数 n，则当样本个数由公式 (6.5)（见问题 1）决定时，极大极小化后的国家损失（见问题 1 公式 (6.4)）超过检查费用的两倍.

现在运用试验风险极小的原则以试图决定 (6.48) 中的系数 α，这时风险由 (6.47) 给出，其中 ρ_3, ρ_4 需由经济工作者定出. 但如没有特别的假定或试验观察的结果，这是不能实现的，因为不知样本价值平均二次离差 s 如何地依赖于样本体积 n. 如果

假定 s 不依赖于 $n\left(\dfrac{\mathrm{d}s}{\mathrm{d}n}=0\right)$, 那么由 (6.47) 得

$$\frac{\partial R}{\partial n}=-\frac{1}{2}\rho_3 Nqsn^{-\frac{3}{2}}+\rho_4 k,$$

解方程 $\dfrac{\partial R}{\partial n}=0$ 最后得知试验风险的极小由

$$n=\frac{\rho_3}{2\rho_4}\left(\frac{Nqs}{k}\right)^{\frac{2}{3}}$$

达到. 但若假定 $\dfrac{\mathrm{d}s}{\mathrm{d}n}=0$ 不合理, 则可用试验的方法来决定 (6.48) 中的系数 α. 试验的方法概括如下: 设从体积为 N 的总体中抽出一些体积不同的样本. 于 (6.47) 中认为 q, k, ρ_3, ρ_4 已知而可对各样本算出 R. 于是得到 R 对 n 的试验相关式. 当这样的试验次数相当大时, 由试验相关式可定出使 R 达到极小的样本体积 n_0, 于是最佳的系数 α 由条件

$$n_0=\alpha\left(\frac{Nqs}{k}\right)^{\frac{2}{3}}$$

决定.

不同的总体的参数——在 (6.48) 中出现的或不出现的——取不同的值, 对这些总体重复上述试验, 可以验证我们为得到 (6.47) 及 (6.48) 而作的假定是否正确.

这样, 我们用一个例子从头到尾说明了应用量度分析理论的大致步骤. 这里应再注意一点, 当用实验方法来估计系数 α 及分析观察结果时, 必须用到概率方法. 因此, 量度分析不完全把概率论从产品抽样检查理论中排除出去, 但只保留了它适当的作用, 这作用像它在其他实验科学中所起的一样.

参考文献

[1] Buckingham E. On physically similar systems. Physical Reviews, 1914, 4: 345.

[2] Drobot S. O analizie wymiarawej. Zastosowania Matematyki, 1954, 1: 233.

[3] Drobot S. On the foundation of dimensional analysis. Studia Mathematica, 1953, 14: 84.

[4] Drobot S. Warmus M., Analiza wymiarowa w badaniu wyrywkowym towarów. Zastosowanie Matematyki, 1954, 2: 1.

[5] Drobot S. Warmus M. Dimensional analysis in sampling inspection of merchandise. Rozprawy Matematyuna V, Warszawa, 1954.

[6] Steinhaus H. Wycena statystyuna jako metoda odbioru towarów produkeji masowej. Studia i Prace Statystyune, 1950, 2.

§7. 统计弹性理论中的一个奇论

近来常试图建立统计弹性理论. 至今在一切静力学计算中, 用的几乎都是决定性的模型, 由于预料中的负荷与材料的法定强度间的离差, 使在静力建筑中必须引进商定的安全系数. 因难于预计使用材料上的浪费, 故对可能的离差估计越精确则设计越经济. 然而就我所知, 虽曾试图建立一种统计理论, 以计算材料的抗力强度, 但至今仍未得到能解决实际问题的结果. 我这里想叙述一个奇论, 它在建立统计弹性理论的开头就碰到. 根据 Steinhaus 所发现的这一奇论, 我再叙述一个理论性问题并指出 Trybuła 所得到的局部解答.

某建筑材料, 例如钢梁的抗力强度是一随机变数. 这可如下了解: 说到钢梁时, 我们想到的一批具体的钢梁, 它们的先天条件是一样的(同样的原料, 在相同的技术条件下由同一工厂制造), 但由于各种不可能精确控制的差别, 在同类的钢梁中, 各别钢梁的抗力强度常彼此不一样.

设有两批钢梁(例如由于所用钢的种类不同, 或技术条件不同, 或横截面的形状不同), 分别以随机变数 x_1 及 x_2 表第一及第二批钢梁的抗力强度. 我们提出下列问题: 什么时候抗力强度为 x_1 的第一批钢梁比抗力强度为 x_2 的第二批好? 如果某一批中全部钢梁的抗力强度大于另一批中全部钢梁的抗力强度, 即有

$$P(x_1 > x_2) = 1 \text{ 或 } P(x_2 > x_1) = 1 \tag{7.1}$$

时, 那么这问题容易回答. 但若(7.1)中无一式成立时, 则实际工作人员必须认定, 哪一批比另一批好, 以便决定在这次建筑

中采用那些钢梁. 当他们提出这问题时, 人们常回答说第一批 x_1 比第二批 x_2 好(记以符号 $x_1 > x_2$). 如果随机地从第一、二批中分别抽出钢梁时, 第一批中的钢梁比第二批中的好的次数多于相反情况的次数, 即

$$x_1 > x_2 \equiv P(x_1 > x_2) > P(x_2 > x_1). \qquad (7.2)$$

关系 $x_1 > x_2$ 的这一定义初看起来是很自然的, 但它实际上却不完善, 因为 $x_1 > x_2$ 不满足推移律, 可以找到三个随机变量 x_1, x_2, x_3, 使 $x_1 > x_2$, $x_2 > x_3$ 及 $x_3 > x_1$. 正是这一事实我在标题中称之为奇论. 现在来造三个随机变量 x_1, x_2, x_3, 使有循环关系

$$x_1 > x_2, \quad x_2 > x_3, \quad x_3 > x_1. \qquad (7.3)$$

令随机变量 x_1 以概率 1 取值 2, $P(x_1 = 2) = 1$; 随机变量 x_2 取两值 1 及 4, 其概率分别 $P(x_2 = 1) = \alpha$, $P(x_2 = 4) = 1 - \alpha$; 最后令 x_3 取值 0 及 3, 概率分别为 $P(x_3 = 0) = \beta$, $P(x_3 = 3) = 1 - \beta$.

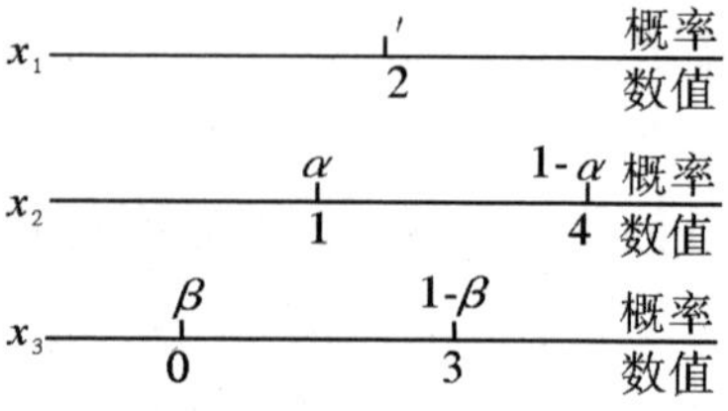

图 7-1

最好把随机变量 x_1, x_2, x_3 所可能取的值及对应的概率用图 7-1 表示, 以便于计算.

今设随机变量互为独立, 则

$$P(x_1 > x_2) = P(x_2 = 1) = \alpha,$$
$$P(x_2 > x_3) = P(x_2 = 4) + P(x_2 = 1)P(x_3 = 0) = 1 - \alpha + \alpha\beta, \qquad (7.4)$$
$$P(x_3 > x_1) = P(x_3 = 3) = 1 - \beta.$$

在 (7.4) 之各式中令 $\alpha = 1 - \beta = \frac{1}{2}(\sqrt{5} - 1)$, 得

$$P(x_1 > x_2) = P(x_2 > x_3) = P(x_3 > x_1) = \frac{1}{2}(\sqrt{5}-1) > \frac{1}{2}.$$

根据定义(7.2)，关系式(7.3)成立。这里有趣的问题是：对一切可能的独立随机变量 x_1，x_2，x_3 而言，(7.4)中全体概率的极大值是否就是 $\frac{1}{2}(\sqrt{5}-1)$？Steinhaus 把这问题用更一般的形式叙述为：对任一自然数 n，试求

$$a_n = \sup_x \min_p \{P(x_1 > x_2), P(x_2 > x_3), \cdots, P(x_n > x_1)\},$$
(7.5)

其中 $\min\limits_p$ 对括号中全体概率而取，$\sup\limits_x$ 对独立随机变量 x_1，x_2，$\cdots$，x_n 的一切可能的分布而取。

当 $n=2$ 时，显然 $a_2 = 0.5$。

当 $n=3$ 时，Trybuła 已证明，上面的例子给出 $a_3 = \frac{1}{2}(\sqrt{5}-1)$。

容易证明，如果像图 7-1 一样定义随机变量 x_1，x_2，x_3，其中 $\alpha = 1-\beta = \frac{1}{2}(\sqrt{5}-1)$，再设以概率 1 有 $x_4 = c_4$，$x_5 = c_5$，$\cdots$，$x_n = c_n$，且实数 c_4，c_5，$\cdots$，c_n 满足条件 $c > c_4 > c_5 > c_6 > \cdots > c_n > 2$，那么得

$$P(x_1 > x_2) = P(x_2 > x_3) = P(x_3 > x_4) = \frac{1}{2}(\sqrt{5}-1),$$
$$P(x_4 > x_5) = P(x_5 > x_6) = \cdots = P(x_n > x_1) = 1.$$

因而当 $n = 4, 5, \cdots$ 时，$a_n \geq a_3 = \frac{1}{2}(\sqrt{5}-1)$。

容易找到 4 个随机变量，使满足

$$P(x_1 > x_2) = P(x_2 > x_3) = P(x_3 > x_4) = P(x_3 > x_1)$$
$$= \frac{2}{3} > \frac{1}{2}(\sqrt{5}-1).$$

为此只需如图 7-1 定义 x_1，x_2，x_3，但此时令 $\alpha = \frac{2}{3}$，$\beta = \frac{1}{2}$，

并如下定义 x_4：$P(x_4=-1)=\dfrac{1}{3}$，$P\left(x_4=\dfrac{5}{2}\right)=\dfrac{2}{3}$. 由此得 $a_4\geqslant\dfrac{2}{3}>a_3$，再像上面证明 $a_n\geqslant a_3(n>3)$ 一样，可得当 $n>4$ 时，$a_n\geqslant\dfrac{2}{3}>a_3$. Steinhaus 认为 a_n 构成一上升数列，以 1 为极限，但这假定尚未证实.

上述的全部结果与思想是 Steinhaus 和 Trybuła 的，它们至今尚未发表.

§8. 根据个体间已知距离，将个体所成的集排序与分类的方法

将个体所成集合排序与分类是一切实验科学方法论中重要问题之一．科学的实验与观察向实验者或观察者提出如何将观察的结果分类的问题，因为任一科学的研究的目的不仅是要收集新的观察的结果，还要将它们分类，这样才能发现自然的法则，才能利用观察的结果来解决实际生活中的问题．许多实验员都是根据主观的分类或自信来解决这些问题．虽然这些来自长期的经验或正确的直觉，常常导致满意的解答，但这种主观方法总是依靠才能多于依靠科学．

设观察结果是一些个体的集合，其中每一个体由几个表示量的数值标志所决定．大多数定性的标志也可以用数值表示（使之数量化）．例如一群人人体测量的结果，是用个体的许多标志的测量数值表示的．这些标志中，可能有一些定性的标志，例如眼睛的颜色，或人的性别．但这些标志仍可数量化，如可以通用的色度来记眼睛颜色，用1记男性，用0记女性．

如果观察员只登记观察对象的一个标志，那么观察对象的排序问题没有任何困难，因只要把表这标志的数字当作直线上点的横坐标，用点来表示个体，就得到一个序．但一般地只按一个数值标志来排序，好处不大，因为它难以完全回答实验员在开始观察时所提出的全部问题（例如根据考古发掘出来的颅骨以决定进化的道路，或研究异种植物间的相似等）．

如有两个或三个标志，直观上可把个体表为平面或三维空间中的点，于是得到一个图形或一模型，但这时个体间已不再

有自然的序,因为二维及三维空间的点是无序的.

但在很多情况下,标志的个数大于三,则由于不可能直观地表示三维以上空间的点,故不可能用模型来表示个体间的相对位置. 此时为了要科学地分析观察结果,必须利用某些分类或排序法. 其中有些是由波兰所创立和应用的,我们现在就来研究这些.

设 N 个个体 A_1,A_2,$\cdots$,A_N 的观察结果是 n 维空间的 N 个点. 这些点也用 A_1,A_2,$\cdots$,A_N 来表示,每一点的坐标是个体的 n 个标志的观察数值:$A_1(c_{11}$,c_{12},$\cdots$,$c_{1n})$,$A_2(c_{21}$,c_{22},$\cdots$,$c_{2n})$,$\cdots$,$A_N(c_{N1}$,c_{N2},$\cdots$,$c_{Nn})$,我们所有的方法建立在个体间距离的观念上. 以 $\rho(A_i, A_k)$ 或简写为 ρ_{ik} 来记 A_i 与 A_k 间的距离. 这里我们不详细讨论如何定义距离 ρ_{ik}. 关于这点以后还要谈到,暂时只指出,并不一定要定义 ρ_{ik} 为类似于 n 维空间常用的距离

$$\rho_{ik} = \sqrt{(c_{i1}-c_{k1})^2+(c_{i2}-c_{k2})^2+\cdots+(c_{in}-c_{kn})^2} \quad (8.1)$$

因为在大多数情况下,采用易于计算的公式作为距离定义更为方便,例如

$$\rho_{ik} = \frac{1}{n}[|c_{i1}-c_{k1}|+|c_{i2}-c_{k2}|+\cdots+|c_{in}-c_{kn}|] \quad (8.2)$$

或

$$\rho_{ik} = \frac{1}{n}[\alpha_1|c_{i1}-c_{k1}|+\alpha_2|c_{i2}-c_{k2}|+\cdots+\alpha_n|c_{in}-c_{kn}|], \quad (8.3)$$

即将 A_i 与 A_k 对应的标志的值取绝对值加权 α_1,α_2,$\cdots$,α_n 后,再取算术平均值.

还可以有其他的距离定义,我们不再多说,只提醒距离应有下列性质:

(i) 距离 ρ_{ik} 是非负数,

$$\rho_{ik} \geqslant 0, \quad i,k=1,2,\cdots,N. \quad (8.4\text{a})$$

(ii) 任一个体 A_i 与自己的距离为 0,

$$\rho_{ii}=0, \quad i=1, 2, \cdots, N. \tag{8.4b}$$

(iii) 对每一对 i, k, ρ_{ik} 是对称函数,

$$\rho_{ik}=\rho_{ki}, \quad i, k=1, 2, \cdots, N. \tag{8.4c}$$

此外距离 ρ_{ik} 还必须有下述性质,若两个个体的距离 ρ_{ik} 甚近,则它们彼此相像,因为相隔甚远的个体不相像.

称每一排列 A_{i_1}, A_{i_2}, $\cdots$, A_{i_N} (i_1, i_2, $\cdots$, i_N 是正整数 1, 2, $\cdots$, N 的任一排列)为个体的线形序. 这序的长定义为序中相邻个体间距离的和

$$\delta(A_{i_1}, A_{i_2}, \cdots, A_{i_N})$$
$$=\rho(A_{i_1}, A_{i_2})+\rho(A_{i_2}, A_{i_3})+\cdots+\rho(A_{i_{N-1}}, A_{i_N}).$$

线形序可直观地解释为 n 维空间中表达这些个体的点间相连的折线形. 如果定义线段 $A_{i_k}A_{i_{k+1}}$ 的长为距离 $\rho_{i_k i_{k+1}}$,而折线形的长为一切折线长的和,那么序的长重合于折线形的长.

在许多实际问题中,可把问题提为:试求观察到的个体间的最短线性序. 例如,若要根据考古发掘以决定进化的概率道路,则可采用下列原则:最可能的进化道路是连接已找到的个体间的最短折线形.

我们不知道如何找最短线性序,就我们所知这问题尚未解决. 当然,可采用计算全部 $N!$ 个线性序的长的办法,因为这只需要多次重复简单的算术运算,所以只要造好程序后再利用电子计算机. 但若 N 不很大,则有经验的人只要顺次试验几次就可找到最短的或接近最短的线性序. 50 年前,著名的波兰人类学家柴肯诺夫斯基创造了专门的方法,来检查已给的线形序是否接近于最短的. 这方法称为差级诊断,常用于各种人类学及其他问题中. 运用这方法的原则如下:设有某线形序,我们看来它可能是接近于最短的,对它造所谓柴肯诺夫斯基表(以后

简称为柴氏表).这是一正方形表,登记了个体间的 N^2 个距离. 在 N 个行与 N 个列的标题中写上这线形序的个体,而在第 i 行与第 j 列的交点上记下此序中第 i 与第 j 号个体间的距离. 由条件(8.4b)与(8.4c)可知柴氏表上主对角线上的元为零,其余的元对主对角线对称. 现在按柴氏表作所谓柴肯诺夫斯基图(以后

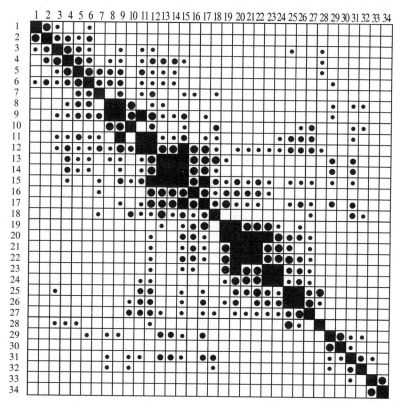

图 8-1 柴肯诺夫斯基图,取自:Perkal J. Analyse morphologique d'un groupe d'équidés, Zoologica Polomiae, 1957, 8. 这里■表长度不超过 7 个单位的距离,●表长度在 8~11 的距离,●表长在 12~15 的距离,●表长在 16~19 的距离. 长度超过 19 个单位的距离用图中白色方块表示.

简称为柴氏图),它不同于柴氏表的是,不用数值的距离而用黑白颜色的记号.用白色正方形来代替最大距离.其他越来越小的距离则用越来越黑的记号来代替,将最短或为零的距离换为黑正方形.如果图中黑色部分沿主对角线集中,那么这线形序就近似于最短的.图 8-1 是一柴氏图,它是毕加略特根据五个标志(颅骨的五个部分)的测量绘制图,用来将 34 匹马排序.

利用柴氏法虽然可得到许多动物学与人类学上的结论,但此法仍有很大缺点.由它不能造最短线性序而只能直观地指出某线性序的性质.此外,当检查一线性序是否接近最短的时候,也无客观判定法则.在批评柴氏法时,可以批评线形序的原则本身.但多数情况下,这法则是合理的,然而有些问题中,例如可能有分支的进化,统计人口迁移等,线形序原则就不能应用.这时应该考虑的,不是如何求最短线形序,而自然地应是求最短的枝形序.所谓个体 $A_1, A_2, \cdots, A_N$ 的枝形序是线段 $A_i A_j$ 的集 D,在 D 中对任一对 A_k, A_l,存在一列而且只有一列线段 $A_{i_1} A_{i_2}, A_{i_2} A_{i_3}, \cdots, A_{i_{s-1}} A_{i_s}$,使 $A_{i_1} = A_k, A_{i_s} = A_l$. 枝形序的长定义为 D 中一切线段长度的和.

寻求个体 $A_1, A_2, \cdots, A_N$ 所成集 z 的最短枝形序的问题完全为伏拉茨拉夫数学小组所解决.根据这方法所创立的生物分类法称为伏拉茨拉夫分类法,

现在指出如何来造某一枝形序 $D(z)$,并证明一定理: $D(z)$ 就是集 z 的最短枝形序.要最短枝形序是唯一的,还要假定集 z 中任两非零的距离不相等,即若 $i, j, k, l (=1, 2, \cdots, n)$ 中至少有三个不同的数,则 $\rho_{ij} \neq \rho_{kl}$.

下面是造 $D(z)$ 的方法.第一步用直线线段把每一个体 A_i 与其最近的个体连起来.这些线段称为第一类线段.它们将这些个体分为一个或更多的(无公共部分的)子集,每一子集与其

中连线共同构成一枝形条. 这些子集称为第一类子集. 第二步是用线段将每一第一类子集与其最邻近的子集连起来(所谓两子集的距离是两集间点的最短距离, 将两子集连起来是指用线段将两子集中最邻近的两点连起来). 联结两个第一类子集的线段称为第二类线段, 这些线段所连接的子集共同构成一个第二类子集. 如果只有一个第二类子集, 那么全体第二类与第二类线段构成一枝形序 $D(z)$, 如果存在两个或更多的第二类子集, 那么可继续造下去, 利用第三类线段来连接第二类子集等等, 直到只得到一个连接 z 中一切点的序为止. 用符号 $D(z)$ 来表这枝形序. 注意 $D(z)$ 中各类线段恰好共有 $N-1$ 条, 因为每一线段将子集个数减少一个(这里认为 z 中每一点是一零类集). 现在举一个具体的例来说明枝形序 $D(z)$ 的造法. 设要造连接七个点的枝形序(见图 8-2). 点间的距离设为平面上普通的(毕达哥拉斯)距离, 图 8-3 上画出了一切第一类线段. 它们将集 z 中的点分成三个第一类子集. 那里有些第一类线段用双线来画, 以便表示, 例如, 最近于 A_1 的是点 A_3, 最近于 A_3 的是 A_1. 在枝形序 $D(z)$ 中单线连接与双线连接并无区别. 图 8-4 中虚线表示第二类线段, 它们与第一类线段共同构成连接集 z 中一切点的枝形序 $D(z)$.

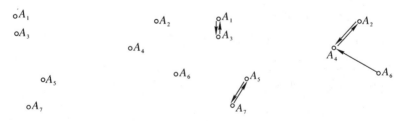

图 8-2 由 7 个个体组成的集 z 的例. 距离 ρ_{ik} 为平面上毕达哥拉斯距离.

图 8-3 造枝形序 $D(z)$ 的第一步. 第一类线段与子集.

现在证明

定理 1 枝形序 $D(z)$ 是个体集 z 的最短枝形序.

证 设 $F(z)$ 是集 z 的最短枝形序. 要证等式 $D(z)=F(z)$ 只要证明 $D(z) \subset F(z)$. 为此任取 $D(z)$ 中一线段 A_iA_j, 今证它也属于 $F(z)$. 设 A_iA_j 在构造 $D(z)$ 时

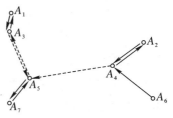

图 8-4 造枝形序 $D(z)$ 的第二步. 虚线表第二类线段.

是一 k 级线段, 它连接某一 $k-1$ 级子集 U 及最近于 U 的另一 $k-1$ 级子集 V. 如 A_iA_j 不属于枝形序 $F(z)$, 则在枝形序 $F(z)$ 中必存在一列线段

$$A_{i_1}A_{i_2}, A_{i_2}A_{i_3}, \cdots, A_{i_{s-1}}A_{i_s}$$

使 $A_{i_1}=A_i$, $A_{i_s}=A_j$. 在这列中存在线段 $A_{i_r}A_{i_{r+1}}$, 它的一端点属于 U 而另一端点不属于 U. 由枝形序 $D(z)$ 的构造得不等式

$$\rho(A_iA_j) < \rho(A_{i_r}A_{i_{r+1}}),$$

这是因为 A_iA_j 是连接 U 与其他子集的最短线段. 因之, 若从 $F(z)$ 中抛去线段 $A_{i_r}A_{i_{r+1}}$, 再添上线段 A_iA_j, 则得一比 $F(z)$ 更短的集 z 的枝形序, 这与 $F(z)$ 是集 z 的最短的枝形序矛盾. 从而得知线段 A_iA_j 必属于枝形序 $F(z)$. 既然 A_iA_j 是 $D(z)$ 中任一线段, 故枝形序 $D(z)$ 的一切线段皆属于 $F(z)$. 由于 $D(z)$ 是集 z 的一枝形序, 而按假设 $F(z)$ 是此集的最短枝形序, 故只有一个可能: $D(z)=F(z)$.

我们用图 8-2, 8-3, 8-4 中的具体例子来说明了最短枝形序的造法, 那里个体用平面上的点来表示, 而其间的距离是普通的毕达哥拉斯距离. 但即使对任意的空间及任一满足条件 (8.4a)(8.4b)(8.4c) 的距离, 最短枝形序的造法仍然十分简单. 为了方便起见可制一个距离表(柴肯诺夫斯基表), 要找距某个

体最近的个体,就只要找某行中最小的数(主对角线上的 0 不算,因它是个体到自身的距离).利用枝形序的拓扑性质,可将任一枝形序绘于平面上并可保持枝形序的全体线段的长度. 我们用一个具体的例子来说明这种图表的制法,这例子登在伏罗茨拉夫数学家宿特卡的文章中. 宿特卡运用伏罗茨拉夫分类法来区分小麦的品种,他根据四个法令那(一种机器的名称——译者)的标志来分类:发育,稠密,弹性与软性,它们刻画了小麦的烤度. 第一与第四个标志对烤度的作用是积极的,而第二与第三个标志的作用是消极的. 作者考虑了 10 种小麦,为使分类更完善,他还引进了两种理想小麦,其一为最佳品种 O,其标志的理想值为 $O(-3,3,3,-3)$,另一为最次品种 P,其标志的理想值为 $P(3,-3,-3,3)$. 真正的品种的标志值如下规范化:任一固定的标志值对 10 个真正品种的平均值为 0,标准离差为 1. 根据 12 个品种的资料宿特卡按公式(8.1)算出了各个距离. 他所得到的柴氏表附于后页,现在来造小麦品种的最短枝形序. 由距离表易见 $1\to 2$, $2\to 5$, $3\to 10$, $4\to 6$, $5\to 6$, $6\to 4$, $7\to 6$, $8\to 4$, $9\to 10$, $10\to 9$, $0\to 3$, $P\to 5$(读为:最近于品种 1 的是品种 2,最近于品种 2 的是品种 5 等). 由此可见第一类线段造成了两个第一类子集,即子集 1,2,4,5,6,7,8,P 及子集 3,9,10,0. 要找连接这些子集的第二类线段,就是要在距离表中,在对应于第一类子集中的品种所在的横行和对应于第二类子集中的品种所在的直行的交点上所标出的数中,找出最小的数. 易见这就是品种 4 与 10 间的距离 215. 现在将这 12 个 4 维空间的点的最短枝形序画在平面上,它只能保持直接相连的点间距离(见图 8-5). 在枝形序的平面图上一切线段的方向都是任意的(图 8-5 中采用的原则是:连接最佳与最次小麦折线铺成直线,而其他的枝线则从两侧与直线垂直). 宿特卡利用这一任意性,在一枝形序上，除保持直接联结的点间的距离外，还得保持了每一个点与最佳小麦间的距离（这一原则已完全决定了枝形序的线段间的夹角），而在另一图上则保持每一点与最次小麦间的距离。根据所得的枝形序宿特卡估计了各种小麦烤成面包的有效度。

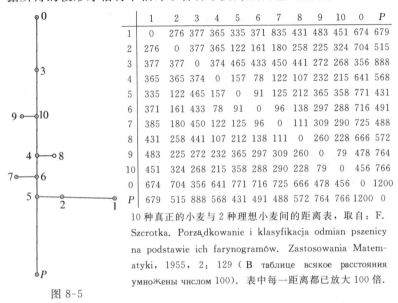

	1	2	3	4	5	6	7	8	9	10	0	P
1	0	276	377	365	335	371	835	431	483	451	674	679
2	276	0	377	365	122	161	180	258	225	324	704	515
3	377	377	0	374	465	433	450	441	272	268	356	888
4	365	365	374	0	157	78	122	107	232	215	641	568
5	335	122	465	157	0	91	125	212	365	358	771	431
6	371	161	433	78	91	0	96	138	297	288	716	491
7	385	180	450	122	125	96	0	111	309	290	725	488
8	431	258	441	107	212	138	111	0	260	228	666	572
9	483	225	272	232	365	297	309	260	0	79	478	764
10	451	324	268	215	358	288	290	228	79	0	456	766
0	674	704	356	641	771	716	725	666	478	456	0	1200
P	679	515	888	568	431	491	488	572	764	766	1200	0

10 种真正的小麦与 2 种理想小麦间的距离表，取自：F. Szcrotka. Porządkowanie i klasyfikacja odmian pszenicy na podstawie ich farynogramów. Zastosowania Matematyki，1955，2；129（В таблице всякое расстояния умножены числом 100）。表中每一距离都已放大 100 倍。

图 8-5

伏罗茨拉夫分类法还应用在许多其他的生物问题中。这里我只想指出运用此法所得的主要结果。考古发掘中所得人的头颅骨（其中有著名的北京人）的枝形序，证实了人类学上关于将这些发掘按时间与依据人类进化道路排序的假设，根据血群频率将人们排序得出了有趣的图案，专家以为它反映了迁徙的道路。特别有趣的是关于太平洋岛上居民的结论。在其他生物问题中，根据所谓对偶枝形序也得到了有趣的结果。我们在造 N 个个体 $A_1 A_2 \cdots A_N$ 的集的枝形序时，把个体看成 n 维空间的点 $A_1(c_{11}, c_{12}, \cdots, c_{1n})$，$A_2(c_{21}, c_{22}, \cdots, c_{2n})$，$\cdots$，$A_N(c_{N1}, c_{N2}, \cdots, c_{Nn})$，故可以根据这些观察的结果造 n 个标志 c_1，

c_2, $\cdots$, c_n 的对偶枝形序,并把这些标志看成为 N 维空间的点 $c_1(c_{11},\ c_{21},\ \cdots,\ c_{N1})$, $c_2(c_{12},\ c_{22},\ \cdots,\ c_{N2})$, $\cdots$, $c_n(c_{1n},\ c_{2n},\ \cdots,\ c_{Nn})$. 对偶枝形序根据被研究的个体的资料,指出各标志间的相似. 我们运用对偶枝形序,例如在生态学的问题中,把观察地点当作个体,植物的不同种类当作标志,因此,简单的枝形序根据出产的植物不同种类的频数,把观察地点排序,而对偶枝形序,则根据在各观察地点出现的各种植物的频数将各类植物排序.

现在来讲标题中和这篇报告开始时所说的第二个问题. 把个体集分类. 这就是说,要把个体的集分为一些子集,使同一个子集中的个体彼此接近,同时它们与其他不属于此子集的个体则相隔甚远. 在许多情形下,若个体所成的集没有这种子集,则可以称此集为齐次的. 但若它们存在,则实验者希望把它们挑选出来,因为非齐次个体集的齐次子集通常对应于个体的不同类型. 要挑出齐次子集,可以利用柴氏图. 在图 8-1 中易见沿主对角线的黑暗地区构成几个黑暗正方形,其中最清楚的有二: 一是标号为 12, 13, 14, 15, 16, 17 的横行与直行的变区,另一是标号为 19, 20, 21, 22, 23, 24 的横行与直行的交区. 这例子表明柴氏表可以指出齐次子集,实践中常用此法来解决问题. 但用伏罗茨拉夫分类法也可达到目的. 还在造最短枝形序 $D(z)$ 时,我们就得到各类子集,它们也指出了个体的一些子集,根据它们可将个体分类. 但这里更方便更符合于实验员的直觉的是下列运用分类长度概念的办法. 设集 z 分成 m 个不相交的子集 $z_k(k=1,\ 2,\ \cdots,\ m)$,

$$z = z_1 + z_2 + \cdots + z_m. \tag{8.5}$$

这分类的长 $\delta(z_1 z_2 \cdots z_m)$ 定义为连接子集中的点的最短枝形序 $D(z_1)$, $D(z_2)$, $\cdots$, $D(z_m)$ 的长的和. 分类 (8.5) 称为将集 z 分

为 m 部分的最佳分类,如果这分类的长达到极小.

要连接各子集 z_k ($k=1, 2, \cdots, m$) 中的点,我们一共需要 $n-m$ 个线段. 反之,每 $n-m$ 个线段,只要它们不构成任一闭折线形,就决定将集 z 分为 m 份的分类. 例如最短枝形序 $D(z)$ 的任意 $n-m$ 根线段决定这样的一个分类. 我们现在证明,把集 z 分为 m 份的最佳分类可由抛去最短枝形序 $D(z)$ 中最长的 $m-1$ 根线段得到. 这一结论是下预备定理的推论.

预备定理 若分类 (8.5) 是将集 z 分为 m 份的最佳分类,则最短枝形序 $D(z_k)$ ($k=1, 2, \cdots, m$) 中的一切线段均属于连接集 z 中的点的最短枝形序 $D(z)$,$\sum_{k=1}^{m} D(z_k) \subset D(z)$.

证 设 $a = A_i A_j \in \sum_{k=1}^{m} D(z_k)$ 为任一线段,它属于某一最短枝形序 $D(z_k)$ ($k=1, 2, \cdots, m$). 今证 $a \in D(z)$. 设 $a = A_i A_j$ 属于枝形序 $D(z_l)$. 则在枝形序 $D(z_l)$ 中,抛去此线段后,子集 z_l 分为两部分 U 及 V. 若设 $a \bar\in D(z)$,则在 $D(z)$ 中可找到一列线段 $A_{i_1} A_{i_2}, A_{i_2} A_{i_3}, \cdots, A_{i_{s-1}} A_{i_s}$ 使 $A_{i_1} = A_i$,$A_{i_s} = A_j$. 在这一列中有线段 $b = A_{i_r} A_{i_{r+1}}$,它的一端点属于 U 而另一端点不属于 U,故 $b \bar\in \sum_{k=1}^{m} D(z_k)$. 按假设 $\rho_{ij} \neq \rho_{kl}$ 故应满足不等式 $b < a$,或不等式 $b > a$. 但第一情形不可能,因若 $b < a$,则从 $\sum_{k=1}^{m} D(z_k)$ 中抛去线段 a 而添上 b,我们就得到 $n-m$ 个线段 $\sum_{k=1}^{m} D(z_k) - a + b$,它决定集 z 分成 m 份的一新分类,其长度短于分类 (8.5) 的长,但按定义分类 (8.5) 是集 z 分成 m 份的最短分类,故这是不可能的. 第二种情形也不可能,因如在 $D(z)$ 中抛去线段 b 再添上 a,我们就得到集 z 的一枝形序 $D(z) - b + a$,它短于枝形序 $D(z)$,而按定义,$D(z)$ 是最短的. 因此由假设

$a \in D(z)$ 我们得到了矛盾. 既然 a 是 $\sum_{k=1}^{m} D(z_k)$ 中任一线段, 预备定理得证.

由此预备定理直接推得下列定理:

定理 2 为了要得到集 z 分为 m 份的最佳分类, 只要从最短枝形序 $D(z)$ 中抛去 $m-1$ 根最长的线段. 其余的线段构成最佳分类(8.5)的各子集的最短线形序 $D(z_k)$.

例如, 运用这分类法于图 8-5 中的小麦得:

若把图 8-5 中 12 种小麦分为 2 类, 则第一类只含最次小麦 P, 其他的小麦构成另一类.

若分这些小麦为 3 类, 则第一类含点 P, 另一类为点 O(最佳小麦), 其余真正的小麦成为第三类.

但若分为 6 份, 则得(1)P, (2)O, (3)1, (4)3, (5)9, 10, (6)2, 4, 5, 6, 7, 8.

最后, 再指出个体集 z 的最短枝形序的一些性质. 这些性质使伏罗茨拉夫分类法的运用受到限制, 但需要好好理解它以便正确的解释枝形序与分类.

枝形序当然依赖于个体间距离的定义, 这定义几乎总是协商性的. 在定义距离时应力求利用实践者在这方面的直觉.

在创立枝形序与分类的理论时, 我们曾假定个体间的距离都不相同($\rho_{ij} \neq \rho_{kl}$). 这一前提是最短枝形序的唯一性所要求的. 实践中在造枝形序 $D(z)$ 时, 常常碰到这种情况, 与某点如 A_1 最近的, 不止一个而有两个或更多的点如 A_i 及 A_j. 从理论上看来, 在构造枝形序以前, 不论取线段 $A_1 A_i$ 或 $A_1 A_j$, 都是一样, 但在实践中, 我们力求在一张图上照顾两种可能而画两根线段. 这种枝形序从拓扑的观点看来已不是枝形序, 可能碰到这种情况, 在平面图上不可能保留一切线段的长.

枝形序及分类不是点间距离的连续函数. 因为二者都由长度最短的线段所构成, 故少许改变线段之长就可能得到完全不同的序与类.

枝形序只能反映其上直接连接的点间的距离. 在枝形序上相近的点在标志空间是相近的(为此, 对非直接相连的个体要假定距离满足三角形性质 $\rho_{ik} \leqslant \rho_{ij} + \rho_{jk}$). 关于枝形序上相距甚远的两个体间的真实距离, 只能够说, 它的长大于枝形序上间接连接此两点的诸线段中的任一线段. 因此, 在标志空间相近的两点在枝形序上可能相距很远. 因之由枝形序上所看出的性质, 必须用距离表来验证, 因为枝形序只是 n 维空间的点在平面点上的简略表现.

参考文献

[1] Feorek K, Łukaszewicz J, Perkal J, Steinhaus H, Zubrzycki S. Sur la eiaison et la divisòn des points d'un ensamblc fini. Colloquium Mathematicum, 1951, 2.

[2] Florek K, Łukaszewicz J, Perkal J, Steinhaus H, Zubrzycki S. Taksouomia wrocŁawska. Przeglad Autropologiozny, 1951, 17.

[3] Perkal J. Analyse morpnologique d'un groupe d'equidés. Zoologica Poloniae, 1957, 8.

[4] Szczatka F. Porządkowanie i klasyfikacja Odmian pszenicy na podstawie ich farynogramów. Zastosowania Matematyki, 1955, 2: 123.

[5] Kelus A, Łukceszewicz J. Porządkowanie popnlacji ludzkich wedŁug częstości grup krwi. Przeglad Antropologiczng, 1954, 20.

[6] Kelus A, Łukaszewicz J. Taksonomia wroct awska, w Zastosowania do Ź agadnieú seroantropologii. Archiwum Immunologii i Terapii Doswiadczalue, 1953, 1.

§9. 平面上点的随机分布与类别存在的统计判定法

在数学的应用中，我们常常遇到这样的问题：如何判断平面上的已给点集是均匀随机分布的结果，还是在它们之间，有相互吸引、排斥或某种特别的分配的趋势？向我们提出这类问题的有天文学家、考古学家和地理学家等．天文学家要求判断，天球上某些星，特别是亮度甚弱的星，是否有集成链带形的趋势？考古学家问，如何根据已发掘古物的出土位置，来发掘地下文物？地理学家肯定，把某区域内的城市绘成地图上的点，则这些点不是随机分布的，相邻的城市有保持一定距离的相互排斥的倾向．这类例子还可举出很多，但我想这已足够说明这些问题常是实际工作者所关心的．在后文中，我想叙述解决这类问题的方法．

先提出一理论问题，所谓某点集在平面上均匀地随机分布是什么意思？为简单计，考虑全平面及无穷点集 z. 定义一区域函数 $N(D)$，对任一区域 D，$n = N(D)$ 表集 z 中属于 D 的点数．在平面上任取一直角坐标系，对每一矢量 (a, b)，每一区域 D，符号 $D(a, b)$ 表将区域 D 中的点按矢量 (a, b) 移动后所得的新区域．

我们说集 z 中的点在平面上随机地分布，如果存在正数 λ，使对任一有限数 n，任一组不相交区域 $D_1, D_2, \cdots, D_n$，及任一组非负整数 $K_1, K_2, \cdots, K_n$，满足条件

$$\lim_{a \to +\infty} \frac{1}{4a^2} \int_{-a}^{a} \int_{-a}^{a} \chi(a, b; z; K_1, K_2, \cdots, K_n; D_1, D_2, \cdots, D_n) \, da \, db$$

$$= \prod_{i=1}^{n} l^{-\lambda|D_i|} \frac{(\lambda \mid D_i \mid)^{K_i}}{K_i!}, \qquad (9.1)$$

这里 χ 在满足条件

$$N[D_1(a, b)] = K_1, \quad N[D_2(a, b)] = K_2, \cdots,$$
$$N[D_n(a, b)] = K_n$$

的点 (a, b) 上取值 $\chi = 1$, 在其他点 (a, b) 上 $\chi = 0$. 又 $|D_i|$ 表区域 D_i 的平面测度(面积).

由定义(9.1)可见，在平面上随机分布的点集是二维有效泊松流.

虽然集 z 在任一有限区域 D 内的点的任一排序, 不影响条件(9.1)的满足, 我们仍要直观地根据条件(9.1)的思想, 创立一判定法, 以判定在有限区域 D 内, z 的有限个点集是否随机分布. 例如, 事先把区域 D 分为 n 个等面积的子域 D_1, D_2, $\cdots$, D_n(所谓事先分割是指这分割不依赖于 z 中的点在 D 中的位置). 如果数 $N(D_1)$, $N(D_2)$, $\cdots$, $N(D_n)$ 显著地与分布为泊松的随机变数的 n 个独立实现(或说 n 个期待数)不同, 这里的泊松分布的参数是 $\frac{1}{n} N(D)$, 而 $N(D)$ 是落于 D 中的 z 集内的点数; 那么抛弃"z 中的点是随机分布"的假定. 观察值与期待值间的离散度可用 χ^2 判别法来估计.

例如, Zubrzycki 运用这方法来研究星际是否构成链带形的问题. 为此他在天球上一球面正方形域(中心点的坐标 $\alpha = 23^h 23^m$, $\delta = +60°$, 边长为 10)内找出 380 个星(星等高于 14^m). 天文学家认为, 这区域内的星显然有构成链带形的趋势. Zubrzycki 将此区域分为 400 正方形, 然后将两邻近的正方形连接而得 200 长方形, 最后又将两长方形连接而得 100 正方形. 对每一分法他都计算了这一区域内星的观察个数与期待个数(设分布为泊松的). 实验结果登记在表 9-1 中. 对于每一分法, 在

直行中用 f_K 表观察到的恰含 K 个星的子域数,用 f'_k 表根据参数是 $\overline{K}$ 的泊松分布所算的这种区域的个数,计算值 $\chi^2=A$ 及概率 $P_s(\chi^2>A)$(这里 s 是自由度数,它等于比较项的项数减 2)后,结果表示,不能抛弃原始假定:在这区域内是随机分布的.

表 9-1 将中心为 $\alpha=23^h23^m$,$\delta=+60°$ 边长为 $1°$ 的天空球面正方形,等分为 400,200 与 100 子域时,对高于 14^m 的星进行观察子域个数(f_k)与期望个数(f'_k)的比较表.

子域个数	400				200				100			
$\overline{k}=$在一子域内星的平均数	0.95				1.90				3.80			
k	f_k	f'_k	$f_k-f'_k$	$\dfrac{(f_k-f'_k)^2}{f'_k}$	f_k	f'_k	$f_k-f'_k$	$\dfrac{(f_k-f'_k)^2}{f'_k}$	f_k	f'_k	$f_k-f'_k$	$\dfrac{(f_k-f'_k)^2}{f'_k}$
0	164	155.3	8.7	0.487	30	29.6	0.4	0.005	11	10.6	0.4	0.002
1	129	148.3	−19.3	2.510	60	56.5	3.5	0.217				
2	77	70.8	6.2	0.543	48	53.9	−5.9	0.646	16	16.0	0.0	0.000
3					32	34.4	−2.4	0.167	17	20.4	−3.4	0.566
4									26	19.5	6.5	2.163
5	30	25.6	4.4	0.757					15	14.9	0.1	0.001
6					30	25.6	4.4	0.756				
7									15	18.6	−3.6	0.698
8												
总共	400	400.0	0.0	4.297	200	200.0	0.0	1.791	100	100.0	0.0	3.430
	$P_2(\chi^2>4.3)=0.12$				$P_3(\chi^2>1.8)=0.62$				$P_4(\chi^2>3.4)=0.50$			

注:见 Zubrzycki. O łańcuszkach gwiezdnych. Zastosowania Matematyki,1954,1:200.

不用在个别子域内观察数与期待数的比较法,Steinhaus 根据周知的事实:对泊松分布 $Ex=E(x-Ex)^2$,建议用另一方法. 他要求计算所谓凝结系数

$$L(n) = \frac{n}{n-1} \frac{\sum_{i=1}^{n}(K_i - \overline{K})^2}{\sum_{i=1}^{n} K_i},$$

其中 n 表子域数，K_i 表在第 i 个子域的点数，$\overline{K}$ 为在一个子域内平均点数.

当在个别子域内星的个数的分布重合于泊松分布时，系数 $L(n)$ 取值 1. 不等式 $L(n) > 1$ 表示在一子域内星数的方差大于泊松分布的方差. 这表示与泊松分布比较时，在观察到的分布中，星数接近于平均值 $\overline{K}$ 的子域出现得较少，而较多地出现的是星凝结得太密或星数不足的子域. 另一情况为 $L(n) < 1$，这时与泊松分布比较时，在一子域内星的个数更为稳定. 第一种后效 Steinhaus 称为星的吸引（若应用此法于其他个体，且以平面上点表这些个体时，则称为点的吸引），第二种后效称为星的（或点的）排斥.

如果后效是吸引的，$L(n)$ 作为 n 的函数，起初随 n（子域个数）上升而上升，在某一点 n_0 到达极大，然后下降. 数 n_0 决定类别的平均个数与点间吸引的距离. 因为通常如有类别，所以点的吸引只在类别中起作用，而类与类间则相互排斥. 对于这种研究，子域的形状应取得不要太长，当 n 上升时，它的直径应按 $n^{-\frac{1}{2}}$ 的比例下降. 为此最好可用一列边长下降的六角形或正方形网.

但上述方法不能最好地符合要求，因为它没有估计到天文学家所提出的是一种特殊的类别，即星的链带. 因而需要特别的方法来判别是否有构成链带的趋势. Zubrzycki 根据伏洛茨瓦夫分类法创造了这种方法. 他在天球的观察域内画出连接全体星的枝形序. 这里用的距离是星际角距，忽略球形的影响后角距就重合于平面上的毕达哥拉斯距离，这影响可以略去是因它

对边为 $1°$ 的球面正方形而言很小. 然后 Zubrzycki 在枝形序 $D(z)$ 中引进了每个星的等级的观念, 一个星的等级定义为在 $D(z)$ 中与该星直接相连的星的个数. 如果星有构成链带的趋势, 那么在 $D(z)$ 中两等星的个数应多于比随机分布的星所成的枝形序中的两等星数. 由于不会从理论上计算随机分布的点的枝形序中点的等级的分布, Zubrzycki 比较了星的等级的经验分布与随机分布点的等级的经验分布. 为此, 他在正方形中造了 500 个随机点的分布图, 每个点的三个坐标数取自随机数表. 对不同的天球区域重复这一实验, Zubrzycki 总是得到正面的肯定: 星的等级分布与随机点所成枝形序中点的等级分布是一致的. Zubrzycki 还改变点的等级的定义, 认为星的链带会更易地影响此量的分布, 如果在它的定义中, 只考虑第一类或第二类线段 (见 §8 中枝形序 $D(z)$ 的构造). 但各次试验都没有发现星的图与随机点的图间有本质差异. 由此只能肯定星的链带只存在于某些天文学家的幻想中.

这里我只用了星的链带的例子, 说明如何利用这些方法来发现平面上点的类别, 但我们还成功地运用这些方法到其他问题中去, 例如本文首段所提出的一些问题.

最后应指出我们虽只对平面上的点提出问题, 但这些方法同样可用来研究 n 维空间的点集, 为此只要在形式上稍微改变定义.

参考文献

[1] Zubrzycki S. O łańcuszkach gwiezdnych. Zastosowania Matematyki, 1954, 1: 197.

[2] von Pahlen E. Lehrbuch der Stellarstatistik. Leipzig, 1937.

[3] Steinhaus H. O wskazni'ku zgęszczenia rozproszonia. Prz̨eglad Geograficzny, 1947, 21.

[4] Kucharczyk i S, Szczepanowski H. Optymalne poletho poszukiwań archeolsgicrnych. Zastosowania Matematyki. 1957, 3.

[5] Floren K, Łukaszewica J, Perkar J, Steinhaus H, Zubrzycki S. Sur la liaison et la division des points d'un eusamble fini. Colloquium Mathematicum, 1951, 2.

§10. 电话总局最佳局址的选定

决定电话总局局址的问题化为要寻找点 Q,使对平面上已给的点 p_1,p_2,$\cdots$,p_n 及已给的一组正数 c_1,c_2,$\cdots$,c_n,满足条件

$$\Phi(Q) = \sum_{i=1}^{n} c_i r_i = 极小值(\text{Minimum}), \quad (10.1)$$

其中 $r_i = \rho(Q, P_i)$ 是点 Q 与 P_i 间的距离.

若 Q 是电话总局局址,p_1,p_2,$\cdots$,p_n 是电话分局局址,数 c_1,c_2,$\cdots$,c_n 正比于把各分局与总局连起来的单位长电线的价值,则条件(10.1)表示:总局的最佳局址保证连接分局与总局的电线网的价值最小. 函数 $\Phi(Q)$ 连续,非负,且当点趋于无穷时 $\Phi(Q)$ 也趋于无穷. 于是满足条件(10.1)的点 Q 总存在. 这个点可用下面的静力类比来决定.

在城市地图上标出点 p_1,p_2,$\cdots$,p_n,将它贴于水平桌面上(见图10-1),在点 p_i 上钻一小孔,每孔用线穿过,每线在桌

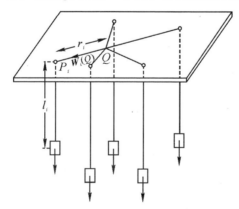

图 10-1 用静力类比法决定电话总局最佳局址的模型

面下的一端系有小砝码,其重与 c_i 成比例,桌面上的一端共同系成一结,自由释放后,结就指出满足条件(10.1)的点 Q 的位置.

为证此注意当各砝码的重心到达最低位置时,上述系统处于平衡状态. 设各线长(自结到砝码重心之长)均为 l. 则在桌面下之线长为 $l_i = l - r_i$,而整个砝码系统重心的高度为(高度自桌面起,沿垂直轴计算,轴的方向朝上)

$$L = -\frac{\sum_{i=1}^{n} c_i l_i}{\sum_{i=1}^{n} c_i} = -\frac{\sum_{i=1}^{n} c_i(l - r_i)}{\sum_{i=1}^{n} c_i} = \frac{\sum_{i=1}^{n} c_i r_i}{\sum_{i=1}^{n} c_i} - l = \frac{\Phi(Q)}{\sum_{i=1}^{n} c_i} - l,$$

当结位于满足条件(10.1)的 Q 点时,它取极小值.

运用静力类比法时,结可能位在某一小孔上或穿入此孔. 这表示最佳局址就是此孔所代表的某分局局址.

满足条件(10.1)的点 Q 称为"铜心",上述静力类比法使我们得到铜心的性质如下:

若铜心 Q 不重合于任一点 P_i(分局局址),则 n 个力

$$W_i(Q) = \frac{c_i}{r_i} \overrightarrow{QP_i} \quad (i = 1, 2, \cdots, n)$$

所成的系统处于平衡,其中位于点 P_i 上的力 $W_i(Q)$ 的方向为 $\overrightarrow{QP_i}$,长为 c_i. 但如铜心重合于某一 P_i,例如 P_n,考虑 $n-1$ 个力

$$W_i(Q) = \frac{c_i}{r_i} \overrightarrow{QP_i} = \frac{c_i}{r_i} \overrightarrow{P_n P_i}, i = 1, 2, \cdots, n-1$$

的和的反作用力,可见其长不超过 c_n.

由此性质可见,沿半直线 $QP_i + \infty$ 任意改变分局局址时,铜心不变. 这说明某些技术人员提出的(见 Лянгер 的书)方法,即把局址放在系统的重心上的方法不能用(这系统由带质量 c_i 的点 P_i 构成,其重心使函数 $\Psi(Q) = \sum_{i=1}^{n} c_i r_i^2$ 达到极小),因为沿诸

半直线 $QP_i+\infty$ 移动 P_i 时，可以任意改变重心，而铜心则不变．

在上述静力类比中，矢量

$$W(Q) = \sum_{i=1}^{n} W_i(Q) = \sum_{i=1}^{n} \frac{c_i}{r_i} \overrightarrow{QP_i}$$

是作用于 Q 点处的结的力，而函数 $\Phi(Q) = \sum_{i=1}^{n} c_i r_i$（我们正要找它的极小值）除差一符号外，与上述力场的势重合．在平面上任选一坐标系，得

$$\Phi(Q) = \Phi(x,y) = \sum_{i=1}^{n} c_i \sqrt{(x_i-x)^2 + (y_i-y)^2},$$

(10.2)

$$W(Q) = W(x,y) = \left[\sum_{i=1}^{n} \frac{c_i}{r_i}(x_i-x), \sum_{i=1}^{n} \frac{c_i}{r_i}(y_i-y)\right]$$

$$= \left[-\frac{\partial \Phi}{\partial x} - \frac{\partial \Phi}{\partial y}\right], \quad (10.3)$$

其中 x，y 是点 Q 的坐标，x_i，$y_i(i=1, 2, \cdots, n)$ 是点 P_i 的坐标．当 $Q=P_i(i=1, 2, \cdots, n)$ 时，$r_i=0$，$W(Q)$ 不存在，因为在这些点上偏导数 $\frac{\partial \Phi}{\partial x}$，$\frac{\partial \Phi}{\partial y}$ 不存在．

(10.2)式右边和中的每一项是一圆锥，其顶点朝下，函数 $z=\Phi(x, y)$ 是这些圆锥的和，本身是一凸向下的曲面，由此可得实用上甚重要的两推论．

(10.1) 若点 P_i 不位在同一直线上，则存在一个且只一个铜心（这表示函数 $\Phi(x, y)$ 有一个且只有一个极小值）．若一切 P_i 均位于同一直线上，则存在一个铜心或这些铜心布满某一闭区间．例如当只有两个分局 P_1，P_2 且 $c_1=c_2$ 时，闭区间 P_1P_2 中每一点都是铜心．关于这点不多讲，因它无实际意义．

(10.2) 对每一常数 c，等势线 $\Phi(x, y)=c$ 是凸的．

由(10.2)可得决定铜心的方法．若通过任一点 $Q \neq P_i$ 作一

直线 $l(Q)$ 垂直于矢量 $W(Q)$，则铜心 Q 位于由直线 $l(Q)$ 及矢量 $W(Q)$ 所决定的半平面内．这是因为垂直于 $W(Q)$ 的直线 $l(Q)$ 是通过 Q 点的等势线的切线．但因等势线是凸的，其内部及函数 $\Phi(x, y)$ 的极小值（应读为：及使函数 $\Phi(x, y)$ 达到极小值的点——译者）位于切线的同一侧，这侧由矢量 $W(Q)$ 指出．如 $Q=P_K$，这性质仍正确，只需在计算矢量 $W(P_K)$ 时，抛去无意义的被加项 $W_K(P_K)$，但这时需记住，当 $|W(P_K)| \leqslant c_K$ 时，铜心重合于点 P_K．

故求出矢量 $W(Q)$ 并对不同的 Q 画出直线 $l(Q)$ 后，我们可以一步步地缩小铜心所在的区域．图 10-2 上表达了这些步骤．在该图上标出了 38 个分局的位置．在表示分局的点旁写上对应于该分局的数 c_i．矢量 $W(Q)$ 的坐标按（10.3）计算，利用图 10-2 上的倒数尺，距离的倒数 $\dfrac{1}{r_i}$ 直接自图中取得．在点 $Q_1(0, 0)$，$Q_2(-5, 10)$，$Q_3(3, 7)$ 上找出矢量 $W(Q)$ 后，我们得到三直找 l_1, l_2, l_3，它们构成 $\triangle ABC$，其中包含铜心．考虑 $\triangle ABC$ 的重心 $Q_4(-4, 4)$ 后，得到四边形 $ADEC$，再考虑此四边形的重心 $Q_5(-1, 6)$，得到更小的四边形 $CFGE$．这里作为下一点 Q_{i+1}，我们总是取凸形的重心，在这凸形中，按以前的估计结果一定包含铜心，这种取法可保证下一凸形域的面积不大于上凸形域面积的 $\dfrac{5}{9}$（因凸形图有下性质：每一通过凸形重心的直线将它几乎分为两等分，并且两块面积之比不小于 $4:5$）．在图 10-2 的例中经过五步以后，得知铜心位于四边形 $CFGE$ 内，现在利用向下凸的曲面总是位于切面之上这一事实，我们可以估计 $\Phi_{\min}$，例如在点 $Q_5(-1, 6)$ 上有

$$|W(Q_5)|=4.9, \quad \Phi(Q_5)=4\,940,$$

但因曲面 $z=\Phi(x, y)$ 在点 Q_5 上的切面沿矢量 $W(Q_5)$ 的方向有

最大的倾斜度，而四边形 $CFGE$ 在这方向的宽度是 $s=6.5$，故
$$\Phi(Q_5)-s|W(Q_5)|<\Phi_{\min}<\Phi(Q_5),$$
代以数值后得
$$4\,908<\Phi_{\min}<4\,940.$$
因之，无论取 $CFGE$ 中哪一点作为局址，我们已经逼近于电线网的最小价值，且相对误差不超过 0.7%。

再来看条件(10.1)。除在点 P_i 外，函数 $\Phi(x,y)$ 有连续偏导数，若铜心不与任一 P_i 重合，则其坐标 (x,y) 满足方程

$$\frac{\partial \Phi(x,y)}{\partial x}=-\sum_{i=1}^{n}\frac{c_i}{r_i}(x_i-x)=0,$$
$$\frac{\partial \Phi(x,y)}{\partial y}=-\sum_{i=1}^{n}\frac{c_i}{r_i}(y_i-y)=0. \quad (10.4)$$

我们不会求方程组(10.4)的精确解，但可用下述累次逼近法：

由观察任取点 $Q^{(0)}$ 作为铜心的 0 次逼近，在方程(10.4)中以 $r_i^{(0)}=|\overrightarrow{Q^{(0)}P_i}|$ 代未知距离 r_i。因之方程(10.4)化为线性方程，取其解 $x^{(1)}$，$y^{(1)}$ 作为铜心的下一次逼近。类似地求出其他的逼近。容易验证，若 $(x^{(K)},y^{(K)})$ 表第 K 次逼近 $Q^{(K)}$ 的坐标，则第 $K+1$ 次逼近 $Q^{(K+1)}$ 的坐标 $x^{(K+1)}$，$y^{(K+1)}$ 为

$$x^{(K+1)}=\frac{\sum_{i=1}^{n}\frac{c_i x_i}{r_i^{(K)}}}{\sum_{i=1}^{n}\frac{c_i}{r_i^{(K)}}},\quad y^{(K+1)}=\frac{\sum_{i=1}^{n}\frac{c_i y_i}{r_i^{(K)}}}{\sum_{i=1}^{n}\frac{c_i}{r_i^{(K)}}}, \quad (10.5)$$

我们不会证明，由公式(10.5)所决定的逼近点收敛于铜心。但几次运用这方法收敛性总是成立，甚至当铜心重合于某一 P_i，即当方程组(10.4)无意义时，也是如此。在图 10-2 上，取点 $Q^{(0)}(0;0)$ 为 0 次逼近，然后得到其余的逼近点 $Q^{(1)}(-0.7;3.2)$，$Q^{(2)}(-0.7;4.5)$ 有

$$\Phi(Q^{(0)})=5\,110,\ \Phi(Q^{(1)})=4\,970,\ \Phi(Q^{(2)})=4\,950,$$

这些点已在图 10-2 绘出。

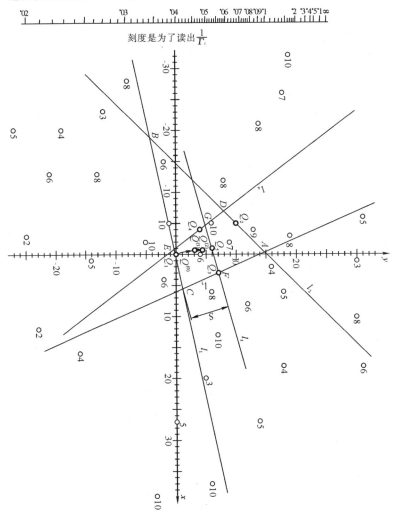

图 10-2 决定电话总局最佳局址的例，此例取自 Łukaszewicz J，Steinhaus H. O wyznacraniu środka miedzi sieci telefonicznej. Lastasowania Matematyki, 1954, 1.

解决最佳电话总局局址的问题时，条件(10.1)中距离 $r_i = \rho(Q, P_i)$ 定义为平面上毕达哥拉斯距离

$$\rho(Q, P_i) = \sqrt{(x_i - x)^2 + (y_i - y)^2}. \qquad (10.6)$$

但在实际问题中，取其他距离 $\rho(Q, P_i)$ 的定义，仍有意义，例如

$$\rho(Q, P_i) = |x_i - x| + |y_i - y|. \qquad (10.7)$$

如果城市的街道系统是垂直的，而且技术条件只容许沿街道架线，那么就该采用上式中的距离定义.

若在条件(10.1)中，距离 $r_i = \rho(Q, P_i)$ 由(10.7)表达，则决定最佳电话总局局址问题化为要个别地决定坐标 x 及 y，因为函数 $\Phi(x, y)$ 这时化为

$$\Phi(x, y) = \sum_{i=1}^{n} c_i |x_i - x| + \sum_{i=1}^{n} c_i |y_i - y|, \qquad (10.8)$$

故 $\Phi(x, y)$ 的极小值可由求(10.8)式右方两和各自的极小值而找到. 容易看出，如果 x 为带权 $c_1, c_2, \cdots, c_n$ 的 $x_1, x_2, \cdots, x_n$ 的中数，y 为带权 $c_1, c_2, \cdots, c_n$ 的 $y_1, y_2, \cdots, y_n$ 的中数时，那么函数取极小值.

到现在为止，在上述讨论中都假定分局局址已知，但实际中常发生更复杂的问题，即已知个别电话用户的地址，而需要同时决定总局与 n 个分局的局址. 在这问题的各种提法中，我这里只叙述一个，看来它是最广泛的.

设已给平面上 N 个点 $\pi_1, \pi_2, \cdots, \pi_N$ 的集 z，对每一点都配有一正数 $r_1, r_2, \cdots, r_N$，要找的是将 z 分为 n 个子集 $z_1, z_2, \cdots, z_n$ 的分割，并要找 $n+1$ 个点 $Q, P_1, P_2, \cdots, P_n$，使函数

$$\begin{aligned}&\Phi(Q, P_1, P_2, \cdots, P_n; z_1, z_2, \cdots, z_n)\\&= \sum_{i=1}^{n} c_i \rho(Q, P_i) + \sum_{i=1}^{n} \sum_{\pi_j \in z_i} r_i \rho(P_i, \pi_j)\end{aligned} \qquad (10.9)$$

(其中正量 c_i 是属于集 z_i 的诸点 π_j 所对应的权 r_j 的已知函数)达到极小.

这问题可用电话来解释. 点 π_j 表个别电话用户(或第二类的电话分局), 它的位置认为已知, 这用户与某分局 P_i 连接的单位长电线的价值 r_j 也已知. 分局又需用电线与总局 Q 连接. 这电线的单位长价值 c_i 是某些 r_j 的已知函数, 这些 r_j 是分局 P_i 与其用户 π_j 间的连线的单位长价值. 函数(10.9)定出整个线路网的价值, 问题在于要求出总局及 n 个分局的最佳地址 Q 及 P_i, 以使线路网的价值极小.

我们不知道这问题的精确解, 只提出下面的逐步逼近法. 由观察任取一点 $Q^{(0)}$ 及集 z 的任一分为 n 份的分割 $z_1^{(0)}$, $z_2^{(0)}$, $\cdots$, $z_n^{(0)}$ 作为解的零次逼近. 然后对每一子集 $z_i^{(0)}$ 求出其中带权 $r_{i,1}$, $r_{i,2}$, $\cdots$, r_{i,n_i} 的点 $\pi_{i,1}$, $\pi_{i,2}$, $\cdots$, π_{i,n_i} 的铜心, 把这铜心取为分局局址 $P_i^{(1)}$, $P_i^{(1)}$ 的权为 $c_i^{(0)} = f(r_{i,1}, r_{i,2}, \cdots, r_{i,n_i})$然后算出权为 $c_i^{(0)}$ 的分局位置 $P_i^{(1)}$ 的铜心 $Q^{(1)}$, 并取此铜心作为总局位置的第一次逼近. 最后把每一点 π_j 与其最邻近的分局 $P_i^{(1)}$ 相连, 于是得到分集 z 为 n 个子集 $z_1^{(1)}$, $z_2^{(1)}$, $\cdots$, $z_n^{(1)}$ 的分割, 以这分割为所求分割第一次逼近. 于是完成了第一步逼近.

我们不知道, 用这种手续造出的逼近法何时收敛于问题的最佳解. 从每一步向下步过渡时, 对函数(10.9)的观察可检查此法可否应用. 我们相信在具体问题中, 用上述逐步逼近法应该得到满意的结果. 只在几个臆造的例中, 我们试用了这方法, 得到了令人满意的收敛性.

参考文献

[1] Langer M. Studien über Aufgaben der Fernsprechtechnik. 1936.

[2] Dietrich H. Zagadnienie najwłaściwszego usytnowania miejskiej centrali telefonicznej. Przegląd Telekomunikacyjny, 1952, 19(25).

[3] Łukaszewicz J, Steinhaus H. O wyznaczaniu środka miedzi sieci telefonicznej. Zastosowania Matematyki, 1954, 1: 299.

§11. 关于个体（看成 n 维空间的点）集合的几个注意

再来研究个体 A_1，A_2，$\cdots$，A_N 所成的集 Z，它可看成 n 维空间的点集. 令每一点的坐标 $A_i(c_{i1}, c_{i2}, \cdots, c_{in})$，$i=1, 2, \cdots, N$，为个体 A_i 的 n 个量度标志 c_1，c_2，$\cdots$，c_N 的测量结果，因此，全部观察的结果构成矩阵

$$c = \begin{pmatrix} c_{11} & c_{12} & \cdots & c_{1n} \\ c_{21} & c_{22} & \cdots & c_{2n} \\ \vdots & \vdots & & \vdots \\ c_{N1} & c_{N2} & \cdots & c_{Nn} \end{pmatrix} \quad (11.1)$$

其中每一横行对应于集 Z 中一个体，而每一直行对应于一标志. 如果两个体 A_i，A_j 满足条件

$$\frac{c_{i1}}{c_{j1}} = \frac{c_{i2}}{c_{j2}} = \cdots = \frac{c_{in}}{c_{jn}}, \quad (11.2)$$

我们就说 A_i，A_j 几何相似. 条件(11.2)表示，n 维空间中的点 A_i，A_j 位于通过坐标原点的同一直线上. 在实验科学中，各种指标定义为某些标志的比（例如人脸的测量指标等于脸宽与脸长之比），这些标志是几何相似的不变数，但伏洛茨瓦夫的数学家 Perkal，经过长期与生物学者合作以后，发现几何相似与生物学上的个体相似的直觉概念并不符合，因而引进了生物相似的新观念. 设 $\sigma_i (i=1, 2, \cdots, n)$ 为标志 c_i 相对全部个体而言的平均二次离差

$$\sigma_i = \sqrt{\frac{1}{N} \sum_{l=1}^{N} (c_{l_i} - \bar{c}_i)^2}, \quad (11.3)$$

其中 $\bar{c}_i = \frac{1}{N} \sum_{l=1}^{N} c_{l_i}$ 是标志 c_i 的算术平均值. 根据 Perkal 的定义, 个体 A_i, A_j 若满足条件

$$\frac{c_{i1} - c_{j1}}{\sigma_1} = \frac{c_{i2} - c_{j2}}{\sigma_2} = \cdots = \frac{c_{in} - c_{jn}}{\sigma_n}, \qquad (11.4)$$

则称 A_i, A_j 为生物相似. 条件 (11.4) 表示向量 $\overrightarrow{A_j A_i}$ 平行于向量 $\boldsymbol{\sigma}(\sigma_1, \sigma_2, \cdots, \sigma_n)$. 常常不考虑矩阵 (11.1) 而考虑矩阵

$$\boldsymbol{\Gamma} = \begin{pmatrix} \gamma_{11} & \gamma_{12} & \cdots & \gamma_{1n} \\ \gamma_{21} & \gamma_{22} & \cdots & \gamma_{2n} \\ \vdots & \vdots & & \vdots \\ \gamma_{N1} & \gamma_{N2} & \cdots & \gamma_{Nn} \end{pmatrix}, \qquad (11.5)$$

其中元 γ_{ij} 系由 (11.1) 中之元经过规范化而得 (使每标志的平均值为 0, 平均二次离差为 1)

$$\gamma_{ij} = \frac{c_{ij} - \bar{c}_j}{\sigma_j} \quad i = 1, 2, \cdots, N; \quad j = 1, 2, \cdots, n.$$

采用记号 (11.5) 后, 使个体 A_i, A_j 生物相似的条件 (11.4) 化为

$$\gamma_{i1} - \gamma_{j1} = \gamma_{i2} - \gamma_{j2} = \cdots = \gamma_{in} - \gamma_{jn}. \qquad (11.6)$$

但 Perkal 建议根据整个集 Z 来描述个体 A_i, 即计算所谓测量指标矩阵, 或如我们已习惯说的, Perkal 指标矩阵

$$\boldsymbol{P} = \begin{pmatrix} P_{11} & P_{12} & \cdots & P_{1n} \\ P_{21} & P_{22} & \cdots & P_{2n} \\ \vdots & \vdots & & \vdots \\ P_{N1} & P_{N2} & \cdots & P_{Nn} \end{pmatrix}, \qquad (11.7)$$

其中

$$P_{ij} = \gamma_{ij} - \frac{1}{n} \sum_{l=1}^{n} \gamma_{il}, \qquad (11.8)$$

这表示已将矩阵 (11.5) 的每一横行规范化, 以使其平均值为 0, 容易证明 Perkal 的各指标是个体生物相似的不变数, 因而两个

生物相似的个体 A_i,A_j 有相同的 Perkal 指标. 由条件(11.6)及公式(11.8)得等式

$$P_{i1}=P_{j1}, \quad P_{i2}=P_{j2}, \quad \cdots, \quad P_{in}=P_{jn}.$$

(11.8)式中的被减数

$$m_i = \frac{1}{n}\sum_{l=1}^{n}\gamma_{il}$$

可称为个体 A_i 的数值,因为它是此个体各标志对此标志关于集 Z 的平均值的平均离差. 但 Perkal 指标指出:是什么标志特别把该个体区分开来,至今我们尚未研究标志 c_1,c_2,$\cdots$,c_n 在个体集合 Z 上是否相关.

我们说标志是相关的,如在对称矩阵

$$S = \begin{pmatrix} \sigma_{11} & \sigma_{12} & \cdots & \sigma_{1n} \\ \sigma_{21} & \sigma_{22} & \cdots & \sigma_{2n} \\ \vdots & \vdots & & \vdots \\ \sigma_{n1} & \sigma_{n2} & \cdots & \sigma_{nn} \end{pmatrix}, \qquad (11.9)$$

中,非对角线上的元 σ_{ij} 不完全为 0,这里二阶中心矩 E_{ij} 定义为

$$\sigma_{ij} = \frac{1}{N}\sum_{l=1}^{N}(c_{li}-\bar{c}_i)(c_{lj}-\bar{c}_j), \quad i,j=1,2,\cdots,n \qquad (11.10)$$

(若 $i=j$,则 $\sigma_{ii}=\sigma_i^2$,σ_i 是由公式(11.3)定出). 因为在许多实际问题中,最好用不相关的标志来刻画个体,故可提出如下问题:试求 n 维空间坐标系的一线性变换,使在新坐标系中,新坐标(我们认为这就是个体的新标志)是不相关的. 容易证明,这一变换由下式定出:

$$k_{ij}=\alpha_{1j}(c_{i1}-\bar{c}_1)+\alpha_{2j}(c_{i2}-\bar{c}_2)+\cdots+\alpha_{nj}(c_{in}-\bar{c}_n), \qquad (11.11)$$
$$i,j=1,2,\cdots,n,$$

其中系数 $\alpha_{lj}(l=1,2,\cdots,n)$ 是新坐标系中第 j 个坐标轴的方向余弦,可由下齐次线性方程组:

$$(\sigma_{11}-\lambda_j)\alpha_{1j}+\sigma_{12}\alpha_{2j}+\cdots+\sigma_{1n}\alpha_{nj}=0,$$
$$\sigma_{21}\alpha_{1j}+(\sigma_{22}-\lambda_j)\alpha_{2j}+\cdots+\sigma_{2n}\alpha_{nj}=0,$$
$$\cdots$$
$$\sigma_{n1}\alpha_{1j}+\sigma_{n2}\alpha_{2j}+\cdots+(\sigma_{nn}-\lambda_j)\alpha_{nj}=0, j=1,2,\cdots,n$$
$$(11.11)'$$

在附加条件
$$\alpha_{1j}^2+\alpha_{2j}^2+\cdots+\alpha_{nj}^2=1, j=1,2,\cdots,n$$

下解出. 方程组 $(11.11)'$ 中的数 $\lambda_j(j=1,2,\cdots,n)$ 是下方程

$$\begin{vmatrix} \sigma_{11}-\lambda & \sigma_{12} & \cdots & \sigma_{1n} \\ \sigma_{21} & \sigma_{22}-\lambda & \cdots & \sigma_{2n} \\ \vdots & \vdots & & \vdots \\ \sigma_{n1} & \sigma_{n2} & \cdots & \sigma_{nn}-\lambda \end{vmatrix}=0 \quad (11.12)$$

的根. 个体 $A_1, A_2, \cdots, A_n$ 的新标志 $K_1, K_2, \cdots, K_n$ 的值 $K_{ij}(i=1,2,\cdots,n; j=1,2,\cdots,n)$ 按公式 (11.12) 计算, 新标志的二阶中心矩矩阵为

$$\boldsymbol{R}=\begin{pmatrix} \lambda_1 & 0 & \cdots & 0 \\ 0 & \lambda_2 & \cdots & 0 \\ \vdots & \vdots & & \vdots \\ 0 & 0 & \cdots & \lambda_n \end{pmatrix},$$

从而可见这些标志不相关.

若 λ_1 是方程 (11.13) 的最大根, 则轴 K_1 称为由标志 $c_1, c_2, \cdots, c_n$ 所构成的 n 维空间的 Z 集轴. 集轴有下列性质: 集 Z 中全体点与直线 l 距离的平方和, 当 l 重合于 Z 集轴时达到极小. Perkal 还给出了集 Z 的个体相似的另一定义: 如果向量 $\overrightarrow{A_iA_j}$ 平行于集轴, 那么说 $A_i, A_j V-$ 相似.

Perkal 在各实际问题中, 成功地研究了不相关的标志. 例如一定年龄的小孩的集 Z, 由两个标志决定: $c_1=$ 身长; $c_2=$ 体

重；对它可以找到不相关的标志 K_1，K_2，它们的生物学解释为发育与失常．在运用此法的另一例中，我们观察一定年龄的某种树木所成的 N 个森林，观察结果构成集 Z．这里测量了两个相关的标志：在该森林中树木的平均高度与平均粗度（距地面 130 cm 的树干直径）．这时不相关的标志由林业工作者解释为整齐性与环境的影响．

参考文献

[1] Perkal J. O Pewnych korelacjach obszarowych. Časopis pro Pěstowania Matematyki a Fysiki. 1949，75.

[2] Perkal J. O wskaźnikach actropologicznych. Przegląd Antropologiczny，1953，19.

[3] Nowakowski T K，Perkal J. Nowe metody badania zaleznosci między wzrostem，waga, a wielkiem młodzieży. Przcgląd Antropologiczny. 1952，18.

[4] Perkal J，Battek J. Proba oceny rozwojk drzewictanów Sylwan. 1955，49.

§12. 关于地质矿藏参数的估计

在这篇报告里,我们假定研究的对象是某一地质参数,例如煤矿的厚度或地面某一点下锌矿的纯锌含量。以符号 $y(p)$ 记此参数,p 为某平面区域 D 内的点。设 $y(P)$ 是平面上点的平稳迷向随机过程,有不变的数学期望

$$E[y(p)] = m, \quad p \in D \qquad (12.1)$$

及不变的方差

$$E[y(p) - m]^2 = S^2 \quad p \in D. \qquad (12.2)$$

它的相关函数只依赖于点 p_1, p_2 间的毕达哥拉斯距离 $d(p_1, p_2)$

$$R\{y(p_1), y(p_2)\} = \frac{1}{S^2} E\{[y(p_1) - m][y(p_2) - m]\}$$
$$= f(d(p_1, p_2)), \qquad (12.3)$$

其中

$$p_1 \in D, \quad p_2 \in D.$$

我们的问题是要求在区域 D 中,参数 $y(p)$ 的平均值 η

$$\eta = \eta(D) = \frac{1}{|D|} \iint_D y(p) \mathrm{d}p \qquad (12.4)$$

($|D|$ 表区域 D 的面积,$\mathrm{d}p$ 表面积的微分)或积分

$$Y = Y(D) = \iint_D y(p) \mathrm{d}p \qquad (12.5)$$

的最佳估值,当然,这时要假定我们已知区域 D 中若干点 p_1, p_2, $\cdots$, p_k 上参数的值:

$$y_1 = y(p_1), \quad y_2 = y(p_2), \quad \cdots, \quad y_k = y(p_k).$$

这是地质学中主要问题之一。不少的作者给出了积分 (12.5) 或积分平均值的各种估值方法。例如 Смирнов 有 8 个不

同的解，其中有已知值 y_1, y_2, $\cdots$, y_k 的数学平均值与几何平均值。这里我只叙述伏洛茨瓦夫数学家 Zubrzycki 的一些结果，他用随机过程中的术语提出这一问题并得到了一些有趣的结果。

假定我们已知由公式(12.1)～(12.3)定义的参数 m, s 及相关函数 $f(d)$，容易求出由公式(12.5)所定义的积分 Y 的下列估值。

（Ⅰ）Zubrzycki 的最一般的估值 $\hat{Y}_1$ 由线性函数定义

$$\hat{Y}_1 = c_0 + c_1 y_1 + \cdots + c_k y_k, \quad (12.6)$$

它满足条件

$$S_1^2 = E(c_0 + c_1 y_1 + \cdots + c_k y_k - Y)^2$$
$$= \min_{r_0, r_1, \cdots, r_n} E(r_0 + r_1 y_1 + \cdots + r_k y_k - Y)^2. \quad (12.7)$$

S_1 称为估值 1 的误差。

公式(12.6)中系数 c_1, c_2, $\cdots$, c_k 可由解下方程组而得

$$w(y_1, y_1)c_1 + \cdots + w(y_1, y_k)c_k = w(y_1, Y),$$
$$w(y_2, y_1)c_1 + \cdots + w(y_2, y_k)c_k = w(y_2, Y),$$
$$\cdots \qquad\qquad\qquad\qquad\qquad\qquad (12.8)$$
$$w(y_k, y_1)c_1 + \cdots + w(y_k, y_k)c_k = w(y_k, Y),$$

而常数 c_0 则由方程

$$c_0 + c_1 E(y_1) + \cdots + C_k E(y_k) = E(Y) \quad (12.9)$$

决定。方程(12.8)中 $w(x, y)$ 为数学期望 $E[(x-Ex)(y-Ey)]$，它是随机变量的二阶混合中心矩。这里它可由下公式计算

$$w(y_i, y_j) = s^2 f[d(p_i, p_j)], \quad (i, j = 1, 2, \cdots, k)$$
$$(12.10)$$

（当 $i=j$ 时得 $w(y_i, y_j) = s^2$，此因 $d(p_i, p_j) = 0$ 而相关函数有性质 $f(0) = 1$）及

$$w(y_i, Y) = \iint_D w[y_i, y(p)] dp$$

$$= s^2 \iint_D f[d(p_i, p)] dp \quad (i = 1, 2, \cdots, k).$$

(12.11)

注意要决定系数 c_0，c_1，$\cdots$，c_k 不必要知道方差 s^2，因为可以 s^2 除(12.8)之两边而消去 s^2. 估值 $\hat{Y}_1$ 之误差 S_1 可按下式计算：

$$S_1^2(D; p_1, \cdots, p_k) = \begin{vmatrix} w(y_1, y_1) & \cdots & w(y_1, y_k) & w(y_1, Y) \\ \vdots & & \vdots & \vdots \\ w(y_k, y_1) & \cdots & w(y_k, y_k) & w(y_k, Y) \\ w(Y, y_1) & \cdots & w(Y, y_k) & w(Y, Y) \end{vmatrix} : \begin{vmatrix} w(y_1, y_1) & \cdots & w(y_1, y_k) \\ \vdots & & \vdots \\ w(y_k, y_1) & \cdots & w(y_k, y_k) \end{vmatrix},$$

(12.12)

其中

$$w(Y, Y) = \iint_D \left\{ \iint_D w(p, q) dp \right\} dq$$
$$= s^2 \iint_D \left\{ \iint_D {}' f[d(p, q)] dp \right\} dq.$$

（Ⅱ）Zubrzycki 的另一估值为 $\hat{Y}_\Pi$，由下齐次线性函数定义：

$$\hat{Y}_\Pi = c_1 y_1 + c_2 y_2 + \cdots + c_k y_k \quad (12.13)$$

并满足条件

$$s_\Pi^2 = E(c_1 y_1 + c_2 y_2 + \cdots + c_k y_k - Y)^2$$
$$= \min_{r_1, r_2, \cdots, r_n} E(r_1 y_1 + r_2 y_2 + \cdots + r_k y_k - Y)^2, \quad (12.14)$$

s_Π 称为估值 $\hat{Y}_\Pi$ 的误差.

(12.13)中的系数 c_1，c_2，$\cdots$，c_k 是下方程组的解，

$$c_1 E(y_1 y_1) + c_2 E(y_1 y_2) + \cdots + c_k E(y_1 y_k) = E(y_1 Y),$$
$$c_1 E(y_2 y_1) + c_2 E(y_2 y_2) + \cdots + c_k E(y_2 y_k) = E(y_2 Y),$$

(12.15)

$\cdots$

$$c_1 E(y_k y_1) + c_2 E(y_k y_2) + \cdots + c_k E(y_k y_k) = E(y_k Y),$$

这里，积 $y_i y_j$ 的数学期望及 $E(y_i Y)$ 由下式计算：

$$E(y_i y_j) = w(y_i, y_j) + E(y_i)E(y_j)$$
$$= s^2 f[d(p_i, p_j)] + m^2 \quad (i, j = 1, 2, \cdots, k),$$
$$(12.16)$$
$$E(y_i Y) = w(y_i Y) + E(y_i)E(Y)$$
$$= s^2 \iint_D f[d(p_i, p)] \mathrm{d}p - m^2 |D|, \quad (12.17)$$

因为容易看出 $E(Y) = m|D|$，要完全决定(12.13)中的系数 $c_1, c_2, \cdots, c_k$，只需要知道相关函数 $f(d)$ 及地质参数的数学期望对其平均二次离差的比 $\dfrac{m}{s}$.

估值 $\hat{Y}_{\mathrm{II}}$ 的误差 S_{II} 可按下式求出

$$S_{\mathrm{II}}^2(D; p_1, \cdots, p_k)$$
$$= \begin{vmatrix} E(y_1 y_2) & \cdots & E(y_1 y_k) & E(y_1 Y) \\ \vdots & & \vdots & \vdots \\ E(y_k y_1) & \cdots & E(y_k y_k) & E(y_k Y) \\ E(Y y_1) & \cdots & E(Y y_k) & E(YY) \end{vmatrix} : \begin{vmatrix} E(y_1 y_2) & \cdots & E(y_1 y_k) \\ \vdots & & \vdots \\ E(y_k y_1) & \cdots & E(y_k y_k) \end{vmatrix}.$$
$$(12.18)$$

（Ⅲ）估值 $\hat{Y}_{\mathrm{III}}$ 也是 $y_1, y_2, \cdots, y_k$ 的齐次线性函数

$$\hat{Y}_{\mathrm{III}} = c_1 y_1 + c_2 y_2 + \cdots + c_k y_k, \quad (12.19)$$

但满足两条件. 第一，要求估值 $\hat{Y}_{\mathrm{III}}$ 是 Y 的无偏估值，即

$$E(\hat{Y}_{\mathrm{III}}) = E(c_1 y_1 + c_2 y_2 + \cdots + c_k y_k) = E(Y) = m|D|.$$
$$(12.20)$$

其次要求估值 $\hat{Y}_{\mathrm{III}}$ 的误差 S_{III} 在所有由 $y_1, y_2, \cdots, y_k$ 所给出的齐次线性无偏估值中到达极小值，即

$$S_{\mathrm{III}}^2 = E(c_1 y_1 + c_2 y_2 + \cdots + c_k y_k - Y)^2$$
$$= \min_{r_1, r_2, \cdots, r_k} E(r_1 y_1 + r_2 y_2 + \cdots + r_k y_k - Y)^2 \quad (12.21)$$

$$E(r_1y_1+r_2y_2+\cdots+r_ky_k)=E(Y).$$

(12.19)中的系数 $c_1, c_2, \cdots, c_k$ 可用拉格朗日未定因子法求出. 运用此方法稍经变换后决定 $c_1, c_2, \cdots, c_k$ 的方程化为

$$c_1w(y_1, y_1)+c_2w(y_1, y_2)+\cdots+c_kw(y_1, y_k)+\lambda=w(y_1, Y),$$
$$c_1w(y_2, y_1)+c_2w(y_2, y_2)+\cdots+c_kw(y_2, y_k)+\lambda=w(y_2, Y),$$
$$\cdots \quad (12.22)$$
$$c_1w(y_k, y_1)+c_2w(y_k, y_2)+\cdots+c_kw(y_k, y_k)+\lambda=w(y_k, Y),$$
$$c_1+c_2+\cdots+c_k=|D|.$$

由方程(12.22)及公式(12.10)(12.11),可见要决定(12.19)式中的系数 $c_1, c_2, \cdots, c_k$ 只需知道相关函数 $f(d)$.

估值 $\hat{Y}_{\text{III}}$ 的误差 S_{III} 为

$$S_{\text{III}}^2(D; p_1, p_2, \cdots, p_k)$$
$$=c_1\{c_1w(y_1, y_1)+c_2w(y_1, y_2)+\cdots+c_kw(y_1, y_k)-w(y_1, Y)\}+c_2\{c_1w(y_2, y_1)+c_2w(y_2, y_2)+\cdots+c_kw(y_2, y_k)-w(y_2, Y)\}+\cdots+c_k\{c_1w(y_k, y_1)+c_2w(y_k, y_2)+\cdots+c_kw(y_k, y_k)-w(y_kY)\}-\{w(Y, y_1)+w(Y, y_2)+\cdots+w(Y, y_k)-w(Y, Y)\} \quad (12.23)$$
$$=-\lambda|D|-\{w(Y, y_1)+w(Y, y_2)+\cdots+w(Y, y_k)-w(YY)\}$$

(Ⅳ) 第四个估值 $\hat{Y}_{\text{IV}}$ 也是 $y_1, y_2, \cdots, y_k$ 的齐次线性函数,但所有系数相同,即

$$\hat{Y}_{\text{IV}}=cy_1+cy_2+\cdots+cy_k. \quad (12.24)$$

此外还要求估值 $\hat{Y}_{\text{IV}}$ 满足条件

$$E(\hat{Y}_{\text{IV}})=E(cy_1+cy_2+\cdots+cy_k)=E(Y). \quad (12.25)$$

条件(12.25)等价于要求估值 $\hat{Y}_{\text{IV}}$ 的误差 S_{IV} 极小,即

$$S_{\text{IV}}^2=E(cy_1+cy_2+\cdots+cy_k-Y)^2$$

$$=\min_r E(ry_1+ry_2+\cdots+ry_k-Y)^2 \quad (12.26)$$

或(12.24)中的 c 由条件(12.25)易算得为

$$c=\frac{1}{k}\mid D\mid, \quad (12.27)$$

而误差 S_{IV} 为

$$S_{\text{IV}}^2(D,p_1,p_2,\cdots,p_k)$$
$$=c^2\sum_{i=1}^k\sum_{j=1}^k E(y_iy_j)-c\sum_{i=1}^k E(y_iY)+E(YY). \quad (12.28)$$

估值 S_{IV} 称为算术平均值，因为由它可得 η(参看(12.4))的估值为观察值 y_1, y_2, $\cdots$, y_k 的算术平均值.

(V) 最后 Zubrzycki 考虑了估值 $\hat{Y}_{\text{V}}$，它完全不利用观察的结果，即

$$\hat{Y}_{\text{V}}=c. \quad (12.29)$$

由误差 S_{V} 最小条件得

$$S_{\text{V}}^2(D)=E(c-Y)^2=\min_r E(r-Y)^2, \quad (12.30)$$

易求出

$$c=E(Y)=m\mid D\mid, \quad S_{\text{V}}^2=E(E(Y)-Y)^2=w(Y,Y). \quad (12.31)$$

由估值 $\hat{Y}_{\text{I}}$, $\hat{Y}_{\text{II}}$, $\hat{Y}_{\text{III}}$, $\hat{Y}_{\text{IV}}$ 及 $\hat{Y}_{\text{V}}$ 的定义，它们的误差满足不等式

$$S_{\text{I}}\leqslant S_{\text{II}}\leqslant S_{\text{III}}\leqslant S_{\text{IV}} \text{ 及 } S_{\text{I}}\leqslant S_{\text{V}}. \quad (12.32)$$

在实际中，在点 $p_i(i=1, 2, \cdots, k)$ 所观察到的，常常不是未知参数的精确值 y_i，而是某一数 y_i^*，由于不可避免的测量误差，它不等于 y_i

$$y_i^*=y_i+\varepsilon_i, \quad i=1, 2, \cdots, k. \quad (12.33)$$

关于误差 ε_1, ε_2, $\cdots$, ε_k，我们假定它们是随机变数，数学期望为 0，方差为 S'^2，而且相互独立，也不依赖于随机变量 y_1,

$y_2, \cdots, y_k$. 在这些假定下,根据在点 $p_1, p_2, \cdots, p_k$ 的观察值 $y_1^*, y_2^*, \cdots, y_k^*$, 已知特征

$$E(y_i^*) = m^*, \quad E(y_i^* - m^*)^2 = S^{*2}$$

及相关函数

$$\frac{1}{S^{*2}} E\{[y^*(p_1) - m^*][y^*(p) - m^*]\} = f^*[d(p_1, p)]$$

来研究未知量(12.4)或(12.5). 为此,先来研究具有误差的测量的特征 m^*, S^{*2} 及 $f^*(d)$ 与未知参数精确测量的特征 m, S^2 及 $f(d)$ 间的关系.

由假定测量误差 ε_i 的数学期望为 0 得

$$m^* = m. \tag{12.34}$$

由于误差 ε 与在任一点 p 上的参数 y 值独立,有

$$S^{*2} = S^2 + S'^2. \tag{12.35}$$

利用误差 ε 与参数值的独立性及不同的观察所得的误差的相互独立性得

$$\begin{aligned}
w(y_i^*, y_j^*) &= E[(y_i^* - E(y_i^*))(y_j^* - E(y_j^*))] \\
&= E\{[(y_i - E(y_i)) + (\varepsilon_i - E(\varepsilon_i))][(y_j - E(y_j)) + (\varepsilon_j - E(\varepsilon_j))]\} \\
&= E[(y_i - E(y_i))(y_j - E(y_j))] + E[(y_i - E(y_i))(\varepsilon_j - E(\varepsilon_j))] + \\
&\quad E[(y_j - E(y_j))(\varepsilon_i - E(\varepsilon_i))] + E[(\varepsilon_i - E(\varepsilon_i))(\varepsilon_j - E(\varepsilon_j))] \\
&= E[(y_i - E(y_i))(y_j - E(y_j))] = w(y_i, y_j),
\end{aligned}$$

$$i, j = 1, 2, \cdots, k, \quad i \neq j, \tag{12.36}$$

利用相关函数的定义,当 $d > 0$ 时得

$$f^*(d) = \frac{S^2}{S^{*2}} f(d). \tag{12.37}$$

用得到等式(12.36)的同样方法可得

$$w(y_i^*, Y) = w(y_i, Y), \quad i = 1, 2, \cdots, k, \tag{12.38}$$

由 $w(\alpha, \beta) = E(\alpha\beta) - E(\alpha)E(\beta)$ 及(12.36)(12.38)得

$$E(y_i^* y_j^*) = E(y_i y_j) \quad (i, j = 1, 2, \cdots, k), \tag{12.39}$$

$$E(y_i^* Y) = E(y_i Y) \quad (i=1, 2, \cdots, k). \tag{12.40}$$

现在注意：若在公式(12.6)~(12.28)中，到处都以 y_i^* 代 y_i ($i=1, 2, \cdots, k$)，S^* 代 S，$f^*(d)$ 代 $f(d)$，则得未知量 Y 根据观察值 $y_1^*, y_2^*, \cdots, y_k^*$ 所产生的新估值 $\hat{Y}_I^*, \hat{Y}_{II}^*, \hat{Y}_{III}^*$ 及 $\hat{Y}_{IV}^*$ 的定义及构造方法。现在除 $m, S, f(d)$ 外还要知道 S^*。按定义计算估值 $\hat{Y}_I^*, \hat{Y}_{II}^*, \hat{Y}_{III}^*, \hat{Y}_{IV}^*$ 的误差 $S_I^*, S_{II}^*, S_{III}^*, S_{IV}^*$ 的公式与计算 $S_I, S_{II}, S_{III}, S_{IV}$ 的一样，只要在对应的公式中引入带星号的值。误差 $S_I^*, S_{II}^*, S_{III}^*, S_{IV}^*$ 满足不等式

$$S_I^* \leqslant S_{II}^* \leqslant S_{III}^* \leqslant S_{IV}^* \text{ 及 } S_I^* \leqslant S_V^*, \tag{12.41}$$

这相当于不等式(12.32)，但此外还有不等式

$$S_I \leqslant S_I^*, \quad S_{II} \leqslant S_{II}^*, \quad S_{III} \leqslant S_{III}^*, \quad S_{IV} \leqslant S_{IV}^*, \tag{12.42}$$

这些不等式反映了下列事实：观察值的误差增大了根据这些值所作出的估值的误差。

例 1 Zubrzyeki 运用他的公式来研究下面的例子。在图 12-1 中，已知矩形四边中点 p_1, p_2, p_3, p_4 上纯锌的数量 y_1, y_2, y_3, y_4，或具有误差的数量 $y_1^*, y_2^*, y_3^*, y_4^*$，要估计此矩形中纯锌的数量。

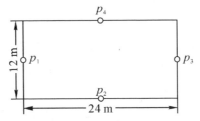

图 12-1 例 1 中的区域 D 及点 p_1, p_2, p_3, p_4 的位置.

在此例中，Zubrzycki 取（用某一单位）

$$m = 10,$$
$$S^2 = 81.7, \tag{12.43}$$

$$f(d) = 10^{-0.1d}.$$

为计算估值 $\hat{Y}_1^*$, $\hat{Y}_2^*$, $\hat{Y}_3^*$ 及 $\hat{Y}_4^*$, 令

$$S^{*2} = 100. \tag{12.44}$$

这些数值接近于 Zubrzycki 在波兰博托城附近锌矿区所求得的实验值. 所得的估值及误差为

$$\hat{Y}_{\text{I}} = 34.397(y_1 + y_3) + 42.224(y_2 + y_4) + 1\,347.58,$$
$$\hat{Y}_{\text{I}}^* = 28.509(y_1^* + y_3^*) + 35.127(y_2^* + y_4^*) + 1\,607.28,$$
$$\hat{Y}_{\text{II}} = 62.684(y_1 + y_3) + 68.789(y_2 + y_4),$$
$$\hat{Y}_{\text{II}}^* = 62.407(y_1^* + y_3^*) + 64.762(y_2^* + y_4^*),$$
$$\hat{Y}_{\text{III}} = 69.143(y_1 + y_3) + 74.857(y_2 + y_4),$$
$$\hat{Y}_{\text{III}}^* = 69.704(y_1^* + y_3^*) + 74.296(y_2^* + y_4^*),$$
$$\hat{Y}_{\text{IV}} = 72(y_1 + y_2 + y_3 + y_4),$$
$$\hat{Y}_{\text{IV}}^* = 72(y_1^* + y_2^* + y_3^* + y_4^*),$$
$$\hat{Y}_{\text{V}} = 2\,880;$$
$$S_{\text{I}}^2 = 8\,475.2 \times 81.7, \quad S_{\text{I}}^{*2} = 9\,608.8 \times 81.7,$$
$$S_{\text{II}}^2 = 12\,654.4 \times 81.7, \quad S_{\text{II}}^{*2} = 16\,111.7 \times 81.7,$$
$$S_{\text{III}}^2 = 13\,577.2 \times 81.7, \quad S_{\text{III}}^{*2} = 18\,352.2 \times 81.7,$$
$$S_{\text{IV}}^2 = 13\,607.7 \times 81.7, \quad S_{\text{IV}}^{*2} = 18\,377.0 \times 81.7,$$
$$S_{\text{V}}^* = 15\,170.8 \times 81.7.$$

例 2 Zubrzycki 再一次运用他的方法来估计上问题中矩形 D 中的纯锌量,只是矩形的边长为以前的一半(6 m 及 12 m);而四点 p_1, p_2, p_3, p_4 则仍是矩形四边的中点,锌矿区的特征也如前一样. 在本例中得估值及误差为

$$\hat{Y}_{\text{I}} = 15.952(y_1 + y_3) + 16.218(y_2 + y_4) + 76.60,$$

$$\hat{Y}_{\text{I}}^* = 13.735(y_1^* + y_3^*) + 14.338(y_2^* + y_4^*) + 158.84,$$

$$\hat{Y}_{\text{II}} = 17.590(y_1 + y_3) + 17.482(y_2 + y_4),$$

$$\hat{Y}_{\text{II}}^* = 16.987(y_1^* + y_3^*) + 16.964(y_2^* + y_4^*),$$

$$\hat{Y}_{\text{III}} = 18.114(y_1 + y_3) + 17.886(y_2 + y_4),$$

$$\hat{Y}_{\text{III}}^* = 18.086(y_1^* + y_3^*) + 17.914(y_2^* + y_4^*),$$

$$\hat{Y}_{\text{IV}} = 18(y_1 + y_2 + y_3 + y_4),$$

$$\hat{Y}_{\text{IV}}^* = 18(y_1^* + y_2^* + y_3^* + y_4^*),$$

$$\hat{Y}_{\text{V}} = 720;$$

$$S_{\text{I}}^2 = 218.92 \times 81.7, \quad S_{\text{I}}^{*2} = 562.27 \times 81.7,$$

$$S_{\text{II}}^2 = 332.24 \times 81.7, \quad S_{\text{II}}^{*2} = 589.4 \times 81.7,$$

$$S_{\text{III}}^2 = 337.52 \times 81.7, \quad S_{\text{III}}^{*2} = 635.61 \times 81.7,$$

$$S_{\text{IV}}^2 = 337.56 \times 81.7, \quad S_{\text{IV}}^{*2} = 635.64 \times 81.7,$$

$$S_{\text{V}} = 1\,955.32 \times 81.7.$$

由这些例中可见,测量误差的存在大大地增加了各估值的误差. 例 1 中精确值的算术平均 $\hat{Y}_{\text{IV}}$ 比带星号的各估值(除 $\hat{Y}_{\text{I}}^*$ 外)都要准确. 在例 2 中,估值 $\hat{Y}_{\text{IV}}$ 比所有带星号的估值准确. 例 1 中,估值 $\hat{Y}_{\text{II}}^*$, $\hat{Y}_{\text{III}}^*$, $\hat{Y}_{\text{IV}}^*$ 比 $\hat{Y}_{\text{V}}$ 的准确性差,而 $\hat{Y}_{\text{V}}$ 不依赖于观察结果. 这种情况在例 2 中不出现,因在例 2 中,观察值与参数在全区域中的值更好地关联着.

在任一情况下,估值 Ⅲ,Ⅳ 的误差无大区别. 以这些估值为例可以观察两种不同影响的对立:其一,在求积分 Y 的估值时,若观察值与参数在全区域中的估值相关越好,则这些观察值就越珍贵;其二,如诸观察值彼此相关,则它们的价值就差些,因为其中的一观察在某种程度上重复另一观察而少给出新的信息. 在例 1 中, 前一影响比后一更强烈, 因为估值 $\hat{Y}_{\text{III}}$ 中 y_2, y_4 (或 $\hat{Y}_{\text{III}}^*$ 中 y_2^*, y_4^*) 的系数超过 y_1, y_3 (或 y_1^*, y_3^*) 的系数. 但 y_2, y_4 (或 y_2^*, y_4^*) 是在矩形长边中点上参数的观察值, 故它们与矩形内参数的值更为相关, 但同时它们彼此间的相关也更密切, 在例 2 中, 区域范围小一半, 后一影响强于前者, 估值 $\hat{Y}_{\text{III}}$ 中 y_2, y_4 (或 $\hat{Y}_{\text{III}}^*$ 中 y_2^*, y_4^*) 的系数小于 y_1, y_3 (或 y_1^*, y_3^*) 的系数.

上述结果只是 Zubrzycki[1] 中的一部分, 那里还研究了与 (12.4)(12.5) 的估值有关的其他问题, 这方面的工作尚未结束, 新的结果将在其他文章中发表.

参考文献

[1] Zubrzycki S. O szacowaniu parametrow złóz geologicznych. Zastosowania Matematyki, 1957, 105.

[2] Смирнов В И. Подсяёт Запасов Минерального Сырья. Москва, 1950.

§13. 关于统计参数的极大中极小估计

极大中极小估计的原则把参数的统计估值问题化为策略论中的问题. 近十年来, 策略论已发展成为一门独立的数学学科, 在一次讨论会上, 甚至连它的基本面貌也不可能叙述. 在这里, 我想叙述 H. Steinhaus 的一个结论, 据我看来, 它是很有意思的. 设在一实验中, 某事件出现的概率未知, 今将此试验独立的重复 n 次, 试根据观察结果来估计此未知概率.

为此, 只需要策略论中的一些基本观念, 我就从这里讲起.

策略论中的主要观念是游戏, 从数学观点看来, 每一游戏由甲、乙两个对立者参加, 并且由两集合 X, Y 及二元函数 $f(x, y)$ 决定, 对每一对元, $x \in X$ 及 $y \in Y$, 有一实数 $w = f(x, y)$ 与之对应. 集 X 中之元 x 称为甲的策略, 集 Y 中之元 y 称为乙的策略, 故 X, Y 分别为甲、乙全部可取策略的集合. 函数 $f(x, y)$ 称为支付函数, 游戏 (x, y, f) 的玩法是, 在每一局中, 两人独立地挑选自己的策略, 若甲的策略为 $x \in X$, 乙的为 $y \in Y$, 则乙付给甲的钱为 $f(x, y)$, 若函数 $f(x, y)$ 取负值, 则表示甲应付给乙的钱为 $|f(x, y)|$. 故当策略各为 x, y 时, $f(x, y)$ 表甲的利润, 它等于乙的亏损. 若集 X, Y 皆有限或可数, 则游戏称为矩形的, 因它完全由矩阵 $\|a_{ij}\|$ 决定, $a_{ij} = f(x_i, y_j)$, 而这游戏的每一局由甲选定矩阵中一横行, 乙独立的选定一直行构成. 横直行的交点上的数即甲的利润.

每一方都想找到最佳的策略. 为此要定义判定策略是否最佳的原则. 在策略论中, 这判别法则为极大中极小的原则. 甲希望获得最大的利润而想找一策略 x_0, 使对任意 $x \in X$ 满足

条件，
$$\min_{y\in X} f(x_0, y) \geq \min_{y\in Y} f(x, y),$$
若这样的策略存在，则用此策略甲在每一局中的利润不小于
$$w = \max_{x\in X} \min_{y\in Y} f(x, y), \tag{13.1}$$
且没有其他的策略能常常使他的利润大于 w. 如这样的 x_0 存在，我们就称它为甲的最佳策略，而数(13.1)称为甲的游戏价值. 若(13.1)中右方的极小中的极大不存在，则甲的游戏价值定义为
$$w = \sup_{x\in X} \inf_{y\in Y} f(x, y) \tag{13.2}$$
且对任 $\varepsilon > 0$，在集 X 中存在策略 $x(\varepsilon)$，它保证甲的利润不小于 $w - \varepsilon_0$. 同样地定义
$$W = \inf_{y\in Y} \sup_{x\in X} f(x, y), \tag{13.3}$$
它称为乙的游戏价值. 对每 $\varepsilon > 0$，在集 Y 中存在一策略 $y(\varepsilon)$，若乙采用此策略，则它的亏损不大于 $W + \varepsilon$. 如存在策略 y_0，它保证乙的亏损不大于 W，我们就称它为对乙最佳的策略.

例如，以矩阵
$$\begin{pmatrix} 0 & -4 & 6 & 2 \\ 5 & 3 & -2 & 4 \\ 3 & 2 & 1 & 1 \end{pmatrix} \tag{13.4}$$
所定义的有限矩形游戏中，集 X 由三个策略构成(矩阵有三横行)，甲的最佳策略为(13.4)中第三横行，集 Y 由四个策略构成(矩阵有四直行)，乙的最佳策略为第二直行. 这里甲的游戏价值为 $w = 1$，乙的游戏价值为 $W = 3$. 若双方都使用最佳策略，则甲的利润为 2.

易证对每一游戏，由(13.2)(13.3)所定义的甲、乙的游戏价值满足不等式
$$w \leq W, \tag{13.5}$$

这直接由显然的事实可见；不可能使甲的利润多于乙的亏损. 使(13.5)化为等式 $w=W$ 的游戏称为闭的，此时 w，W 的公共值称为游戏的价值. 例如矩阵

$$\begin{bmatrix} -2 & 2 & -3 \\ -2 & 0 & 4 \\ -4 & -2 & 3 \end{bmatrix} \quad (13.6)$$

定义一个价值为 -1 的闭游戏. 此时，甲，乙的最佳策略分别为矩阵的第二横行及第一直行. 比较例(13.4)及(13.6)，易见在闭游戏(13.6)中，双方的最佳策略是一对相互反对的最佳策略. 但在开游戏(13.4)中，如乙已知甲使用了最佳策略(即选定(13.4)的第三横行)，他便会放弃自己的最佳策略(第二直行)，因为存在更好的反对第三横行的策略(第三或第四直行)，反之如甲已知乙采用最佳策略，他必放弃其最佳策略，因这最佳策略在反对乙的最佳策略时，不是最佳的，当游戏是开的时，不存在一对相互反对的最佳策略. 易证定理.

定理 1 若在游戏 (X, Y, f) 中存在两策略 $x_1 \in X$ 及 $y_1 \in Y$，使 x_1 是反对 y_1 的最佳策略，即对任一 $x \in X$ 有

$$f(x_1, y_1) \geqslant f(x, y_1), \quad (13.7)$$

而 y_1 是反对 x_1 的最佳策略，即对任一 $y \in Y$ 有

$$f(x_1, y_1) \leqslant f(x_1, y), \quad (13.8)$$

则游戏 (X, Y, f) 是闭的，$f(x_1, y_1)$ 是此游戏的价值，且 x_1，y_1 分别是甲、乙的最佳策略.

证 由假设(13.7)及定义(13.3)得

$$f(x_1, y_1) = \sup_{x \in X} f(x, y) \geqslant \inf_{y \in Y} \sup_{x \in X} f(x, y) = W. \quad (13.9)$$

同样由假定(13.8)及定义(13.2)得

$$f(x_1, y_1) = \inf_{y \in Y} f(x_1, y) \leqslant \sup_{x \in X} \inf_{y \in Y} f(x, y) = w.$$

$$(13.10)$$

由(13.9)(13.10)得
$$w \geqslant f(x_1, y_1) \geqslant W.$$
由不等式(13.5)得等式
$$w = W = f(x_1, y_1).$$
从而得证定理 1.

今设有游戏(X, Y, f),根据此游戏如下定义一新游戏(Θ, H, φ):

Θ 是集 X 上一切概率测度 ξ 的集.

H 是集 Y 上一切概率测度 η 的集.

$\varphi = \varphi(\xi, \eta) = E_{\xi, \eta} f(x, y)$ 是函数 $f(x, y)$ 当 x 的分布为 ξ 而 y 独立地有分布 η 时的数学期望. 这样定义的游戏(Θ, H, φ)称为游戏(X, Y, f)的随机化. 在游戏(Θ, H, φ)中, 甲的策略是取值 $x \in X$ 的随机变量, 乙的策略为取值 $y \in Y$ 的随机变量, 这可如下设想: 甲不是自己从集 X 中去挑选策略 x, 而去造一随机机械, 它按由测度 ξ 所决定的分布去随机挑选策略, 乙也去造一机械, 它不依赖于甲的机械的指示而按 η 所决定的分布, 从集 Y 中去挑选策略 y. 在随机游戏中, 支付函数 $\varphi(\xi, \eta)$ 等于在无限局中甲的平均利润, 这些局都是借助于这二机械进行的.

易证若游戏(X, Y, f)是闭的, 其最佳策略为 x_0 及 y_0, 价值为 V, 则其随机化(Θ, H, φ)也是闭的, 有相同的价值 V, 其最佳策略为 ξ_0 及 η_0, ξ_0 以概率 1 集中于策略 x_0, η_0 以概率 1 集中于策略 y_0.

但对相当广泛的游戏(X, Y, f)的集合, 随机化(Θ, H, φ)甚至于当(X, Y, f)是开的时, 也是闭的, 有一基本定理, 我们不在此证明, 甚至也不严格叙述. 它常使实际工作者, 当每采用某些解, 而它们又化为要玩某一游戏时, 不得不采用最佳随机策略. 这类应用可举战争中的策略为例.

未知参数的统计估值理论中许多问题可用策略论来叙述而得到有趣的结果. 这里我们只考虑最简单的估值问题, 即估计二项分布中未知参数 p, 设独立地重复试验 n 次时, 具有未知出现概率 p 的事件出现了 m 次, 设此参数的估值为 $\hat{p}=\hat{p}(m, n)$. 在多数实际问题中, 此估值的误差 $|p-\hat{p}|$ 常引起某一损失, 后者是此误差的函数. 我们假定损失与误差平方成正比. 现在要决定最佳估值 $\hat{p}$.

在问题的这种提法下, 可以在策略论中找到对应于上述情况的模型, 设游戏的甲方是统计者, 其对手乙方称为自然界. 这游戏的玩法如下. 对确定的 n, 统计者挑选任一估值 $\hat{p}=\hat{p}(m, n)$, 对一切可能的值 $m=0, 1, \cdots, n$, 它对应于区间 $[0, 1]$ 中的一实数, 而自然界则自区间 $[0, 1]$ 中任挑一实数 p. 因此, 统计者一切可取策略的集合 X 是 $n+1$ 维立方体, 它是区间 $[0, 1]$ 的 $n+1$ 次笛卡儿乘积, 而自然界的一切可取策略的集合 Y 是区间 $[0, 1]$. 支付函数 $f[\hat{p}(m, n) p]$ 定义为估值 $\hat{p}$ 的误差平方的数学期望乘 -1, 即

$$f[\hat{p}(m,n), p] = -E(\hat{p}(m,n)-p)^2$$
$$= -\sum_{m=0}^{n}[\hat{p}(m,n)-p]^2 C_n^m p^m(1-p)^{n-m},$$

(13.11)

由定义(13.11)易见统计者的利润常常是非正的.

易证上面定义的游戏是开的. 因为其中没有一对互相反对的最佳策略. 为证此, 设结论不对. 即设游戏是闭的, 且某概率 p_0 是自然界的最佳策略, 则统计者反对此策略 p_0 的最佳策略为 $\hat{p}(m, n)=p_0$, 后者不依赖于观察结果而为参数的准确值 p_0, 这时(当自然界采用策略 p_0 时), 只有这策略才保证统计者

的利润最大，即 $f[p_0, p_0]=0$. 但在反对统计者这一策略时，自然界的策略 $p=p_0$ 是最坏的. 自然界反对策略 $\hat{p}(m,n)=p_0$ 的最佳策略是策略 $p=1\left(\text{如 } p_0\leqslant\frac{1}{2}\right)$ 或策略 $p=0\left(\text{如 } p\geqslant\frac{1}{2}\right)$.

因此，上述问题的完全解要从对随机游戏 (Θ, H, φ) 的研究中去寻求. 这里统计者的随机策略集 Θ 包含 $n+1$ 维立方体中的一切概率测度. 选定某一策略 ξ 后，对任一观察结果 (n, m)，统计者决定的不是一个值 $\hat{p}$，而是估值的整个分布 $\xi(\hat{p}; n, m)$ 同时，自然界的策略集 H 是区间 $[0,1]$ 上所有的分布. 支付函数对任一固定的 n 为

$$\varphi(\xi,\eta) = -E(\hat{p}-p)^2$$
$$= -\int_0^1 \sum_{m=0}^n C_n^m p^m (1-p)^{n-m} \int_0^1 (\hat{p}-p)^2 d\xi(\hat{p};n,m) d\eta(p).$$

(13.12)

今设已知自然界的策略为 $\eta(p)$，试求统计者反对此策略的最佳策略，为此要求下式的极大，

$$I = -\int_0^1 C_n^m p^m (1-p)^{n-m} (\hat{p}-p)^2 d\eta(p). \quad (13.13)$$

因为当 (13.12) 极大化后，统计者可不依赖于每一 m 而决定自己的策略. 令 $\dfrac{dI}{d\hat{p}}=0$，易见当 $\hat{p}$ 取值

$$\hat{p}_0(m,n,\eta(p)) = \frac{\int_0^1 p^{m+1}(1-p)^{n-m} d\eta(p)}{\int_0^1 p^m (1-p)^{n-m} d\eta(p)} \quad (13.14)$$

时，(13.13) 取极大. 值 (13.14) 确为使 (13.13) 取极大的点，因 I 为 $\hat{p}$ 的连续、非正函数，而且当 $\hat{p}$ 无限下降或上升时它趋于 $-\infty$. 由于对每一组 n, m 及 $\eta(p)$，(13.13) 在点 (13.14) 上有唯一的极大，可见统计者反对自然界的策略 $\eta(p)$ 的最佳策略为 $\xi(\hat{p};$

n，m），它对每一 n 及 m 以概率 1 等于由（13.14）定义的估值 $\hat{p}_0(m, n, \eta(p))$. 因此，反对自然界的任一策略 $\eta(p)$ 的统计者的最佳策略是非随机策略. 故若 $(\Theta, H, \varphi(\xi, \eta))$ 是闭的，则 $(X, H, \psi(\hat{p}, \eta))$ 也是闭的，其中支付函数 $\psi(\hat{p}, \eta)$ 由下式定义，

$$\psi(\hat{p}, \eta) = -E(\hat{p} - p)^2$$
$$= -\int_0^1 \sum_{m=0}^n C_n^m p^m (1-p)^{n-m} [\hat{p}(n,m) - p]^2 \mathrm{d}\eta(p). \quad (13.15)$$

例如自然界采取策略 $\eta_1(p)$，它是区间 $[0, 1]$ 上的均匀分布（这前提等价于贝叶斯假定）. 则由（13.14）可算出统计者反对此策略的最佳策略为

$$\hat{p}_1(m, n) = \frac{m+1}{n+2}. \quad (13.16)$$

拉普拉斯曾提出此式以估计未知参数 p，但他的后继者批判了它. 若统计者采用策略（13.16），则支付函数为

$$\psi(\hat{p}_1, \eta) = -\int_0^1 \sum_{m=0}^n C_n^m p^m (1-p)^{n-m} \left[\frac{m+1}{n+2} - p\right]^2 \mathrm{d}\eta(p)$$
$$= \frac{-1}{(n+2)^2} \int_0^1 \sum_{m=0}^n C_n^m p^m (1-p)^{n-m} [m^e + 2m(1 - 2p - np) + 1 -$$
$$2p(n+2) + p^2(n+2)^2] \mathrm{d}\eta(p) \quad (13.17)$$
$$= \frac{-1}{(n+2)^2} \int_0^1 [(n-4)p(1-p) + 1] \mathrm{d}\eta(p).$$

在上式最后一步里，我们用了

$$\sum_{m=0}^n C_n^m p^m (1-p)^{n-m} = 1,$$
$$\sum_{m=0}^n m C_n^m p^m (1-p)^{n-m} = np,$$
$$\sum_{m=0}^n m^2 C_n^m p^m (1-p)^{n-m} = n^2 p^2 + np(1-p).$$

由（13.17）式易见，当 $n=4$ 时，被积函数不依赖于 p，且对自

然界的每一策略，支付函数为

$$\varphi(\hat{p}_1, \eta) = \frac{-1}{(n+2)^2} = -\frac{1}{36}. \tag{13.18}$$

因当 $n=4$ 及 $\hat{p}=\hat{p}_1$ 时，支付函数 $\varphi(\hat{p}_1, \eta)$ 不依赖于 η，故在 $n=4$ 的情况下，自然界的每一策略都同样地好，而这表示自然界的每一策略在反对统计者的策略 $\hat{p}=\hat{p}_1$ 时，都是最佳的．特别，策略 $\eta_1(p)$ 是反对 $\hat{p}_1$ 的最佳策略．但因策略 $\hat{p}_1$ 是反对自然界的策略 $\eta_1(p)$ 的最佳策略，故我们找到了一对相互最佳的策略，由定理1，这表示当 $n=4$ 时 $(X, H, \psi(\hat{p}, \eta))$ 是闭的，$\hat{p}_1$ 及 η_1 是统计者及自然界的最佳策略，而(13.18)表游戏的价值．

今试求游戏 $(X, H, \psi(\hat{p}, \eta))$ 当 n 任意时的一般解．为此，不用策略(13.16)而考虑统计者的策略 $\hat{p}_2$，它由更一般的函数表达，即

$$\hat{p}_2(m, n) = \frac{m+a}{n+b}, \tag{13.19}$$

其中 a, b 不依赖于 m，但可能与 n 有关．

重复得到公式(13.17)时所用的计算，可见当统计者利用策略 $\hat{p}_2$ 而自然界采用任一策略 $\eta(p)$ 时，支付函数为

$$\psi(\hat{p}_2, \eta) = \frac{-1}{(a+b)^2} \int_0^1 [(b^2-n)p^2 + (n-2ab)p + a^2] d\eta(p). \tag{13.20}$$

由(13.20)易见当 $a = \frac{1}{2}\sqrt{n}$，$b = \sqrt{n}$ 时，被积函数不依赖于 p，因之函数 $\psi(\hat{p}_2, \eta)$ 不依赖于 η，故得到结论：自然界的每一策略 $\eta(p)$ 在反对统计者的策略

$$\hat{p}_2(m, n) = \frac{m + \frac{\sqrt{n}}{2}}{n + \sqrt{n}} \tag{13.21}$$

时，都是最佳的．要证明估值(13.21)是统计者的最佳策略，必须求出自然界的某一策略，在反对此策略时，策略(13.21)是最好的．为此需求出一分布 $\eta_2(p)$，以此分布代入(13.14)式后，其左方为策略(13.21)． H. Steinhaus 找到了此分布 $\eta(p)$，它的密度为

$$d\eta(p) = c[p(1-p)]^s dp,$$

其中

$$s > -1, \quad \frac{1}{c} = \int_0^1 [p(1-p)]^s dp. \quad (13.22)$$

故由(13.14)得

$$\hat{p}_0(m, n, \eta(p)) = \frac{\frac{1}{c}\int_0^1 p^{s+m+1}(1-p)^{s+n-m} dp}{\frac{1}{c}\int_0^1 p^{s+m}(1-p)^{s+n-m} dp} = \frac{m+s+1}{n+2s+2}.$$

$$(13.23)$$

后一等式像等式(13.16)一样由下式而得

$$\int_0^1 p^\alpha (1-p)^\beta dp = \frac{\Gamma(\alpha+1)\Gamma(\beta+1)}{\Gamma(\alpha+\beta+2)}, \text{及 } \Gamma(\alpha+1) = \alpha\Gamma(\alpha).$$

由(13.23)可见，如想得到策略(13.21)，必须令 $s = \frac{1}{2}\sqrt{n} - 1$．因此，策略(13.21)是统计者反对自然界的策略 $\eta_2(p)$ 的最佳策略，对于 $\eta_2(p)$，p 的概率密度为

$$d\eta_2(p) = [p(1-p)]^{\frac{1}{2}\sqrt{n}-1} \frac{\Gamma(\sqrt{n})}{\left[\Gamma\left(\frac{\sqrt{n}}{2}\right)\right]^2} dp. \quad (13.24)$$

因此我们找到了一对相互最佳的策略，于是证明了，对每一 n，游戏 $(X, H, \psi(\hat{p}, \eta))$ 是闭的，而估值(13.21)是统计者的最佳策略．由公式(13.20)可见游戏 $(X, H, \psi(\hat{p}, \eta))$ 的价值是

$$\psi(\hat{p}_2, \eta) = \frac{-a^2}{(n+b)^2} = \frac{-1}{4(1+\sqrt{n})^2},$$

解上述问题时，H. Steinhaus 不知道它早已解决．J. L. Hodges, Jr. E. L. Lehmann[3] 承认 Herman Rulin 早已解决这问题，但我在此详细地叙述了 H. Steinhaus 的解，因他只用了最基本的方法，而且还可用此法来解一些至今尚未解决的更一般的问题．

较广泛的已解决的问题是要估计未知参数 $p_1, p_2, \cdots, p_k$ ($\sum_{i=1}^{k} p_i = 1$). 设某实验的结果是不相容事件 $A_1, A_2, \cdots, A_k$ 中之一．经 n 次独立实验后，事件 A_1 出现 m_1 次，A_2 出现 m_2 次，$\cdots$，A_k 出现 m_k 次 ($\sum_{i=1}^{k} m_i = n$). 根据观察值 $m_1, m_2, \cdots, m_k$ 我们希望估计未知参数 $p_1, p_2, \cdots, p_k$. 这里我们不想精确地定义对应于此问题的游戏，设估值 $\hat{p}_1, \hat{p}_2, \cdots, \hat{p}_k$ 所引起的损失与

$$E\left[\sum_{i=1}^{k}(\hat{p}_i - p_i)^2\right] \tag{13.25}$$

成正比．若自然界的策略已随机化，则像 $k=2$ 时一样，游戏是闭的．统计者的最佳策略是估值

$$\hat{p}_i = \frac{m_i + \frac{1}{k}\sqrt{n}}{n + \sqrt{n}}, \quad i=1, 2, \cdots, k. \tag{13.26}$$

自然界的一切策略在反对统计者这一策略时都是最好的．统计者的策略(13.26)，在反对自然界的分布密度为

$$\frac{\Gamma(\sqrt{n})}{\left[\Gamma\left(\frac{\sqrt{n}}{k}\right)\right]^k}(p_1 p_2 \cdots p_k)^{\frac{\sqrt{n}}{k} - 1}$$

的策略时是最好的．如统计者采用最佳策略(13.26)，他的利润（恒非正）到达极小中的极大

$$-E\left[\sum_{i=1}^{k}(\hat{p}_i - p_i)^2\right] = \frac{-n}{(n+\sqrt{n})^2} \frac{k-1}{k},$$

它不依赖于自然界的策略，而且是游戏的价值．

参考文献

[1] Steinhaus H. Über einige prinzipielle Fragen der mathematishen statistik. Tagung über Wahrscheinlichkeitsrechnung und Mathematische Statistik. Berlin, 1954, Deutscher Verlag der wissenschaften.

[2] Steinhaus H. The problem of estimation. Ann. Math. Statist., 1957, 28: 633-648.

[3] Hodges J L, Lehmann Jr E L. Some problems in minimax point estimation. Ann. Math. Statist., 1950, 21(2): 182-197.

[4] Savage L J. The foundations of statistics. John Wiley and Sons, New York, 1954 (see 13.4).

[5] Mckinsey J C C. Introduction to the theory of games. The Rand Corporation, New York, 1952.

[6] Girshick M A, Blackwell P. Theory of games and statistical decisions. John Wiley and Sons, New York, 1954.

§14. 关于平面上经验曲线的长度

曲线长度是某曲线近邻中的无界泛函. 两条相当邻近的曲线, 其长度的差可以任意的大. 因此当我们只知道某一曲线的近似, 而要测量它的长度时, 便发生很大困难. 这类问题, 例如当地理学家要测量海岸长度时便发生. 这时甚至很难精确定义什么叫"海岸", 这种曲线实际上没有, 虽然在各种地图上都用曲线来表示"海岸". 如果在各种地图上来测量这些曲线的长, 可能发现当地图的规模增大时, 岸长无限上升. 再看另一个例子. 设地理学家希望比较两条河流的长度, 但他有的只是两张地图, 第一张上绘有第一条河, 另一张上绘有另一条. 如这些图的大小不一样, 地理学家便不会比较它们的长度, 而请求数学家的帮助. 这里我想叙述曲线观念的两种推广. 其一为 H. Steinhaus 的 p 级长的观念, 另一为 J. Perkal 的 ε-长观念.

设有一距离测定器, 它由画有一族平行直线(两相邻线间的距离为 d)的透明薄板构成. 现在要测量图上曲线 s 的长度. 可用以下办法: 把测定器随机地放在曲线 s 上, 算得这曲线与测定器上各直线的交点总数为 n_1, 然后把测定器转移角度 $\frac{\pi}{k}$, 再算得交点数为 n_2; 于是得一组数 $n_1, n_2, \cdots, n_k$ 及数 $N = \sum_{i=1}^{k} n_i$. 曲线 s 的近似长度为

$$L = \frac{1}{2k} N d\pi. \tag{14.1}$$

这近似长度的精确度依赖于 d 及 k. 公式小还不能消除上述长度的奇怪现象, 但可改用下法, 不用 $n_1, n_2, \cdots, n_k$, 而

用数 n_1', n_2', $\cdots$, n_k'，其定义为
$$n_i' = \begin{cases} n_i, & n_i \leqslant p, \ i=1, 2, \cdots, k. \\ p, & n_i > p, \ p > 0. \end{cases}$$
数值
$$L_p = \frac{1}{2k} N' \mathrm{d}\pi, \tag{14.2}$$
其中 $N = n_1' + n_2' + \cdots + n_k'$，逼近于某一泛函，证此泛函为 $L_p(s)$，并称为 p 级长。

J. Perkal 的方法在于化曲线长度的测量为某区域面积的测量。对任一 $\varepsilon > 0$，以 $A_\varepsilon(s)$ 表与 s 的距离不超过数 ε 的全体点集，并称之为曲线 s 的 ε-近邻，即
$$A_\varepsilon(s) = \underset{x}{E}\{p(x, s) \leqslant \varepsilon\}. \tag{14.3}$$

H. Minkowski 定义曲线 s 的长为极限
$$L(s) = \lim_{\varepsilon \to 0} \frac{\alpha_\varepsilon(s)}{2\varepsilon}, \tag{14.4}$$
其中 $\alpha_\varepsilon(s)$ 表区域 $A_\varepsilon(s)$ 的面积。但 J. Perkal 不用极限 (14.4) 而引进曲线 s 的 ε-长 $L^{(\varepsilon)}(s)$ 的观念
$$L^{(\varepsilon)}(s) = \frac{\alpha_\varepsilon(s) - \pi\varepsilon^2}{2\varepsilon}, \tag{14.5}$$
这里要自 $\alpha_\varepsilon(s)$ 中减去半径为 ε 的圆的面积 $\pi\varepsilon^2$，以避免系统误差。为了实际地计算 ε-长 (14.5)，J. Perkal 建议采用圆形距离测定器，它由画有一族圆的透明薄板构成。把测定器放在此曲线的图上后，为要计算 $\alpha_\varepsilon(s)$，就只要计算与曲线 s 有公共部分的圆的个数。

采用 H. Steinhaus 的 p 级长的观念或 J. Perkal 的观念后，各种比较实验曲线长度的问题可化为完全可实现的比较这些曲线的 p 级长或 ε-长的问题。但当长度的严格观念没有意义时（例如海岸长），ε-长的观念可能非常宝贵并能很好的符合于地理学家的直觉。

参考文献

[1] Steinhaus H. On the length of empirical curves. Časopis Pro pestovdni Matematiky a Fysiky, 1949, 74.

[2] Steinhaus H. Length, shape and area. Colloquium Mathematicum, 1954, 3: 1-13.

[3] Perkal J. On the ε-length. Bulletin de l'Academie Polonaise des Sciences, 1956, 4: 399-403.

[4] Minkowski H. Gesammelte Abhandlungen. Leipzig-Berlin, 1911, 2.

§15. 某些生物问题中贝叶斯公式的应用

在每本概率论教程中都可找到贝叶斯公式

$$P(A_i \mid B) = \frac{P(A_i)P(B \mid A_i)}{p(A_1)P(B \mid A_1)+P(A_2)P(B_2 \mid A_2)+\cdots+P(A_n)P(B \mid A_n)}. \quad (15.1)$$

由它可以计算当某事件 B 发生后，一组不相容的且唯一可能的原因 A_1，A_2，$\cdots$，A_n 中任一个的后验概率 $P(A_i \mid B)$，这时只要利用已知的先验概率 $P(A_i)$ 及在原因 A_i 下事件 B 的条件概率 $P(B \mid A_i)$.

这公式在实际中很少应用，因为在多数情况下原因 A_i（$i=1$，2，$\cdots$，n）的先验概率 $P(A_i)$ 未知，这时，常采用贝叶斯假定：一切原因的先验概率相同. 这假定在未经验证前，可能产生错误的结论. 由于贝叶斯假定常被严厉批判，某些作者便把这些批判理解为对公式(15.1)的批判. 但后者是概率论中公理系的推论，在未引进贝叶斯假定以前，它完全是正确的. 例如 W. Feller 在其名著[1]中只用小字列出公式(15.1)及对它的批评. 在讲问题 4 时我们曾多次运用连续型的公式(15.1)，并指出在某些理论中，按作者的意见认为已克服未知先验分布及其他有关的困难，但实际上这些理论等价于贝叶斯假定. 这篇报告里我们说明，在我们的应用数学工作中，曾多次碰到公式(15.1)完全可合理地应用的情况.

第一个应用贝叶斯公式的例是根据血清分析的结果来研究父子关系. H. Steinhaus 第一个指出在某些情况下可计算父子

关系的概率. 这方法已由我研究清楚并在解决父子关系的具体审判事务中得以应用. 结果发现, 概率论与贝叶斯公式在这些事务中给出了客观的判决方法. 此外根据血清分析的资料及法院的判决可以统计地估计判决与资料的符合程度. 这些结果在法律学者中引起了长久热烈的讨论, 因为看来运用概率论于审判中还是第一回.

第二个例子是应用贝叶斯公式来研究孪生的胚胎发生问题(双生、三生或四生)在我离开波兰的前些日子曾将我与伏洛茨瓦夫的医生 T. K. Nowakowski(他是热心于把数学方法运用于医学的人)合作的文章投去付印, 在文章中指出了计算某双生, 三生或四生由一个卵子或多个卵子发展起来的概率, 在具体情况下, 这些概率是根据许多兄弟或姊妹与其父母的形态学上的标志来计算的. 这时公式(15.1)中的先验概率是从双生, 三生及四生根据男、女性别的各种结合的频率观察的统计资料中计算出来的, 而条件概率则根据所观察标志的嫡系遗传法则来计算. 利用我们的方法, 可以断定, 例如, 1954 年在伏洛茨瓦夫诞生的四生以概率 0.964 是由一卵发展起来的.

参考文献

[1] Feller W. An introduction to probability theory and its applications, Vol. 1. John Wiley and Sons, New York, 1950.

[2] Steinhaus H. O dochodreniu ojcostwa. Zastosowania Matematyki, 1954, 1.

[3] Steinhaus H. The establishment of paternity. Prace wrocławskiego Towarzystwa Naukowego A. 1954, 32.

[4] Łukaszewicz J. O dochodzeniu ojcostwa. Zastosowania Matematyki, 1956, 2.

[5] Łukaszewicz J, Nowakwski T K. Obliczanie prawdopodobieństwa jednojajowości, Czworaczków. Zastosowanta Matematyki, 1960, 5(1): 119-139.

南开大学学报(自然科学版),1964,5(5)

扩散过程在随机时间替换下的不变性

§1. 扩散过程的随机时间替换

采用[1]中的定义与记号. 我们的目的是研究在随机时间替换下,扩散过程仍然变为扩散过程的充分必要条件,并讨论联系于典范扩散过程的微分方程及特征算子方程的一些性质. 扩散过程的定义见[1,P217].

设(E, F)为光滑流形,维数为l[1,P214][①],$X=(x_t, \zeta, \mathcal{M}_t, P_x)$为$(E, F)$中的扩散过程(因而自然是马氏过程),不失一般性,可以假定X是完全的[1,P123]. X的特征算子记为$\mathcal{A}$[1,P203],导出微分算子记为D[1,P221]. 任取$x \in E$,以$\mathcal{D}(x)$表示一切在点x的邻域内有定义而且二次连续可微的函数的集,以F_x表示一切定义域包含x的坐标系$\psi \in F$的集,则对任意$f \in \mathcal{D}(x)$,任意$\psi = (x^1, x^2, \cdots, x^l) \in F_x$,有[1,P215]

---

① 这里光滑流形的定义见[1,P214].

扩散过程在随机时间替换下的不变性

$$Df(x) = \frac{1}{2}\sum_{i,j=1}^{l} a^{ij}(x)\frac{\partial^2 f(x)}{\partial x^i \partial x^j} + \sum_{i=1}^{l} b^i(x)\frac{\partial f(x)}{\partial x^i} - c(x)f(x),$$
(1.1)

其中对任意 x，矩阵 $(a^{ij}(x))$ 非负定，$c(x) \geqslant 0$；换句话说，D 是非负定的二阶微分算子．

对 D，引进条件：对任意 $x \in E$，

$$\sum_{i,j=1}^{l}|a^{ij}(x)| + \sum_{i=1}^{l}|b^i(x)| + |c(x)| > 0. \quad (1.2)$$

现在考虑 X 的齐次连续可加泛函 $\varphi_t^s(\omega)$ $(0\leqslant s\leqslant t<\zeta(\omega))$，[1，第 6 章]，假定

$$\varphi_s^s(\omega) = 0, \quad 0\leqslant s<\zeta(\omega), \quad (1.3)$$

$$\varphi_t^s(\omega) > 0, \quad 0\leqslant s<t<\zeta(\omega). \quad (1.4)$$

令 $\widetilde{\zeta}(\omega) = \varphi_{\zeta-0}(\omega)$，$\tau_t(\omega) = \sup(u: \varphi_u^0(\omega)\leqslant t)$，如 $0\leqslant t<\widetilde{\zeta}(\omega)$．已经知道 [1，定理 10.10]，$\widetilde{X} = (x_{\tau_t}, \widetilde{\zeta}, \mathscr{M}_{\tau_t}, P_x)$ 是 E 中的齐次马氏过程，称 $\widetilde{X}$ 为自 X 经 φ_t^s 变换得来，以下简记 φ_t^0 为 $\varphi(t)$．

试研究何时 $\widetilde{X}$ 也是扩散过程．

定理 1.1 设 X 为完全的扩散过程，满足 (1.2)；$\varphi_t^s(\omega)$ $(0\leqslant s\leqslant t<\zeta(\omega))$ 是 X 的齐次连续可加泛函，满足 (1.3)(1.4)；$\widetilde{X}$ 为自 X 经 φ_t^s 变换得来的马氏过程．则 $\widetilde{X}$ 是扩散过程的充分必要条件是

$$C-\lim_{U\downarrow x}\frac{E_x\varphi(\eta_U)}{E_x\eta_U} = a(x) > 0, \quad \text{一切 } x \in E \quad (1.5)$$

(U 为含 x 的开集，其闭包 $\overline{U}$ 紧，η_U 为首出 U 的时间 [1，P152)][①]．此时 $\widetilde{X}$ 的特征算子 $\widetilde{\mathscr{A}}$ 在点 x 的定义域 $\mathscr{D}_{\widetilde{\mathscr{A}}}(x)$ 包含 $\mathscr{A}$ 在点 x 的定义域 $\mathscr{D}_{\mathscr{A}}(x)$，而且对 $f \in \mathscr{D}_{\mathscr{A}}(x)$，有

[①] 因 X 连续，η_U 重合于 $0+$ 后首出 U 的时刻；(1.5)中极限的详细意义见[1，P201]，(1.5)中 C 表 E 中的拓扑．

$$\widetilde{\mathcal{O}} f(x) = \mathcal{O} f(x)/a(x). \tag{1.6}$$

若补设(1.5)中的极限 $a(x) < +\infty$,则

$$\mathcal{D}_{\widetilde{\mathcal{O}}}(x) = \mathcal{D}_{\mathcal{O}}(x). \tag{1.7}$$

证 由 X 的连续性,$E_x \eta_U > 0$ $(x \in U)$. 任取坐标系 $\psi = (x^1, x^2, \cdots, x^l) \in F_{x_0}$,令

$$\Delta^{ij}(x) = (x^i - x_0^i)(x^j - x_0^j), \quad \Delta^i(x) = x^i - x_0^i,$$

则有

$$a^{ij}(x_0) = \mathcal{O} \Delta^{ij}(x_0) = C - \lim_{U \downarrow x_0} \frac{\int_{U'} \Delta^{ij}(x) \pi_U(x_0, \mathrm{d}x)}{E_{x_0} \eta_U}, \tag{1.8}$$

$$b^i(x_0) = \mathcal{O} \Delta^i(x_0) = C - \lim_{U \downarrow x_0} \frac{\int_{U'} \Delta^i(x) \pi_U(x_0, \mathrm{d}x)}{E_{x_0} \eta_U}, \tag{1.9}$$

$$c(x_0) = -\mathcal{O} 1(x_0) = C - \lim_{U \downarrow x_0} \frac{1 - \pi_U(x_0, U')}{E_{x_0} \eta_U}, \tag{1.10}$$

其中 $U' = \overline{U} \cap (E \setminus U)$ 是 U 的边界,而

$$\pi_U(x_0, \Gamma) = P_{x_0}(x_{\eta_U} \in \Gamma), \quad \Gamma \subset U'.$$

由(1.2)(1.8)(1.9)(1.10)知:对 x_0 的充分小邻域 V(精确些,即存在含 x_0 的开集 U,$\overline{U}$ 紧,使对一切开集 V, $x_0 \in V \subset U$,$\overline{V}$ 紧),有

$$E_{x_0} \eta_V < +\infty (x_0 \in E).$$

以下对 $\widetilde{X}$ 的量都于其上标以"～"号,对每固定的 ω,$\tau_t(\omega)$ 把 $[0, \widetilde{\zeta}(\omega)]$ 连续地变到 $[0, \zeta(\omega)]$ 上,故由 X 的连续性得 $\widetilde{X}$ 的连续性,而且[1, P447]

$$\pi_U(x, A) = \widetilde{\pi}_U(x, A).$$

因此,若 $f \in \mathcal{D}_{\widetilde{\mathcal{O}}}(x)$,则

$$\widetilde{\mathcal{O}} f(x) = C - \lim_{U \downarrow x} \frac{\int_{U'} f(y) \pi_U(x, \mathrm{d}x) - f(x)}{E_x \widetilde{\eta}_U}, \tag{1.11}$$

其中由 $\widetilde{X}$ 的连续性知 $E_x\widetilde{\eta}_U > 0$ $(x \in U)$，回忆

$$\mathcal{O}\!f(x) = C - \lim_{U \downarrow x} \frac{\int_{U'} f(y) \pi_U(x, \mathrm{d}y) - f(x)}{E_x \eta_U}. \quad (1.12)$$

今对任意固定的 $x_0 \in E$，定义函数

$$g(x) = \begin{cases} \sum_{i,j=1}^{l} e_{ij}(x^i - x_0^i)(x^j - x_0^j), & \sum_{i,j=1}^{l} |a^{ij}(x_0)| > 0, \\ \sum_{i=1}^{l} e_i(x^j - x_0^i), & \sum_{i,j=1}^{l} |a^{ij}(x_0)| = 0, \\ & \text{但} \sum_{i=1}^{l} |b^i(x_0)| > 0, \\ -1, & \sum_{i,j=1}^{l} |a^{ij}(x_0)| + \sum_{i=1}^{l} |b^i(x_0)| = 0 \\ & （此时由(1.2) c(x_0) > 0）, \end{cases}$$

(1.13)

其中 e_{ij}（及 e_i）为任意常数，满足

$$\sum_{i,j=1}^{l} e_{ij} a^{ij}(x_0) \neq 0, e_{ij} = e_{ji} \left(\text{及} \sum_{i=1}^{l} e_i b^i(x_0) \neq 0 \right).$$

由于 X 是扩散而且 $g(x) \in \mathscr{D}(x_0)$，显然

$$g(x) \in \mathscr{D}_{\mathcal{O}}(x_0), \quad \mathcal{O}g(x_0) \neq 0 \quad (1.14)$$

（后一式由 $g(x)$ 的定义直接算出）。由此及(1.12)可见：对一切含 x_0 的充分小的邻域 U，有

$$\int_{U'} \pi_U(x_0, \mathrm{d}y) g(y) - g(x_0) \neq 0.$$

今设 $\widetilde{X}$ 扩散。由扩散过程的定义，函数 $1, \Delta^i(x), \Delta^{ij}(x)$ 都属于 $\mathscr{D}_{\widetilde{\mathcal{O}}}(x_0)$，所以 $g(x) \in \mathscr{D}_{\widetilde{\mathcal{O}}}(x_0)$，而且

$$0 \leqslant C - \lim_{U \downarrow x_0} \frac{E_{x_0} \widetilde{\eta}_U}{E_{x_0} \eta_U}$$

$$= C-\lim_{U\to x_0}\frac{\left[\iint_{U'}\pi_U(x_0,dy)g(y)-g(x_0)\right]\div E_{x_0}\eta_U}{\left[\iint_{U'}\pi_U(x_0,dy)g(y)-g(x_0)\right]\div E_{x_0}\tilde{\eta}_U}$$

$$=\frac{\mathcal{O}l\ g(x_0)}{\widetilde{\mathcal{O}l}\ g(x_0)}\leqslant +\infty. \tag{1.15}$$

(1.15)中极限不可能等于 0，实际上，(1.14)中已证 $\mathcal{O}l\ g(x_0)\neq 0$；由于 $g\in\mathscr{D}_{\widetilde{\mathcal{O}l}}(x_0)$，故 $\widetilde{\mathcal{O}l}\ g(x_0)$ 有穷，从而(1.15)中 $\frac{\mathcal{O}l\ g(x_0)}{\widetilde{\mathcal{O}l}\ g(x_0)}\neq 0$. 其次，(1.15)中的 $\frac{E_{x_0}\tilde{\eta}_U}{E_{x_0}\eta_U}\geqslant 0$，故得证

$$0<a(x_0)=C-\lim_{U\downarrow x_0}\frac{E_{x_0}\tilde{\eta}_U}{E_{x_0}\eta_U}\leqslant +\infty \quad (\text{一切}\ x_0\in E). \tag{1.16}$$

注意，由于 X，φ_t^i 都连续，而且满足(1.4)，故

$$E_{x_0}\tilde{\eta}_U=E_{x_0}\varphi(\eta_U) \quad (x_0\in U), \tag{1.17}$$

由此及(1.17)即得证第一结论的必要性部分.

今设对一切 x，存在极限(1.5)，则 $\widetilde{X}$ 必扩散，为此，只要证 $1,\ x^i,\ x^i x^j\in\mathscr{D}_{\widetilde{\mathcal{O}l}}(x)$ 对某一坐标系成立. 但由假定 X 扩散，故对任一坐标系[1，P218]，$1,\ x^i,\ x^i x^j\in\mathscr{D}_{\mathcal{O}l}(x)$，因而只要证 $\mathscr{D}_{\mathcal{O}l}(x)\subset\mathscr{D}_{\widetilde{\mathcal{O}l}}(x)$，任取 $f\in\mathscr{D}_{\mathcal{O}l}(x)$，有

$$C-\lim_{U\downarrow x}\frac{\int_{U'}\pi_U(x,dy)f(y)-f(x)}{E_x\tilde{\eta}_U}$$

$$=C-\lim_{U\downarrow x}\frac{\int_{U'}\pi_U(x,dy)f(y)-f(x)}{E_x\eta_U}\times\frac{E_x\eta_U}{E_x\tilde{\eta}_U}. \tag{1.18}$$

注意(1.17)及(1.5)，即得上式中的极限为 $\frac{\mathcal{O}l\ f(x)}{a(x)}$，即 $f\in D_{\widetilde{\mathcal{O}l}}(x)$ 而且(1.6)式成立，从而第一、第二结论都完全证明了.

最后设 $a(x)<+\infty$. 为证(1.7)，只要证 $\mathscr{D}_{\widetilde{\mathcal{O}l}}(x)\subset\mathscr{D}_{\mathcal{O}l}(x)$.

任取 $g \in \mathscr{D}_{\widetilde{\mathrm{o}}}(x)$，以此 g 代入 (1.18) 中的 f，并在 (1.18) 中将 $\widetilde{\eta}_U$，η_U 互换，即得
$$\mathrm{o}\!\!\!/\, g(x) = \widetilde{\mathrm{o}\!\!\!/}\, g(x) \cdot a(x),$$
故 $g \in \mathscr{D}_{\mathrm{o}\!\!\!/}(x)$，并且附带证明了：$X$，$\widetilde{X}$ 的导出微分算子的系数成比例. ∎

注 1.1 由 (1.5) 及定理 1.1 第三结论可见，如果两可加泛函都对同一 $a(x)$ 满足 (1.5)，$0 < a(x) < +\infty$，一切 x，那么它们产生的两 $\widetilde{X}$ 具有相同的 $\widetilde{\mathrm{o}\!\!\!/}$，因此如固定一个 $a(x)$，可加泛函 φ 的选择有很大的自由性，特别，若 $a(x)$ 连续，则只要取 φ 为积分型泛函
$$\varphi_t^s = \int_s^t a(x_u) \mathrm{d}u.$$

实际上，由
$$\frac{E_x[a(x) - \varepsilon]\tau_U}{E_x \tau_U} \leqslant \frac{E_x \int_0^{\tau_U} a(x_u) \mathrm{d}u}{E_x \tau_U} \leqslant \frac{E_x[a(x) + \varepsilon]\tau_U}{E_x \tau_U},$$
可见 $\quad 0 < a(x) = C - \lim\limits_{U \downarrow x} \dfrac{E_x \varphi(\tau_U)}{E_x \tau_U} < +\infty.$

注 1.2 设 X 为 Wiener 过程，φ 为对应于测度 μ 的 S 泛函 [1, P374]，这时 E 为 l 维欧氏空间，
$$C - \lim_{U \downarrow x} \frac{E_x \varphi(\tau_U)}{E_x \tau_U} = C - \lim_{U \downarrow x} \frac{\int_U g(x,y) \mu(\mathrm{d}y)}{\int_U g(x,y) \mathrm{d}y}.$$
$g(x, y)$ 为过程 X 在有界开集 U 上的格林函数 [1, P566].

§2. Dirichlet 问题与特征算子方程

以下设 E 为 l 维欧氏空间，在 E 中考虑微分算子
$$D = \sum_{i,j=1}^{l} a^{ij}(x) \frac{\partial^2}{\partial x^i \partial x^j} + \sum_{i=1}^{l} b^i(x) \frac{\partial}{\partial x^i}. \tag{2.1}$$
假定 D 的系数满足条件

(i) $a^{ij}(x)$, $b^i(x)$ ($i, j = 1, 2, \cdots, l$) 有界而且在 E 中满足具同一指数 $\lambda > 0$ 的 Hölder 条件；

(ii) 存在常数 $\gamma > 0$，使对任一 $x \in E$，任一组实数 λ_1, $\lambda_2, \cdots, \lambda_l$，有
$$\sum_{i,j=1}^{l} a^{ij}(x) \lambda_i \lambda_j \geqslant \gamma \sum_{i=1}^{l} \lambda_i^2,$$
则必存在唯一的不断的[①]扩散过程 X_D，转移密度 $p(t, x, y)$ 是 $\frac{\partial u}{\partial t} = Du$ 的基本解，而且导出微分算子是 D，称此 X_D 为典范扩散过程[1, P238]，它的特征算子记为 $\mathscr{A}$，样本函数记为 $x_i(\omega)$, $t \geqslant 0$。以下所谓 "G 有规则边界" 是对 X_D 而言[1, P537]。

定理 2.1 设 D 满足(i)(ii)，对任意具有规则边界的区域 G（使 $\overline{G} \subset E$）及 $G' [= \overline{G} \cap (E \setminus G)]$ 上的任意连续函数 $f(x)$，外 Dirichlet 问题
$$\begin{cases} Du = 0, & x \in E \setminus G, \\ u = f, & x \in G', \end{cases} \tag{2.2}$$
有唯一有界解的充分必要条件是
$$\lim_{t \to +\infty} |x_t| = 0 \ (P_x\text{-几乎，一切 } x \in E). \tag{2.3}$$

[①] 唯一性见[1, P799]，不断性由赵昭彦证明。

证 设 U 为任一非空有界开集，τ 表示首出 U 的时刻，由于对 X_D，$p(t, x, y) > 0$ 对一切 $t > 0$ 及 $x, y \in E$ 成立，故 $p_x(\tau = +\infty) < 1$ (一切 $x \in E$). 既然 X_D 还是连续的强 Feller 过程，故可用 [1, 定理 13.7] 而知 $E_x \tau$ 有界，于是更有

$$P_x(\tau < +\infty) = 1, \quad 一切 \ x \in E. \tag{2.4}$$

今取 $U_n = [x : |x| < n]$，n 为正整数，以 τ_n 表首出 U_n 时刻，由 (2.4)，$P_x(\tau_n < +\infty) = 1$. 不妨设 X_D 为标准过程，由标准性知 [1, P161] $P_x(\tau_n \to +\infty) = 1$. 既然 $|x_{\tau_n}| = n$，得

$$\varlimsup_{t \to +\infty} |x_t| = +\infty \quad (P_x\text{-几乎，一切 } x \in E).$$

由此易知条件 (2.3) 与 X_D 的常返性等价 [2]，因而也与外 Dirichler 问题有唯一有界解等价. ∎

注 2.1 由上面所证的 $P_x(\tau_n \to +\infty) = 1$ 及 [3, 引理 4.1]，知方程

$$\mathscr{A} u - u = 0$$

在 E 中无正有界解.

定理 2.2 设开集 G 有紧闭包而且有规则边界，$V(x)$ ($x \in G$) 是有界非负连续函数，则存在 $\beta > 0$，使对一切正数 $\lambda < \beta$，方程

$$\begin{cases} \mathscr{A} u + \lambda V_u = 0, & x \in G, \\ u = 1, & x \in G', \end{cases} \tag{2.5}$$

在 $\bar{G}$ 中有正解.

证 由 (2.4) 及 [1, P544]，存在 $t > 0$ 及 $\alpha > 0$，使

$$\sup_{x \in E} P_x(\tau > t) = \alpha < 1. \tag{2.6}$$

τ 为首出 G 的时刻 (因而它不大于 $0+$ 后首出 G 时刻)，其次令 $\xi = \int_0^\tau V(x_s) ds$，试证

$$F_\lambda(x) = E_x e^{\lambda \xi} < +\infty, \quad 0 \leqslant \lambda < \beta, \tag{2.7}$$

其中 β 是某正常数. 实际上，如 $\|V\| = \sup_{x \in G} |V(\lambda)| = 0$，结

论显然正确. 否则令 $s = \|V\| t$, 有

$$E_x e^{\lambda \xi} \leq \sum_{n=0}^{+\infty} e^{\lambda(n+1)s} P_x(\xi\tau > ns)$$

$$\leq \sum_{n=0}^{+\infty} e^{\lambda(n+1)s} P_x(\|V\|\tau > ns)$$

$$\leq \sum_{n=0}^{+\infty} e^{\lambda(n+1)s} P_x(\tau > nt)$$

$$\leq \sum_{n=0}^{+\infty} e^{\lambda\|V\|(n+1)t} \alpha^n$$

$$= \frac{e^{\lambda\|V\|t}}{1 - e^{\lambda t\|V\|}\alpha} \equiv A_\lambda,$$

故如 $e^{\lambda t\|V\|}\alpha < 1$ 即 $\lambda < -\dfrac{\ln \alpha}{t\|V\|} = \beta$ 时,

$$E_x e^{\lambda \xi} \leq A_\lambda < +\infty \quad (\text{一切 } x).$$

既然 X_D 是强 Feller 过程, 根据[4, 定理1], 知由(2.7)定义的 $F_\lambda(x)$ 是(2.5)的正解. ∎

最后, 附带指出典范扩散过程 X 的一个性质. 设 $f(x)$ ($x \in E$) 是有界 Borel 可测函数, 令

$$R_\lambda f(x) = \int_0^{+\infty} e^{-\lambda t} T_t f(x) dt \quad (\lambda > 0), \tag{2.8}$$

T_t 为过程的半群, 则 $R_\lambda f(x)$ 是 x 的有界连续函数; 又若 $f \in \hat{C}$, 则 $R_\lambda f \in \hat{C}$.

实际上, 因 X 是强 Feller 的, 故 $T_t f(x)$ 对 x 连续, 由于 $|e^{-\lambda t} T_t f| \leq \|f\| e^{-\lambda t}$, 而后者对 t 可积, 故可在(2.8)中的积分号下取极限而知 $R_\lambda f(x)$ 对 x 连续, 它显然还有界:

$$\|R_\lambda f\| \leq \frac{\|f\|}{\lambda}.$$

其次, 因 X 也是 $\hat{C}$ 过程, 故如 $f \in \hat{C}$, 则 $T_t f \in \hat{C}$, 于(2.8)中令 $x \to +\infty$ 即得 $R_\lambda f(x) \to 0$, 既然已证 $R_\lambda f$ 有界连续, 故 $R_\lambda f \in \hat{C}$.

换言之, X 满足 Hunt 条件(C)[5].

参考文献

[1] Дынкин Е Б. Марковские процессы. Москва: Госуд. нздат. Физмат. Литер，1963(英译本见本书下篇的文献[6])．

[2] 王梓坤. 常返马尔可夫过程的某些性质. 数学学报，1966，16(2)：166-178.

[3] Хасьминский Р З. Эргодические свойства возвратных диффузионных процессов и стабнлизация решений задачн Коши для параболических уравнений. Теоря Вероятн. и ее Примен，1960，5：196-214.

[4] Хасьминский Р З. О положительных решениях уравнения $\mathscr{A} u+Vu=0$. Теоря Вероятн. и ее Примен，1959，4：332-341.

[5] Hunt G A. Markov processes and potentials. Illinois J. Math.，1957，1：44-93，316-369，1958，2：154-213.

生灭过程的遍历性与 0-1 律

§1. 基本概念与特征数

设 $X=\{x(t),\ t\geqslant 0\}$ 是定义在概率空间 $(\Omega,\mathscr{B},P)$ 上的具有可列多个状态的齐次马氏过程，$\Omega=(\omega)$，相空间为 $E=\mathbf{N}$，转移概率 $P(t)=(P_{ij}(t))(i\geqslant 0,\ i,\ j\in E)$ 满足条件：对任一 $i\in E$ 有

$$P_{ij}(t)\geqslant 0,\ \sum_{j\in E}P_{ij}(t)=1, \tag{1.1}$$

$$\sum_{k\in E}P_{ik}(t)P_{kj}(s)=P_{ij}(t+s),\ t\geqslant 0, s\geqslant 0, \tag{1.2}$$

$$\lim_{t\to 0}P_{ii}(t)=1. \tag{1.3}$$

由此知存在极限

$$\lim_{t\to 0}\frac{P_{ij}(t)-\delta_{ij}}{t}=q_{ij},\quad i,j\in E, \tag{1.4}$$

其中 $\delta_{ii}=1$，$\delta_{ij}=0\ (i\neq j)$. 以后恒设

$$0<\sum_{j\neq i}q_{ij}=-q_{ii}\equiv q_i<+\infty,\ i\in E, \tag{1.5}$$

称 $Q=(q_{ij})$ 为密度矩阵，而 X（或 $P(t)$）称为 Q 过程，以表示它

与 Q 有(1.4)的关系,特别,如

$$\begin{cases} q_{i,i+1}=b_i\ (>0),\quad q_{i,i-1}=a_i\ (>0),\\ q_{ii}=-(a_i+b_i),\quad q_{ij}=0\ (\mid i-j\mid >1) \end{cases} \quad (1.6)$$

(补定义 $a_0=0$),称这种 Q 过程为生灭过程.

我们的目的是:设已给(1.6)形的矩阵 Q,试讨论以此 Q 为密度矩阵的全体 Q 过程的遍历性、0-1 律以及过分函数的性质.

令 N^t 为由 $\{x(u),u\geqslant t\}$ 所产生的 σ 代数,$\Pi=\bigcap\limits_{t\geqslant 0} N^t$. 对 $i\in E$,令

$\mathcal{O}\!\!\!I_i=(A:A\in\Pi$,而且对任一 $t\geqslant 0$,有 $P_i(\theta_t A\triangle A)=0)$,其中 θ_t 表过程的推移算子,$\triangle$ 表示对称差,P_i 表示开始分布集中在 i 的条件概率,以后 E_i 表示关于 P_i 的数学期望. 最后令 $\mathcal{O}\!\!\!I=\bigcap\limits_{i\in E}\mathcal{O}\!\!\!I_i$,$\mathcal{O}\!\!\!I$ 中的集称为不变集. 如对任一不变集 A 及任一 $i\in E$,有 $P_i(A)=0$ 或 1,就说对 X,0-1 律成立.

不妨设 X 是完全可分的 Borel 可测过程,令

$$\eta_i(\omega)=\inf(t:t>\tau(\omega),\ x(t,\omega)=i), \quad (1.7)$$

其中 $\tau(\omega)$ 为第一个跳跃点,即

$$\tau(\omega)=\inf(t:x(t,\omega)\neq x(0,\omega)), \quad (1.8)$$

状态 i 称为常返的,如 $P_i(\eta_i<+\infty)=1$;称为遍历的,如 $E_i\eta_i<+\infty$;如一切状态都常返(遍历),就说过程 X 常返(遍历). 显然遍历必常返.

引进数字特征

$$m_i=\frac{1}{b_i}+\sum_{k=0}^{i-1}\frac{a_i a_{i-1}\cdots a_{i-k}}{b_i b_{i-1}\cdots b_{i-k}b_{i-k-1}},\ i\geqslant 0; \quad (1.9)$$

$$e_i=\frac{1}{a_i}+\sum_{k=0}^{+\infty}\frac{b_i b_{i+1}\cdots b_{i+k}}{a_i a_{i+1}\cdots a_{i+k}a_{i+k+1}},\ i>0; \quad (1.10)$$

$$R=\sum_{i=0}^{+\infty}m_i,\quad S=\sum_{i=1}^{+\infty}e_i; \quad (1.11)$$

$$z_0 = 0, z_n = 1 + \sum_{k=1}^{n-1} \frac{a_1 a_2 \cdots a_k}{b_1 b_2 \cdots b_k}, z = \lim_{n \to +\infty} z_n. \quad (1.12)$$

这些数字的概率意义见[1]，粗略地说，R 是质点沿过程的轨道自零出发首次到达 $+\infty$ 的平均时间，而 S 是自 n 出发首次到达 0 的平均时间(改造 n 为反射壁)且当 n 趋于 $+\infty$ 时的极限．

如果 $R=+\infty$，在[2]中已证明：X 常返的充分必要条件是 $z=+\infty$；X 遍历的充分必要条件是 $z=+\infty$，$e_1<+\infty$．因此我们只要研究 $R<+\infty$ 时 X 的常返性与遍历性，结果证明了：如 $R<+\infty$，一切生灭过程都是遍历的，因而都是常返的．其次证明了：对一切生灭过程(不论 R 是否等于 $+\infty$)，0-1 律成立．最后考虑生灭过程的过分函数 $\{f_n\}$[①]，结果发现：必存在极限 $\lim_{n \to +\infty} f_n$，特别，若 $R<+\infty$，或 $R=+\infty$ 且 $z=+\infty$，则 $f_n = f$(常数)．这样，生灭过程的遍历性、常返性和 0-1 律成立的问题得以完满解决．

---

① 由于 E 可列，定义在 E 上的函数是一序列．

§2. 常返性和遍历性

由(1.6)知 $P_{ij}(t)>0$(一切 $t>0$, $i,j\in E$),即一切状态互通,故为证 X 常返(遍历),只要证某一状态常返(遍历)就够了.

以后无特别声明时,总设已给的 Q 是满足(1.6)的矩阵.

定理 2.1 若 $R<+\infty$,则一切 Q 过程 X 常返.

证 由于 $R<+\infty$,Q 过程不唯一,例如 Doob 过程就是一种 Q 过程(Doob 过程的定义见[1]或[3],它由 Q 及 E 上一概率分布 $\pi=(\pi_0,\pi_1,\pi_2,\cdots)$ 所决定,故记为 (Q,π) 过程.全体 Q 过程的集记为 $\{A\}$,全体 Doob 过程记为 $\{D\}$.

先设 $X\in\{D\}$,试证 X 常返.实际上,由于 $R<+\infty$,存在一列飞跃点[1] $\tau_1<\tau_2<\cdots<\tau_n\to+\infty$.不妨设 $P(x(0)=i)=1$.在 $[\tau_k,\tau_{k+1}]$ 中没有 i 区间(定义见[3])的概率为 $\sum_{j>i}\pi_j(1-c_{ji})<1$,其中 $c_{ji}=\dfrac{z-z_j}{z-z_i}<1$(参看[1]).因此,在 $[\tau_1,\tau_{n+1}]$ 中没有 i 区间的概率为 $\left[\sum_{j>i}\pi_j(1-c_{ji})\right]^n$,当 $n\to+\infty$ 时,此概率趋于 0,故自 i 出发,回到 i 的概率为 1,于是知 X 常返.

今设 $X\in\{A\}$,$P(x(0)=i)=1$,由[1]知 X 是一列 Doob 过程 X_n 的极限,而且 X_n 的任一 i 区间保留为 X 的 i 区间,故 X 也常返.

定理 2.2 若 $R<+\infty$,则一切 Q 过程 X 遍历.

证 分两种情况考虑:先设 $S<+\infty$.以 m_{ij} 表示自 i 出发

首次到达 j 的平均时间[①],即 $m_{ij}=E_i\eta_j$,则

$$m_{ij}=\frac{1}{a_i+b_i}+\frac{a_i}{a_i+b_i}m_{i-1,i}+\frac{b_i}{a_i+b_i}m_{i+1,i}.$$

若 $X\in\{D\}$,不难看出 $m_{i-1,i}<R$,$m_{i+1,i}\leqslant R+S$,则

$$m_{ii}\leqslant\frac{1}{a_i+b_i}+R+S<+\infty,$$

即 m_{ii} 对一切 $X\in\{D\}$ 有界. 若 $X\in\{A\}$,取一列 $X_n\in\{D\}$,使 X_n 在[1]的意义下收敛于 X[②],对 X_n 及 X 分别用(1.7)式定义的随机变量记为 $\eta_i^{(n)}$ 及 η_i,则 $\eta_i^{(n)}\uparrow\eta_i$,由单调收敛定理得

$$\lim_{n\to+\infty}m_{ii}^{(n)}=\lim_{n\to+\infty}E_i\eta_i^{(n)}=E_i\eta_i,$$

既然 $m_{ii}^{(n)}\leqslant\frac{1}{a_i+b_i}+R+S$,故 $E_i\eta_i<+\infty$ 而 X 遍历.

次设 $S=+\infty$,这时任一 X 的遍历性已在[1]中证明,参看[1]中(8.10)式.

注 2.1 定理 2.1 的结论已含于定理 2.2 中,为了说明证明的方法(见[1,P168 iv[③]]),我们单独予以证明.

---

① m_{ii} 应理解为自 i 出发,离开 i 后首次回到 i 的平均时间.
② 即,以概率 1,X_n 的样本函数 $x_n(t,\omega)$ 关于勒贝格测度对几乎一切 t 收敛于 X 的样本函数 $x(t,\omega)(n\to+\infty)$.
③ 见本卷第 5 篇 §9:(iv). ——编者

§3. 过分函数极限的存在性和 0-1 律

称 Q 过程 X 为规则的,如果它完全可分、Borel 可测,而且它的样本函数在任一有限 t 区间内以概率 1 只有有限多个跳跃点. 因而规则 Q 过程的样本函数以概率 1 是跳跃函数. 跳跃点列设为 $\{\xi_n\}$, $\xi_0 \equiv 0$,则 $\xi_n \uparrow +\infty$ 的概率为 $1(n \to +\infty)$. 令
$$y_n = x(\xi_n + 0), \tag{3.1}$$
即 $y_n = \lim_{t \downarrow \xi_n} x(t)$,则 $\{y_n\}$, $n \in \mathbf{N}$,是一马氏链,称之为 X 的嵌入马氏链.

称非负序列 $\{f_i\}(i \in E)$ 为 X(或 $P(t)$)的过分函数(可取 ∞ 为值),如对任一 $t \geqslant 0$,任一 $i \in E$,有
$$\sum_{j \in E} P_{ij}(t) f_j \leqslant f_i. \tag{3.2}$$
(在一些文献里,过分函数的定义中还要求 $\lim_{t \to 0} \sum_{j \in E} P_{ij}(t) f_j = f_i$,我们这里无需此条件.)

以下所谓"任意生灭过程"是指"任意(1.6)形的 Q 及任意 Q 过程".

定理 3.1 设 X 为任意生灭过程,$\{f_i\}$ 为 X 的任一过分函数,则必存在极限
$$\lim_{i \to +\infty} f_i = f. \tag{3.3}$$
如 $R < +\infty$,或 $R = +\infty$ 且 $z = +\infty$,则 $f_i \equiv f$(常数).

证 若对某 j,$f_j = +\infty$,则因 $P_{ij}(t) > 0$ 对一切 $i, j \in E$ 及 $t > 0$ 成立,故由(2.2)知 $f_i \equiv +\infty$. 因而不妨设 $\{f_i\}$ 是有限的过分函数.

如 $R < +\infty$ 或 $R = +\infty$ 且 $z = +\infty$,由定理 2.1 及[2],知

X 常返(以下恒不妨设 X 完全可分 Borel 可测),但在[5]中证明了:若 X 是具可列多个状态的互通的常返马氏过程,则 X 的任一过分函数恒等于常数,由此知 $f_i \equiv f$.

剩下一种情形是 $R = +\infty$ 且 $z < +\infty$. 这时 X 非常返,然而是规则的. 以 N_t 表示由 $\{x(u), 0 \leqslant u \leqslant t\}$ 所产生的 σ 代数,则 $\{f_{x(t)}, N_t, P\}$ 是可分的半 Martingale,于是 P 几乎地存在极限

$$\lim_{t \to +\infty} f_{x(t,\omega)} = f(\omega). \qquad (3.4)$$

既然 X 规则、非常返,故对几乎一切 ω 及任一正数 N,存在 $T(\omega) > 0$,使当 $t \geqslant T(\omega)$ 时,有 $x(t, \omega) \geqslant N$,换句话说[①],

$$P(\lim_{t \to +\infty} x(t, \omega) = +\infty) = 1. \qquad (3.5)$$

由此及 X 的规则性,可见对 X 的嵌入马氏链 $\{y_n\}$,也有 $P(y_n(\omega) \to +\infty) = 1$. 注意对生灭过程,自 i 出发,经一步跳跃后只能到 $i+1$ 或 $i-1$,故对几乎一切 ω,$y_n(\omega)$ 必取一切正整数(除有穷多个外)而趋于 $+\infty$. 于是由(3.4)知以概率 1,

$$\lim_{n \to +\infty} f_n = \lim_{n \to +\infty} f_{y_n(\omega)} = \lim_{t \to +\infty} f_{x(t,\omega)} = f(\omega). \qquad (3.6)$$

这说明 $f(\omega)$ 以概率 1 等于某常数 f,而且(3.3)成立. ∎

定理 3.2 对一切生灭过程 X,0-1 律成立.

证 不妨设 X 完全可分 Borel 可测. 如 X 常返,由[4]中一般结果知对 X 0-1 律成立.

设 X 非常返,即 $R = +\infty$,$z < +\infty$,此时 X 还规则,从而(3.5)成立,不论开始分布如何,任取不变集 $A \in \mathscr{A}$,由定义及马氏性知对任意 $t \geqslant 0$,

$$P_i(A) = P_i(\theta_t A) = \int_\Omega P_i(\theta_t A \mid N_t) P_i(\mathrm{d}\omega)$$

$$= \int_\Omega P_{x_t}(A) P_i(\mathrm{d}\omega) = \sum_{j \in E} P_{ij}(t) P_j(A), \qquad (3.7)$$

---

① 如换 P 为 P_i,(2.5)(2.6)式同样正确.

所以作为 i 的函数 $\{P_i(A)\}$ 是 X 的过分函数,由定理 3.1,(3.6),存在常数 f,使
$$f = \lim_{i \to +\infty} P_i(A) = \lim_{t \to +\infty} P_{x_t}(A). \tag{3.8}$$
注意 $A \in \mathcal{A}$,$P_i(A \Delta \theta_t A) = 0$,故上式右方等于
$$\lim_{t \to +\infty} P_i(\theta_t A \mid N_t) = \lim_{t \to +\infty} P_i(A \mid N_t)$$
$$= P_i(A \mid N_{+\infty}) = \chi_A(\omega) \quad (P_i \text{—几乎}). \tag{3.9}$$
$\chi_A(\omega)$ 是 A 的特征函数,由(3.8)及(3.9)得
$$P_i(\chi_A(\omega) = f) = 1,$$
因而 $P_i(A) = 0$ 或 1. ∎

注 3.1 其实证明了更强的结论:对固定的 $A \in \mathcal{A}$,或 $P_i(A) \equiv 1$,或 $P_i(A) \equiv 0$. 这由(3.7)及 $P_{ij}(t) > 0 (t > 0)$ 推出.

作为一例,考虑生灭过程 X 的任一齐次非负可加泛函 $\varphi \equiv \varphi_t^s(\omega)(0 \leq s \leq t < +\infty)$(定义见[6],第 6 章),由于 $\varphi_{+\infty}^0 = \varphi_t^0 + \theta_t \varphi_\infty^0$, $E_i \varphi_\infty^0 \geq E_i E_{x_t} \varphi_\infty^0$,可见函数 $f_i = E_i \varphi_{+\infty}^0$ 对 X 过分,由定理 3.1,极限 $\lim_{i \to +\infty} E_i \varphi_{+\infty}^0$ 存在.

最后,我们顺便指出特征数间的一关系.

设 $R < +\infty$,则 $S < +\infty$ 的充分必要条件是 $e_1 < +\infty$.

实际上,由 $e_1 \leq S$ 立得**必要性**. 其次,易见
$$e_{n+1} = \frac{a_n}{b_n}\left(e_n - \frac{1}{a_n}\right)$$
$$= \frac{a_1 \cdots a_n}{b_1 \cdots b_n} e_1 - \frac{a_2 \cdots a_n}{b_1 b_2 \cdots b_n} - \frac{a_3 \cdots a_n}{b_2 b_3 \cdots b_n} - \cdots - \frac{a_n}{b_{n-1} b_n} - \frac{1}{b_n}.$$

以之代入 $S = \sum_{n=1}^{\infty} e_n$,得一个二重级数,按对角线求和,并注意
$$Z = 1 + \sum_{k=1}^{+\infty} \frac{a_1 a_2 \cdots a_k}{b_1 b_2 \cdots b_k},$$
$$R = \sum_{n=0}^{+\infty} \left(\frac{1}{b_n} + \frac{a_{n+1}}{b_n b_{n+1}} + \frac{a_{n+1} a_{n+2}}{b_n b_{n+1} b_{n+2}} + \cdots\right),$$

即得

$$S = e_1 z - \left(R - \frac{1}{b_0} z\right) = \left(e_1 + \frac{1}{b_0}\right) z - R.$$

由假定 $R < +\infty$，$e_1 < +\infty$，又 $z < +\infty$，故 $S < +\infty$.

参考文献

[1] 王梓坤. 生灭过程构造论. 数学进展，1962，5(2)：137-170.

[2] Karlin S, Mcgregor J L. The classification of birth and death processes. Trans. Am. Math. Soc.，1957，86：366-400.

[3] Chung K L. Markov chains with stationary transition probabilities. 1st ed，1960. 2nd ed. 1967.

[4] 王梓坤. 马尔可夫过程的0-1律. 数学学报，1965，15(3)：342-353.

[5] 王梓坤. 常返马尔可夫过程的若干性质. 数学学报，1966，16(2)：166-178.

[6] Дынкин Е В. Марковские процессы. Москва：Госуд. издат. Физмат. Литер. 1963. (英译本：Dynkin E B. Markov Processes. Berlin：Springer-Verlag，1965).

On Zero-One Laws for Markov Processes[1]

The object of this paper is the investigation of necessary and sufficient conditions for the validity of zero-one laws for Markov processes, and the presentation of some conveniently applied sufficient conditions. The zero-one law for sequences of independent random variables and its importance are well known [7]. In recent research on Markov processes, one frequently encounters events whose probability can only be either zero or one. These can be broadly divided into two kinds, i. e. , infinitely near and infinitely far; for examples of the former and latter sort, see Theorem 3 of § II. 11 and Theorem 4 of § II. 10, respectively, in [6]. However, the existing results along these lines have mostly been proved individually by resorting to special conditions; hence there is a need for a general treatment. Apparently, the first general result concerning infinitely near zero-one laws for homogeneous Markov processes appears in [3], where a sufficient condition for the validity of such a

---

[1] Received: 1963-04-08; Revised: 1964-11-27.

law is given. This result is restated, in slightly modified form, in [8], and extended to the nonhomogeneous case (see Theorem 2 below). Various applications of zero-one laws may be found in [1][2][4][5][9][10].

The Markov processes considered in the present paper have arbitrary state spaces, and in general need not be homogeneous. The paper is divided into three sections. In the first section, we obtain two necessary and sufficient conditions for the validity of an infinitely near zero-one law. On the basis of these, we then derive several simple and convenient sufficient conditions; for example, we prove that this law is necessarily valid for a Markov process having only enumerably many states, provided only that the process satisfies the usual standard condition (1.17) or (1.17)$'$; subsequently, we discuss the connection between this law and the strong Markov property. In the second section, we find a necessary and sufficient condition for the validity of the infinitely far zero-one law, and characterize the totality of tail events. In the third section, we specialize the discussion to homogeneous Markov processes, and prove that the infinitely far zero-one law is valid for all the recurrent Markov processes ordinarily encountered in the literature; subsequently, we apply the theory of Blackwell to the study of conditional zero-one laws.

Some simple examples are appended to the first and the third sections, including a counter example to a result in [6].

§ 1. Infinitely near zero-one laws

We adopt the definition of a Markov process given in [8], but consider only honest Markov processes. Let there be given a space $\Omega = (\omega)$ of elementary events, a measurable space $(E, \mathcal{B})$, where the σ-algebra $\mathcal{B}$ contains all singletons, and a function $x(t, \omega) \equiv x_t(\omega)$, defined for all $\omega \in \Omega$, $t \geqslant 0$, taking values in E; for all $0 \leqslant s \leqslant t$, let there be given σ-algebras $\mathfrak{M}_t^s$ and $\mathfrak{M}^s$ in Ω, such that $\mathfrak{M}^s \supseteq \mathfrak{M}_t^s$, $t \geqslant s$; finally, for each $s \geqslant 0$, $x \in E$, let there be given a probability measure $P_{s,x}$ on $\mathfrak{M}^s$. We say that these entities define a Markov process $X = (x_t, \mathfrak{M}_t^s, P_{s,x})$, provided that

(i) for any $0 \leqslant s \leqslant t \leqslant u$, $\Gamma \in \mathcal{B}$, we have
$$\{x_t \in \Gamma\} \in \mathfrak{M}_t^s \subseteq \mathfrak{M}_u^s;$$

(ii) if the transition function is defined by
$$p(s, x; t, \Gamma) = P_{s,x}(x_t \in \Gamma), \quad 0 \leqslant s \leqslant t, \quad \Gamma \in \mathcal{B}, \quad (1.1)$$
then $p(s, x; s, E \setminus x) = 0$ and $p(s, x; t, \Gamma)$ is a $\mathcal{B}$-measurable function of x (under these circumstances, we have
$$p(s, x; s, \{x\}) = 1);$$

(iii) the Markov property holds: for $0 \leqslant s \leqslant t \leqslant u$, $x \in E$, $\Gamma \in \mathcal{B}$, we have
$$P_{s,x}\{x_u \in \Gamma \mid \mathfrak{M}_t^s\} = p(t, x_t; u, \Gamma) \quad (P_{s,x}), \quad (1.2)$$
where the notation $(P_{s,x})$ means almost everywhere with respect to the measure $P_{s,x}$.

Let $\mathcal{F}\{x_u, u \in T\}$ denote the smallest σ-algebra of Ω con-

taining all the sets $\{x_u \in \Gamma\} (u \in T, \Gamma \in \mathscr{B})$, and let
$$N_t^s = \mathscr{F}\{x_u, s \leqslant u \leqslant t\}, \quad N_t = N_t^0,$$
$$N^s = \mathscr{F}\{x_u, s \leqslant u\}.$$

For any finite measure μ on $\mathscr{B}$, the integral
$$P_{s,\mu}(A) = \int_E P_{s,x}(A)\mu(\mathrm{d}x), A \in N^s \quad (1.3)$$
defines a measure $P_{s,\mu}$ on N^s; let $A \in \overline{N}^s$, provided that, for every finite measure μ on $\mathscr{B}$, there exist $A_1 \in N^s$, $A_2 \in N^s$ such that $A_1 \subseteq A \subseteq A_2$ and $P_{s,\mu}(A_1) = P_{s,\mu}(A_2)$. We have $\overline{N}^s \supseteq N^s$. For $A \in \overline{N}^s$, define $P_{s,\mu}(A) = P_{s,\mu}(A_1)$, thereby extending $P_{s,\mu}$ to $\overline{N}^s$.

Let $A \in \mathfrak{M}_{t+0}^s$, provided that $A \in \mathfrak{M}_v^s$ for all $v > t$. Clearly,
$$\mathfrak{M}_t^s \subseteq \mathfrak{M}_{t+0}^s = \bigcap_{v>t} \mathfrak{M}_v^s.$$

The σ- algebra $\overline{N}^s \cap \mathfrak{M}_{s+0}^s$ of Ω will play an important role here. Intuitively speaking, it may be regarded as the σ- algebra formed by the events of the infinitely near future realtive to s.

We say that the infinitely near zero-one law holds for the Markov process X, provided that, for any $s \geqslant 0$, $x \in E$ and $A \in \overline{N}^s \cap \mathfrak{M}_{s+0}^s$, we have $P_{s,x}(A) = 0$ or 1.

Theorem 1 Any one of the following two conditions is necessary and sufficient for the validity of the infinitely near zero-one law for the Markov process $X = (x_t, \mathfrak{M}_t^s, P_{s,x})$:

(i) For any $0 \leqslant s < u$, $\Gamma \in \mathscr{B}$, there exists a sequence $\{t_n\}$, $t_n \downarrow s$, such that, for all $x \in E$, we have[①]

---

① The notation $P_{s,x}$ lim means convergence in the measure $P_{s,x}$. The sufficiency of condition (ii) has already been mentioned in § 5.21 of [8].

$$P_{s,x}\lim_{t_n \downarrow s} p(t_n, x_{t_n}; u, \Gamma) = p(s, x; u, \Gamma).$$

(ii) $\tilde{X} = (x_t, \mathfrak{M}_{t+0}^s, P_{s,x})$ is a Markov process.

Proof For any $0 \leqslant s \leqslant v_n < u$, $\Gamma \in \mathscr{B}$, $x \in E$, we have

$$P_{s,x}(x_u \in \Gamma \mid \mathcal{M}_{v_n}^s) = p(v_n, x_{v_n}; u, \Gamma) \quad (P_{s,x}). \quad (1.4)$$

Choose any sequence $\{v_n\}$ such that $v_n \downarrow r \geqslant s$. By a martingale convergence theorem [6], it is evident that the limit

$$\lim_{v_n \downarrow r} P_{s,x}(x_u \in \Gamma \mid \mathcal{M}_{v_n}^s) = P_{s,x}(x_u \in \Gamma \mid \mathcal{M}_{r+0}^s) \quad (P_{s,x})$$

exists, hence, it follows by (1.4) that, for any sequence $\{v_n\}$, $v_n \downarrow r$, the limit $\lim_{v_n \downarrow r} p(v_n, x_{v_n}; u, \Gamma)$ exists, and moreover,

$$P_{s,x}(x_u \in \Gamma \mid \mathcal{M}_{r+0}^s) = \lim_{v_n \downarrow r} p(v_n, x_{v_n}; u, \Gamma) \quad (P_{s,x}). \quad (1.5)$$

For any $\varepsilon > 0$ and $0 \leqslant s \leqslant r \leqslant t \leqslant u$, we have

$$P_{s,x}(\mid p(t, x_t; u, \Gamma) - p(r, x_r; u, \Gamma) \mid > \varepsilon)$$

$$= \int_\Omega P_{s,x}(\mid p(t, x_t; u, \Gamma) - p(r, x_r; u, \Gamma) \mid > \varepsilon \mid \mathcal{M}_r^s) P_{s,x}(dw)$$

$$= \int_\Omega P_{r,x_r}(\mid p(t, x_t; u, \Gamma) - p(r, x_r; u, \Gamma) \mid > \varepsilon) P_{s,x}(dw)$$

$$= \int_E P_{r,y}(\mid p(t, x_t; u, \Gamma) - p(r, y; u, \Gamma) \mid > \varepsilon) P(s, x; r, dy).$$

(1.6)

Now suppose that (i) holds. For any $r < u$, $\Gamma \in \mathscr{B}$, there exists a sequence $\{r_n\}$, $r_n \downarrow r$, such that, for any $y \in E$, we have

$$\lim_{r_n \downarrow r} P_{r,y}(\mid p(r_n, x_{r_n}; u, \Gamma) - p(r, y; u, \Gamma) \mid > \varepsilon) = 0.$$

Let t in (1.6) take the values r_n; then, by (6) and the dominated convergence theorem, we obtain

$$P_{s,x}\lim_{r_n \downarrow r} p(r_n, x_{r_n}; u, \Gamma) = p(r, x_r; u, \Gamma). \quad (1.7)$$

Since the $\{v_n\}$ in (1.5) can be any sequence satisfying $v_n \downarrow r$, we may, in particular, choose $v_n = r_n$; comparing (1.5) and (1.7), and bearing in mind the uniqueness of limits for convergence in measure, we see that

$$P_{s,x}(x_u \in \Gamma \mid \mathfrak{M}_{r+0}^s) = p(r, x_r; u, \Gamma) \quad (P_{s,x}) \quad (1.8)$$

holds for any $0 \leqslant s \leqslant r < u$, $\Gamma \in \mathscr{B}$, $x \in E$. if $r = u$, formula (1.8) remains valid, since in that case both sides are equal to $\chi_\Gamma(x_\Gamma)$, $(P_{s,x})$; here, $\chi_\Gamma(y)$ denotes the characteristic function of Γ, i.e., $\chi_\Gamma(y)$ is equal to 1 or 0 according as $y \in \Gamma$ or $y \overline{\in} \Gamma$. This proves (ii).

As for the proof that (ii) implies the infinitely near zero-one law, this is given in the corollary to Theorem 5.11 of [8], and will not be repeated here.

We now assume that the infinitely near zero-one law is valid and proceed to prove (i). Let $N_{s+0}^s = \bigcap_{t > s} N_t^s$; obviously, $N_{s+0}^s \subseteq \overline{N}^s \cap \mathfrak{M}_{s+0}^s$. When proving (1.5), we showed that, for any sequence $t_n \downarrow s$, $s < u$, there exists $\lim_{t_n \downarrow s} p(t_n, x_{t_n}; u, \Gamma)(P_{s,x})$; if we extend its domain of definition by assigning the value zero to those points for which the limit is undefined, the result is clearly $N_{t_n}^s$-measurable for all n, hence measurable with respect to $N_{s+0}^s = \bigcap_n N_{t_n}^s$, and therefore measurable with respect to $\overline{N}^s \cap \mathfrak{M}_{s+0}^s$. Since by assumption $\overline{N}^s \cap \mathfrak{M}_{s+0}^s$ contains only sets whose measure $P_{s,x}$ is 0 or 1, it follows that there exists a constant c, independent of w, such that

$$\lim_{t_n \downarrow s} p(t_n, x_{t_n}; u, \Gamma) = c \quad (P_{s,x}). \quad (1.9)$$

It only remains to prove that $c = p(s, x; u, \Gamma)$. To do this, we use the Markov property and (1.9), obtaining

$$p(s, x; u, \Gamma) = P_{s,x}(x_u \in \Gamma) = P_{s,x}(x_s = x, x_u \in \Gamma)$$
$$= \int_{(x_s = x)} p(t_n, x_{t_n}; u, \Gamma) P_{s,x}(dw)$$
$$\to \int_{(x_s = x)} c P_{s,x}(dw) = c.$$

Hence, $p(s, x; u, \Gamma) = c$.

In actual fact, we have proved even more. It is apparent from (1.5) and (1.7) that, for $s \leqslant r < u$, $\Gamma \in \mathscr{B}$, and any sequence $v_n \downarrow r$, we have

$$\lim_{v_n \downarrow r} p(v_n, x_{v_n}; u, \Gamma) = p(r, x_r; u, \Gamma) \quad (P_{s,x}). \quad (1.10)$$

In particular, setting $r = s$, it follows that condition (i) is equivalent to

(i′) For any sequence $v_n \downarrow s$, the relation

$$\lim_{v_n \downarrow s} p(v_n, x_{v_n}; u, \Gamma) = p(s, x; u, \Gamma) \quad (P_{s,x}) \quad (1.11)$$

holds for all $x \in E$, $\Gamma \in \mathscr{B}$, $u > s$.

If $\{x_t, t \geqslant 0\}$ is an independent stochastic process, then it is also a Markov process, and furthermore $p(s, x; u, \Gamma)$ is independent of s, x ($u > s$), hence (i) is always satisfied and the infinitely near zero-one law holds.

We now specialize the phase space to a topological measurable space $(E, C, \mathscr{B})$, where (E, C) is a topological space and $(E, \mathscr{B})$ is a measurable space. A topological measurable space is called a D-space provided that $\mathscr{B}$ is the σ-algebra generated by some subsystem of the system C of open sets, and moreover that for every $U \in \mathscr{B} \cap C$, there exists a $\mathscr{B}$-measurable continuous function $f(x)$ such that $f(x) \neq 0$ if and only if $x \in U$.

A Markov process X taking values in a D-space is said to be almost right continuous, provided that, for any $0 \leqslant s \leqslant r$,

$x \in E$, there exists a sequence $t_n \downarrow r$ such that, almost everywhere,

$$\lim_{t_n \downarrow r} x_{t_n}(\omega) = x_r(\omega) \quad (P_{s,x}). \tag{1.12}$$

Let $H_1(H_2)$ denote the totality of bounded $\mathscr{B}$ - measurable functions (bounded continuous $\mathscr{B}$ - measurable functions) defined on $(E, C, \mathscr{B})$; obviously, $H_2 \subseteq H_1$. For $f \in H_1$, define

$$F(r, y) = \int_E p(r, y; u, dz) f(z). \tag{1.13}$$

Of course, $F(r, y)$ depends upon u and f. If the transition function $p(s, x; u, \Gamma)$ is homogeneous, i. e., satisfies

$$p(s, x; s+t, \Gamma) = p(0, x; t, \Gamma) \equiv p(t, x, \Gamma), \tag{1.14}$$

then we define

$$F(y) = \int_E p(t, y, dz) f(z), \quad f \in H_1. \tag{1.15}$$

The homogeneous transition function $p(t, x, \Gamma)$ is said to be strongly Feller (or Feller), provided that, for any $f \in H_1$ (or $f \in H_2$), we have $F(y) \in H_2$.

Corollary 1 The infinitely near zero-one law is valid provided that one of the following two conditions is satisfied:

(i) X is almost right continuous, and, for any $f \in H_2$,

$$\lim_{\substack{y \to x \\ r \downarrow s}} F(r, y) = F(s, x); \tag{1.16}$$

(ii) X is almost right continuous, and the transition function is Feller.

Proof We need only verify that, under either of the assumptions (i) and (ii), condition (ii) in Theorem 1 is satisfied. If X is right continuous, Theorem 5.11 of [8] shows that (ii) in Theorem 1 holds, but the method of proof given there is still applicable in the almost right continuous case; therefore, the proof is omitted here. Note that, if E is the space of real numbers, then, for a process to be almost right continuous, it suffices that it be right stochastically continuous (i. e. , replace the limit in condition (1. 12) by $P_{s,x}\lim$), hence the condition "almost right continuous" is substantially weaker than the condition "right continuous".

The following theorem shows that processes having a countable number of states, which are often encountered in the literature, all satisfy the infinitely near zero-one law.

Theorem 2 Let E be a countable set, $\mathscr{B}$ the σ-algebra of all subsets of E, and suppose that, for any $s \geqslant 0$, we have
$$\lim_{t \downarrow s} p_{yy}(s, t) = 1, \text{ for all } y \in E, \tag{1.17}$$
where $p_{xy}(s, t) = p(s, x; t, \{y\})$, $\{y\}$ being the singleton containing y. Then, the infinitely near zero-one law is valid.

Note. If $p_{xy}(s, t)$ is homogeneous, then condition (1. 17) reduces to the usual so-called standard condition (see [6], § II. 2):
$$\lim_{t \downarrow 0} p_{yy}(t) = 1, \text{ for all } y \in E. \tag{1.17}'$$

Proof of Theorem 2 First we show that, under condition (1. 17), $p_{xy}(s, t)$ is a right continuous function of $s \in [0, t)$. In fact, if $t > r > s$, then
$$p_{xy}(r, t) - p_{xy}(s, t)$$
$$= p_{xy}(r, t)[1 - p_{xx}(s, r)] - \sum_{z \neq x} p_{xz}(s, r) p_{zy}(r, t) \tag{1.18}$$
and since each term on the right does not exceed $1 - p_{xx}(s, r)$,
$$|p_{xy}(r, t) - p_{xy}(s, t)| \leqslant 1 - p_{xx}(s, r) \to 0 \quad (r \downarrow s). \tag{1.19}$$

Next, for any $\varepsilon > 0$, $s \leqslant r < t$, we have

$$p_{s,x}(|x_t - x_r| > \varepsilon) = \sum_y p_{r,y}(|x_t - x_r| > \varepsilon) p_{xy}(s,r)$$
$$\leq \sum_y [1 - p_{yy}(r,t)] p_{xy}(s,r). \quad (1.20)$$

Since the right-hand side is dominated by the convergent series $\sum_y p_{xy}(s,r) = 1$, it follows from (1.17) that
$$\lim_{t \downarrow r} P_{s,x}(|x_t - x_r| > \varepsilon) = 0.$$

Hence there exists a sequence $t_n \downarrow r$ such that
$$\lim_{t_n \downarrow r} x_{t_n} = x_r \quad (P_{s,x}).$$

Since the points of E are isolated, it follows from the above relation that, for almost all $\omega(P_{s,x})$, there exists a positive integer $N(\omega)$ such that, when $n \geq N(\omega)$, then $x_{t_n}(\omega) = x_r(\omega)$. Hence, by (1.19), we obtain
$$\lim_{t_n \downarrow r} p_{x_{t_n},y}(t_n, u) = p_{x_r,y}(r, u) \quad (P_{s,x}). \quad (1.21)$$

The choice of $\{t_n\}$ depended upon x; however, by virtue of the countability of E, it is possible by a diagonal method to select a sequence $\{t_n\}$, $t_n \downarrow r$, such that (1.21) holds for all of the measures $P_{s,x}(x \in E)$. Consequently, condition (i) of Theorem 1 is satisfied (to see this, take $r = s$ in (1.21), and notice that, in this case, (i) needs only hold for singletons Γ).

As an example of a process for which the infinitely near zero-one law is not valid, consider an X having but three states, $E = (0, 1, 2)$ and a homogeneous transition function given by $p_{00}(0) = 1$, $p_{01}(t) = p_{02}(t) = \frac{1}{2}(t > 0)$; $p_{11}(t) \equiv p_{22}(t) \equiv 1 \ (t \geq 0)$. Suppose that the initial distribution is concentrated at 0; one can choose $x_t(\omega)$, $t \geq 0$, such that all sample functions are continuous for $t > 0$. Let

$$T(\omega) = \inf\ (t;\ t>0,\ x(t,\ \omega)=1),$$

Then $A \equiv (\omega; T(\omega) = 0) = (x_\varepsilon(\omega) = 1) \in N_\varepsilon^0$, where $\varepsilon > 0$ is arbitrary, hence $A \in N_{0+0}^0 \subseteq \overline{N^0} \cap \mathfrak{M}_{0+0}^0$, however, $P_0(A) = p_{01}(\varepsilon) = \frac{1}{2}$, where $P_x = P_{0,x}$. It is easily seen that, in this case (i′) is not satisfied; in fact, for a homogeneous transition function, if we take $s=0$, relation (1.11) becomes

$$\lim_{v_n \downarrow 0} p(u-v_n,\ x_{v_n},\ \Gamma) = p(n,\ x,\ \Gamma) \quad (p_x), \quad (1.22)$$

where $u>0$. If we choose $x=0$, $\Gamma = \{1\}$, then, in the present example, the right-hand side of the above formula becomes $p_{01}(u) = \frac{1}{2}$; but the left-hand side is, with P_0-probability $\frac{1}{2}$, equal to $\lim_{v_n \downarrow 0} p(u-v_n,\ 1,\ \{1\}) = 1$, and, with P_0-probability $\frac{1}{2}$, equal to $\lim_{v_n \downarrow 0} p(u-v_n,\ 2,\ \{1\}) = 0$, thus (1.22) does not hold in this case.

We may ask what relation the strong Markov property may have with the validity of the infinitely near zero-one law. This depends upon the way in which the strong Markov property is defined. If the definition is that given in [8], then there is no direct implication relation between them, because this version of the strong Markov property only requires that the Markov property hold for times more numerous than constant times (i.e., so-called "independent of the future" times), whereas, according to (ii) in Theorem 1, the validity of the infinitely near zero-one law requires that the Markov property hold for constant times, but for a larger σ - algebra $\mathfrak{M}_{t+0}^s$. In [4, 10], a strong Markov property is also defined for homogeneous Markov processes $X = (x_t, N_t^0, P_x)$; it requires that the Markovian property hold, not only for more times, but for a larger σ-algebra as well, and is therefore stronger than the strong Markov property of [8], that is, with reference to the special case $X = (x_t, N_t^0, P_x)$. In particular, from the strong Markov property of [4, 10], one can deduce that $\tilde{X} = (x_t, N_{t+0}^0, P_x)$ is a Markov process, hence $P_x(A) = 0$ or 1 for all $A \in \overline{N^0} \cap N_{0+0}^0 = N_{0+0}^0$.

Notice that, for $X = (x_t, \mathfrak{M}_t^s, P_{s,x})$, if $N_{t+0}^s \subseteq \mathfrak{M}_t^s$, then $(x_t, N_{t+0}^s, P_{s,x})$ is also a Markov process. It follows by Theorem 1 that $P_{s,x}(A) = 0$ or 1 for any $A \in N_{s+0}^s$, where $s \geqslant 0$, $x \in E$ are arbitrary.

§ 2. Infinitely far zero-one laws

We now proceed to study infinitely far zero-one laws for $X = (x_t, \mathfrak{M}_t^s, P_{s,x})$. Let $\prod = \bigcap_{t \geqslant 0} N^t$, and let $\overline{\prod}_{s,x}$ denote the completed σ-algebra of $\prod$ with respect to $P_{s,x}$. If, for all $A \in \overline{\prod}_{s,x}$, we have $P_{s,x}(A) = 0$ or 1, then we say that the $P_{s,x}$-infinitely far zero-one law holds; if the $P_{s,x}$-infinitely far zero-one law holds for all $s \geqslant 0$, $x \in E$, then we say that the infinitely far zero-one law holds. Intuitively speaking, the sets in $\overline{\prod}_{s,x}$ may be called tail events.

In what follows, $B \stackrel{.}{=} C(P_{s,x})$ means that $P_{s,x}(B \triangle C) = 0$, where $B \triangle C = (B \setminus C) \cup (C \setminus B)$.

Theorem 3 Arbitrarily fix $A \in \overline{\prod}_{s,x}$. Then

(i) A can be represented as

$$A \stackrel{.}{=} \bigcap_{m=1}^{+\infty} \bigcup_{n=m}^{+\infty} (x_{t_n} \in E_n) \stackrel{.}{=} \bigcup_{m=1}^{+\infty} \bigcap_{n=m}^{+\infty} (x_{t_n} \in E_n) \quad (P_{s,x}), \quad (2.1)$$

where $\{t_n\}$ is any sequence of constants such that $t_n \uparrow +\infty$, and $E_n \in \mathscr{B}$.

(ii) For $P_{s,x}(A) = 0$ or 1, it is necessary and sufficient that there exist a sequence of constants $\{t_n\}$, $t_n \uparrow +\infty$ such that

$$\lim_{n \to +\infty} P_{t_n, x_{t_n}}(A) = C \text{ (constant)} \quad (P_{s,x}) \quad (2.2)$$

(whence it follows that this holds for any such sequence of constants $\{t_n\}$), in which case we must have $P_{s,x}(A) = C$.

(iii) If the process $\{P_{t,x_t}(A), t \geqslant s\}$ defined on $(\Omega, N^s,$

$P_{s,x}$) is separable, then a necessary and sufficient condition for $P_{s,x}(A)=0$ or 1 is

$$\lim_{t \to +\infty} P_{t,x_t}(A) = C \ (P_{s,x}). \tag{2.3}$$

Proof We need only prove these assertions for $A \in \prod$. By the Markov property,

$$P_{s,x}(A \mid N_t^s) = P_{t,x_t}(A) \ (P_{s,x}). \tag{2.4}$$

Letting t go to infinity through any sequence $\{t_n\}$, we get

$$P_{s,x}(A \mid N^s) = \lim_{n \to +\infty} P_{t_n,x_{t_n}}(A) \ (P_{s,x}).$$

Since $\prod \subseteq N^s$, we have

$$\chi_A(\omega) = \lim_{n \to +\infty} P_{t_n,x_{t_n}(\omega)}(A) \ (P_{s,x}). \tag{2.5}$$

Choose any constant α, $0<\alpha<1$, and set

$$E_n = (y: \ P_{t_n,y}(A) > \alpha) \in \mathscr{B}. \tag{2.6}$$

Then, for this $\{E_n\}$, relation (2.1) holds. In fact, let Ω_0 denote the set of ω for which (2.5) holds; then $P_{s,x}(\Omega_0)=1$. If $\omega \in A$, then $\chi_A(\omega)=1$. Therefore, it follows from (2.5) that, for all sufficiently large n, we have $P_{t_n,x_{t_n}}(\omega)(A) > \alpha$, and thus $x_{t_n}(\omega) \in E_n$, hence $\omega \in \bigcup_{m=1}^{+\infty} \bigcap_{n=m}^{+\infty} (x_{t_n} \in E_n)$. Conversely, if $\omega \in \Omega_0 - A$, then $\chi_A(\omega) = 0$. It follows from (2.5) that, for all sufficiently large n, we have $P_{t_n,x_{t_n}}(\omega)(A) \leqslant \alpha$, and thus $x_{t_n}(\omega) \overline{\in} E_n$, hence $\omega \overline{\in} \bigcap_{m=1}^{+\infty} \bigcup_{n=m}^{+\infty} (x_{t_n} \in E_n)$.

Consequently, we have proved that, except for at most of $P_{s,x}$-zero measure, we have $\bigcap_{m=1}^{+\infty} \bigcup_{n=m}^{+\infty} (x_{t_n} \in E_n) \subseteq A \subseteq \bigcup_{m=1}^{+\infty} \bigcap_{n=m}^{+\infty} (x_{t_n} \in E_n)$; however, the set on the left-hand side clearly contains the set on the right-hand side, whence (2.1) is proved.

Note that the $\{E_n\}$ of (2.6) depends upon $\{t_n\}$ and α.

To prove (ii), we need only show that there exists a sequence $t_n \uparrow +\infty$ such that (2.2) constitutes a sufficient condition for $P_{s,x}(A)=0$ or 1, and that the validity of (2.2) for any sequence $t_n \uparrow +\infty$ is a necessary condition.

Assume that (2.2) holds for some sequence $t_n \uparrow +\infty$. Comparing (2.2) and (2.5), it is seen that $\chi_A(\omega) = C$, $(P_{s,x})$, hence C must be 0 or 1, and $P_{s,x}(A)=C$. Conversely, if $P_{s,x}(A)=0$ or 1, then either $\chi_A = 0$ $(P_{s,x})$ or $\chi_A = 1$ $(P_{s,x})$. It follows from (2.5) that, for any sequence $t_n \uparrow +\infty$, we have either $\lim_{n \to +\infty} P_{t_n, x_{t_n}}(A) = 0$ $(P_{s,x})$ or $\lim_{n \to +\infty} P_{t_n, x_{t_n}}(A) = 1$ $(P_{s,x})$. In the former case we take $C=0$, in the latter case we take $C=1$; thus, we have shown that (2.2) holds for any sequence $t_n \uparrow +\infty$.

Finally, if $\{P_{t,x_t}(A), t \geqslant s\}$ is a separable process, then (iii) can be deduced from (ii) and separability ([7], Chapter 2, Theorem 2.3).

As a simple consequence of Theorem 3, consider the process $\{x_t, t \geqslant 0\}$, with independent random variables. Since, in this case, $P_{s,x}(A)$ is independent of s, x, the E_n in (2.6) is equal to E or to $\varnothing$ (the empty set), thus, (2.1) becomes $A \doteq \Omega$ or $A \doteq \varnothing (P)$, where P is identical with all of the $P_{s,x}$. Moreover, (2.3) always holds, hence the zero-one law is valid; this result is well known for sequences of independent random variables.

Another consequence is as follows. If the only bounded solutions of the integral equation

$$f(s,x) = \int_E p(s,x;t,\mathrm{d}y)f(t,y) \qquad (2.7)$$

are $f(s, x) \equiv C$(constant), then, for any

$$A \in \bigcap_{\substack{s>0 \\ x \in E}} \prod_{s,x},$$

we have $P_{s,x}(A) = 0$ or 1; moreover, if $P_{s,x}(A) = 0$ holds for some pair (s, x), then it holds for any pair (s, x). In fact, we see from

$$P_{s,x}(A) = \int_E p(s,x;t,\mathrm{d}y) P_{t,y}(A) \qquad (2.8)$$

that $\{P_{s,x}(A)\}$ is a solution of (2.7); but, by assumption, $P_{s,x}(A) \equiv C$, hence (2.2) is satisfied, and the required conclusion follows at once from Theorem 3.

Theorem 3 also implies the following important fact. Let $\prod_{s,x}^{\{t_n\}}$ denote the completed σ-algebra of $\bigcap_{m=0}^{+\infty} \mathscr{F}\{x_{t_n}, n > m\}$ with respect to $P_{s,x}$, where $t_n \uparrow \infty$ is arbitrary; clearly, $\prod_{s,x} \supseteq \prod_{s,x}^{\{t_n\}}$, whence (2.1) shows that $\prod_{s,x} = \prod_{s,x}^{\{t_n\}}$. Consequently, the study of the $P_{s,x}$-infinitely far zero-one law for the process X reduces to the corresponding problem for the sequence $\{x_{t_n}, n \geq 0\}$.

§ 3. The infinitely far zero-one law for homogeneous Markov processes

In the following, we shall study the homogeneous Markov process $X=(x_t, \mathfrak{M}_t, P_x, \theta_t)$ (for the definition, see § 2.8 of [8]). By virtue of the invariance of the transition function with respect to shift transformations, it is naturally more appropriate, in this case, to consider a σ-subalgebra $\mathfrak{A}_x$ of $\prod_x (= \prod_{0,x})$, rather than $\prod_x$ itself. If we let $A \in \mathfrak{A}_x$ provided that $A \in \prod$, then for any $t \geqslant 0$ we have $P_x(\theta_t A \Delta A) = 0$. Define $\mathfrak{A} = \bigcap_{x \in E} \mathfrak{A}_x$. We call the sets in $\mathfrak{A}$ invariant sets. Let $A \in \mathfrak{A}$; then, since

$$P_x(A \mid \mathcal{M}_t) = P_x(\theta_t A \mid \mathcal{M}_t) = P_{x_t}(A) \quad (P_x),$$

the relation (3.1)

$$\chi_A = \lim_{n \to +\infty} P_{x_{t_n}}(A) \quad (P_x)$$

holds for any sequence of constants $t_n \uparrow +\infty$ and any $x \in E$. Then, just as Theorem 3 is proved from (2.5), we can similarly prove

Theorem 4 Arbitrarily fix $A \in \mathfrak{A}$ and $x \in E$. Then,

(i) The set A can be represented as

$$A \doteq \bigcap_{m=1}^{+\infty} \bigcup_{n=m}^{+\infty} (x_{t_n} \in e) \doteq \bigcup_{m=1}^{+\infty} \bigcap_{n=m}^{+\infty} (x_{t_n} \in e) \quad (p_x), \quad (3.2)$$

where $e = (y: P_y(A) > \alpha) \in \mathcal{B}$, $0 < \alpha < 1$, and $t_n \uparrow +\infty$ is arbitrary.

(ii) A necessary and sufficient condition for $P_x(A) = 0$ or 1 is that there exist a sequence of constants $\{t_n\}$, $t_n \uparrow +\infty$ such that

$$\lim_{n \to +\infty} P_{x_{t_n}}(A) = C \text{ (constant)} \quad (P_x) \qquad (3.3)$$

(whence it follows that this holds for any such sequence of constants), in which case we must have $P_x(A) = C$.

(iii) If the process $\{P_{x_t}(A), t \geqslant 0\}$ defined on (Ω, N^0, P_x) is separable, then a necessary and sufficient condition for $P_x(A) = 0$ or 1 is

$$\lim_{t \to +\infty} P_{x_t}(A) = C \quad (P_x). \qquad (3.4)$$

Note that the e in (3.2) depends only upon α, hence is appropriately denoted by e_α.

From Theorem 4, we directly obtain

Corollary 2 Arbitrarily fix $A \in \mathfrak{A}$ and $x \in E$. Then

(i) A necessary and sufficient condition for $P_x(A) = 0$ ($P_x(A) = 1$) is: for almost all ω (with respect to P_x), there exists a sequence of positive integers $\{k_n\}$, $k_n = k_n(\omega) \uparrow +\infty$, and some α, $0 < \alpha < 1$, such that $x_{k_n}(\omega) \in E \setminus e_\alpha (x_{k_n}(\omega) \in e_\alpha)$ for all n.

(ii) If the process $\{P_{x_t}(A), t \geqslant 0\}$ on (Ω, N^0, P_x) is separable, then a necessary and sufficient condition for $P_x(A) = 0$ ($P_x(A) = 1$) is: for almost all ω (with respect to P_x), there exists a sequence of positive numbers $\{t_n\}$, $t_n = t_n(\omega) \uparrow +\infty$ and some α, $0 < \alpha < 1$, such that $x_{t_n}(\omega) \in E \setminus e_\alpha (x_{t_n}(\omega) \in e_\alpha)$ for al n.

Proof If $P_x(A) = 0$, then $\chi_A(\omega) = 0$ (P_x); choosing $t_n = n$ in (3.1), we get $\lim_{n \to +\infty} P_{x_{t_n}}(A) = 0$, hence, for almost all $\omega(P_x)$, there exists a positive integer $N = N(\omega)$ such that, for

$n \geqslant N$, we have $P_{x_n(\omega)}(A) \leqslant \alpha$, i. e., $x_n(\omega) \in E \setminus e_\alpha$. Then, choosing $k_n(\omega) = N(\omega) + n$, we obtain the necessity of (i).

Conversely, suppose that $\{k_n(\omega)\}$ fulfills the requirement of (a). Relation (3.1) is true for any sequence $t_n \uparrow +\infty$, hence

$$\lim_{n \to +\infty} P_{x_n}(A) = \chi_A(P_x).$$

Since, for almost all ω (P_x), there exists a subsequence $\{P_{x_{k_n}(\omega)}(A)\}$ of $\{P_{x_n(\omega)}(A)\}$, such that $P_{x_{k_n}(\omega)}(A) \leqslant \alpha$, it follows from the above formula that

$$0 = \lim_{n \to +\infty} P_{x_n}(A) = \chi_A(P_x).$$

This means that $P_x(A) = 0$. The argument for the case $P_x(A) = 1$ is similar.

If $\{P_{x_t}(A), t \geqslant 0\}$ is separable, then (3.1) can be replaced by

$$\lim_{t \to +\infty} P_{x_t}(A) = \chi_A(P_x),$$

whence, repeating the above argument, we obtain (ii).

By applying the above results, it can be proved that the infinitely far zero-one law is valid for all the recurrent type homogeneous Markov processes commonly encountered in the literature. To this end, we first give the general definition of recurrence. Let X be a homogeneous Markov process taking values in a D-space. The state x is called a recurrent state, provided that, for any open set U containing x, there exists a sequence of real numbers $t_n = t_n(\omega) \uparrow +\infty$, such that

$$P_x(x_{t_n}(\omega) \in U, \text{ all } n) = 1. \tag{3.5}$$

Theorem 5 Let x be a recurrent state of a strong Feller process, and suppose that the process $\{P_{x_t}(A), t \geqslant 0\}$ is separable,[①] and $A\in\mathfrak{A}$. Then $P_x(A)=0$ or 1.

Proof Bearing in mind that $P_y(A)$ is a bounded $\mathscr{B}$-measurable function of y, it follows from the strong Feller property and the relation

$$P_y(A) = E_y P_{x_t}(A) = \int_E p(t,y,\mathrm{d}z) P_z(A),$$

that $P_y(A)$ is a continuous function of y. Suppose that $0 < P_x(A) = c < 1$. Then

$$U = \left(y: \frac{c}{2} < P_y(A) < \frac{1+c}{2}\right) \quad (3.6)$$

is an open set containing x. By the recurrence of x, there exists $t_n(\omega) \uparrow +\infty$ such that (3.5) holds for the U of (3.6). Since

$$U \subseteq (y: P_y(A) > \frac{c}{2}) = e_{\frac{c}{2}},$$

$$P_x(x_{t_n}(\omega) \in e_{\frac{c}{2}}, \text{ all } n) = 1.$$

Hence, by Corollary 2(ii) we get $P_x(A)=1$. On the other hand, from $U \subseteq E \setminus e_{\frac{1+c}{2}}$ and $P_x(x_{t_n}(\omega) \in E \setminus e_{\frac{1+c}{2}}, \text{ all } n)=1$ we get $P_x(A)=0$. But the constant $P_x(A)$ cannot be equal to both zero and one; this contradiction shows that $P_x(A)=0$ or 1.

We now proceed to discuss, one by one, various types of recurrent homogeneous processes, starting from the simplest case.

(i) Let $X=(x_n, N_n, P_x)$ be a countable Markov chain, $n\in\mathbf{N}$, $E=\mathbf{N}$. Let $p_{xy}^{(n)}=p(n, x, \{y\})$. If E is given the discrete topology, then this process is strong Feller. Clearly, for $A\in\mathfrak{A}$, $\{P_{x_n}(A), n \geqslant 0\}$ is separable. Hence, if x is recurrent, then $P_x(A)=0$ or 1. Note that the definition of recur-

---

① To be precise, the process $\{P_{x_t}(A), t \geqslant 0\}$ on (Ω, N^0, P_x) is separable.

rence coincides in this case with the definition given in the theory of Markov chains ([6], § 1.4); a necessary and sufficient condition for the recurrence of x is $\sum_{n=1}^{+\infty} p_{xx}^{(n)} = +\infty$.

Suppose that x and y communicate, i.e., that there exist two positive integers m, n such that $p_{xy}^{(n)} p_{yx}^{(m)} > 0$; then y is also recurrent. Let $\eta(\omega) = \inf(n: x_n(\omega) = y)$; as is well known, $P_x(\eta(\omega) < +\infty) = 1$. Since the set of parameters is discrete, X must have the strong Markov property, hence, if $A \in \mathfrak{A}$ and $\theta_t A = A$ (all $t \geqslant 0$), then

$$P_x(A) = \int_{(\eta < +\infty)} P_x(A \mid N_\eta) P_x(d\omega)$$
$$= \int_{(\eta < +\infty)} P_{x(\eta)}(A) P_x(d\omega)$$
$$= P_y(A) P_x(\eta < +\infty) = P_y(A),$$

where the σ-algebra N_η is defined as in ([8], § 5.1). This shows that $P_x(A)$, $P_y(A)$ are either both 0 or both 1.

(ii) Let $X = (x_t, N_t, P_x)$ be a countable Markov process, with E as in (α) above. This process, again, is strongly Feller. Suppose that the transition function $P_{xy}(t)$ is a Lebesgue measurable function of t for all $x, y \in E$. In [6](§ 11.10) it is proved that $\int_0^{+\infty} P_{xx}(t) dt = +\infty$ is equivalent to $\sum_{n=1}^{+\infty} P_{xx}(n) = +\infty$. Observe that $P_{xy}(n)$ is the n-step transition function of the Markov chain $\tilde{X} = (\tilde{x}_n, \tilde{N}_n, \tilde{P}_x)$ imbedded in X, where $\tilde{x}_n = x_n$, $\tilde{N}_n = \mathscr{F}\{\tilde{x}_m, 0 \leqslant m \leqslant n\} \subseteq N_n$, and $\tilde{P}_x$ is the restriction of P_x to $\mathscr{F}\{\tilde{x}_m, m \geqslant 0\}$, so that, if B belongs to the completion $\tilde{\mathscr{F}}_x\{\tilde{x}_m, m \geqslant 0\}$ of the latter σ-algebra with respect to $\tilde{P}_x$, then

$\widetilde{P}_x(B) = P_x(B)$. Now suppose that $\int_0^{+\infty} p_{xx}(t)\,dt = +\infty$, whence $\sum_{n=0}^{+\infty} p_{xx}(n) = +\infty$, so that x is a recurrent state of $\widetilde{X}$, and hence is also a recurrent state of X. Consider the invariant set A of X; by Theorem 4(i), A is also an invariant set of $\widetilde{X}$. Therefore, by (i), we obtain $P_x(A) = \widetilde{P}_x(A) = 0$ or 1. Furthermore, if x and y communicate in X, i.e., if there exist $t_1 > 0$, $t_2 > 0$ such that $p_{xy}(t_1) p_{yx}(t_2) > 0$, then it follows by [6](§ II. 10) that x and y also communicate in $\widetilde{X}$. Consequently, $P_y(A) = \widetilde{P}_y(A) = \widetilde{P}_x(A) = P_x(A)$ holds for all $A \in \mathfrak{A}$ such that $\theta_t A = A$ (all $t \geqslant 0$).

(iii) Let $X = (x_t, \mathfrak{M}_t, P_x)$ be a continuous strong Feller process, with $E = (r_1, r_2)$ a finite or infinite one-dimensional interval. Define

$$\tau_1(a) = \inf(t: x_t \leqslant a), \quad \tau_2(a) = \inf(t: x_t \geqslant a),$$

and assume that the following condition is satisfied:

R: for any $a, b \in E$, $a < b$, we have

$$P_a(\tau_2(b) < +\infty) P_b(\tau_1(a) < +\infty) = 1.$$

Under this condition, it has been shown that (see [11], § 15.5)

$$P_x(\lim_{t \uparrow +\infty} x_t = r_1, \overline{\lim}_{t \uparrow +\infty} x_t = r_2) = 1 \quad (\text{all } x \in E). \quad (3.7)$$

It follows from (3.7) and the continuity of the process that every state x is recurrent. Moreover, by the strong Feller property and continuity, it also follows that $\{P_{x_t}(A), t \geqslant 0\}$ is a continuous process, $A \in \mathfrak{A}$. Hence by Theorem 5 we obtain $P_x(A) = 0$ or 1, $x \in E$. As in (i), we deduce that, if $\theta_t A = A$, for any $t \geqslant 0$, then $P_x(A) = P_y(A)$ for all $x, y \in E$.

(iv) A generalization of the preceding example is as follows. Let X be a right continuous Feller process taking values in a D-space, x a recurrent state. Then, since $P_y(A)$ is continuous in y for $A \in \mathfrak{A}$, the function $\{P_{x_t}(A), t \geqslant 0\}$ is right continuous in t. By Theorem 5, we obtain at once $P_x(A) = 0$ or 1.

Summarizing the above discussion, we obtain

Theorem 6 Let X be a homogeneous Markov process, $x \in E$, $A \in \mathfrak{A}$. Under any one of the following, we have $P_x(A) = 0$ or 1.

(i) X is a countable Markov chain, $\sum_{n=1}^{+\infty} p_{xx}^{(n)} = +\infty$; if x and y communicate, and, for all integers $t \geqslant 0$, $\theta_t A = A$, then $P_x(A) = P_y(A) = 0$ or 1;

(ii) X is a countable Markov process, $p_{xy}(t)$ is a Lebesgue measurable function of t (for all $x, y \in E$), and $\int_0^{+\infty} p_{xx}(t) dt = +\infty$, if x and y communicate, and, for all $t \geqslant 0$, $\theta_t A = A$, then $P_x(A) = P_y(A) = 0$ or 1;

(iii) X is a continuous strong Feller process, $E = (r_1, r_2)$, and condition R is satisfied; if $\theta_t A = A$ for all $t \geqslant 0$, then
$$P_x(A) = P_y(A) = 0 \text{ or } 1 \text{ (all } x, y \in E);$$

(iv) X is a right continuous strong Feller process taking values in a D-space, and x is recurrent.

In the nonrecurrent case, the infinitely far zero-one law need not be valid. For example, let X be a Markov chain with $p_{01} = \frac{1}{2}$, $p_{02} = \frac{1}{2}$, $p_{11} = p_{22} = 1$; then $A = \bigcup_{m=1}^{+\infty} \bigcap_{n=m}^{+\infty} (x_n = 1) \in \mathfrak{A}$, but $P_0(A) = \frac{1}{2}$. As another example, let X be a right continuous Markov process with $E=(0, 1, 2)$, $p_{00}(t)=\frac{1}{2}(1+e^{-t})$, $p_{01}(t)=p_{02}(t)=\frac{1}{4}(1-e^{-t})$, $p_{11}(t)=p_{22}(t)=1$, $t\geqslant 0$. Let $s_1(\omega)=(t: x_t(\omega)=1)$, $A=(\omega: s_1(\omega) \text{ unbounded})$; then $A\in\mathfrak{A}$, but $P_0(A)=P_0(x(\tau+0)=1)=\frac{1}{2}$, where $\tau=\inf(t: x_t(\omega)\neq 0)$. This example shows that the second conclusion of Corollary 2 in [6](§ Ⅱ. 10) is false (it is asserted there that $P_0(A)=0$ or 1).

Consider the imbedded Markov chain $\widetilde{X}=(\widetilde{x}_n, \widetilde{N}_n, \widetilde{P}_x)$ of any homogeneous Markov process $X=(x_t, \mathfrak{M}_t, P_x)$ such that E is countable; $\widetilde{X}$ is defined exactly as in (ii) above. In accordance with the theory of Blackwell ([6], § 1.17), one can, relative to $\widetilde{X}$ and $\widetilde{P}_x$, decompose E into the union of a finite or countable number of almost closed sets:

$$E=E_0\cup E_1\cup E_2\cup\cdots, \tag{3.8}$$

where $E_i E_j=\varnothing$ $(i\neq j)$, the E_i $(i>0)$ are atomic sets and E_0 is a completely nonatomic set. Here, and in the following theorem, $x\in E$ is fixed.

Theorem 7 A necessary and sufficient condition for $P_x(A)=0$ or 1 for all $A\in\mathfrak{A}$ is that $E=E_i(i>0)$, i.e., that the decomposition (3.8) consists of just a single atomic set.

Proof For an arbitrary $B\in\mathscr{B}$, let $\mathscr{L}(B)$ denote any set of ω satisfying the following two relations:

$$\mathscr{L}(B)\doteq\bigcap_{m=1}^{+\infty}\bigcup_{n=m}^{+\infty}(x_n\in B) \quad (P_x), \tag{3.9}$$

$$\mathscr{L}(B)\in\mathfrak{A}. \tag{3.10}$$

We know by (3.2) that, for any $A\in\mathfrak{A}$, there exists a $B\in\mathscr{B}$

such that $\mathscr{L}(B) \doteq A(P_x)$; in particular, $\mathscr{L}(E) \doteq \Omega(P_x)$. By hypothesis, $E = E_i$, hence $\mathscr{L}(E_i) \doteq \Omega(P_x)$. Therefore,
$$P_x(A) = P_x(\mathscr{L}(B)) = P_x(\mathscr{L}(B) \cap \mathscr{L}(E_i)) = P_x(\mathscr{L}(B \cap E_i)),$$
But $E_i = E$ is an atomic set, hence $P_x(\mathscr{L}(B \cap E_i))$ is either equal to 0, or equal to $P_x(\mathscr{L}(E_i)) = P_x(\Omega) = 1$. This proves sufficiency.

Conversely, assume that $P_x(A) = 0$ or 1 for all $A \in \mathfrak{A}$. If we suppose that there exist distinct E_i and E_j, both nonempty, then $\mathscr{L}(E_i)$ and $\mathscr{L}(E_j)$ both belong to $\mathfrak{A}$, moreover, $P_x(\mathscr{L}(E_i)) > 0$, $P_x(\mathscr{L}(E_j)) > 0$; hence their values are both less than 1 (since $\sum_{n=0}^{+\infty} P_x(\mathscr{L}(E_n)) = 1$), which contradicts our hypothesis. Thus there can exist but one E_i. It remains to prove that this E_i is atomic. If this were not so, i.e., $E_i = E_0$, then E_0 would contain two disjoint subsets B_1, $B_2 \in \mathscr{B}$, such that $P_x(\mathscr{L}(B_1)) > 0$, $P_x(\mathscr{L}(B_2)) > 0$, whence $P_x(\mathscr{L}(B_1))$ could not be equal to either 0 or 1, contrary to hypothesis.

Incidentally, we point out the following fact. If we assume that $E = E_0$, we then have another extreme, diametrically opposite to the zero-one law, i.e., in this case, for any constant c, $0 < c < 1$, there is a $\Lambda \in \mathfrak{A}$ such that $P_x(\Lambda) = c$.

In fact, one may define $\widetilde{\mathfrak{A}}$ for $\widetilde{X} = (\widetilde{X}_n, \widetilde{N}_n, \widetilde{P}_x)$ in the same manner[1] that $\mathfrak{A}$ was defined for X. Setting $t_n = n$ in (3.2), it is evident that $\mathfrak{A} \subseteq \widetilde{\mathfrak{A}}$. On the other hand, if $A \in \widetilde{\mathfrak{A}}$, then, for any $y \in E$, we have

---

[1] Merely replace $t \geqslant 0$ by integer values $n \geqslant 0$.

$$\chi_A = \lim_{n \to +\infty} P_{x_n}(A) \quad (P_y),$$

$$\theta_t \chi_A = \lim_{n \to +\infty} P_{x_{n+t}}(A)$$

$$= \lim_{n \to +\infty} P_y(A \mid N_{n+t}) = P_y(A \mid N_{+\infty}) = \chi_A \quad (P_y),$$

where $t \geq 0$ is arbitrary. Hence, $P_y(A \triangle \theta_t A) = 0$ and $A \in \mathfrak{A}$. Thus, we have proved that $\mathfrak{A} = \widetilde{\mathfrak{A}}$.

If $E = E_0$, then by [6](§1.17) we know that, for any c, $0 < c < 1$, there exists a $\Lambda \in \widetilde{\mathfrak{A}} = \mathfrak{A}$ such that $P_x(\Lambda) = c$.

In the foregoing discussion, we have made use of the imbedded Markov chain $\widetilde{X} = (\widetilde{x}_n, \widetilde{N}_n, \widetilde{P}_x)$ to study the zero-one law for X. By virtue of the arbitrariness of $\{t_n\}$, $t_n \uparrow +\infty$ in (3.2), one can, naturally, make use of the more general imbedded Markov chain $\widetilde{X}_h = (\widetilde{x}_{nh}, \widetilde{N}_{nh}, \widetilde{P}_x)$, where $h > 0$ is arbitrary and $\widetilde{P}_x$ is the restriction of P_x to $\mathscr{F}_x(\widetilde{x}_{nh} \equiv x_{nh}, n \geq 0)$. If we consider the decomposition (3.8) of E relative to $\widetilde{X}_h$, and $\widetilde{P}_x$, then Theorem 7 and the above-mentioned fact are also valid for this decomposition.

The following conditional zero-one law is sometimes required. Let $A \in \mathfrak{A}$, then

$$P_x(A \mid \mathscr{L}(E_i)) = 0 \text{ or } 1, \quad i > 0. \tag{3.11}$$

In fact, since $A \in \mathfrak{A} = \widetilde{\mathfrak{A}}$, there exists $B \in \mathscr{B}$ such that $P_x(A \triangle \mathscr{L}(B)) = 0$. From $P_x(\mathscr{L}(E_i)) > 0$, we obtain

$$P_x(A \mid \mathscr{L}(E_i)) = \frac{P_x(A \cap \mathscr{L}(E_i))}{P_x(\mathscr{L}(E_i))}$$

$$= \frac{P_x(\mathscr{L}(B \cap E_i))}{P_x(\mathscr{L}(E_i))}.$$

By the atomicity of E_i, $P_x(\mathscr{L}(B \cap E_i))$ can only be equal to 0 or $P_x(\mathscr{L}(E_i))$, which proves (3.11) (for the discrete parameter case, see Lemma 2.2 of [1]).

As an application, let $f(x)$ be a bounded $\mathscr{B}$ - measurable function defined on E, and let the $\{x_t,\ t \geq 0\}$ in $X = (x_t,\ \mathfrak{M}_t,\ P_x,\ \theta_t)$ be measurable [7]. From

$$\left(\omega: \int_0^{+\infty} f(x_t) dt = +\infty\right) = \left(\omega: \int_s^{+\infty} f(x_t) dt = +\infty\right)$$
$$= \theta_s \left(\omega: \int_0^{+\infty} f(x_t) dt = +\infty\right)$$

it is seen that $A \equiv \left(\omega: \int_0^{+\infty} f(x_t) dt = +\infty\right) \in \mathfrak{A}$. Hence, if one of the conditions in Theorem 6 is satisfied, then $P_x(A) = 0$ or 1. If there exists an E_i ($i > 0$), then the conditional zero-one law (3.11) is valid. These results can be extended without difficulty to general additive functionals.

Using the results of the present paper, it can be proved that both kinds of zero-one laws are valid for any birth and death process (see [2] for the definition), irrespective of whether or not the process is recurrent; see [12] for the details.

Bibliography

[1] Yang Chaoqun (Yang Ch'ao-ch'ün). Integral type functionals of countable Markov processes and boundary properties of two-sided birth and death processes. Progress in Math., 1964, 7: 397-424 (Chinese).

[2] Wang Zikun (Wang Tzu-k'un). On distributions of functionals of birth and death processes and their applications in the theory of queues. Sci. Sinica, 1961, 10: 160-170.

[3] Blumenthal R M. An extended Markov property.

Trans. Amer. Math. Soc. , 1957, 85: 52-72.

[4] Blumenthal R M, Getoor R K and McKean Jr H P. Markov processes with identical hitting distributions. Illinois J. Math. , 1962, 6: 402-420.

[5] Volkonskiĭ V A. Additive functionals of Markov processes. Trudy Moskov. Mat. Obšc, 1960, 9: 143 — 189; English transl. , Select. Transl. Math. Statist. and Prob. , Inst. for Math. Statist. and Amer. Math. Soc. , Providence R I. 1965, 5: 127-178.

[6] Kai Lai Chung. Markov chains with stationary transition probabilities. Die Grundlehren der Mathematischen Wissenschaften, Bd. 104, Springer-Verlag, Berlin, 1960.

[7] Doob J L. Stochastic processes. Wiley, New York, 1953.

[8] Dynkin E B. Foundations of the theory of Markov processes. Chinese Transl. , Science Press, Peking, 1962.

[9] Hunt G A. Markov processes and potentials. I, Illinois J. Math, 1957, 1: 44-93.

[10] H P McKean Jr and Tanaka H. Additive functionals of the Brownian Path. Mem. Coll. Sci. Univ. Kyoto Ser. A, 1960/61, 33: 479-506.

[11] Dynkin E B. Markov processes. Moscow, 1963 (Russian).

[12] Wang Zikun (Wang Tzu-k'un). Ergodicity and zero-one laws for birth and death processes. Nankai Univ. J. (Nat. Sci.), 1964, 5(5): 89-94 (Chinese).

Translated by: E. J. Brody

The Martin Boundary and Limit Theorems for Excessive Functions[①]

Abstract As proved in [8], for birth and death processes (with a conservative density matrix) every excessive function $u(i)$ has a limit $\lim_{i\to+\infty} u(i)$. It is natural to investigate the limit theorems for excessive functions of the general Markov chains. The purpose of this paper is to solve this problem by applying the Martin boundary theory which was first introduced by Doob[3] for the Markov chains and later developed by Hunt and Watanabe[5][6]. In Section 1 we give a sketch of the Martin boundary theory and prove some lemmas that we shall use. In Section 2 it is shown that for any nonrecurrent Markov chain[②] every finite excessive function has a limit in the sense of F-convergence and the exact value of this limit is found. For a discussion of the relation between the F-convergence and the

---

① Received: 1965-01-13.

② If the chain is recurrent and the states communicate with each other, then every finite excessive function is a constant.

usual convergence we introduce the concept of atomic kernel. In the case of atomic kernel the $F-$convergence is reduced to the usual convergence. For example, the imbedded Markov chain of bilateral birth and death process has at most two atomic kernels. In Section 3 our results are applied to denumerable Markov processes with continuous time parameters, chiefly to integral functionals of these processes; moreover, the bilateral birth and death processes are considered in detail.

§ 1. The martin boundary

§ 1.1. Generality

Let I be a denumerable set, $p=(p(i, j))$, a transition matrix on I, i.e., a matrix satisfying the following conditions:
$$p(i,j) \geqslant 0, \quad \sum_{j \in I} p(i,j) \leqslant 1.$$
A nonnegative function $h(i)(i \in I)$ is called excessive if
$$ph \cdot (i) \equiv \sum_{j \in I} p(i,j)h(j) \leqslant h(i), \ i \in I.$$
If this inequality is replaced by equality, and $0 \leqslant h < +\infty$, then h is called harmonic. A measure μ on I satisfying
$$\mu p \cdot (j) \equiv \sum_{i \in I} \mu(i) p(i,j) \leqslant \mu(j), \ j \in I$$
is called an excessive measure.

Let h be an excessive function. If $I^h = (i: 0 < h(i) < +\infty)$ is nonempty, we define
$$p^h(i, j) = \frac{p(i, j)h(j)}{h(i)}, \ i, \ j \in I^h; \qquad (1.1)$$

The Martin Boundary and Limit Theorems for Excessive Functions

then $p^h = (p^h(i, j))$ is a transition matrix on I^h. Similarly, we can define the h-excessive function (i.e., an excessive function on I^h for p^h), h-harmonic function, etc. A Markov chain $(x_n(\omega), \beta(\omega))$ with transition matrix p^h and state space I^h is called an h-chain, where $\beta(\omega)$ is the stopping time, $0 \leqslant n \leqslant \beta(\omega)$ (or $0 \leqslant n < +\infty$ if $\beta(\omega) = +\infty$), $\omega \in \Omega$. If $\beta(\omega) \equiv +\infty$, we write $\{x_n(\omega)\}$ or $\{x_n\}$ for $(x_n(\omega), \beta(\omega))$. Notice that the 1-chain is a Markov chain with transition matrix p.

Let $p_n(i, j)$ be the n-th step transition probability for the 1-chain, $p_0(i, j) = \delta_{ij}$ (the Kronecker symbol), and

$$G(i,j) = \sum_{n=0}^{+\infty} p_n(i,j). \tag{1.2}$$

In the following we shall always assume that the 1-chain is non-recurrent, i.e.,

$$G(i, j) < +\infty \quad (i, j \in I). \tag{1.3}$$

Fix a measure γ on I such that

$$\gamma(i) > 0 \ (i \in I); \quad \sum_{i \in I} \gamma(i) = 1.$$

Define a function

$$k(i, j) = \frac{G(i, j)}{\zeta(j)}, \quad i, j \in I,$$

where $\zeta(j) = \sum_{i \in I} \gamma(i) G(i,j)$ (without loss of generality we can assume that $\zeta(j) > 0$). In I a metric d can be introduced such that in this metric a point sequence $\{j_n\} \subset I$ is a Cauchy sequence if and only if either $\{j_n\}$ contains only finitely many different elements, or it contains infinitely many different elements, and $\{k(i, j_n)\}$ is a Cauchy sequence of real numbers for each fixed $i \in I$. The completion of I by d is denoted by I^*. The set $B = I^* \setminus I$ is called the Martin boundary, which depends on p and γ. Let $\mathscr{F}$ be the minimal σ-algebra containing all open sets of I^*.

Let $\xi \in B$. Define an excessive function $k(\cdot, \xi)$ by
$$k(i, \xi) = \lim_{\substack{j \to \xi \\ d}} k(i, j). \tag{1.4}$$

We say that an excessive function h is minimal if from equality $h = h_1 + h_2$ (h_1, h_2 are excessive) one can deduce that h_1, h_2 are proportional to h. A boundary point $\xi \in B$ is called minimal if $k(i, \xi)$ is minimal, harmonic, and satisfying
$$\sum_{i \in I} \gamma(i) k(i, \xi) = 1. \tag{1.5}$$
The set of all minimal points is denoted by B_e.

We shall use the following results of [5]:

(i) Let function h be excessive and γ-integrable; then for the h-chain (x_n, β), almost everywhere (a.e.) either $x_\beta \in I$ when β is finite, or x_n tends to some point $x_\beta \in B_e$ in the topology of I^* when β is infinite.

(ii) For each h in (i) there is a unique measure μ^h on $I \cup B_e$ such that
$$h(i) = \int_{I \cup B_e} k(i, \xi) \mu^h(d\xi); \tag{1.6}$$
the measure μ^h has a probability interpretation as follows: Suppose that the h-chain has an initial distribution γ^h ($\gamma^h(i) = \gamma(i) h(i)$); then the final state x_β has a distribution μ^h, i.e.,
$$\mu^h(c) = P(x_\beta \in c), \quad c \in \mathscr{F}. \tag{1.7}$$
The measure μ^h concentrates on B_e if and only if h is harmonic.

(iii) By a suitable choice of basic space Ω, for $\xi \in B_e$ we can consider the h-chain in (ii) under the condition $x_\beta = \xi$. The conditional chain is denoted by $\{x_{n_\xi}\}$. For μ^h almost all ξ, ($\mu^h -$

ξ), $\{x_{n_\xi}\}$ is a $k(\cdot, \xi)$-chain (and hence it is Markov) with a state space $I_\xi = (i: k(i, \xi) > 0)$ and an initial distribution $\gamma^{k(\cdot, \xi)}$. Since $k(\cdot, \xi)$ is harmonic,

$$\sum_{j \in I_\xi} \frac{p(i,j)k(j,\xi)}{k(i,\xi)} = 1 (i \in I_\xi),$$

this chain is not stopped, i. e., $\beta(\omega) \equiv +\infty$.

§ 1.2. Some lemmas

Lemma 1 Let $\xi \in B_e$ and $\{x_n\}$ be a $k(\cdot, \xi)$- chain with any initial distribution α ($\sum_i \alpha(i) = 1$). We have $P(x_\beta = \xi) = 1$.

Proof The probability P depends on the initial distribution α. It is best to denote P by P_α. If $\alpha = \gamma^{k(\cdot, \xi)}$, from (iii) and (1.5)

$$1 = P_\alpha(x_\beta = \xi) = \sum_{i \in I_\xi} \gamma(i) k(i,\xi) P_i(x_\beta = \xi), \quad (1.8)$$

where P_i is the probability generated by the $k(\cdot, \xi)$- chain with the initial distribution concentrating on state i. Since $\gamma(i) k(i, \xi) > 0$, from (1.5) and (1.8), $P_i(x_\beta = \xi) = 1 (i \in I_\xi)$. Hence for any initial distribution α (which must concentrate on I_ξ), we have

$$P_\alpha(x_\beta = \xi) = \sum_{i \in I_\xi} \alpha(i) P_i(x_\beta = \xi) = \sum_{i \in I_\xi} \alpha(i) = 1.$$

Let

$$\mu_i^h(c) = P_i(x_\beta \in c), \quad c \in \mathcal{F};$$

it is the distribution of the final state for the h - chain with the initial distribution concentrating on i. From (1.6)(1.7) we have seen that μ^h plays an important role. But how to find the values of $\mu_i^h(c)$ and $\mu^h(c)$? In a special case, as c is a one-point set $j \in I^h$, from (2.21) in [5] we get

$$\begin{cases} \mu_i^h(j) = \dfrac{G(i,j)[h(j) - ph \cdot (j)]}{h(i)}, & i, j \in I^h, \\ \mu^h(j) = \sum_{i \in I^h} \gamma(i) G(i,j)[h(j) - ph \cdot (j)], & j \in I^h. \end{cases}$$

(1.9)

As to the boundary set c, we have

Theorem 1 Suppose that $c \subset B_e$, $c \in \mathscr{F}$; then

$$\mu_i^h(c) = \frac{h_c(i)}{h(i)}, \quad i \in I^h,$$

$$\mu^h(c) = \sum_{i \in I^h} \gamma(i) h_c(i),$$

where $h_c(i)$ is the reduite① of h at i for c.

Proof Take an arbitrary element $s \bar{\in} I$, and define $p(s, i) = \gamma(i)$, $p(i, s) = 0 (i \in I)$. Then the domain of the definition of $p(i, j)$ is extended from I to $I \cup \{s\}$. Since $\gamma(i) > 0$, the extended p has a centre② s. Hence we may use the formula (4.20) of [6]:

$$h_c(i) = \int_c k(i, \xi) \mu^h(d\xi).$$

(1.10)

From this we obtain (cf. (2.19) in [5])

$$\mu_i^h(c) = \int_c \frac{k(i, \xi)}{h(i)} \mu^h(d\xi) = \frac{h_c(i)}{h(i)}, \quad i \in I^h.$$

(1.11)

From (1.7) and (1.11) follows

$$\mu^h(c) = \sum_{i \in I^h} \gamma^h(i) \mu_i^h(c) = \sum_{i \in I^h} \gamma(i) h_c(i).$$

On minimality we have the following simple results; for

---

① For definition, see [6]
② In terms of [6]

completeness we also give their proofs.

Lemma 2 (i) The constant function 1 is minimal harmonic if and only if there is a bounded, harmonic function not identical with zero, and any such function is a constant.

(ii) If $\xi \in B_e$ and h is a bounded $k(\cdot, \xi)$-harmonic function, then h is identical with some constant.

Proof (i) The following fact is useful[5,6]: For any minimal, harmonic function h, there is a unique point $\xi \in B_e$, on which concentrates the measure μ^h. Therefore, if 1 is minimal harmonic, μ^1 concentrates on some point $\xi \in B_e$, i.e., $P(x_\beta = \xi) = 1$, where x_β is the final state of the 1-chain and P is the probability generated by p and the initial distribution γ. Since $\gamma(i) > 0$, we have $P_i(x_\beta = \xi) = 1 (i \in I)$. But any bounded, harmonic function u is expressed by $u(i) = E_i f(x_\beta)$, where f is a boundary function defined on B_e, and E_i is the expectation with measure P_i.[5] Hence

$$u(i) = E_i f(x_\beta) = f(\xi)$$

is a constant, not depending on i.

Conversely, take a non-trivial bounded harmonic function u. By hypothesis, u is identical with a constant $c > 0$. This means that c is harmonic, and therefore 1 is also harmonic. If the harmonic function $v \leqslant 1$, by hypothesis v is a constant, so that v is proportional to 1. It follows that 1 is minimal, harmonic.

(ii) Since $\mu^{k(\cdot, \xi)}$ concentrates on point ξ, by an argument similar to that given in the proof (i), we know that any bounded $k(\cdot, \xi)$-harmonic function h is a constant.

§ 2. The limit theorems for excessive functions

§ 2.1. The existence of F - limits

In the investigation of limit theorems for excessive functions the F - convergence plays an important role. Let $\xi \in B_e$. According to Doob[3], a subset D of I is called an F - neighbourhood of ξ if for the $k(\cdot, \xi)$ - chain $\{x_n(\omega)\}$ with the initial distribution $\gamma^{k(\cdot, \xi)}$ (hence by Lemma 1, with any initial distribution on I_ξ), there is a positive integer $N \equiv N(\omega)$ such that for almost all ω, $x_n(\omega) \in D$ holds for all $n \geqslant N(\omega)$.

We say that a function $v(i)(i \in I)$ has an F - limit b (which may be infinite) at the point $\xi \in B_e$ in symbol $F \lim\limits_{i \to \xi} v(i) = b$, if for any neighbourhood G of b in the extended line $(-\infty, +\infty)$, there is an F - neighbourhood D of ξ such that $v(i) \in G$ for all $i \in D$.

The following Lemma 3 is pointed out in [3] without proof.

Lemma 3 The F - limit $F \lim\limits_{i \to \xi} v(i) = b$ exists if and only if for the $k(\cdot, \xi)$ - chain $\{x_n\}$ with the initial distribution $\gamma^{k(\cdot, \xi)}$ (and hence with any initial distribution on I_ξ) we have

$$\lim_{n \to +\infty} v(x_n) = b \quad \text{a. e.}.$$

Proof Suppose that $F \lim\limits_{i \to \xi} v(i) = b$; then for any neighbourhood G of b, there is an F - neighbourhood D of ξ such that $v(i) \in G$ for all $i \in D$. For fixed D, by definition, for the

The Martin Boundary and Limit Theorems for Excessive Functions

$k(\cdot, \xi)$-chain $\{x_n\}$ with initial distribution $\gamma^{k(\cdot,\xi)}$, we can find $N \equiv N(\omega) > 0$ such that $x_n \in D$ if $n \geq N$; therefore $v(x_n) \in G$ a. e..

Conversely, suppose that $\lim_{n \to +\infty} v(x_n) = b$ a. e.. By a suitable choice of values on a set with measure zero, b is measurable with the invariant σ-algebra [1, § I.17] for $\{x_n\}$. But by Lemma 2 (ii) and P113 in [1] this σ-algebra contains only the sets with measure 0 or 1; hence b is a constant a. e.. Let

$$\Omega_0 = (\omega: \lim_{n \to +\infty} v(x_n(\omega)) = b, \ b\text{ - const}).$$

Then $P(\Omega_0) = 1$. For any neighbourhood G of b, if $\omega \in \Omega_0$, there is a positive integer $N \equiv N(\omega)$ such that for all $n \geq N$

$$v(x_n(\omega)) \in G. \tag{2.1}$$

Take $D = (x_n(\omega): \omega \in \Omega_0, n \geq N(\omega))$, i.e., D is a set of states $x_n(\omega)$ satisfying the two conditions in brackets; $D \subset I$. By definition, D is an F-neighbourhood of ξ. From (2.1) it follows that $v(i) \in G$ for all $i \in D$.

In the following it is assumed that $\lim_{n \to \beta} v(x_n)$ always exists and is equal to $v(x_\beta)$ if $\beta < +\infty$. The measure μ restricted on B_e is denoted by μ_B. Recall that "$\mu_B - \xi$" means "for μ_B - almost all".

Theorem 2 Let (x_n, β) be the 1-chain with initial distribution γ. If $\lim_{n \to +\infty} v(x_n)$ exists a. e., then there exists limit

$$F \lim_{i \to \xi} v(i) = b(\xi) (\mu_B - \xi).$$

Proof By Lemma 3 it is sufficient to prove that for $\mu_B - \xi$ there exists $\lim_{n \to +\infty} v(y_n)$, where $\{y_n\}$ is a $k(\cdot, \xi)$-chain with initial distribution $\gamma^{k(\cdot,\xi)}$. According to (iii) in Section 1, by suitable choice of basic space Ω, we need only prove that

$\lim\limits_{n\to+\infty} v(x_{n\xi})$ exists a. e.. Let P^ξ be the probability induced by the Markov chain $\{x_{n\xi}\}$; then by hypothesis

$$1 = P(\text{exists } \lim_{n\to\beta} v(x_n)) = P(\beta < +\infty) + P(\text{exists } \lim_{n\to+\infty} v(x_n))$$

$$= \mu(I) + \int_{B_e} P^\xi(\text{exists } \lim_{n\to+\infty} v(x_{n\xi})) \mu_B(d\xi).$$

In virtue of $\mu(I) + \mu_B(B_e) = \mu(I \cup B_e) = 1$, there exists $b(\xi)$ such that

$$P^\xi(\lim_{n\to+\infty} v(x_{n\xi}) = b(\xi)) = 1 \quad (\mu_B - \xi).$$

§ 2.2. The exact values of F – limits

How to find the value of $F \lim\limits_{i\to\xi} v(i)$? This question can be solved if v is a finite excessive function.

Lemma 4[①] Suppose that for the Markov chain $\{x_n\}$ 1 is a minimal harmonic function corresponding to the point $\xi (\in B_e)$. If v is a finite excessive function of this chain, then

$$F \lim_{i\to\xi} v(i) = \inf_{i\in I} v(i).$$

Proof By Lemma 3 it is sufficient to prove

$$\lim_{n\to+\infty} v(x_n) = \inf_{i\in I} v(i) \quad \text{a. e.}.$$

The left limit exists and is finite since v is a finite excessive function[5]. In order to prove the equality, we first suppose that v is bounded. Consider the Riesz decomposition of v

$$v(i) = g(i) + h(i),$$

where g is a non-negative potential and h is a bounded harmonic function. By Lemma 2, $h(i) \equiv c$ (const). By an argument sim-

---

① "h – excessive (harmonic)" and "excessive (harmonic) for the h – chain" are synonyms.

ilar to that given in Theorem 12.8 of [4], it is not difficult to show that $\lim_{n\to+\infty} g(x_n)=0$ a. e.. Since $g(i) \geqslant 0$, we have
$$\lim_{n\to+\infty} v(x_n) = c = \inf_{i\in I} v(i) \quad \text{a. e..}$$
For a general v, put $a(\omega) = \lim_{n\to+\infty} v(x_n(\omega))$. As mentioned above, we can assume that $a(\omega)$ is finite. Define a function
$$v_m(i) = \min(m, v(i)).$$
Then v_m is bounded and excessive. Therefore
$$\lim_{n\to+\infty} v_m(x_n(\omega)) = \inf_{i\in I} v_m(i).$$
In virtue of $v_m \uparrow v$ and the finiteness of $a(\omega)$, there is a positive integer $N \equiv N(\omega)$ such that for all $n \geqslant N$,
$$v(x_n(\omega)) \leqslant a(\omega) + 1 \quad \text{a. e..}$$
Take any positive integer $M \equiv M(\omega) \geqslant a(\omega) + 1$; then
$$v(x_n(\omega)) = v_m(x_n(\omega)) \quad (n \geqslant N),$$
for all $m \geqslant M$. Note that for $m \geqslant M_1$ (M_1 being some positive integer), we have
$$\inf_{i\in I} v_m(i) = \inf_{i\in I} v(i).$$
Therefore
$$\lim_{n\to+\infty} v(x_n(\omega)) = \lim_{n\to+\infty} v_{M+M_1}(x_n(\omega)) = \inf_{i\in I} v(i) \quad \text{a. e..}$$

Theorem 3 Let u be a finite excessive function; then for $\mu_B - \xi$, we have

(i) $$F \lim_{i\to\xi} \frac{u(i)}{h(i)} = \inf_{i\in H} \frac{u(i)}{h(i)}, \qquad (2.2)$$

where $\xi \in B_e$, and h is a non-trivial minimal harmonic function corresponding to ξ, $H = (i: h(i) > 0)$;

(ii) if h is bounded, then

$$F \lim_{i\to\xi} u(i) = \sup_{i\in H} h(i) \cdot \inf_{i\in H} \frac{u(i)}{h(i)}; \qquad (2.3)$$

(iii) suppose that v is also a finite excessive function, h is bounded and $F\lim_{i\to\xi} v(i)>0$; then

$$F\lim_{i\to\xi}\frac{u(i)}{v(i)} = \inf_{i\in H}\left(\frac{u(i)}{h(i)}\right) \div \inf_{i\in H}\left(\frac{v(i)}{h(i)}\right). \qquad (2.4)$$

Proof Since h is minimal harmonic for p, 1 is minimal harmonic for p^h. The final distribution of the h-chain concentrates on the point ξ corresponding to h; therefore, the constant 1, as a minimal harmonic function for h - chain, corresponds also to ξ. Moreover, $\frac{u}{h}$ is h-excessive, finite in H. By Lemma 4 we get (2.2).

Taking $u\equiv 1$ in (2.2), we have

$$F\lim_{i\to\xi} h(i) = \sup_{i\in H} h(i) > 0. \qquad (2.5)$$

If h is bounded, this limit is finite. Substituting it in (2.2), we have

$$F\lim_{i\to\xi} u(i) = F\lim_{i\to\xi} h(i) \cdot F\lim_{i\to\xi}\frac{u(i)}{h(i)} = \sup_{i\in H} h(i) \cdot \inf_{i\in H}\frac{u(i)}{h(i)}.$$

This proves (2.3). Applying (2.3) to v and dividing (2.3) by the formula just obtained, we get (2.4) by the hypothesis

$$F\lim_{i\to\xi} v(i) > 0.$$

Remark 1 For a given boundary point $\xi\in B_e$, a possible choice of h is $k(\cdot, \xi)$. According to the equality (see (2.19) in [5]),

$$P_i(x_\beta=\xi) = k(i, \xi)\mu\{\xi\}. \qquad (2.6)$$

We see that $k(\cdot, \xi)$ must be bounded if $\mu\{\xi\}>0$. Since any minimal harmonic function corresponding to h must be proportional to $k(\cdot, \xi)$, h is bounded if $k(\cdot, \xi)$ is bounded.

§ 2.3. The case of atomic kernel

The F-convergence of finite excessive functions has been thoroughly studied. But when is the F-convergence reduced to the usual convergence?

We give a simple but sufficient condition which is useful in practice. First of all we introduce the concept of atomic kernel.

Let $\{x_n\}$ be a Markov chain without stopping. Suppose that the initial distribution is γ. Consider the Blackwell decomposition of state space for this chain:

$$I = E_0 \cup E_1 \cup E_2 \cup \cdots, \qquad (2.7)$$

where E_i are disjoint, almost closed sets, E_0 is completely non-atomic, while $E_j (j>0)$ are atomic sets. We call $\xi_j (\in B_e)$ an atomic boundary point if $\mu\{\xi_j\} > 0$. In [2] it is shown that there exists a one-to-one correspondence between all atomic almost closed sets and all atomic boundary points. Let E_j correspond to ξ_j.

Given a subset ε_j of $E_j (j>0)$, we say that ε_j is an atomic kernel if

$$P(\mathscr{L}(E_j)) = P(\mathscr{L}(\varepsilon_j)), \qquad (2.8)$$

and if for any infinite subset A of ε_j

$$P(\mathscr{L}(\varepsilon_j - A)) = 0, \qquad (2.9)$$

where $\mathscr{L}(\varepsilon) = \bigcup_{m=1}^{+\infty} \bigcap_{n=m}^{+\infty} (x_n \in \varepsilon)$.

Lemma 5 Any unbounded[①] sequence $\{j_n\}$ of ε_j F-converges to ξ_j as $n \to +\infty$.

Proof Given any F-neighbourhood C of ξ_j. For the

---

① A sequence $\{j_n\}$ is called unbounded if for any finite subset D of I, there is $N > 0$ such that $j_n \overline{\in} D$ for all $n \geq N$.

$k(\cdot, \xi_j)$-chain $\{y_n\}$ there are a positive integer $N \equiv N(\omega)$ and an ω-set Ω_0, $P(\Omega_0) = 1$ such that for any $\omega \in \Omega_0$, we have $y_n(\omega) \in C$ for all $n \geqslant N$. Let

$$Y = (y_n(\omega) : \omega \in \Omega_0, n \geqslant N(\omega)) \subset C.$$

Taking $N(\omega)$ sufficiently large, one can assume that $Y \subset E_j$; by (2.8) one can assume further that $Y \subset \varepsilon_j$. From this follows that all (but finitely many) $j_n \in Y$, because otherwise there would be an infinite subset A of $\{j_n\}$ such that

$$P(\mathscr{L}(\varepsilon_j - A)) \geqslant P(\mathscr{L}(Y - A)) = P(\mathscr{L}(Y)) = \mu\{\xi_j\} > 0,$$

which contradicts (2.9). Hence

$$j_n \in Y \subset C$$

for all $n \geqslant M$. Here M is a constant.

Theorem 4 Suppose that $\{x_n\}$ is a Markov chain with the initial distribution γ and u is a finite excessive function for this chain. If ε_j is an atomic kernel and $\{j_n\}$ is any unbounded sequence of ε_j, then there exists a finite limit $\lim_{n \to +\infty} u(j_n)$. Moreover,

$$\lim_{n \to +\infty} u(j_n) = \sup_{i \in H} h(i) \cdot \inf_{i \in H} \frac{u(i)}{h(i)}, \quad (2.10)$$

where h is a minimal harmonic function corresponding to ξ_j, $H = (i : h(i) > 0)$.

Proof Since ξ_j corresponds to the atomic almost closed E_j, $\mu\{\xi_j\} > 0$. By remark 1, h is bounded. From (2.3) follows that the value on the right side of (2.10) is equal to $F \lim_{i \to \xi_j} u(i)$. But by Lemma 5

$$\lim_{n \to +\infty} u(j_n) = F \lim_{i \to \xi_j} u(i).$$

This concludes the proof of (2.10).

§ 2. 4. The limit theorems for excessive measures

By duality it is not difficult to obtain the corresponding results for excessive measures. Suppose that for the transition matrix p there is a strictly positive excessive measure $\alpha(i)>0 (i\in I)$ (If p is irreducible, i. e., if each state communicates with every other state, such α exists). Let

$$q_{ji} = \frac{p(i, j)\alpha(i)}{\alpha(j)}. \qquad (2.11)$$

Suppose that β is a finite excessive measure for p. Define

$$\beta^*(i) = \frac{\beta(i)}{\alpha(i)}, \ i \in I;$$

then β^* is a finite excessive function for $Q=(q_{ji})$. Let ξ^* be a minimal boundary point for the Q-chain and F^* be its F-convergence. From (2.2) we get

$$F^* \lim_{i \to \xi^*} \frac{\beta(i)}{\alpha(i) h^*(i)} = \inf_{i \to H^*} \frac{\beta(i)}{\alpha(i) h^*(i)},$$

where h^* is a minimal harmonic function for the Q-chain and corresponds to ξ^*, $H^* = (i : h^*(i) > 0)$.

§ 3. Application to denumerable Markov processes

§ 3.1. The continuous time-parameter case

Let $X \equiv \{x(t, w), 0 \leqslant t < \tau(w)\}$ be a denumerable homogeneous Markov process whose transition matrix satisfies the condition:

$$\lim_{t \to 0} p_{ii}(t) = 1, \quad i \in I. \tag{3.1}$$

Suppose that the sample functions of this process are right continuous. Let $Q = (q_{ij})$ be its density matrix, where

$$q_{ij} = \lim_{t \to 0} \frac{p_{ij}(t) - \delta_{ij}}{t}.$$

In the following we always assume that

$$0 < q_i \equiv -q_{ii} = \sum_{j \neq i} q_{ij} < +\infty, \quad i \in I. \tag{3.2}$$

Denote by $\tau_n(\omega)$ the n-th jump of sample functions, i.e.,

$\tau_0(\omega) \equiv 0$,

$\tau_n(\omega) = \inf(t : t > \tau_{n-1}(\omega), x(t, \omega) \neq x(\tau_{n-1}(\omega), \omega))$.

Moreover, it is assumed that the stopping time is given by

$$\tau(\omega) = \lim_{n \to +\infty} \tau_n(\omega), \tag{3.3}$$

so that $X \equiv \{x(t, \omega), 0 \leqslant t < \tau(\omega)\}$ is the so-called minimal process. The Markov chain

$$y_n(\omega) = x(\tau_n(\omega), \omega), \quad n \in \mathbf{N} \tag{3.4}$$

is called the imbedded chain of process X. In order to define the Martin boundary for $\{y_n\}$, we suppose that the transition matrix of $\{y_n\}$ satisfies (1.3). The symbol $F \lim_{i \to \xi}$ used below is relative to this boundary.

A non-negative function $u(i)$ $(i \in I)$ is called excessive for X, or simply X-excessive if for any $t \geqslant 0$,

$$\sum_{j \in I} p_{ij}(t) u(j) \leqslant u(i), \quad i \in I. \tag{3.5}$$

It was shown in [6] that the function u is X-excessive if and only if it is excessive for the imbedded chain $\{y_n\}$. Therefore, the results of Section II can be applied to finite X-excessive functions.

§ 3.2. A limit theorem

For a given function $f(i) \geqslant 0$ $(i \in I)$ satisfying the condition

$$u(i) \equiv E_i \int_0^\tau f(x(t, \omega)) dt < +\infty, \quad i \in I, \tag{3.6}$$

it is easy to show that the function u defined above is finite and excessive[①] for X.

Assume that the imbedded chain $\{y_n\}$ has the initial distribution γ, and ε is an atomic kernel corresponding to the boundary point ξ for this chain. Clearly, $\mu\{\xi\} > 0$. By (2.6) we know that

$$\mu_i\{\xi\} \equiv P_i(y_\beta = \xi)$$

is a minimal harmonic function corresponding to ξ for $\{y_n\}$. Take any unbounded sequence $\{j_n\} \subset \varepsilon$. From (2.10) follows

$$\lim_{n \to +\infty} E_{j_n} \int_0^\tau f(x(t, \omega)) dt = \sup_{i \in H} \mu_i\{\xi\} \cdot \inf_{i \in H} \frac{E_i \int_0^\tau f(x(t, \omega)) dt}{\mu_i\{\xi\}}, \tag{3.7}$$

where $H = (i: \mu_i\{\xi\} > 0)$.

Especially in (3.7) taking first $f(i) = \delta_{li}$ and then $f(i) =$

---

① See Sections 10.15 and 12.3 in [4].

$\delta_{mi}(l, m \in I)$, we obtain two equalities. Dividing the first equality by the second, if[①]

$$M_m \equiv \inf_{i \in H} \frac{\int_0^{+\infty} p_{im}(t)\,dt}{\mu_i\{\xi\}} > 0, \qquad (3.8)$$

we get the following formula about the integral ratio for transition probabilities of minimal process

$$\lim_{n \to +\infty} \frac{\int_0^{+\infty} p_{j_n l}(t)\,dt}{\int_0^{+\infty} p_{j_n m}(t)\,dt} = \frac{M_l}{M_m}. \qquad (3.9)$$

§ 3.3. F-convergence reduced to usual convergence for the bilateral birth and death process X

A denumerable homogeneous Markov process is called a bilateral birth and death process if its density matrix $Q = (q_{ij})$ satisfies the conditions:

$$q_{ij} = 0, \text{ if } |i-j| > 1;$$
$$q_{i,i-1} = a_i > 0, \quad q_{i,i+1} = b_i > 0, \quad q_i \equiv -q_{ii} = a_i + b_i,$$

$i, j \in I = \mathbf{Z}$, I being the set of all integers. Introduce the following characteristic numbers:

$$x_i = -b_0\left(1 + \frac{b_{-1}}{a_{-1}} + \cdots + \frac{b_{-1}b_{-2}\cdots b_{i+1}}{a_{-1}a_{-2}\cdots a_{i+1}}\right), \quad \text{if } i < -1,$$

$$x_{-1} = -b_0, \quad x_0 = 0, \quad x_1 = a_0,$$

$$x_i = a_0\left(1 + \frac{a_1}{b_1} + \cdots + \frac{a_1 a_2 \cdots a_{i-1}}{b_1 b_2 \cdots b_{i-1}}\right), \quad \text{if } i > 1,$$

$$z_1 = \lim_{i \to -\infty} x_i, \quad z_2 = \lim_{i \to +\infty} x_i.$$

In [7] it is shown that if c_{in} is the probability of reaching n

---

[①] Note that by (3.7) $\lim_{n \to +\infty} \int_0^{+\infty} p_{j_n m}(t)\,dt = M_m \cdot \sup_{i \in H} \mu_i\{\xi\}$.

by finite steps from i for the imbedded chain $\{y_n\}$, then

$$c_{in} = \begin{cases} \dfrac{z_2 - x_i}{z_2 - x_n}, & \text{if } n < i, \\ \dfrac{x_i - z_1}{x_n - z_1}, & \text{if } n > i \end{cases} \qquad (3.10)$$

$\left(\dfrac{+\infty}{+\infty}=1\right)$; and the condition (1.3) is not satisfied for $\{y_n\}$ if and only if $z_1 = -\infty$, $z_2 = +\infty$①. Hence we should only consider the cases where at least one z_i is finite.

Case A $z_1 = -\infty$, $z_2 < +\infty$. In this case there is only one atomic almost closed set I; $\varepsilon = (n, n+1, \cdots)$ is an atomic kernel[7]. Notice

$$k(i,j) = \frac{G(i,j)}{\sum_{v \in I} \gamma(v) G(v,j)} = \frac{c_{ij} G(j,j)}{\sum_{v \in I} \gamma(v) c_{vj} G(j,j)} = \frac{c_{ij}}{\sum_{v \in I} \gamma(v) c_{vj}}. \qquad (3.11)$$

If $j > i$, by (3.10) $c_{ij} = 1$. Hence

$$k(i,j) = \frac{1}{\sum_{v<j} \gamma(v) + \sum_{v \geq j} \gamma(v) c_{vj}}. \qquad (3.12)$$

Let $+\infty$ be the minimal boundary point to which corresponds ε. Since $\sum_{v \in I} \gamma(v) = 1$, we have

$$k(i, +\infty) = \lim_{j \to +\infty} k(i, j) = 1. \qquad (3.13)$$

By the proof of Lemma 2 (i), $\mu\{+\infty\} \equiv P(y_\beta = +\infty) = 1$.

Besides $+\infty$ there is another boundary point $-\infty$. Similarly,

---

① If $z_1 = -\infty$, $z_2 = +\infty$, then $\{y_n\}$ is recurrent. Hence each excessive function u for $\{y_n\}$ (or for the minimal process X) is a constant. Notice that u is either finite everywhere or identical with ∞ since $p_{ij}(t) > 0$, $t > 0$, $i, j \in I$.

$$k(i, -\infty) = \frac{z_2 - x_i}{\sum_{v \in I} \gamma(v)(z_2 - x_v)}.$$

But since $\mu\{-\infty\} = 0$, this boundary is trivial.

Suppose that u is a finite excessive function for the minimal process X. By (2.10) and taking $h(i) = k(i, +\infty) = 1$ there, we get

$$\lim_{i \to +\infty} u(i) = \inf_{i \in I} u(i).$$

Case B $z_1 > -\infty$, $z_2 < +\infty$. There are two atomic almost closed sets, each of which has an atomic kernel $\varepsilon_1 = (\cdots, -n-1, -n)$ or $\varepsilon_2 = (n, n+1, \cdots)$. Let $-\infty$, $+\infty$ be two minimal boundary points to which correspond ε_1, ε_2 respectively. Then

$$k(i, -\infty) = \frac{z_2 - x_i}{\sum_{v \in I} \gamma(v)(z_2 - x_v)}, \quad k(i, +\infty) = \frac{x_i - z_1}{\sum_{v \in I} \gamma(v)(x_v - z_1)}.$$

For any finite excessive function of the minimal process X, we have

$$\lim_{i \to -\infty} u(i) = (z_2 - z_1) \cdot \inf_{i \in I} \frac{u(i)}{z_2 - x_i},$$

$$\lim_{i \to +\infty} u(i) = (z_2 - z_1) \cdot \inf_{i \in I} \frac{u(i)}{x_i - z_1}.$$

We omit the case $z_1 > -\infty$, $z_2 = +\infty$ since it can be considered in the same way as in Case A.

References

[1] Chung K L. Markov Chains with Stationary Transition Probabilities. 1960, Springer, Berlin: Gottinger, Heidelberg.

[2] Chung K L. On the boundary theory for Markov chains. Acta Mathematica, 1963, 110(1-2): 9-77.

［3］Doob J L. Discrete potential theory and boundaries. J. Math. and Mech., 1959, 8: 433-458.

［4］Дынкин Е Ъ. Марковские процессы. Физматгиз, 1963.

［5］Hunt G A. Markov chains and Martin boundaries. Illinois J. Math., 1960, 4: 313-340.

［6］Watanabe T. On the theory of Martin boundaries induced by countable Markov processes. Mem. Coll. Sci. Univ. Kyoto, Series A, Math, 1960, 33: 39-108.

［7］Yang Chao-chun. The integral functionals of denumerable Markov processes and boundary properties of bilateral birth and death processes. Progress in Math., 1964, 7(4): 397-424(Chinese).

［8］Wang Tzu-kwen. Ergodic property and zero-one law for birth and death processes. Acta Scientiarum Naturalium Universitatis Nankaiensis (Math.), 1964, 5(5): 89-94.

Martin 的边界和过分函数的极限定理

摘要 利用 Martin 边界，证明了对非常返马尔可夫链，有限过分函数在 F-收敛下有极限，并求出了此极限．为讨论 F-收敛与通常收敛之关系，引进了原子核概念．在原子核情况，F-收敛化为通常收敛，然后应用这些结果以研究可列马尔可夫过程的积分型泛函．

Some Properties of Recurrent Markov Processes[①]

§ 1. Introduction[②]

We use the definitions and notations of [1]. Let $X=(x_t, \zeta, \mathfrak{M}_t, p_x)$ be a Markov process on the state space $(E, \mathscr{B})$. By a process in this paper, we mean a Markov process, since we shall confine our attention to Markov processes.

One of the important types of process is the recurrent process. In Markov chains where both time and E are discrete, the study of the recurrent property was made quite early [2]; however, the study of general recurrent processes is a matter of recent years [3][4]. Our aim is to discuss certain properties of

---

　　① Received: 1964-01-24; Revised: 1965-01-27.
　　② The author is grateful to the audience in the discussion class, who gave suggestions for improvements, and to Shi Ren-jie (Shih Jen-chieh), who has carefully read through the paper.

Some Properties of Recurrent Markov Processes

general recurrent processes. The following is a summary of the paper.

In a recurrent process, the first entrance (contact) time into a certain set plays an important role. In [1], the definition of the first entrance time after time $s+0$ is given by the limit of the first entrance time after time $u(>s)$ as $u \downarrow s$. In §2 we first give an equivalent definition of the first entrance (or first contact) time into a set Γ after time s (or $s+0$), and in the new definition, we write down explicitly an expression for the first entrance time after time $s+0$, and consequently we can see its obvious difference from the first entrance time after time s. It is the same in the case of first contact times. Finally we discuss the relation between these times in fundamental, countable processes. In §3, we discuss necessary and sufficient conditions for recurrent processes, and we can obtain some results similar to those in Markov chains after we add some restrictions to the processes. For example: the recurrent property can be expressed in terms of transition probabilities in a canonical diffusion process. In §4, we study two problems: the excessive functions for recurrent processes and the strong zero-one law at infinity. We have proved that any continuous excessive function for a right-continuous recurrent process is identically equal to a constant; in the case of a fundamental, countable process or one-dimensional regular, continuous, strong Markov process, a necessary and sufficient condition for the recurrent property is that any finite excessive function is a constant. In [5], we have proved that the zero-one law holds at infinity for a nonterminating right continuous recurrent strong Feller process. Here we have more satisfactory results for standard processes: a necessary and sufficient condition for the zero-one law to hold is that any non-negative, bounded harmonic function on E is a constant. Finally, in §5 we apply our results to second-order differential equations of several variables, and discuss the relation between the fundamental solution of the equation $\frac{\partial u}{\partial t}=Du$ and the Dirichlet problem for the equation $Du=0$.

As we have said earlier, all the definitions and notations in this paper are adopted from [1]; we will point out the pages where they appear, and will not repeat them here except when necessary. Otherwise it would make our description too long.

§ 2. First entrance and first contact times

Let $X=(x_t,\ \zeta,\ \mathfrak{M}_t,\ p_x)$ be a Markov process defined on a state space $(E,\ \mathfrak{B})$. For $\omega \in (\zeta(\omega)>t)$, let $F_t^s = F_t^s(\omega)$ be the range of the sample function during the period $[s,\ t]$, i. e.

$$F_t^s(\omega) = (x_u(\omega):\ s \leqslant u \leqslant t). \qquad (2.1)$$

For the set $\Gamma \subseteq E$ and any real number $s \geqslant 0$, define the function

$$\eta_s^{(1)}(\omega) = \begin{cases} \sup(t:\ F_t^s \cap \Gamma = \varnothing) & \text{if } \zeta(\omega)>s,\ \text{and the } t\text{-set inside} \\ \sup(t:\ F_t^s \subseteq E \setminus \Gamma), & \text{the bracket is nonempty;} \\ s, & \text{if } \zeta(\omega)>s,\ \text{and the above} \\ & t\text{-set is empty;} \\ s, & \text{if } \zeta(\omega) \leqslant s. \end{cases}$$

$$(2.2)$$

In [1], $\eta_s^{(1)}(\omega)$ is defined as the first entrance time into Γ after time s. Obviously, if $s \leqslant u$, then $\eta_s^{(1)}(\omega) \leqslant \eta_u^{(1)}(\omega)$, hence the limit

$$\eta_s^{(2)}(\omega) = \lim_{u \downarrow s} \eta_u^{(1)}(\omega), \qquad (2.3)$$

exists. In [1], $\eta_s^{(2)}(\omega)$ is defined as the first entrance time into Γ after time $s+0$.

Definition (2.3) is hard to handle, and we give an equivalent definition for $\eta_s^{(1)}$ and $\eta_s^{(2)}$.

Lemma 1 Let

$$\tau_s^{(1)}(\omega) = \begin{cases} \inf(t: t \geqslant s, \\ \quad x_t(\omega) \in \Gamma), & \text{if } \zeta(\omega) > s, \text{ and the } t\text{-set inside} \\ & \text{the bracket is nonempty;} \\ \zeta(\omega), & \text{if } \zeta(\omega) > s, \text{ and the above } t\text{-set is} \\ & \text{empty;} \\ s, & \text{if } \zeta(\omega) \leqslant s. \end{cases}$$

(2.4)

$$\tau_s^{(2)}(\omega) = \begin{cases} \inf(t: t > s, \\ \quad x_t(\omega) \in \Gamma), & \text{if } \zeta(\omega) > s, \text{ and the } t\text{-set inside} \\ & \text{the bracket is nonempty;} \\ \zeta(\omega), & \text{if } \zeta(\omega) > s, \text{ and the above } t\text{-set is} \\ & \text{empty;} \\ s, & \text{if } \zeta(\omega) \leqslant s. \end{cases}$$

(2.5)

Then[①]

$$\tau_s^{(1)}(\omega) = \eta_s^{(1)}(\omega); \tag{2.6}$$

$$\tau_s^{(2)}(\omega) = \eta_s^{(2)}(\omega). \tag{2.7}$$

(2.6) was pointed out in [8], appendix § 2.8; its proof is easy and so we omit it here. The following proves (2.7).

Replace s in (2.4) by u ($>s$) and denote the new formula by (2.4)′; this is the definition of $\tau_u^{(1)}(\omega)$. Consider the three cases in (2.5) separately; if $\zeta(\omega) \leqslant s$ ($< u$). and since $\zeta(\omega) < u$, then we have

$$\tau_s^{(2)}(\omega) = s = \lim_{u \downarrow s} u = \lim_{u \downarrow s} \tau_u^{(1)}(\omega)$$
$$= \lim_{u \downarrow s} \eta_u^{(1)}(\omega) = \eta_s^{(2)}(\omega), \tag{2.8}$$

---

① Note that $\tau_s^{(1)}$ is equivalent to $\tau_s^{(2)}$ if we change $t \geqslant s$ inside the bracket in (2.4) to $t > s$ in (2.5). In the case of (2.2), we can obtain the same result no matter whether we interpret $\sup(t: F_t^s \cap \Phi = \varnothing)$ as $\sup(t: t \geqslant s, F_t^s \cap \Gamma = \varnothing)$ or $\sup(t: t > s, F_t^s \cap \Gamma = \varnothing)$. This goes for Lemma 2 too.

for the last two equalities, we use (2.6) and (2.3). If the second case of (2.5) occurs, since $\zeta(\omega) > s$, hence if $u(>s)$ is sufficiently close to s, then $\zeta(\omega) > u$ and
$$(t: t \geq u, \ x_t(\omega) \in \Gamma) \subseteq (t: t > s, \ x_t(\omega) \in \Gamma) = \emptyset.$$
Hence from $(2.4)'$, we know $\tau_u^{(1)}(\omega) = \zeta(\omega) = \tau_s^{(2)}(\omega)$, and therefore
$$\tau_s^{(2)}(\omega) = \lim_{u \downarrow s} \tau_u^{(1)}(\omega) = \lim_{u \downarrow s} \eta_u^{(1)}(\omega) = \eta_s^{(2)}(\omega). \qquad (2.9)$$

Finally consider the first case. If we can show there exists a sequence $\{u_n\}$, $u_n \downarrow s$, such that
$$\lim_{n \to +\infty} \tau_{u_n}^{(1)}(\omega) = \tau_s^{(2)}(\omega), \qquad (2.10)$$
and since $\lim_{u \downarrow s} \tau_u^{(1)}(\omega)$ exists, then we have
$$\tau_s^{(2)}(\omega) = \lim_{n \to +\infty} \tau_{u_n}^{(1)}(\omega) = \lim_{u \downarrow s} \tau_u^{(1)}(\omega) = \lim_{u \downarrow s} \eta_u^{(1)}(\omega) = \eta_s^{(2)}(\omega),$$
hence (2.7) is proved. Therefore we only have to prove (2.10).

First we let $\tau_s^{(2)} = s$. From (2.5) we have for every $\frac{1}{n} > 0$, there exists u_n, such that $x_{u_n} \in \Gamma$ and $0 < u_n - \tau_s^{(2)} < \frac{1}{n}$; when n is sufficiently large $\zeta > u_n$. From the definition of $\tau_{u_n}^{(1)}$ we obtain $\tau_{u_n}^{(1)} = u_n$; hence
$$0 < \tau_{u_n}^{(1)} - \tau_s^{(2)} < \frac{1}{n}, \qquad \lim_{n \to +\infty} \tau_{u_n}^{(1)} = \tau_s^{(2)}.$$

Next we let $\tau_s^{(2)} = a > s$. Therefore there is no point t in the interval (s, a) satisfying the condition inside the bracket in (2.5). Hence when n is suffciently large, $\zeta > s + \frac{a-s}{n}$, and also
$$\tau_s^{(2)} = \inf\left(t: \ t \geq s + \frac{a-s}{n}, \ x_t \in \Gamma\right) = \tau_{s+\frac{a-s}{n}}^{(1)};$$

let $u_n = s + \dfrac{a-s}{n}$, then we have proved (2.10).

Now we assume that the state space is a topological and measurable space $(E, \mathfrak{C}, \mathfrak{B})$ ([1], P789), and denote by $\hat{F}_t^s$ the closure of F_t^s in (2.1). Define the function

$$\eta_s^{(3)}(\omega) = \begin{cases} \sup(t: \hat{F}_t^s \cap \Gamma = \varnothing) \\ \sup(t: \hat{F}_t^s \subseteq E \setminus \Gamma), \\ \qquad \text{if } \zeta(\omega) > s, \text{ and the } t\text{-set inside the bracket is nonempty;} \\ s, \qquad \text{if } \zeta(\omega) > s, \text{ and the above } t\text{-set is empty;} \\ s, \qquad \text{if } \zeta(\omega) \leqslant s. \end{cases}$$

(2.11)

In [1], $\eta_s^{(3)}(\omega)$ is defined as the first contact time with Γ after time s. Obviously, if $s \leqslant u$, then $\eta_s^{(3)}(\omega) \leqslant \eta_u^{(3)}(\omega)$, hence there exists the limit:

$$\eta_s^{(4)}(\omega) = \lim_{u \downarrow s} \eta_u^{(3)}(\omega). \qquad (2.12)$$

In [1], $\eta_s^{(4)}(\omega)$ is defined as the first contact time with Γ after time $s+0$.

Lemma 2 If we let[①]

$$\tau_s^{(3)}(\omega) = \begin{cases} \inf(t: t \geqslant s, \text{ there exists } s_n, s \leqslant s_n \uparrow t, \text{ so that } \lim_{n \to +\infty} x_{s_n} \in \Gamma), \\ \qquad \text{if } \zeta(\omega) > s, \text{ and the } t\text{-set inside the bracket is nonempty;} \\ \zeta(\omega), \qquad \text{if } \zeta(\omega) > s, \text{ and the above } t\text{-set is empty;} \\ s, \qquad \text{if } \zeta(\omega) \leqslant s. \end{cases}$$

(2.13)

---

① The s_n in both (2.13) and (2.14) can be identically equal.

$$\tau_s^{(4)}(\omega) = \begin{cases} \inf(t: t>s, \text{ there exists } s_n, \ s \leqslant s_n \ t, \text{ so that} \\ \quad \lim_{n \to +\infty} x_{s_n} \in \Gamma), \text{ if } \zeta(\omega) > s, \text{ and the } t\text{-set inside the} \\ \quad \quad \quad \quad \quad \quad \quad \text{bracket is nonempty;} \\ \zeta(\omega), \quad \quad \quad \quad \text{if } \zeta(\omega) > s, \text{ and the above } t\text{-set is} \\ \quad \quad \quad \quad \quad \quad \quad \text{empty;} \\ s, \quad \quad \quad \quad \quad \quad \text{if } \zeta(\omega) \leqslant s. \end{cases}$$

(2.14)

then

$$\tau_s^{(3)}(\omega) = \eta_s^{(3)}(\omega); \quad (2.15)$$

$$\tau_s^{(4)}(\omega) = \eta_s^{(4)}(\omega). \quad (2.16)$$

Proof We shall omit the proof of (2.15). To prove (2.16), repeat the proof of Lemma 1 sentence by sentence up to (2.10), and we only have to show for the first case in (2.14) the following: there exists a sequence $\{u_n\}$, $u_n \downarrow s$, such that

$$\lim_{n \to +\infty} \tau_{u_n}^{(3)}(\omega) = \tau_s^{(4)}(\omega). \quad (2.17)$$

Replace s in (2.13) by u ($>s$) and denote the new formula by (2.13)′, this is the definition of $\tau_u^{(3)}(\omega)$.

First let $\tau_s^{(4)} = s$. From the definition, for every positive integer n, there exists $u_n > s$, such that $0 < u_n - \tau_s^{(4)} < \frac{1}{n}$, and also there exists $s_m^{(n)}$, such that $s \leqslant s_m^{(n)} \uparrow u_n (m \to +\infty)$, at the same time $\lim_{m \to +\infty} x_{s_m^{(n)}} \in \Gamma$. We may let $u_{n-1} > u_n > \tau_s^{(4)}$, $u_n \downarrow \tau_s^{(4)} = s$. From the definition of $\tau_{u_n}^{(3)}$ in (2.13)′, when n is sufficiently large, we have the first case in (2.13)′, and also $\tau_{u_n}^{(3)} \leqslant u_{n-1}$; hence

$$0 < \tau_{u_n}^{(3)} - \tau_s^{(4)} \leqslant u_{n-1} - \tau_s^{(4)} < \frac{1}{n-1} \to 0 \quad (n \to +\infty).$$

Next let $\tau_s^{(4)} = a > s$, hence there is no point t in interval (s, a) satisfying the condition inside the bracket in (2.14); therefore when n is sufficiently large, we have $\zeta > s + \dfrac{a-s}{n}$, and also

$$\tau_s^{(4)} = \inf\left(t: t \geqslant s + \frac{a-s}{n}, \text{ there exists } s_n,\right.$$
$$\left. s + \frac{a-s}{n} \leqslant s_n \uparrow t, \text{ so that } \lim_{n \to +\infty} x_{s_n} \in \Gamma\right)$$
$$= \tau_{s+\frac{a-s}{n}}^{(3)};$$

let $u_n = s + \dfrac{a-s}{n}$, and we have proved (2.17).

Now let $E = \mathbf{N}$ be the set of non-negative integers, $\bar{E} = E \cup \{+\infty\}$, $X = (x_t, +\infty, N_t, p_x)$ is defined to be a fundamental, countable process, if and only if it has the following properties (i)~(v):

(i) The transition probability $P_{xy}(t) = p(t, x, \{y\})$ satisfies

$$\lim_{t \to 0} p_{xx}(t) = 1 \quad (\text{every } x \in E). \tag{2.18}$$

(ii) X is completely separable: For any sequence of points $R = \{r_n\}$ dense in $[0, +\infty)$, and any $\omega \in \Omega$, and any $t \in [0, +\infty)$, there exists a subsequence $\{s_n\} \subseteq R$, such that we have

$$\lim_{n \to +\infty} s_n = t, \quad \lim_{n \to +\infty} x_{s_n}(\omega) = x_t(\omega)$$

simultaneously.

(iii) X is lower right semicontinuous if and only if for every $\omega \in \Omega$, and for every $t \geqslant 0$, we have

$$\lim_{s \downarrow t} x_s(\omega) = x_t(\omega).$$

(iv) X is a measurable process([1], P143).

(v) $\overline{\mathfrak{N}_t} = \overline{\mathfrak{N}_{t+0}} = \bigcap_{u>t} \overline{\mathfrak{N}_u}$.

It has been proved in [6]: for every set of transition

probability[1] ($p_{xy}(t)$), x, $y \in E$, satisfying (2.18), there always exists a process X satisfying (ii)$\sim$(v) such that the transition probabilities of X coincide with those given, i.e. ($p_{xy}(t)$). Because of separability, it is possible that $x_t(\omega) = +\infty$, but for every fixed $t \geq 0$,

$$p_x(x_t = +\infty) = 1 - \sum_{y \in E} p_{xy}(t) = 0. \qquad (2.19)$$

Hereafter we sometimes denote $\tau_s^{(i)}$ by $\tau_s^{(i)}(\Gamma)$, $i = 1, 2, 3, 4$. If $s = 0$, then we simply write $\tau_0^{(i)}$ as $\tau^{(i)}$, and call $\tau^{(1)} = \tau^{(1)}(\Gamma)$, $\tau^{(3)} = \tau^{(3)}(\Gamma)$ the first entrance time into Γ, and the first contact time with Γ respectively.

Lemma 3[2] For every fundamental, countable process, if $\Gamma \subseteq E$, then

(i) $\tau_s^{(1)} = \tau_s^{(2)} = \tau_s^{(3)} = \tau_s^{(4)}$;

(ii) there exists a sequence of finite sets $\Gamma_1 \subseteq \Gamma_2 \subseteq \cdots \subseteq \Gamma$, such that $\tau_s^{(i)}(\Gamma_n) \downarrow \tau_s^{(i)}(\Gamma)$, $i = 1, 2, 3, 4$;

(iii) $(\tau^{(i)} > t) \in \mathfrak{N}_{t+0} (t \geq 0)$, $i = 1, 2, 3, 4$.

Proof (i) We first prove $\tau_s^{(1)} \geq \tau_s^{(2)}$. If $x_s(\omega) \in \Gamma$, obviously we have $\tau_s^{(1)}(\omega) = \tau_s^{(2)}(\omega)$; therefore we only have to consider the case $x_s(\omega) = j \in \Gamma$. Since $\Gamma \subseteq E$, therefore $j \neq +\infty$, from 3° and the discrete nature of E, for any arbitrary $\varepsilon > 0$, there exists $s < t < s + \varepsilon$ such that $x_t(\omega) = j$, hence $\tau_s^{(2)}(\omega) \leq s + \varepsilon$; since ε is arbitrary, we obtain

---

[1] We assume $\sum_{y \in E} P_{xy}(t) = 1$ for every $x \in E$, $t \geq 0$.

[2] In general a fundamental, countable process is not the standard process defined in [1], hence Lemma 3 is not a special case of Theorem 4.3 in [1]. However, part of the proof is similar.

$$\tau_s^{(2)}(\omega) = s \leqslant \tau_s^{(1)}(\omega).$$

To show $\tau_s^{(3)} \geqslant \tau_s^{(1)}$: we only have to consider the case $\tau_s^{(3)} < +\infty$. From the definition of $\tau_s^{(3)}$, for $\varepsilon > 0$, there always exists $t \in [\tau_s^{(3)}, \tau_s^{(3)} + \varepsilon)$, $s \leqslant s_n \uparrow t$, such that $x_{s_n} \to j \in \Gamma$. When n is sufficiently large, $x_{s_n} = j$, then $\tau_s^{(1)} \leqslant s_n \leqslant t \leqslant \tau_s^{(3)} + \varepsilon$.

Finally we note, from the definitions of $\tau_s^{(i)}$ and Lemmas 1, 2 that

$$\tau_s^{(2)} \geqslant \tau_s^{(1)}, \quad \tau_s^{(3)} \leqslant \tau_s^{(1)}, \quad \tau_s^{(3)} \leqslant \tau_s^{(4)} \leqslant \tau_s^{(2)}.$$

Combining the above inequalities we obtain the proof of (i).

(ii) Let $\tau_s^{(1)}(\Gamma) < +\infty$. From the definition of $\tau_s^{(1)}(\Gamma)$, for every arbitrary $\varepsilon > 0$, there exists $t \in [\tau_s^{(1)}(\Gamma), \tau_s^{(1)}(\Gamma) + \varepsilon)$, such that $x(t, \omega) = j \in \Gamma$. Arrange the members of Γ in increasing order, denote the set having the first n members by Γ_n; when n is sufficiently large, $j \in \Gamma_n$, hence

$$\tau_s^{(1)}(\Gamma) \leqslant \tau_s^{(1)}(\Gamma_n) \leqslant t \leqslant \tau_s^{(1)}(\Gamma) + \varepsilon.$$

Obviously $\tau_s^{(1)}(\Gamma_n)$ is not increasing, and we have the proof of (ii). If $\tau_s^{(1)}(\Gamma) = +\infty$, then obviously $\tau_s^{(1)}(\Gamma_n) = +\infty$.

(iii) Since X is completely separable, E is discrete, and $+\infty \overline{\in} \Gamma$, then we have

$(\tau^{(1)} < t) = ($there exists a rational number r,

$$0 \leqslant r < t, \ x_r(\omega) \in \Gamma) \in \mathfrak{N}_t \quad (t > 0),$$

$$(\tau^{(1)} \leqslant t) = \lim_{n \to +\infty} \left(\tau^{(1)} < t + \frac{1}{n}\right) \in \mathcal{N}_{t+0} \quad (t \geqslant 0),$$

combining this with (i) we obtain (iii).

If we introduce in $E = \mathbf{N}$ the distance $\rho(x, y) = 0(x, y)$, $\rho(x, y) = 1(x \neq y)$, and let $\mathfrak{B}$ be the $\mathfrak{S}$ algebra formed by all the subsets of E, then E is a measurable, metric space.

§ 3. Necessary and sufficient conditions for recurrent processes

Hereafter we always assume $X=(x_t,\ \zeta,\ \mathfrak{M}_t,\ p_x)$ to be

(i) a right continuous Markov process of a fundamental, countable process on a topological, measurable state space $(E, \mathfrak{S}, \mathfrak{B})$;

(ii) $\qquad p_x(\zeta>0)=1$, every $x\in E$. $\qquad$ (3.1)

These two conditions hold throughout the paper, and we will not repeat them.

Take any two points $x,\ y\in E$; we say X is recurrent from x to y if and only if for every neighborhood U of y, we have

p_x(there exist infinitely many $t_n=t_n(\omega)\uparrow\zeta(\omega)$, such that

$$x_{t_n}(\omega)\in U,\ n\in\mathbf{N}^*\ )=1;\qquad (3.2)$$

X is a recurrent process if and only if for any two points $x,\ y\in E$, X is recurrent from x to y.

Theorem 1 Let X be a standard process ([1], P150);[①] then a necessary and sufficient condition for X to be recurrent is: for any point $x\in E$ and any nonempty open set U, we have

$$p_x(\tau_U<\zeta)=1. \qquad (3.3)$$

Here τ_U is the first entrance time into U.

Proof From (3.2) we immediately obtain (3.3), hence

[①] If the state space is a locally compact Hausdorff space with countable base, we do not have to consider the trivial case of E having only one point, since by (3.1) X is recurrent. We shall always assume that E contains at least two different points.

we only have to prove sufficiency. Take any two nonempty open sets, G, F, such that $\overline{G} \cap \overline{F} = \varnothing$, $\overline{G}$ denotes the closure of G. Define

$$\begin{cases} g = \inf(t: t \geqslant 0, x_t \in G) & f = \inf(t: t \geqslant 0, x_t \in F), \\ \quad = \zeta \quad \text{(if the above set} & \quad = \zeta \quad \text{(if the above set} \\ \qquad \text{is empty);} & \qquad \text{is empty);} \\ g_1 = g; & f_1 = g_1 + \theta_{g_1} f, \text{ if } g_1 < \zeta, \\ & \quad = \zeta, \text{ otherwise;} \\ g_n = f_{n-1} + \theta_{f_{n-1}} g, \text{ if } f_{n-1} < \zeta, & f_n = g_n + \theta_{g_n} f, \text{ if } g_n < \zeta, \\ \quad = \zeta, \text{ otherwise;} & \quad = \zeta, \text{ otherwise.} \end{cases}$$

(3.4)

Hence g_1 is the first entrance time into G, f_1 is the first entrance time into F after first entrance time into G, etc. We want to show for any $x \in E$ and for any positive integer n, we have

$$p_x(g_n < \zeta) = 1, \quad p_x(f_n < \zeta) = 1. \tag{3.5}$$

Actually, from (3.3) we obtain, for any $x \in E$,

$$p_x(g < \zeta) = 1, \quad p_x(f < \zeta) = 1. \tag{3.6}$$

hence

$$p_x(g_1 < \zeta) = p_x(g < \zeta) = 1. \tag{3.7}$$

Combining (3.6) (3.7) and the strong Markov property, we obtain

$$\begin{aligned} p_x(f_1 < \zeta) &= p_x(g_1 < \zeta, g_1 + \theta_{g_1} f < \zeta) \\ &= p_x(g_1 < \zeta, \theta_{g_1}(f < \zeta)) \\ &= \int_{(g_1 < \zeta)} p_{x_{g_1}}(f < \zeta) p_x(d\omega) \\ &= p_x(g_1 < \zeta) = 1. \end{aligned} \tag{3.8}$$

Thus if $n=1$, (3.5) is proved. Now assume when $n=m$ (3.5) holds. Similar to (3.8), we have

$$p_x(g_{m+1}<\zeta) = p_x(f_m<\zeta, f_m+\theta_{f_m}g<\zeta)$$
$$= p_x(f_m<\zeta, \theta_{f_m}(g<\zeta))$$
$$= \int_{(f_m<\zeta)} p_{x_{f_m}}(g<\zeta) p_x(d\omega)$$
$$= p_x(f_m<\zeta) = 1.$$

Similarly, we have $p_x(f_{m+1}<\zeta)=1$. Hence (3.5) is proved.

From the right continuity of the process, we obtain

$$x_{g_n} \in \overline{G}, \quad x_{f_n} \in \overline{F}. \tag{3.9}$$

Now for any given points x, $y \in E$, and any neighborhood U of y, we can take two open sets G, F, such that $\overline{G} \subseteq U$, $\overline{G} \cap \overline{F} = \varnothing$. For this pair of G, F, we use the two sequences $\{g_n\}$, $\{f_n\}$ defined in (3.4). We shall prove: take $t_n = g_n$, then (3.2) holds.

In fact, since X is right continuous and $\overline{G} \cap \overline{F} = \varnothing$, hence $\zeta > g_{n+1} > g_n$ almost everywhere in p_x, and there exists the limit

$$g_{+\infty} = \lim_{n \to +\infty} g_n \leq \zeta; \tag{3.10}$$

and also from (3.9)

$$x_{g_n} \in \overline{G} \subseteq U. \tag{3.11}$$

Therefore, if we can prove

$$p_x(g_{+\infty}=\zeta)=1, \tag{3.12}$$

combining (3.10)~(3.12) we immediately obtain (3.2). To prove (3.12) let $A=(g_{+\infty}<\zeta)$; note $g_1<f_1<g_2<f_2<\cdots$, a.e. in p_x, hence we know that the three sequences $\{g_n\}$, $\{f_n\}$, $\{g_n, f_n\}$ have a common limit $g_{+\infty}$. Since X is a standard process, we have

$$\lim_{n\to+\infty} x_{g_n} = x_{g_{+\infty}}, \quad \lim_{n\to+\infty} x_{f_n} = x_{g_{+\infty}} \quad (A, \text{ a. e. in } p_x).$$

With this and (3.9), we obtain $x_{g_{+\infty}} \in \overline{G}$, $x_{g_{+\infty}} \in \overline{F}$; however $\overline{G} \cap \overline{F} = \emptyset$, this implies $p_x(A) = 0$, hence (3.12) is proved.

Remark 1 Let X be a fundamental, countable process. A necessary and

$$p_x(\tau_y < +\infty) = 1;$$

here τ_y is the first entrance time into the single point set $\{y\}$.

In fact, we can proceed as in the proof of Theorem 1 up to (3.11) (at this point, let $G = U = \{y\}$, $F = \{z\}$, z is any state different from y, and use lower right semicontinuity instead of right continuity). This leaves (3.12) to be proved, but (3.12) is proved in [7], P205.

Theorem 2 Let X be a canonical diffusion process ([1], P238):

(i) If X is recurrent, then for every point $x \in E$, and for any nonempty open set G with a compact closure, we have

$$\int_0^{+\infty} p(t, x, G) dt = +\infty. \qquad (3.13)$$

(ii) Conversely, if X is nonterminating ([1], P117), and if (3.13) holds for a certain point x and a certain open set G with a compact closure, then X is recurrent.

Proof (i) We may let X be a standard process ([1], P534). Let η_G be the first exit time from G (i. e. $\eta_G = \tau_{E\setminus G}$ is the first entrance time into $E \setminus G$). From the openness of G and the continuity of X, we have

$$M_x \eta_G > 0, \text{ every } x \in G. \qquad (3.14)$$

Because $M_x \eta_G$ is harmonic on G ([1], P544), therefore $M_x \eta_G$ is lower semicontinuous on G ([1], P534). Take any open set

Some Properties of Recurrent Markov Processes

$V \subseteq G$, such that $\overline{V} \subseteq G, \overline{V}$ compact, and we shall show

$$\inf_{y \in \overline{V}} M_y \eta_G = a > 0. \quad (3.15)$$

In fact, if $a=0$, then there exists $y_n \in \overline{V}$, such that $M_{y_n}\eta_G \to 0$ $(n \to +\infty)$. From the compactness of $\overline{V}$, we may let $y_n \to y \in \overline{V}$. From the lower semicontinuity of $M_x \eta_G$, we obtain

$$M_y \eta_G \leqslant \varliminf_{n \to +\infty} M_{y_n} \eta_G = 0,$$

but $y \in \overline{V} \subseteq G$, therefore the above equation contradicts (3.14), and (3.15) is proved.

G can be made smaller if necessary (note if (3.13) holds for smaller G, it certainly holds for large G), so that there exists an open set F with $\overline{G} \cap \overline{F} = \emptyset$, hence $\overline{V} \cap \overline{F} = \emptyset$. Replace G, F in (3.4) by V, F and use the two sequences $\{v_n\}$, $\{f_n\}$, $v_n < f_n$ defined in (3.4). Let

$$\tau_n = \inf(t: t \geqslant 0, \ x_{v_n + t} \in E \setminus G), \quad n \in \mathbf{N}^*;$$

obviously $x_s \in G$ for every $s \in [v_n, v_n + \tau_n)$. Define

$$s_G(\omega) = (t: t \in [0, \zeta(\omega)), \ x_t(\omega) \in G).$$

Since X is continuous, for every ω fixed, $s_G(\omega)$ is a Borel measurable set on the real line. Let μ be the Lebesgue measure; then

$$M_x[\mu(s_G)] \geqslant M_x\Big(\sum_{n=1}^{+\infty} \tau_n\Big) = \sum_{n=1}^{+\infty} M_x \tau_n = \sum_{n=1}^{+\infty} M_x M_{x_{v_n}} \tau_1.$$

$$(3.16)$$

But from (3.15), for every $y \in \overline{V}$, we have

$$M_y \tau_1 = M_y(\eta_G - \eta_v) = M_y \theta_{\eta_v} \eta_G$$
$$= M_y M_{x_{\eta_v}} \eta_G \geqslant a > 0. \quad (3.17)$$

Since $x_{v_n} \in \overline{V}$, from the above expression and (3.16), we obtain

$$\int_0^{+\infty} p(t, x, G) dt = M_x[\mu(s_G)] \geqslant \sum_{n=1}^{+\infty} a = +\infty.$$

(ii) Let (3.13) hold for a certain x and a certain nonempty open set G with a compact closure, then there exists a compact set k such that

$$p_y(\text{there exists } t\geqslant 0 \text{ such that } x_t \in \bar{k}) = 1 \quad (3.18)$$

for every $y \in E$, otherwise from [3], Lemma 3.1 we have

$$\int_0^{+\infty} p(t,x,G)\mathrm{d}t \leqslant \int_0^{+\infty} p(t,x,\bar{G})\mathrm{d}t < +\infty,$$

which contradicts our assumption. Note that a canonical diffusion process is a strong Feller process and $p(t, y, U) > 0$ for every $y \in E$, $t > 0$ and for every nonempty open set ([1], P232 and P799), hence from (3.18) and [3], P198, we know (3.3) is satisfied. As we said earlier, we may assume X to be a standard process, hence by applying Theorem 1 we prove that X is recurrent.

From the proof of part (i) of the above theorem, we know for a recurrent, canonical diffusion process X,

$$M_x \zeta \geqslant M_x[\mu(s_G)] = +\infty \quad (\text{every } x \in E) \quad (3.19)$$

holds.

Remark 2 If the transition probabilities of a fundamental, countable process satisfy the inequality $P_{xy}(t) > 0$ (for every x, $y \in E$, $t > 0$), then either one of the following conditions is a necessary and sufficient condition for X to be recurrent:

(i) $\int_0^{+\infty} p_{xx}(t)\mathrm{d}t = +\infty$, for every $x \in E$;

(ii) $\int_0^{+\infty} p_{xx}(t)\mathrm{d}t = +\infty$, for a certain $x \in E$.

Proof Refer to [7] § Ⅱ.10.

Example 1 Let $X = (x_t, +\infty, \mathfrak{M}_t, p_x)$ be an n-dimensional Wiener process, hence when $n>2$, X is not recurrent; when $n\leqslant 2$, X is recurrent. In fact, if $n>2$, then $\lim\limits_{t\to+\infty} |x_t| = +\infty$ (a. e. in p_x every x) ([1], P590). It is obvious that (3.2) does not hold for any bounded open set U. If $n\leqslant 2$, for any nonempty open set G, and any $x\in E$, (3.13) holds ([1], P585), since X is a canonical diffusion process, from Theorem 2 (ii), we know that X is recurrent.

§ 4. Excessive functions and strong zero-one law

Let the transition probability of a process X be $p(t, x, A)$, $x \in E$, $A \in \mathfrak{B}$. $f(x)$ ($x \in E$) is said to be an excessive function for X if and only if $f(x)$ is excessive for $p(t, x, A)$ ([1], P493). The assumptions on the state space in §3 hold here.

Theorem 3 (i) Any continuous[①] excessive function of a right-continuous recurrent process is a constant.

(ii) Any excessive function of a recurrent fundamental, countable process is a constant.

Proof (i) Let X be a right-continuous, recurrent Markov process, and let $f(x)$ be any continuous excessive function for X. Let
$$\widetilde{\mathcal{M}}_t = (A+B: A \in \mathcal{M}_t, B \in \mathcal{M}[\zeta \leqslant t]),$$
then $\widetilde{\mathcal{M}}_t$ is the $\mathfrak{S}$-algebra on Ω, and for every $x \in E$, the process $(f(x_t), \widetilde{\mathcal{M}}_t, p_x)$ is a semi-martingale ([1], P506~507); from the right-continuity of X and continuity of f, $f(x_t)$ can be shown to be right-continuous. Hence from the convergence theorem of martingales, there exists an ω-set $\widetilde{\Omega}$, of full probability, i.e. a set which satisfies $p_x(\widetilde{\Omega}) = 1$ (for every $x \in E$), such that when $\omega \in \widetilde{\Omega}$ there exists limit
$$\xi(\omega) = \lim_{t \uparrow \zeta} f(x_t(\omega)). \tag{4.1}$$
Take any two points $x, y \in E$, for every $\varepsilon > 0$, define U_ε to be

---

[①] The assumption of continuity naturally implies the assumption of finiteness.

the neighborhood of y
$$U_\varepsilon = (a: |f(a)-f(y)| < \varepsilon). \qquad (4.2)$$
From the recurrent property of X, (3.2) holds for the points x, y and $U=U_\varepsilon$, i. e.
$$p_x(|f(x_{t_n}(\omega))-f(y)| < \varepsilon, n \in \mathbf{N}^*) = 1, \qquad (4.3)$$
where the sequence $\{t_n\}$ is the same sequence defined in (3.2). But from (4.1), we have $\lim_{n \to +\infty} f(x_{t_n}(\omega)) = \xi(\omega)$, hence from (4.3) we obtain
$$p_x(|\xi(\omega)-f(y)| \leqslant \varepsilon) = 1.$$
Since ε is arbitrary, therefore we have
$$p_x(\xi(\omega)=f(y))=1. \qquad (4.4)$$
Now for any point $z \in E$, similarly we have $p_x(\xi(\omega)=f(z))=1$, with this result and (4.4) we immediately obtain $f(y)=f(z)$, therefore f is identically equal to a constant.

(ii) Now let us assume X is a recurrent fundamental, countable process, to avoid the difficulty caused by $x_t(\omega)$ possibly taking the value $+\infty$, we only consider the process on $t \in \mathbf{N}$, i. e. we only consider $\{x_n(\omega), n \in \mathbf{N}\}$. From (2.19) and the countability of the parameter set (set of non-negative integers), there exists a set $\widetilde{\Omega}$ of full probability, when $\omega \in \widetilde{\Omega}$, $x_n(\omega) \neq +\infty$ (for all n). If necessary we may reduce the space of fundamental events to $\widetilde{\Omega}$, and obtain a Markov chain $\widetilde{X}=(x_n, \widetilde{\mathfrak{N}}_n, p_x)$ not taking the value $+\infty$, where $\widetilde{\mathfrak{N}}_n$ is the $\mathfrak{S}$ algebra on $\widetilde{\Omega}$ generated by $\{x_i(\omega), i \leqslant n\}$. According to [7], § Ⅱ. 10, the fact that X is recurrent immediately implies that $\widetilde{X}$ is also recurrent, and it is easy to see that if f is an excessive function for X, it is also an excessive function for $\widetilde{X}$.[①] Hence if f is fi-

nite, then for every $x \in E(f(x_n), \widetilde{\mathfrak{N}}_n, p_x)$ is a semi-martingale, and there exists limit
$$\lim_{n \to +\infty} f(x_n(\omega)) = \xi(\omega)) \quad \text{(a. e. in } p_x) \qquad (4.5)$$
From the recurrent property of $\widetilde{X}$, for any $x, y \in E$, p_x(there exists a sequence of positive integers $t_n = t_n(\omega) \uparrow +\infty$,

such that $x_{t_n}(\omega) = y, n \in \mathbf{N}^*) = 1$,

hence from (4.5) we obtain $p_x(\xi(\omega) = f(y)) = 1$, returning to (4.4) and we have proved that if f is finite, then f is a constant. Now let $f(y) = +\infty$, for a certain $y \in E$, from the recurrent property of X, it is easy to see that $p_{xz}(t) > 0$ (all x, $z \in E$, $t \geqslant 0$), since f is excessive, we have
$$f(x) \geqslant \sum_{z \in E} p_{xz}(t) f(z) \geqslant p_{xy}(t) f(y) = +\infty \quad (x \in E).$$

Theorem 4 Let X be a standard process whose intrinsic topology C_0 ([1], P168) coincides with the topology C on E; or let X be a fundamental, countable process; then a necessary and sufficient condition for X to be recurrent is that any finite excessive function is a constant.

Proof Necessity. We have proved this in Theorem 3 for a fundamental, countable process. Let X be a standard process and $C_0 = C$; because any excessive function f is C_0-continuous ([1], P501), therefore f is continuous (i.e. C-continuous). From Theorem 3, the condition follows.

Sufficiency For any nonempty open set U, consider

---

① f is excessive for $\widetilde{X}$ if and only if f is any non-negative function satisfying $\sum_{y \in E} p_{xy} \cdot f(y) \leqslant f(x)$ (every x) such that $p_{xy} = p_{xy}(1)$.

the function
$$\pi(x) = p_x(\tau_U^{(2)} < \zeta), \qquad (4.6)$$
where $\tau_U^{(2)}$ is the first entrance time into U after time $0+$. If X is a standard process, it is proved in [1], P498, that $\pi(x)$ is an excessive function; if X is a fundamental, countable process, similarly we can prove $\pi(x)$ to be excessive. According to the assumptions, $\pi(x) \equiv C$(constant). However, using the right continuity or lower right semicontinuity, for any $x \in U$, we have $\pi(x) = 1$. Hence $C = 1$; and $p_x(\tau_U^{(2)} = \tau_U) = 1$, where τ_U is the first entrance time into U. Thus we have shown: $p_x(\tau_U < \zeta) = 1$, for all $x \in E$. From Theorem 1 and Remark 1 we prove that X is recurrent.

Corollary 1 Let X satisfy the assumptions in Theorem 4, and let $\phi_t^s(\omega)$ be the W-functional of X ([1], P261); if X is recurrent, and the function $M_x \phi_{+\infty}^0$ is finite, then it is identically equal to a constant (no assumption of finiteness is necessary for a fundamental, countable process).

In face, it is easy to show $f(x) = M_x \phi_{+\infty}^0$ is an excessive function for X.

Corollary 2 Let X be a regular, continuous strong Markov process ([1], P657) on a one-dimensional interval (open or closed), then a necessary and sufficient condition for X to be recurrent is that any finite excessive function is a constant.

In fact, for such X, $C_0 = C$([1], P673).

Now we shall study the zero-one law. Let $X = (x_t, +\infty, \mathfrak{M}_t, p_x)$ be a Markov process on general state space $(E, \mathfrak{B})$, let $\prod = \bigcap_{t \geq 0} \mathfrak{N}^t$, where $\mathfrak{N}^t$ denotes the $\mathfrak{S}$-algebra generated by

(x_u, $u \geqslant t$). Let
$$\Lambda = (A : A \in \prod, \text{ and } \theta_t A = A, \text{ all } t \geqslant 0).$$
Obviously, Λ is a sub-$\mathfrak{S}$-algebra of $\prod$ on Ω. We say the strong zero-one law at infinity (or simply strong zero-one law) for X holds if and only if for every $A \in \Lambda$, $p_x(A) \equiv 0$ or $p_x(A) \equiv 1$ (all $x \in E$).

For a study of zero-one law, refer to [5]. Here we shall give a satisfactory result for a continuous standard process. By a harmonic function we mean a harmonic function for X on E ([1], P524).

Theorem 5 A necessary and sufficient condition for the strong zero-one law to hold in the case of nonterminating,[①] continuous process is that any non-negative, bounded harmonic function for X is a constant.

Proof Sufficiency Let $A \in \Lambda$; then $p_x(A)$ as a function of x, is non-negative, bounded and $\mathfrak{B}$-measurable, and for any Markov time τ, because of $\theta_\tau A = A$, we have
$$p_x(A) = p_x(\theta_\tau A) = M_x M_{x_\tau} \chi_A = M_x p_{x_\tau}(A),$$
where χ_A is the characteristic function of A; therefore $p_x(A)$ is harmonic on E. According to our assumption $p_x(A) \equiv C$ (C constant) and also $A \in \Lambda$, so that
$$p_x(A \mid \mu_t) = p_x(\theta_t A \mid \mu_t) = p_{x_t}(A) \quad (\text{a. e. in } p_x),$$
$$\chi_A = \lim_{n \to +\infty} p_{x_n}(A) = C \quad (\text{a. e. in } p_x).$$
Hence $C = 0$ or 1.

---

① In fact, in the definition of strong zero-one law, we assume that X is nonterminating.

Some Properties of Recurrent Markov Processes

Necessity Let X be a continuous standard process; then any non-negative bounded harmonic function $f(x)$ for X is an excessive function; moreover, it is an I-function ([1], P507). In fact, from [1], Theorem 12.4, we know that f is an excessive function, and also from the definitions of harmonic function and I-function, we know f is also an I-function. Hence from [1], Theorem 12.7, we obtain

$$f(x) = M_x(\chi_{\Omega'} \lim_{t \uparrow +\infty} f(x_t)) \quad \text{(for all } x \in E\text{)}, \quad (4.7)$$

where $\Omega' = \bigcap_n \bigcup_{r \in \Lambda}(x_r \in E \setminus \Gamma_n)$, Λ is a set of non-negative rational numbers, and Γ_n is any sequence of compact sets,[①] $\Gamma_n \uparrow E$. Let $\eta = \chi_{\Omega'} \lim_{t \uparrow \infty} f(x_t)$, then

$$\theta_s \eta = \theta_s \chi_{\Omega'} \lim_{t \uparrow +\infty} f(x_{s+t}) = \chi_{\Omega'} \lim_{t \uparrow \infty} f(x_t) = \eta,$$

$$\theta_s(\eta \leqslant a) = (\theta_s \eta \leqslant a) = (\eta \leqslant a) \quad (s \geqslant 0),$$

hence $(\eta \leqslant a) \in \Lambda$ and η is Λ-measurable. According to the assumption that the strong zero-one law holds, Λ contains only sets with measure zero or one. Therefore

$$\eta(\omega) \equiv C \quad \text{a. e. } p_x.$$

This constant C may depend on x, hence it is better to denote it by C_x, though it is actually independent of x. If we assume the contrary, say $C_x < d < C_y$, then

$$p_x(\eta \leqslant d) = 1, \quad p_y(\eta \leqslant d) = 0,$$

this result contradicts the assumption that $p_z(\eta \leqslant d)$ pertaining to z is identically equal to 0 or one. Thus we have proved

$$\eta(\omega) \equiv C \quad \text{(a. e. } p_x, \text{ all } x \in E\text{)}.$$

---

① Such choice of $\{\Gamma_n\}$ has no relation to Ω', in fact, $\Omega' = (\omega$: any compact set outside the range of $x_t(\omega))$.

Finally we use (4.7) and we immediately obtain
$$f(x)=M_x\eta=M_x C=C, \text{ all } x\in E.$$

Remark 3 It can be shown from Corollary 2 and Theorem 5 that a necessary and sufficient condition for a regular, continuous standard process on a one-dimensional interval to be recurrent is that any finite excessive function is a constant; and a necessary and sufficient condition for the strong zero-one law to hold is that any bounded, non-negative harmonic function is a constant.

Example 2 An example of a nonrecurrent process on which the strong zero-one law holds can be given by the $n(\geqslant 3)$-dimensional Wiener process. In fact, we have shown in Example 1 that this process is nonrecurrent. Next, from [1], P581 we know that any non-negative harmonic function for this process is a constant. Since we may let this be a standard process, it follows from Theorem 5 that the strong zero-one law holds.

§ 5. Applications in differential equations

Let the state space of a process $X = (x_t, \zeta, \mathfrak{M}_t, p_x)$ be $(E, \mathfrak{B})$. We call the point x an instantaneous entrance point into G if for the set $G \in \mathfrak{B}$, $p_x(r_G^{(2)} > 0) = 0$. Here $r_G^{(2)}$ is the first entrance time into G after time 0_+.

Lemma 4 Let X be a standard process or a fundamental, countable process; then a necessary and sufficient condition for x to be an instantaneous entrance point into G is

$$\lim_{t \to 0} p_x(x_u \overline{\in} G, \text{ all } u \in (0, t]) = 0; \qquad (5.1)$$

a sufficient condition is

$$\overline{\lim_{t \to 0}} p(t, x, G) > 0. \qquad (5.2)$$

Proof Let $q(t, x, G) = p_x(x_u \overline{\in} G, \text{ all } u \in (0, t])$, $q(t, x, G)$ has a meaning for the two kinds of process in the lemma. Obviously, it is a nonincreasing function in t, therefore there exists limit $\lim_{t \to 0} q(t, x, G) \geq 0$.

Next let $A_t = (x_u \overline{\in} G, \text{ all } u \in (0, t])$, then $\lim_{t \to 0} A_t = \bigcup_{t > 0} A_t$, hence

$$\lim_{t \to 0} q(t, x, G) = p_x(\bigcup_{t > 0} A_t) = p_x(\text{there exists } t = t(\omega) > 0,$$

such that $x_u(\omega) \overline{\in} G$, all $u \in (0, t]) = p_x(r_G^{(2)} > 0)$.

This proves the first result. Now let (5.2) hold, then there exists a sequence $t_n \downarrow 0$ such that $\lim_{n \to +\infty} p(t_n, x, G) = a > 0$. Let $C = \bigcap_{m=1}^{\infty} \bigcup_{n=m}^{\infty} (x_{t_n} \in G)$, then

$$p_x(C) = \lim_{m \to \infty} p_x(\bigcup_{n=m}^{\infty} (x_{t_n} \in G)) \geq \lim_{n \to +\infty} p(t_n, x, G) = a > 0.$$

If $\omega \in C$, then $\tau_G^{(2)} = 0$, therefore $p_x(\tau_G^{(2)} > 0) \leqslant 1-a < 1$. For the two kinds of process in the lemma, $(\tau_G^{(2)} > 0) \in \overline{\mathfrak{N}}_{0+0}$, $\overline{\mathfrak{N}}_{0+0} = \overline{\mathfrak{N}}_0$. Hence from the infinitely close zero-one law ([1] P124) we have $p_x(\tau_G^{(2)} > 0) = 0$.

We call a point x a regular boundary point of G if $x \in G' = \overline{G} \cap (E \setminus G)$, and x is an instantaneous entrance point into $E \setminus G$. This definition is consistent with the definition given in [1], P536. If all the points in G' are regular, then we say that G has a regular boundary.

Let E be an l-dimensional Euclidean space, and D a second order differential operator on E:

$$Df(x) = \sum_{i,j=1}^{l} a^{ij}(x) \frac{\partial^2 f(x)}{\partial x_i \partial x_j} + \sum_{i=1}^{l} b^i(x) \frac{\partial f(x)}{\partial x_i}, \quad (5.3)$$

$x = (x_1, x_2, \cdots, x_l) \in E$, we shall always assume that

(i) the functions $a^{ij}(x)$, $b^i(x)$ ($i, j = 1, 2, \cdots, l$) are bounded, and satisfy the Hölder condition with the same exponent $\lambda > 0$ on E;

(ii) there exists a constant $\gamma > 0$, such that for any $x \in E$, and set of real numbers $\lambda_1, \lambda_2, \cdots, \lambda_l$, we have

$$\sum_{i,j=1}^{l} a^{ij}(x) \lambda_i \lambda_j \geqslant \gamma \sum_{i=1}^{l} \lambda_i^2.$$

According to [1], Theorem 5.11, there exists a diffusion $\hat{C}$-process X_D whose transition density $p(t, x, y)$ is the fundamental solution to the equation $\frac{\partial u}{\partial t} = Du$, having generator D. This process is unique (for uniqueness refer to [1], P799) and nonterminating.① In the next theorem, we assume G has a regular boundary for the process X_D.

Theorem 6 Let the differential operator D satisfy conditions (i) (ii) and

(iii) $a^{ij}(x)$, $b^i(x)(i, j=1, 2, \cdots, l)$ are differentiable three times; then the following two conditions are equivalent:

i) the fundamental solution $p(t, x, y)$ ([1], P798) of the equation

$$\frac{\partial u}{\partial t} = Du \tag{5.4}$$

for a certain $x \in E$ and a certain open set G, $\overline{G}$ compact (hence for any $x \in E$, any nonempty open set G, such that $\overline{G}$ is compact) has the property:

$$\int_0^{+\infty} \int_G p(t,x,y) \, dy \, dt = +\infty. \tag{5.5}$$

ii) for any region G with a regular boundary, such that $\overline{G} \subset E$, and for any continuous function $f(x)$ on G', the Dirichlet problem

$$\begin{cases} Du=0, & x \in E \setminus G, \\ u=f, & x \in G' \end{cases} \tag{5.6}$$

has an unique solution.

Proof If we can prove that conditions i), ii) are equivalent to the recurrent property of the process X_D, then the theorem is proved.

Note that $p(t, x, y)$ is the transition density of X_D, $p(t, x, G) = \int_G p(t,x,y) \, dy$, hence from Theorem 2, we know that

---

① Zhao Zhao-yan (Chao Chao-yen) has given the proof that X_D is nonterminating.

i) is equivalent to the recurrent property of X_D.

As for X_D, if we use the proof in [1], P537, it is not difficult to show: if x is a regular boundary point of G, then x is also regular as defined by [3], P205. According to [3], Lemma 5.2, condition ii) is equivalent to the condition:

iii) there exists a compact set k, such that for every $x \in E$, we have
$$p_x(\text{there exists } t \geq 0, \text{ such that } x_t \in k) = 1.$$
Therefore we have only to show that condition iii) is equivalent to the recurrent property of X_D. We may let X_D be a standard process. If X_D is recurrent, then for every nonempty open set U, if $\overline{U}$ is compact, from Theorem 1, we have $p_x(\tau_U < \zeta) = 1$, all $x \in E$. But $x(\tau_U) \in \overline{U}$, so let $k = \overline{U}$, and we have (iii). Conversely, if iii) holds, from the strong Feller property of X_D, and also note that $p(t, x, G) > 0$ for X_D (for all $t > 0$, $x \in E$, G a nonempty open set), and use the proof in [3], p.198, we prove that X_D is recurrent.

Theorem 7 Let D satisfy (i)(ii) and the corresponding X_D be nonterminating; then the following two conditions are equivalent:

i) any non-negative, bounded, twice-continuous differentiable solution to the equation $Du = 0 (x \in E)$ is a constant;

ii) the strong zero-one law holds for X_D.

Proof We may let X_D be a standard process. The entire set of non-negative, bounded, twice-continuous differentiable solutions to the equation $Du = 0$ ($x \in E$) coincides with the entire set of non-negative, bounded harmonic functions ([1], P549); therefore the proof follows from Theorem 5.

Bibliography

[1] Dynkin E B. Markov processes. Fizmatgiz, Moscow, 1963 (Russian).

[2] Kolmogorov A N. Markov chanis with a countable number of possible states. Bjull. Moskov. Gos. Univ, 1937, 1 (3): 1-16 (Russian).

[3] Has´minskiĭ R Z. Ergodic properties of recurrent diffusion processes and stabilization of the solution of the Cauchy problem for parabolic equations. Teor. Verojatnost. i Primenen, 1960, 5: 196-214 (Russian).

[4] Tadashi Ueno. On recurrent Markov processes. Kodai Math. Sem. Rep, 1960, 12: 109-142.

[5] Wang Zi-kun(Wang Tzu-k'un). On zero-one laws for Markov processes. Acta Math. Sinica, 1965, 15 (3): 342-353=Chinese Math. Acta, 1965, 7: 41-54.

[6] Shi Ren-jie(Shih Jen-chieh). Trans formations of stochastic times in countable Markov process. Nankai Univ. J. (Nat. Sci.), 1964, 5: 51-88 (Chinese).

[7] Chung K L. Markov Chains with Stationary Transition Probabilities. Springer-Verlag, Berlin, 1960.

[8] Dynkin E B. Theory of Markov processes. Fizmatgiz, Moscow, 1959; English transl., Prentice Hall, Englewood Cliffs N J, and Pergamon Press, Oxford, 1961; Chinese Transl., Science Press, Peking, 1962.

Translated by:
W. F. Chung

常返马尔可夫过程的若干性质

摘要 对标准过程，给出了常返性的充分必要条件．对某些过程证明了：过程常返的充分必要条件是它的每一有限过分函数是常数；强 0-1 律成立的充分必要条件是每一有界调和函数是常数．然后应用于椭圆型偏微分方程之研究中．

地震迁移的统计预报[①]

§1. 引 言

本文利用随机转移的思想来研究发震地区的迁移问题．根据历史资料，计算出一重、二重及小地区的转移概率，然后以这三个因素作为预报因子构造出预报测度，并讨论了我国中部南北带地区 5 级以上地震以及全国五个主要地区 6 级以上地震的迁移及预报方法．

先简单介绍一下随机转移的概念．设有一个随机地（即偶然性地）运动的质点 A，它每经过一单位时间就作一次随机的转移．它所可能处的状态（或称地区）假定为 $1, 2, \cdots, m$（即第一区、第二区，$\cdots$，第 m 区）．譬如说，它现在处于状态 3，那么下一次可能转移到 1，也可能转移到 2，也可能转移到第 m 个状态．事先不能准确地预言它到底转移到哪里，只能说它转移到某状态的可能性有多大．表示这种可能性大小的数字叫概率．

---

① 本文与王启明，孙惠文，徐道一，朱成熹，李漳南，胡龙桥合作．

设在第 n 个时刻,已观察到 A 处于状态 j. 在这个条件下,下一次(即第 $n+1$ 个时刻)它转移到状态 k 的概率用 nP_{jk} 来表示,称 $nP_{jk}(j, k=1, 2, \cdots, m; n\in \mathbf{N}^*)$ 为一重转移概率. 如果 $nP_{jk}=P_{jk}$,即与 n 无关,那么称 (nP_{jk}) 是平稳的. 对于平稳情形,如果对另一状态 i,有 $P_{ji}\leqslant P_{jk}$,那么下一次转移到 k 的可能性比转移到 i 的可能性大或相等. 若上面的不等式不仅对一个状态 i,而是对一切 i 都成立,即有

$$P_{jk}=\max_{i} P_{ji}, \tag{1.1}$$

那么,假若转移的统计规律完全由一重转移概率决定,则下一次转移到状态 k 的可能性最大. 这个简单想法是本文的一个出发点.

但实际上影响转移的统计规律的,除了一重转移概率外,还可能有其他的因素,如二重转移概率以及小地区因素等.

二重转移概率 nP_{ijk} 的意义是:它是质点 A 于第 $n-1$ 个时刻位于状态 i,第 n 个时刻位于状态 j,在这些已知条件下,下一个时刻转移到状态 k 的概率 $(i, j, k=1, 2, \cdots, m)$.

如果状态代表地区,有时会发现,一个地区又可能再细分成几个小区,而每个小区的转移情况有明显的差别. 虽然同属于一个状态,从这一小区往往较集中地转移到某一状态,而从另一小区则大都转到另一状态. 因此要考虑小区的影响.

现在将上述概念应用到地震的迁移和预报中,把质点 A 设想为地震,把状态 $1, 2, \cdots, m$ 设想为可能发生地震的地区;A 在这些状态上转移就表示地震在这些地区中转移. 本文的基本思想就是根据一重、二重转移概率和小地区三个因素,构造一个预报测度 M,它给每个状态 k 一个数量 $M(k)$,后者可以衡量状态 k 出现的可能性的大小. 每发生一次地震,就可以根据这次地震及以往地震的资料来计算一次预报测度的值. 如果

状态 k_0 的测度 $M(k_0)$ 最大,那么下次最大可能发震的地区就是 k_0. 根据这个最大值还可算出地震发生在 k_0 地区的概率. 若 $M(k_1)$ 次大,则 k_1 就是下次地震发生的第二可能地区;若 $M(k_2)$ 最小,则 k_2 就是发震的最小可能地区.

贯彻本文始终的一个基本假定是:地震迁移的随机性是二重相依的. 就是说,如果已知这次和上次的发震地点,那么下次的发震地点就不依赖于更早的地震的地点. 因此,在预报下次地震时,只需用上两次的地震资料就够了. 我们所以如此假定,一方面是从过去的统计资料中看出,下次地震与两次以前的地震的相依性很小;另一方面也是由于地震资料不够多,如考虑多重转移,势必使资料更加分散,影响统计效果.

1972 年 11 月至 1973 年 8 月,我们对南北地震带(见图 1-1)共实报了六次,除一次错报外,其余五次基本正确. 对发震时间我们也作了试报,尚在改进中.

本文方法还可用以预报气象中的问题,如降水量等.

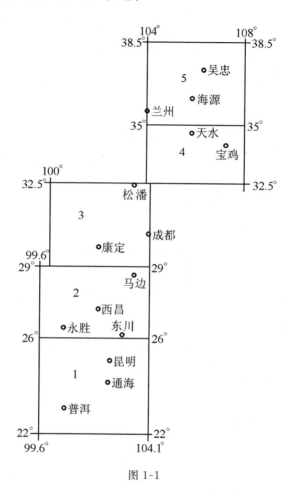

图 1-1

§2. 中国中部南北地震带的地震迁移

§2.1. 状态的划分

如何把南北地震带（包括宁夏、甘肃、四川、云南的大部地区）分成状态是首先碰到的重要问题．分几个？如何划分？这对预报效果有很大影响．根据相关分析统计的结果（参看[1]），考虑到南北带地震活动特点及地质构造上一些明显特点，我们将本地带划分为五个状态，它们的范围见图1-1同时我们将本地区1950～1970年五级以上地震资料按状态的迁移画成图（见图2-1）.

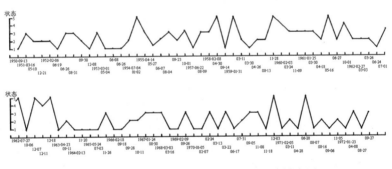

图 2-1 南北地震带 $M_L \geqslant 5$ 级地震迁移样本图

§2.2. 年段的区分

对 1950～1970 年中 $M_s \geqslant 5$ 的地震进行分析后，发现这 21 年中的震中迁移特性随着年份不同有较大差异．可以把 21 年分成三段：①1950～1957 年；②1958～1962 年；③1963～1970 年，即 8 年、5 年、8 年三段．用数理统计中 χ^2 检验发现：两个 8 年段（共 16 年）的统计性质没有显著差异，而中间的 5 年确与 16 年有显著差异，中华人民共和国成立前也有类似的现象．从图 2-1 上亦可看出，5 年段中 4，5 状态发震较多，而 8 年段中则极少．1971 年开始，似又应处于 5 年段时期．

这样，21 年作为整体来看是非平稳的．但是就 8 年段、5 年段内部而言，我们考虑的链是近似平稳的．经过平稳性统计检验也可相信是平稳的，这为使用平稳马尔可夫链进行地震迁移规律研究提供了理论基础．

§2.3. 一重转移概率

如何计算一重转移概率 P_{jk}？我们用频率来近似代替概率．鉴于上述 5 年段与 8 年段的统计性质有差异，分别对两个年段计算转移频率．

先考虑 5 年段．为了计算 P_{jk}，先数出在状态 j 发震的总次数，譬如说共 12 次；再计算这 12 次中转移到 k 状态的几次，譬如说 7 次，则得 $P_{(jk)}^{(5)}=\dfrac{7}{12}$，上标(5)表示 5 年段．8 年段照此办理．这样计算出两组一重转移概率 $P_{jk}^{(5)}$ 及 $P_{jk}^{(8)}$（$j, k=1, 2, \cdots, 5$）(表 2-1)．

表 2-1　一重转移频数表

j \ k	1	2	3	4	5	年段(年)
1	0	1	1	0	2	5
	8	5	8	0	0	8
2	2	3	0	1	2	5
	3	3	4	0	1	8
3	0	1	4	0	2	5
	9	3	4	0	0	8
4	0	1	1	0	1	5
	0	0	0	0	0	8
5	3	2	0	2	0	5
	0	0	1	0	0	8

上面已经提到，目前很可能处于 5 年段(1971～1975 年)，因此主要应使用 $P_{jk}^{(5)}$，但 8 年段也有一定的影响. 所以采用两者的加权平均值

$$P_{jk}=a_j P_{jk}^{(5)}+b_j P_{jk}^{(8)}, \qquad (2.1)$$

其中 a_j, b_j 是权，用下述得分方法确定.

先用 $P_{jk}^{(5)}$ ($k=1, 2, \cdots, 5$) 来报 1958～1962 年的地震，设

$$P_{jk_0}^{(5)} \geqslant P_{jk_1}^{(5)} \geqslant P_{jk_2}^{(5)} \geqslant P_{jk_3}^{(5)} \geqslant P_{jk_4}^{(5)}.$$

这表示：如果这次发震区为 j，以 $P_{jk}^{(5)}$ 为准则来预报时，下次发震地区最大可能为 k_0，次大可能为 k_1，第三可能为 k_2. 再看看下次地震实际发生在哪里. 如果在 k_0，就人为地规定 j 得 5 分；若出现在 k_1，则得 3 分；若出现在 k_2，则得 1 分；若出现在 k_3 或 k_4 则不得分. 假若出现 $P_{jk_0}^{(5)}=P_{jk_1}^{(5)}$ 的情况，地震又发生在 k_0 或 k_1，则得 4 分，余类推. 这样，对应于 1958～1962 年每一次地震计算一次分数. 所得总分记为 f.

同样，用 $P_{jk}^{(8)}$ 来报 1958～1962 年地震，总分记为 g. f, g 累积方法见表 2-2. 令

$$a_j=\frac{f}{f+g}, \quad b_j=\frac{g}{f+g}, \qquad (2.2)$$

代入(2.1)，即得 P_{jk}. 要注意的是权 a_j, b_j 是依赖于 j 的，即依赖于转移前的状态.

上面两次报的都是同 15 年段，那是因为要求预报的时间(1971 年至今)看来正处于一个新的 5 年段，从而采用了偏重于 $P_{jk}^{(5)}$ 的加权方法.

前面只用了 1970 年前的资料来计算 $P_{(jk)}^{(5)}$, $P_{(jk)}^{(8)}$, a_j, b_j 及 P_{jk}，1970 年后的地震用作预报检验，在预报过程中也要随时利用 1970 年后发生并已经过检验的地震. 因此要把新资料累积起来，把它增加到组成 P_{jk} 的数据中去. 自 1971 年起，每发生

一次地震,例如是从 2 转移到 5,就把有关资料加到原来的统计表中去,办法是在原表中 2 转移到 5 那一栏中增加 1. 然后重新计算一次 $P_{jk}^{(5)}$ 及 a_j, b_j, 得到一个新的 P_{jk}, 这就是累积方法. 表 2-2 中第一行 f 及 g 各值,是分别用 $P_{ik}^{(5)}$ 及 $P_{ik}^{(8)}$ ($i=1$, $2, \cdots, 5$) 预报 1958 年至 1962 年底这 5 年段的地震时,各状态得分的累积数.

表 2-2 一重转移 5 年、8 年得分累积统计表

状态 时间\年段分数	1		2		3		4		5	
	5(f)	8(g)	5(f)	8(g)	5(f)	8(g)	5(f)	8(g)	5(f)	8(g)
1958 年至 1962 年底(前一个 5 年段)	14	5	23	10	27	13	9	0	23	0
1970-12-03	19	5	23	10	27	13	9	0	23	0
1971-02-05	19	5	23	10	27	13	9	0	28	0
1971-03-11	21	6	23	10	27	13	9	0	28	0
1971-04-28	21	6	25	12	27	13	9	0	28	0
1971-06-28	26	6	25	12	27	13	9	0	28	0
1971-08-07	26	6	25	12	27	13	9	0	28	0
1971-08-16	29	7	25	12	27	13	9	0	33	0
1971-09-14	29	7	29	14	27	13	9	0	33	0
1971-11-05	32	8	29	14	27	13	9	0	33	0
1972-01-23	32	8	34	16	27	13	9	0	33	0
1972-04-08	33	12	34	16	27	13	9	0	33	0
1972-08-27	33	12	34	16	27	18	9	0	33	0
1972-09-27	34	16	34	16	27	18	9	0	33	0

第一行状态 1 对应的 $f=14$, $g=5$, 在第二行里所以变为 $f=19$, $g=5$, 是因为已知 1970 年 11 月底以前,最后一次地震是在 1 区发生的,如用 $P_{1k}^{(5)}$ ($k=1, 2, \cdots, 5$) 来预报下次地

震，即 1970 年 12 月 3 日的地震，其第一可能状态又报中了状态 5，故状态 1 应得 5 分，即 f 增加 5 分．若用 $P_{(1k)}^{(8)}$（$k=1$，2，…，5）来报下次地震，则所报的前三个可能状态都没有报中，故状态 1 不得分，因而 g 值不变（$P_{(1k)}^{(5)}$ 及 $P_{(1k)}^{(8)}$ 的值参见表 2-1）．其余各行仿此累积．

表 2-3 及表 2-4 的 f，g 值也仿此积累而得．

表 2-3 二重转移 5 年、8 年段得分累积统计表

状态	1		2		3		4		5	
时间 \ 5年,8年段分数	5(f)	8(g)	5(f)	8(g)	5(f)	8(g)	5(f)	8(g)	5(f)	8(g)
1958 年至 1962 年年底（前一个 5 年段）	18	5	32	7	27	8	13	0	31	0
1970-12-03	18	5	32	7	27	8	13	0	31	0
1971-02-05	18	5	32	7	27	8	13	0	31	0
1971-03-11	18	10	32	7	27	8	13	0	31	0
1971-04-28	18	10	32	8	27	8	13	0	31	0
1971-06-28	18	10	32	8	27	8	13	0	31	0
1971-08-07	18	10	32	8	27	8	13	0	34	0
1971-08-16	21	15	32	8	27	8	13	0	34	0
1971-09-14	21	15	36	9	27	8	13	0	34	0
1971-11-05	24	15	36	9	27	8	13	0	34	0
1972-01-23	24	15	41	10	27	8	13	0	34	0
1972-04-08	26	20	41	10	27	8	13	0	34	0
1972-08-27	26	20	41	10	27	13	13	0	34	0
1972-09-27	26	22	41	10	27	13	13	0	34	0

表 2-4　地区在 5 年及 8 年段得分累积统计表

状态　时间	1		2		3		4		5	
5年,8年段分数	5(f)	8(g)	5(f)	8(g)	5(f)	8(g)	5(f)	8(g)	5(f)	8(g)
1958 年至 1962 年年底(前一个 5 年段)	14	3	26	21	30	12	14	0	23	0
1970-12-03	17	3	26	21	30	12	14	0	23	0
1971-02-05	17	3	26	21	30	12	14	0	28	0
1971-03-11	19	3	26	21	30	12	14	0	28	0
1971-04-28	19	6	29	26	30	12	14	0	28	0
1971-06-28	24	6	29	26	30	12	14	0	28	0
1971-08-07	24	6	29	26	30	12	14	0	32	0
1971-08-16	28	9	29	26	30	12	14	0	32	0
1971-09-14	28	9	34	31	30	12	14	0	32	0
1971-11-05	28	12	34	31	30	12	14	0	32	0
1972-01-23	28	12	39	36	30	12	14	0	32	0
1972-04-08	28	17	39	36	30	12	14	0	32	0
1972-08-27	28	17	39	36	30	17	14	0	32	0
1972-09-27	30	22	39	36	30	17	14	0	32	0

§2.4. 二重转移概率的计算

计算方法与一重的计算方法相似. 先求 $P^{(5)}_{(ijk)}$，若 5 年段中共有 7 次从 i 转移到 j，而这 7 次中又有 3 次是从 j 转移到 k，则取 $P^{(5)}_{ijk} = \frac{3}{7}$. 同样方法求 $P^{(8)}_{ijk}$ ($i, j, k = 1, 2, \cdots, 5$)（表 2-5）. 加权平均为

$$P_{ijk} = c_j P^{(5)}_{ijk} + d_j P^{(8)}_{ijk}, \qquad (2.3)$$

权数 c_j, d_j 仍用上述的记分法来求（表 2-3），在外推预报 1970 年后地震时同样要进行累积.

表 2-5 二重转移频数表

i, j \ k	1	2	3	4	5	年段(年)
1, 1	0	0	0	0	0	5
	3	2	3	0	0	8
1, 2	0	1	0	0	0	5
	1	1	2	0	1	8
1, 3	0	0	1	0	1	5
	5	1	1	0	0	8
1, 4	0	0	0	0	0	5
	0	0	0	0	0	8
1, 5	0	1	0	1	0	5
	0	0	0	0	0	8
2, 1	0	1	1	0	0	5
	1	0	2	0	0	8
2, 2	1	1	0	0	1	5
	1	1	1	0	0	8
2, 3	0	0	0	0	0	5
	1	1	2	0	0	8
2, 4	0	1	0	0	0	5
	0	0	0	0	0	8
2, 5	0	1	0	1	0	5
	0	0	1	0	0	8
3, 1	0	0	0	0	0	5
	4	2	2	0	0	8
3, 2	0	0	0	0	1	5
	1	1	1	0	0	8
3, 3	0	1	2	0	1	5
	2	1	1	0	0	8
3, 4	0	0	0	0	0	5
	0	0	0	0	0	8
3, 5	2	0	0	0	0	5
	0	0	0	0	0	8
4, 1	0	0	0	0	0	5
	0	0	0	0	0	8

续表

i, j	k 1	2	3	4	5	年段（年）
4, 2	0	1	0	0	0	5
	0	0	0	0	0	8
4, 3	0	0	1	0	0	5
	0	0	0	0	0	8
4, 4	0	0	0	0	0	5
	0	0	0	0	0	8
4, 5	1	0	0	0	0	5
	0	0	0	0	0	8
5, 1	0	0	0	0	2	5
	0	1	0	0	0	8
5, 2	1	0	0	1	0	5
	0	0	0	0	0	8
5, 3	0	0	0	0	0	5
	1	0	0	0	0	8
5, 4	0	0	1	0	1	5
	0	0	0	0	0	8
5, 5	0	0	0	0	0	5
	0	0	0	0	0	8

§2.5. 小地区因素

当仔细考虑转移情况时，我们发现 1，2，3 三个状态分别还可细分成几个小地区，例如，状态 1 可分为两个小地区 1_A 和 1_B，在 1_A 中发生的地震，若属于 8 年段，则大都转移到状态 3；在 1_B 中发生的地震，8 年段中则大部分转到 1. 细分后总共得到 9 个地区 1_A，1_B，2_A，2_B，3_A，3_B，3_C，4，5，其范围见图 2-2. 对于小地区 1_A，可以把它在 5 年段中转移到 5 个状态的频数分别登记下来，并以频率当作小地区的转移概率 $D^{(5)}_{1_A, k}$，即

$$D^{(5)}_{1_A, k} = \frac{5 \text{ 年段中自 } 1_A \text{ 转移到状态 } k \text{ 的次数}}{5 \text{ 年段中在 } 1_A \text{ 内总共发震次数}}.$$

类似得到 5 年段中的自小地区 r 到状态 k 的转移频率 $D_{rk}^{(5)}$。同样可求出 $D_{rk}^{(8)}$（表 2-6），然后再用得分方法算出加权平均：

$$D_{rk}=U_r D_{rk}^{(5)}+V_r D_{rk}^{(8)}, \qquad (2.4)$$

这里 $r=1_A$，1_B，$\cdots$，3_C，4，5；$k=1$，2，$\cdots$，5（参看表 2-6）。外推或预报时也要累积。

表 2-6 地区转移频数表

λ, φ \ k	1	2	3	4	5	年段（年）
1_A	0	0	0	0	1	5
	0	2	8	0	0	8
1_B	0	1	1	0	1	5
	8	3	0	0	0	8
2_A	1	0	0	1	1	5
	3	0	2	0	0	8
2_B	1	3	0	1	1	5
	0	3	2	0	1	8
3_A	0	0	0	0	2	5
	1	3	3	0	0	8
3_B	0	0	0	0	0	5
	4	0	0	0	0	8
3_C	0	0	4	0	0	5
	4	0	2	0	0	8
4	0	2	1	0	1	5
	0	0	0	0	0	8
5	3	2	0	2	0	5
	0	0	1	0	0	8

§ 2.6. 预报测度

利用上面求出的 P_{jk}，P_{ijk}，D_{rk} 的线性组合，可以构造预报测度 $M(k)$：

$$M(k)=AE_j P_{ijk}+BP_{jk}+CD_{rk}, \qquad (2.5)$$
$$k=1, 2, \cdots, 5$$

(注意最后一项中的 r 由最后那次地震落在 j 中哪个小区而定). 这里 A, B, C 仍是分别用 P_{ijk}, P_{jk} 及 D_{rk} 来报 1958~1962 年中地震时的得分总数,不过为计算简单计,只记录最大可能得分,即如预报中的最大可能状态与实际发震地点相符就得 1 分,否则不得分. 统计结果得 $A=13$, $B=11$, $C=16$. (2.5)中所以出现 $E_j = \dfrac{d_1}{d_2}$ (d_2 是 5 年段内 j 中的地震总数, d_1 是这些地震中自 i 转来的地震总数, $E_j \leqslant 1$) 是因为资料少,故 P_{ijk} 如不为 0,就容易取很大的数值;用 $E_j = \dfrac{d_1}{d_2}$ 乘 P_{ijk}, 就相当于对 P_{ijk} 进行了"压缩".

预报测度 $M(k)$ 中的 k 遍历 5 个状态,即 $k=1, 2, 3, 4, 5$. $M(k)$ 其实依赖于最后两次发震地区 i, j 及最后那次所处 j 中的小地区,它还与地震的次数 n 有关,因为(2.5)中右方 A, B, C, E_j, P_{ijk}, P_{jk}, D_{rk} 等都由于累积而与 n 有关.

为了要预报下一次(设为第 $n+1$ 次)发震于哪一状态,先查明上两次发生的两个状态 i, j 等于什么,譬如说 $i=2$, $j=1$, 即上两次是从 2 转到 1;再查上次发震的经纬度 (λ, φ) 是落在哪一小地区,譬如说落在 1_A 中(图 2-2), 有了这三个数据就可以从一直累积到第 n 次地震的有关表中查出 A, B, C, E_j, 还可以由表 2-3、表 2-1、表 2-5 算出 $P_{21,1}$, $P_{21,2}$, $\cdots$, $P_{21,5}$, P_{11}, P_{12}, $\cdots$, P_{15} 以及 $D_{1A,1}$, $D_{1A,2}$, $\cdots$, $D_{1A,5}$ 的值,代入(6)就得出 $M(1)$, $M(2)$, $\cdots$, $M(5)$, 譬如说:$[M(1), M(2), M(3), M(4), M(5)] = [1.4, 6.7, 8.5, 0, 18.7]$, 这里 $M(5) = 18.7$ 最大,所以下次地震出现于状态 5 的可能性最大,它的概率是 $P(5) = \dfrac{M(5)}{S} = \dfrac{18.7}{35.3} = 52.9\%$, 其中, $S = M(1) + M(2) + \cdots + M(5)$. 其次可能的地区是状态 3, 它出现的概率为 $P(3) = \dfrac{8.5}{35.3} = 24\%$, 而状态 4 基本上不会出现. 如果我们预报

两个状态,可以说有76.9%的把握下次地震会发生在状态5或3中.

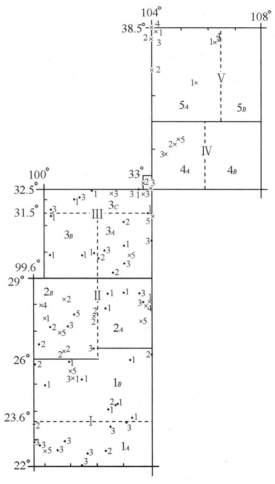

图 2-2 南北地震带"地区"划分示意图

1. 资料取自1950至1970年11月18日,共78个样本
2. ·3表示转移到状态3的"8年段"样本

   ×3表示转移到状态3的"5年段"样本,其余类推

§2.7. 实际效果

我们根据 1950～1970 年的资料用上述方法来外推 1971～1972 年 5 级以上地震的地区时,所得结果见表 2-7.

由表 2-7 可以看出:两年中共发生 13 次 5 级以上的地震,错报 2 次,半对(即实际发震于预报中第二可能地区)2 次,全对(即实际发震地区与预报中的最大可能地区一致)8 次.

表 2-7 如下法算出:

第一行中 (3, 1) 表示 1970 年 11 月底以前最后两次地震发生在第 3 区及第 1 区,而且其中最后一次是在小区 1_B 中,其东经及北纬的度数 (λ, φ) 为 $(25.2°, 101.9°)$. 现在要预报下次发震的地点. 为此,先算 $P_{ij,k}$,查表 2.3 的 (3, 1) 行知

$$(P_{31,1}^{(5)}, P_{31,2}^{(5)}, P_{31,3}^{(5)}, P_{31,4}^{(5)}, P_{31,5}^{(5)}) = (0, 0, 0, 0, 0),$$

$$(P_{31,1}^{(8)}, P_{31,2}^{(8)}, P_{31,3}^{(8)}, P_{31,4}^{(8)}, P_{31,5}^{(8)}) = \left(\frac{4}{8}, \frac{2}{8}, \frac{2}{8}, 0, 0\right).$$

再由表 2-4 的第一行知,状态 1 在 5 年段及 8 年段的得分数各为 18 和 5,其和为 23,故由 (2.3) 得

$$(P_{31,1}, P_{31,2}, P_{31,3}, P_{31,4}, P_{31,5})$$
$$= \frac{18}{23} \times (0, 0, 0, 0, 0) + \frac{5}{23} \times \left(\frac{4}{8}, \frac{2}{8}, \frac{2}{8}, 0, 0\right)$$
$$= (0.11, 0.05, 0.05, 0, 0). \tag{2.6}$$

类似地由表 2-1,表 2-2 的第一行及 (2.1) 算得

$$(P_{11}, P_{12}, P_{13}, P_{14}, P_{15}) = (0.10, 0.25, 0.28, 0, 0.37). \tag{2.7}$$

由表 2-5 的 1_B 行,表 2-6 的第一行及 (2.4) 算得

$$(D_{1_B,1}, D_{1_B,2}, D_{1_B,3}, D_{1_B,4}, D_{1_B,5}) \tag{2.8}$$
$$= (0.13, 0.32, 0.28, 0, 0.28).$$

最后根据公式 (2.5) 算出预报测度 $M(k)$ 为

地震迁移的统计预报

表 2-7 南北带预报检验情况表 ($M_s \geq 5$)

i,j	已知参数			二重转移概率 $P_{i,j,k}$	权 $A \times E_l$	一重转移概率 $P_{j,k}$	权 B	地区转移概率 $D_{j,k}$	权 C	预报测度 $M(k)$	和 S	预测最大可能 状态	概率/%	预测次大可能 状态	概率/%	实发 状态	经纬度	效果
	时间	经纬度	地区															
3,1	1970-11-18	25.2 101.9	1_B	(0.11,0,0.05,0,0.05,0,0)	$13 \times \frac{0}{4}$	(0.10,0,0.25,0,0.28,0,0.37)	11	(0,0.13,0.32,0.28,0,0.28)	16	(3,1,7,9,7.5,0,8.5)	27	5	31	2	29	5	35.5 105.3	好
1,5	1970-12-03	35.5 105.3	5	(0,0,5,0,0.5,0)	$13 \times \frac{4}{7}$	(0,0.43,0,0.29,0,0.29,0)	12	(0,0.43,0,0.29,0,0,0.29)	16	(12.9,0.9,0,9,9,0)	31.8	1	36	2	27	1	25.4 99.6	好
5,1	1971-02-05	25.4 99.6	1_B	(0,0,0.22,0,0,0,0.78)	$13 \times \frac{2}{5}$	(0,0.08,0.21,0.24,0,0.24,0,0.48)	13	(0,0.11,0.25,0.21,0,0,0.43)	17	(2,8,2.6,7,0,17,4)	34.3	5	50.7	2	23.6	2	29 103.6	中
1,2	1971-03-11	29 103.6	2_A	(0.04,0,0.82,0.07,0,0,0.04)	$13 \times \frac{2}{5}$	(0.26,0,0.34,0.11,0,0.09,0.2)	13	(0,0.45,0.15,7.4,6.3,2.5,5.8)	30.4	(11.5,1,7.4,6.3,2.5,5.8)	30.4	1	36.5	5	19.1	1	22.8 101.1	好
2,1	1971-04-28	22.8 101.1	1_A	(0.12,0,0.32,0,0.56,0,0)	$13 \times \frac{2}{6}$	(0.09,0,0.31,0,0.21,0,0.39)	13	(0,0,0.05,0,0.19,0,0.76)	18	(0,6,6.3,8,6,0,18.7)	34.2	5	54.7	3	25.1	5	37.8 106.3	好
1,5	1971-06-28	37.8 106.3	5	(0.33,0,0.33,0,0,0,0.33,0)	$13 \times \frac{3}{7}$	(0,0.50,0,0.25,0,0.25,0)	14	(0,0,0.50,0,0.25,0,0.25,0)	19	(18,2,0,10,0,10,0)	38.2	1	47.7	2	26.1	2	23.9 103.2	好
5,1	1971-08-07	23.9 103.2	1_B	(0,0,0.57,0,0,0,0.43)	$13 \times \frac{2}{7}$	(0,0.31,0,0.19,0,0,0.47)	15	(0,0.14,0.37,0,0.16,0,0.32)	20	(3.9,14,8,6,1,0,15,9)	40.7	5	39.1	2	26.1	2	29 103.5	中
1,2	1971-08-16	29 103.5	2_A	(0,0.44,0.44,0,0.08,0,0,0.04)	$14 \times \frac{2}{9}$	(0,0,0.31,0,0.12,0,0.08,0,0.18)	15	(0,0.54,0,0.19,0,0.13,0,0.13)	21	(17,6,6,1,5,7.3,9.5,5.6)	38.8	1	45.3	2	15.6	1	23.1 100.8	好
2,1	1971-09-14	23.1 100.8	1_A	(0.14,0,0.19,0,0.47,0,0,0.19)	$14 \times \frac{3}{8}$	(0,0.07,0.35,0,0.18,0,0.40,0)	15	(0,0,0.05,0,0.19,0,0.76)	21	(1,8,1.8,7,3.9,4,0,23.7)	42.3	5	56	3	22	5	28.8 103.5	好
2,1	1971-11-05	28.8 103.5	2_A	(0.57,0,0.31,0,0.08,0,0,0.04)	$14 \times \frac{3}{10}$	(0,0.37,0.29,0,0.07,0,0.08,0,0.18)	15	(0,0.59,0,0.19,0,0,0.10,0.10)	22	(20.9,5.6,5.6,3,4.5,1)	40.6	1	51.5	2	13.8	1	23.5 103	好
2,1	1972-01-23	23.5 103	1_A	(0.13,0,0.31,0,0.41,0,0,0.15)	$15 \times \frac{4}{8}$	(0,0.76,0,0.40,0,0.17,0,0.36)	16	(0,0,0.24,0,0.24,0,0.52)	23	(2,1,12,1.8,4,0,17.9)	40.5	5	44.2	2	29.9	3	29.5 101.2	差
1,2	1972-04-08	29.5 101.2	1_A	(0.22,0,0.11,0,0.11,0,0.57)	$16 \times \frac{2}{7}$	(0.18,0,0.16,0,0.47,0,0.19)	16	(0,0.24,0,0.24,0,0,0.52)	24	(10,4,2,7.9,4,0,4.4,8)	27.3	1	34.4	2	31	3	22.5 100.2	差
3,1	1972-08-27	22.5 100.2	1_A	(0.16,0,0.03,0,0.42,0,0.39)	$16 \times \frac{1}{10}$	(0,0.10,0.35,0,0.25,0,0.29)	16	(0,0.29,0,0,0,0,0)	24	(0,0,0.20,0,0.43,0,0.37)	41.6	5	35			3	30.2 101.6	好
1,3	1972-09-27	30.2 101.6	3_B	(0.45,0,0.05,0,0.27,0,0.22)	$16 \times \frac{3}{8}$	(0.30,0,0.15,0,0.40,0,0.15)	16	(1,0,0,0,0)	25	(32,5,2,7,8,0,0,3,7)	46.9	1	69	3	17.1	3		中

369

$$[M(1), M(2), M(3), M(4), M(5)] = 13 \times \frac{0}{4} \times (0.11, 0.05, 0.05, 0, 0) + 11 \times (0.10, 0.25, 0.28, 0, 0.37) + 16 \times (0.13, 0.32, 0.28, 0, 0.28)$$

$$= (3.1, 7.9, 7.5, 0, 8.5),$$

这里的权数 $A=13, B=11, C=16$ 是由统计得出的, 而 $E_j = E_1 = \frac{d_1}{d_2}$, 这里 d_2 是 5 年段中在状态 1 内所发生的地震总数, 由表 2-1 第一行, $d_2 = 1+1+2 = 4$, d_1 是这些地震中由状态 3 转来的次数, 由表 2-3 的 $(3, 1)$ 所对应的 5 年段行知 $d_1 = 0$, 故 $E_1 = \frac{0}{4}$, 又 $S = 3.1 + 7.9 + 7.5 + 0 + 8.5 \approx 27$,

故预报概率为 $\left(\frac{3.1}{27}, \frac{7.9}{27}, \frac{7.5}{27}, 0, \frac{8.5}{27}\right).$

其中 $\frac{8.5}{27}$ 最大, 它是下次地震发生在状态 5 中的概率, 故下次应首先报第 5 区, 结果于 1970 年 12 月 3 日实际在第 5 区发震.

表 2-7 第二行表示第二次预报的情况. 这时 $i=1, j=5$, 计算方法与第一次同, 不过应吸取最近在 5 区有震所带来的新讯息, 而应累积新资料. 为此, 首先表 2-1 中的 $j=1$ 及 $k=5$ 所对应的 5 年段的数 2 应改为 3, 这是因为最近又发生了一次由状态 1 到 5 的转移. 类似地, 表 2-3 中的 $(i, j) = (3, 1)$ 及 $k=5$ 所对应的 5 年段的 0 也应改为 1, 表 2-5 中 $(\lambda, \varphi) \in 1_B$, $k=5$ 所对应的 1 应改为 2. 其次计算得分数时应采用表 2-2, 表 2-4 及表 2-6 中的第二行, 即 1970.12.03 所对应的那一行. 在应用 (2.5) 式时, 由于用 (2.7) 式中的 $P_{1k}(k=1, 2, \cdots, 5)$ 也能报中状态 5, 因而一重转移概率又得 1 分, 故系数 B 应加上 1, 但用 (2.6) 式或 (2.8) 式都不能报中状态 5, 故 A, C 不变. 此外, $E_j = E_5 = \frac{2}{7}$, 利用这些, 就可以作出第二次预报. 以后各次预报仿此.

§3. 全国5个主要地震区的地震迁移

应用上述方法,对全国主要地震区大地震迁移规律进行分析,也得到了一些初步结果. 全国主要地震区除西藏、台湾外可分成5个状态:华北为第一状态、云南为第二状态、四川为第三状态、甘肃宁夏为第四状态、新疆为第五状态. 各区的范围见图 3-1. 选取 1900～1969 年的 $M_s \geq 6$ 地震共 115 个(可识别的余震不包括在内).

5 个状态表示 5 个大区,除华北大区(1 状态)外,其余 4 个大区又根据局部地区转移的集中性不同,分别划出 3～5 个不等的小区,其范围见图 3-1.

在求权数的过程中,这里没有能分出类似南北带的 5 年段或 8 年段的短周期,因此仍用 1900～1969 年的全部材料,其作法与上面一样,计分方法与南北带 5 年段和 8 年段的计分方法一样. 二重转移、一重转移和地区转移的得分分别为 $a=353$,$b=293$,$c=388$,与(2.5)类似得出预报测度为

$$M(k) = a \times \frac{d_1}{d_2} P_{ijk} + b \times P_{jk} + c \times D_{rk}, \quad k=1, 2, \cdots, 5,$$

其中 P_{ijk},P_{jk},D_{rk} 分别为二重、一重和地区的转移概率.

对 1970～1972 年 9 月底发生的 14 个地震作了检验预报,其结果见表 3-1. 报第一可能命中 8 个,占 57.1%,报第二可能命中 5 个,占 35.7%,合计占 92.8%. 上述结果表明此法对全国 5 个震区迁移情况的预报有一定效果. 但是需要指出的,由于华北 1900 年后很少大震,甘肃、宁夏在预报时间内大震也较少,这两个区在转移概率中占比例较少,较难报准,这是一个今后要进一步解决的问题.

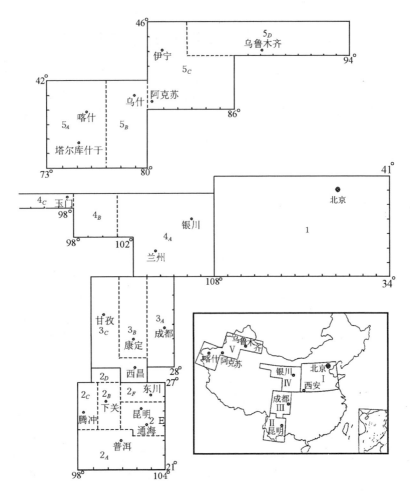

图 3-1 全国 5 个主要地震区位置示意图

地震迁移的统计预报

表 3-1　全国 5 个主要地震区预报检验情况表 ($M_s \geq 6$)

已知参数			权 $a \times \dfrac{d_1}{d_2}$	二重转移概率 $P_{i,i,k}$	权 b	一重转移概率 $P_{j,k}$	权 c	地区转移概率 $D_{j,k}$	预报测度 $M(k)$	预测状态					实际发震状态	地震效果
i,j	时间	经纬度								第一可能 基率(%)	基率	第二可能 基率(%)	基率	第三可能 基率(%)	基率	
5.1	1969-07-18	38.2 119.4	1	$353 \times \dfrac{1}{6}$	293	$(1,0,0,0,0)$	388	$\left(\dfrac{1}{6}, \dfrac{2}{6}, \dfrac{2}{6}, 0, \dfrac{1}{6}\right)$	$(172.1, 226.9, 226.9, 0, 113.5)$	30.6	2,3		1	23.2	2	好
1.2	1970-01-05	24.0 102.7	1_E	$353 \times \dfrac{10}{35}$	297	$\left(\dfrac{1}{2}, \dfrac{1}{2}, 0,0,0\right)$	392	$\left(\dfrac{2}{35}, \dfrac{11}{35}, \dfrac{7}{35}, \dfrac{5}{35}, \dfrac{10}{35}\right)$	$(27.1, 215.4, 59.4, 154.4, 252.8)$	35.5	5	30.3	4	21.7	2	中
2.2	1970-02-07	23.0 102.2	2_E	$357 \times \dfrac{11}{36}$	302	$\left(0, \dfrac{5}{11}, \dfrac{4}{11}, 0, \dfrac{2}{11}\right)$	394	$\left(\dfrac{2}{36}, \dfrac{12}{36}, \dfrac{7}{36}, \dfrac{5}{36}, \dfrac{10}{36}\right)$	$(16.8, 238.8, 246.1, 101.1, 211.6)$	30.2	3	29.3	5	26	3	好
2.3	1970-02-24	30.6 103.2	3_A	$362 \times \dfrac{6}{28}$	303	$\left(0, \dfrac{2}{6}, \dfrac{2}{6}, 0, \dfrac{2}{6}\right)$	399	$\left(\dfrac{9}{28}, \dfrac{12}{28}, \dfrac{5}{28}, 0, \dfrac{2}{28}\right)$	$(0, 255.4, 134.3, 121.1, 266.2)$	34.2	5	32.8	3	17.2	2	中
3.2	1970-03-13	23.8 102.9	2_E	$365 \times \dfrac{10}{37}$	306	$\left(0, \dfrac{3}{10}, \dfrac{4}{10}, 0, \dfrac{3}{10}\right)$	403	$\left(\dfrac{2}{37}, \dfrac{12}{37}, \dfrac{9}{37}, \dfrac{5}{37}, \dfrac{10}{37}\right)$	$(16.5, 243.9, 76.0, 176.2, 285.0)$	35.7	5	30.5	4	22.0	3	好
2.5	1970-07-29	40.0 78.0	5_B	$369 \times \dfrac{10}{29}$	309	$\left(0, \dfrac{3}{10}, \dfrac{4}{10}, 0, \dfrac{3}{10}\right)$	405	$\left(\dfrac{1}{29}, \dfrac{4}{29}, \dfrac{8}{29}, \dfrac{5}{29}, \dfrac{11}{29}\right)$	$(22.5, 223.4, 145.1, 255.7, 90.1)$	30.6	4	26.6	3	17.3	2	中
5.2	1970-02-05	25.4 99.6	2_B	$371 \times \dfrac{8}{38}$	312	$\left(0, \dfrac{1}{8}, \dfrac{3}{8}, 0, \dfrac{4}{8}\right)$	407	$\left(\dfrac{2}{38}, \dfrac{12}{38}, \dfrac{8}{38}, \dfrac{5}{38}, \dfrac{11}{38}\right)$	$(25.9, 117.5, 65.7, 141.0, 444.4)$	55.9	5	17.7	4	14.8	5	好
5.2	1971-05-25	41.5 79.3	5_B	$376 \times \dfrac{11}{30}$	315	$\left(0, \dfrac{4}{11}, \dfrac{4}{11}, 0, \dfrac{3}{11}\right)$	412	$\left(\dfrac{2}{30}, \dfrac{12}{30}, \dfrac{8}{30}, \dfrac{5}{30}, \dfrac{5}{30}\right)$	$(89.7, 280.3, 211.7, 189.8, 88.9)$	32.5	2	24.6	4	22.0	2	差
2.3	1971-04-28	22.8 101.1	3_A	$380 \times \dfrac{10}{39}$	319	$\left(0, \dfrac{2}{10}, \dfrac{5}{10}, 0, \dfrac{3}{10}\right)$	416	$\left(\dfrac{1}{39}, \dfrac{12}{39}, \dfrac{9}{39}, \dfrac{5}{39}, \dfrac{12}{39}\right)$	$(25.9, 209.6, 250.3, 106.1, 228.6)$	30.5	3	27.8	3	25.5	3	好
2.5	1971-06-26	41.4 79.3	5_B	$385 \times \dfrac{12}{31}$	323	$\left(0, \dfrac{1}{12}, \dfrac{4}{12}, 0, \dfrac{7}{12}\right)$	420	$\left(\dfrac{1}{31}, \dfrac{6}{31}, \dfrac{9}{31}, \dfrac{4}{31}, \dfrac{11}{31}\right)$	$(90.8, 375.1, 141.9, 192.1, 89.2)$	42.1	2	21.6	3	15.9	4	好
2.5	1971-08-16	28.8 103.6	3_A	$385 \times \dfrac{9}{29}$	326	$\left(0, \dfrac{2}{9}, \dfrac{6}{9}, 0, \dfrac{1}{9}\right)$	424	$\left(0, \dfrac{4}{29}, \dfrac{7}{29}, \dfrac{4}{29}, \dfrac{10}{29}\right)$	$(0, 315.2, 122.4, 143.6, 277.2)$	36.6	5	32.2	4	16.7	2	好
3.5	1971-09-14	23.0 101.1	3_A	$392 \times \dfrac{13}{40}$	330	$\left(0, \dfrac{5}{13}, \dfrac{5}{13}, 0, \dfrac{3}{13}\right)$	427	$\left(\dfrac{1}{40}, \dfrac{5}{40}, \dfrac{13}{40}, \dfrac{5}{40}, \dfrac{10}{40}\right)$	$(16.5, 212.5, 245.2, 102.8, 272.7)$	32.1	5	28.8	3	25.0	4	好
2.5	1972-01-16	40.5 78.8	5_B	$397 \times \dfrac{13}{32}$	335	$\left(0, \dfrac{5}{13}, \dfrac{5}{13}, 0, \dfrac{3}{13}\right)$	432	$\left(\dfrac{1}{32}, \dfrac{5}{32}, \dfrac{14}{32}, \dfrac{2}{32}, \dfrac{10}{32}\right)$	$(81.9, 347.8, 225.8, 174.4, 88.5)$	37.8	2	24.5	3	18.9	2	好
2.3	1972-08-27	22.3 100.1	3_A	$400 \times \dfrac{14}{41}$	339	$\left(0, \dfrac{2}{14}, \dfrac{10}{14}, 0, \dfrac{2}{14}\right)$	432	$\left(\dfrac{1}{41}, \dfrac{8}{41}, \dfrac{13}{41}, \dfrac{3}{41}, \dfrac{11}{41}\right)$	$(26.0, 196.8, 223.2, 100.6, 320.4)$	36.9	5	25.7	4	22.6	3	好
2.3	1972-09-27	29.5 107.3	3_B	$400 \times \dfrac{7}{30}$	340	$\left(0, \dfrac{2}{7}, \dfrac{5}{7}, 0, \dfrac{2}{7}\right)$	436	$\left(0, \dfrac{11}{30}, \dfrac{9}{30}, \dfrac{3}{30}, \dfrac{7}{30}\right)$	$(0, 163.3, 342.9, 34.0, 325.9)$	29.3	3	27.8	5	13.9		中

§4. 存在问题

统计预报的方法主要的根据是以往的地震资料，如果某地区的资料少，预报的效果就差，如果完全没资料，那么就根本不能用此法预报．因此，这种方法的主要缺点在于对历史上从未发生过或很少发生过地震的地区不能预报，对从未发生过转移的也难以预报．其次，这个方法只能预报某地区下一次发震的概率的大小，而不能绝对肯定发生或不发生地震，即使算得某地发生地震的概率高达90%，也不能绝对肯定发生，因为还存在10%的概率没有地震．所以，最好把这一方法与其他方法结合起来，才能取得更好的效果．

本文中的方法有待长期实践的检验，我们希望通过今后的预报实践来不断改进．

这里只考虑了时间、地点和震级三个因素，其他因素，如震源深度、震源力学特性等均未考虑．在报地点时只用了地点，连时间和震级的信息都未完全用上，这自然是一个缺点．

如何更好地划分状态，也存在问题．我们主要是根据历史上地震迁移和相关性来分区．虽然也考虑了一些地质构造条件，但还很粗糙．对小区的划分也存在类似的问题．

下一次地震发生在什么时间？这个问题我们虽然作了一些初步的探讨，但还需要进一步努力．我们希望以后能在这方面做些工作．

附言

本文中1972年10月完稿，随后我们对(2.4)中的系数 U_r，V_r 及预报测度(2.5)中的系数 A, B, C 做了改进，即与计算

(2.1)的系数 a_j，b_j 一样，使它们与状态有关，以便更反映实际．同时把绝对概率（或称零重转移概率）作为预报因子加进公式(2.5)．改进后的预报测度可表为

$$M(k)=\alpha_{ij}P_{ijk}+\beta_j P_{jk}+r_j P_k+\delta_j D_{rk},$$

其中 $P_k(k=1,2,\cdots,5)$ 是状态 k 的绝对概率，r 是状态 j 内震中所处的那个小区，求系数的方法仍用记分法．

其次，由于每个状态的范围很大，在确定下次地震落在某一状态后，我们进一步搜索最大可能发震的更小地区（此区越小越好）作为重点预报区．

参考文献

[1] 李洪吉，孙惠文．中国中部南北地震带的统计分区．地质科学，1973，8(2)：162-166．

[2] 复旦大学数学系．概率论与数理统计．上海：上海科学技术出版社，1961．

[3] Harbaugh J W et al. Computer Simulation in Geology. New York：John Wiley & Sons，1970：575．

预测大地震的一种数学方法[①]

摘要 本文的目的是预测今后 4 个月内我国大陆是否有大地震发生. 根据我国大陆以外地方最近半年的地震建立一个判断量 X, 如果 $X > 0.53$, 那么认为将有大震发生, 进一步的分析还可大致确定发震的地区.

§1. 基本思想

我们的目的是预测最近 4 个月内, 我国大陆是否有大地震发生. 根据的资料是全球(主要是环太平洋带)1900～1965 年大震目录, 希望利用世界大震来预报我国大陆的大震. 这里所谓的大震, 是指震级 $\geqslant 7$ 级的地震.

方法的基本思想如下:

(i) 找出与我国大陆大震相关密切的地区共 11 块, 记为 A_1, A_2, $\cdots$, A_{11}. 这些地区大致分布在阿富汗、土耳其、缅甸、苏门答腊、苏拉威西、所罗门群岛、菲律宾、斐济、汤加、

[①] 南开大学数学系统计预报科研小组.

美国、墨西哥、中南美、阿留申、日本、千岛群岛等地．每块地区又包含若干块小地区．所谓相关密切，是指每一块地区 A_i 发震后，我国大陆在半年（或 7 个月）以内接着发震的频率较高．这些地区称为**相关区**，它所包含的每一小地区则称为**相关小区**.

(ii) 在 1900～1965 年中，在地区 A_i 内总共发生地震的次数设为 n_i，其中有 m_i 次各在 7 个月内引起我国大震．令

$$a_i = \frac{m_i}{n_i}, \tag{1.1}$$

它是该地区中的地震在 7 个月内转移到我国的频率．这个数越大，表示该区与我国大震的关系越密切．将 a_i 称为 A_i 的**相关频率**.

(iii) 全球在某指定的 7 个月（例如 1950 年 1～7 月）中发生的大地震的总数设为 γ，其中计有 S 次分别落在相关区 A_{i_1}，A_{i_2}，…，A_{i_s} 中（A_{i_1} 可能重合于 A_{i_2} 等），它们对应的相关频率是：a_{i_1}，a_{i_2}，…，a_{i_s}，作线性组合，给出预报判别函数

$$X = d_1 a_{i_1} + d_2 a_{i_2} + \cdots + d_s a_{i_s}, \tag{1.2}$$

权数 d_i 的取法见后．称 X 为这 7 个月的**判别量**.

(iv) 根据 X 的大小，就可预报这 7 个月后的 4 个月中（在上例中为 1950 年 8～11 月）我国大陆是否有大震发生：

若 $X \geq 0.53$ 则报有；若 $X \leq 0.5$ 则报无；

若 $0.5 < X < 0.53$ 则不作结论，继续观察.

§2. 相关区的选择

选择相关区的做法是：在世界大震目录上，先标出我国大陆的各次大震，然后把每次大震（例如其中一次为1950年8月15日西藏察隅大震）之前7个月内全球所发生的各次大震的震中标出，例如1950年2~8月内共发生10次大震，震中分别为[①]（22N，100E），（46N，144E），（21S，169E），（13N，143.5E），（20.3S，169.3E），（47S，15W），（20.3S，169.3E），（8S，70.8W），（6.5S，155E），（27.3S，62.5W）. 其中2次发生在前3个月即1950年2~4月中，其余8次发生在后4个月内，用红点记在坐标纸上．挑出这种红点密集的地区，其中一块 A 例如

$$A_{10} = \begin{pmatrix} 0\text{—}34\text{S} \\ 62\text{—}91\text{W} \end{pmatrix},$$

即经度为 $0°\text{S} \sim 34°\text{S}$，纬度为 $62°\text{W} \sim 91°\text{W}$ 的那块地区. 设1900~1965年中，共含有 m 个红点，且在 A 区内总共发生大震 n 次，如果

$$\frac{m}{n} > 74\%,$$

即 $\frac{m}{n}$ 相当大时，就取 A 为一块相关区. $\frac{m}{n}$ 即 A 的相关频率.

我们选择的相关区如下：

第1区　伊朗、阿富汗等地

$$\begin{pmatrix} 37\text{N}\text{—}47\text{N} \\ 16\text{E}\text{—}29.5\text{E} \end{pmatrix} \quad \begin{pmatrix} 35\text{N}\text{—}43\text{N} \\ 32\text{E}\text{—}55\text{E} \end{pmatrix} \quad \begin{pmatrix} 38\text{N}\text{—}42\text{N} \\ 67\text{E}\text{—}73\text{E} \end{pmatrix}$$

$$\begin{pmatrix} 24.5\text{N}\text{—}35\text{N} \\ 62\text{E}\text{—}67.5\text{E} \end{pmatrix} \quad \begin{pmatrix} 36\text{N}\text{—}37.5\text{N} \\ 0\text{—}20\text{W} \end{pmatrix}$$

---

① S表示南纬，N表示北纬；E表示东经，W表示西经.

预测大地震的一种数学方法

第 2 区　缅甸、苏门答腊等地

$$\begin{pmatrix} 5N—18N \\ 90E—98E \end{pmatrix} \begin{pmatrix} -1S—5N \\ 76E—102E \end{pmatrix} \begin{pmatrix} -9S—2.5S \\ 90E—105.5E \end{pmatrix}$$

第 3 区　苏拉威西、所罗门群岛等地

$$\begin{pmatrix} 3S—8.5S \\ 118.5E—130.5E \end{pmatrix} \begin{pmatrix} 1S—3N \\ 121E—123E \end{pmatrix} \begin{pmatrix} 1S—2N \\ 125E—128E \end{pmatrix}$$

$$\begin{pmatrix} 4.5N—7.5N \\ 125.5E—128E \end{pmatrix} \begin{pmatrix} 8N—14N \\ 123E—126.5E \end{pmatrix} \begin{pmatrix} 7N—12N \\ 127.5E—129E \end{pmatrix}$$

$$\begin{pmatrix} 0—7S \\ 132E—138.5E \end{pmatrix} \begin{pmatrix} 3S—7S \\ 141E—143E \end{pmatrix} + \begin{pmatrix} 4S—8S \\ 143E—147E \end{pmatrix}$$

$$\begin{pmatrix} 4S—8S \\ 148.5E—150E \end{pmatrix} + \begin{pmatrix} 6S—8S \\ 150E—151.5E \end{pmatrix} \begin{pmatrix} 4.5S—6S \\ 153E—155E \end{pmatrix}$$

$$\begin{pmatrix} 6.5S—8S \\ 154.5E—167E \end{pmatrix}$$

第 4 区　我国台湾、菲律宾等地

$$\begin{pmatrix} 12N—18N \\ 138E—148E \end{pmatrix} \begin{pmatrix} 18.5N—22N \\ 145E—147E \end{pmatrix}$$

$$\begin{pmatrix} 26N—32N \\ 130.5E—133E \end{pmatrix} + \begin{pmatrix} 26N—30N \\ 135E—141E \end{pmatrix}$$

$$\begin{pmatrix} 16N—21N \\ 120E—122E \end{pmatrix} \begin{pmatrix} 21.5N—24N \\ 120E—121E \end{pmatrix}$$

第 5 区　斐济、汤加等地

$$\begin{pmatrix} 13S—17S \\ 160E—167E \end{pmatrix} \begin{pmatrix} 19.5S—23S \\ 167E—171E \end{pmatrix} + \begin{pmatrix} 18S—20S \\ 169.5E—171E \end{pmatrix}$$

$$\begin{pmatrix} 28S—30S \\ 175W—179W \end{pmatrix} \begin{pmatrix} 20S—23S \\ 176W—178W \end{pmatrix} + \begin{pmatrix} 20.8S—23.5S \\ 178W—180W \end{pmatrix}$$

$$\begin{pmatrix} 24S—26S \\ 172E—177W \end{pmatrix} \begin{pmatrix} 19S—23S \\ 172W—174W \end{pmatrix} \begin{pmatrix} 15S—19S \\ 173W—180W \end{pmatrix}$$

$$\begin{pmatrix} 32.5S—39.5S \\ 177E—178W \end{pmatrix} \begin{pmatrix} 43S—60S \\ 114E—166E \end{pmatrix}$$

第6区　日本、千岛群岛等地

$$\begin{pmatrix}32N—36N\\134.5E—136E\end{pmatrix}\begin{pmatrix}33N—36N\\138E—142E\end{pmatrix}\begin{pmatrix}38N—39.5N\\139.5E—145E\end{pmatrix}$$

$$\begin{pmatrix}41N—46N\\143E—144.5E\end{pmatrix}\begin{pmatrix}43N—49N\\135E—142.5E\end{pmatrix}\begin{pmatrix}43N—48N\\146E—148E\end{pmatrix}$$

$$\begin{pmatrix}43N—45N\\148E—150E\end{pmatrix}\begin{pmatrix}45N—49.5N\\150E—157E\end{pmatrix}\begin{pmatrix}53N—58N\\154E—164E\end{pmatrix}$$

第7区　白令海峡等地

$$\begin{pmatrix}45N—49N\\176E—170W\end{pmatrix}\begin{pmatrix}50N—55N\\170E—180E\end{pmatrix}\begin{pmatrix}51.5N—53N\\173W—176W\end{pmatrix}$$

$$\begin{pmatrix}53N—55.5N\\156.5W—170W\end{pmatrix}\begin{pmatrix}57N—62N\\149W—156W\end{pmatrix}+\begin{pmatrix}61N—65N\\146W—148W\end{pmatrix}$$

第8区　美国、墨西哥等地

$$\begin{pmatrix}20N—28N\\107W—113W\end{pmatrix}\begin{pmatrix}30N—41N\\117W—142W\end{pmatrix}$$

第9区　中美[①]

$$\begin{pmatrix}14N—20N\\60W—70W\end{pmatrix}\begin{pmatrix}6N—19N\\78W—91W\end{pmatrix}\begin{pmatrix}12N—20N\\92W—97W\end{pmatrix}$$

$$\begin{pmatrix}18N—20N\\98W—106W\end{pmatrix}\begin{pmatrix}7N—11N\\36W—44W\end{pmatrix}\text{左方三角形}$$

第10区　中南美

$$\begin{pmatrix}8S—13S\\72W—78W\end{pmatrix}\begin{pmatrix}14S—18S\\66W—71W\end{pmatrix}+\begin{pmatrix}18S—22S\\68.5W—71W\end{pmatrix}$$

$$\begin{pmatrix}22S—30S\\62W—68W\end{pmatrix}\begin{pmatrix}28S—34S\\66W—72W\end{pmatrix}\begin{pmatrix}0—4S\\79W—81W\end{pmatrix}$$

第11区　南奥克尼群岛等地

$$\begin{pmatrix}60S—62S\\16W—40W\end{pmatrix}\begin{pmatrix}54S—61S\\58W—60W\end{pmatrix}$$

---

① 中美指的是：危地马拉、伯利兹、洪都拉斯、萨尔瓦多、尼加拉瓜、哥斯达黎加、巴拿马、古巴、海地、多米尼加等. 见本套书第7卷第333页.

§3. 判别量 X 的精确公式

现在把公式精确化. 直观地想，一般说来前 3 个月中的地震与后 4 个月中的地震对未来的地震影响是不一样的，因此有必要把它们区别开来. 其次，还要考虑到这半年内全球地震的活动性. 例如：全球只发生 4 次，其中有 2 次落入相关区；或全球共发生 20 次，其中也只有 2 次落入相关区. 落入次数虽都是 2，但这两种情况是大不相同的，应考虑落入次数与总数之比.

设我们现在处于时刻 t_0，要预测未来的 4 个月中 (t_0, t_4) 我国大陆有无大震. 为此，利用过去 7 个月中 (t_{-7}, t_0) 的地震记录，将它分成两段：前 3 个月 (t_{-7}, t_{-4}) 及后 4 个月 (t_{-4}, t_0). 今定义

$$X = \frac{s_1}{\gamma_1} \cdot \frac{1}{s_1} \sum_{j=1}^{s_1} d^{(j)} a^{(j)} + \frac{s_2}{\gamma_2} \cdot \frac{1}{s_2} \sum_{j=1}^{s_2} D^{(j)} A^{(j)}$$

$$= \frac{1}{\gamma_1} \sum_{j=1}^{s_1} d^{(j)} a^{(j)} + \frac{1}{\gamma_2} \sum_{j=1}^{s_2} D^{(j)} A^{(j)}, \quad (3.1)$$

这里，前一项对应于 (t_{-7}, t_{-4})，后一项对应于 (t_{-4}, t_0). 其中，γ_1 为 (t_{-7}, t_{-4}) 中所发生的全球大震总次数；

s_1 为其中落入相关区的次数 $(s_1 \leqslant \gamma_1)$；

$a^{(j)}$ 为 s_1 次中第 j 次大震所在相关区的相关频率，设为

$$a^{(j)} = \frac{m^{(j)}}{n^{(j)}},$$

$m^{(j)}$，$n^{(j)}$ 的意义与 (1) 中的 m_j，n_j 一样. 设这 $m^{(j)}$ 次大震中，有 $h^{(j)}$ 次是经过 4 个月后才引起我国大陆地震的大震次数，则令

$$d^{(j)} = \frac{h^{(j)}}{m^{(j)}}, \quad (3.2)$$

类似地利用 (t_{-4}, t_0) 中的资料，可定义 γ_2，s_2，$A^{(j)}$，$D^{(j)}$，区别只是在

$$D^{(j)} = \frac{H^{(j)}}{M^{(j)}} \qquad (3.3)$$

中，$H^{(j)}$ 是不到 4 个月就引起我国大陆地震的大震次数.

以 (3.2)(3.3) 代入 (3.1)，可得 X 的另一表达式为

$$\begin{aligned} X &= \frac{1}{\gamma_1} \sum_{j=1}^{s_1} \frac{h^{(j)}}{m^{(j)}} \cdot \frac{m^{(j)}}{n^{(j)}} + \frac{1}{\gamma_2} \sum_{j=1}^{s_2} \frac{H^{(j)}}{M^{(j)}} \cdot \frac{M^{(j)}}{N^{(j)}} \\ &= \frac{1}{\gamma_1} \sum_{j=1}^{s_1} \frac{h^{(j)}}{n^{(j)}} + \frac{1}{\gamma_2} \sum_{j=1}^{s_2} \frac{H^{(j)}}{N^{(j)}}, \end{aligned} \qquad (3.4)$$

此式的直观意义是明显的.

具体应用 (3.1) 时，我们一劳永逸地对每块相关区 A_j，算出 $d_j a_j$ 与 $D_j A_j$，算法与求 $d^{(j)} a^{(j)}$，$D^{(j)} A^{(j)}$ 一样，只是应以"第 j 个相关区"来代替上面的"第 j 个大震所在的相关区 $A^{(j)}$".

§4. 举 例

例1 1950年8月15日西藏察隅大震(8.5级)前7个月全球大震落入各相关区情况如表4-1.

表 4-1

地区号	1	2	3	4	5	6	7	8	9	10	11	落入次数 s	$\sum d^{(j)} \cdot a^{(j)}$	$\sum D^{(j)} \cdot A^{(j)}$	总次数 γ
$d_j a_j$	0.368	0.311	0.293	0.429	0.224	0.267	0.395	0.154	0.183	0.383	0.214				
$D_j A_j$	0.447	0.551	0.421	0.371	0.517	0.489	0.395	0.769	0.550	0.511	0.571				
前3个月落入次数					1	1						2	0.491		2
后4个月落入次数			1	1	3				1			6		2.854	8

此表说明：前3个月内落入区域5及6的各有1次，故 $s_1 = 2$，落入非相关区的没有，故 $\gamma_1 = s_1 = 2$，类似算得 $s_2 = 6$，$\gamma_2 = 8$，代入公式(3.1)得

$$X = \frac{1}{2}(0.224 + 0.267) + \frac{1}{8}(0.421 + 0.371 + 3 \times 0.517 + 0.511)$$
$$= 0.246 + 0.357 = 0.603 > 0.53.$$

例2 考虑1939年1月25日至1939年8月25日这段时间内，前3个月落入3区与5区各1次，落入非相关区5次，后4个月内落入3区1次、5区2次，落入非相关区2次. 故 $s_1 = 2$，$\gamma_1 = 7$，$s_2 = 3$，$\gamma_2 = 5$，代入公式(3.1)得

$$X = \frac{1}{7}(0.293 + 0.224) + \frac{1}{5}(0.421 + 2 \times 0.517)$$
$$= \frac{1}{7} \times 0.517 + \frac{1}{5} \times 1.455$$
$$= 0.074 + 0.291 = 0.365 < 0.5,$$

故得1939年9~10月我国大陆无大震，与实际相符.

§5. 实践检验

(i) 1900～1965 年，我国大陆（包括与缅甸、印度、尼泊尔、苏联等地边界附近地区）共发生大震 61 次，对每次大震计算一次 X，共得 61 个 X 值，这 61 个值除 3 个外，最小的是 0.53，详细的分布情况是

$X \geqslant 0.8$, 12 次， $0.7 \leqslant X < 0.8$, 20 次，
$0.6 \leqslant X < 0.7$, 14 次， $0.53 \leqslant X < 0.6$, 12 次，
$0.335 \leqslant X < 0.5$, 3 次.

(ii) 此外，我们又抽选了间隔为 6～7 个月的 109 个数（其后 4～5 个月内大陆没有大震），也算出 109 个 X 值，除 2 个外，它们的最大值是 0.5. 详细分布情况是

$X \leqslant 0.4$, 101 次， $0.4 < X \leqslant 0.5$, 6 次，
$0.5 < X \leqslant 0.6$, 1 次， $0.6 < X$, 1 次.

(iii) 上面分析的是 1900～1965 年的资料．我们再用此法来检验 1965 年以后我国的 6 次大陆大震，算得

第 1 次 1966 年 3 月 22 日邢台大震前 7 个月内全球只发生 1 次大震，不便计算.

第 2 次 1969 年 7 月 18 日渤海地震，$X = 0.679$.

第 3 次 1970 年 1 月 4 日云南通海地震，$X = 0.76$.

第 4 次 1973 年 2 月 6 日四川甘孜地震，$X = 0.623$.

第 5 次 1974 年 5 月 11 日云南昭通地震，$X = 0.554$. 此次之前为 1974 年 5 月 9 日的日本地震，两者相隔不到两昼夜，而日本地震消息传来时，昭通地震已发生. 日本地震前 $X = 0.427$，加上日本地震后 $X = 0.554$.

§6. 预报地区

所预报的大震将在哪里发生？现将解答这一问题的方法简述如下.

(i) 将我国大陆分作 3 区：Ⅰ区，包括新疆、青海、甘肃、宁夏等地；Ⅱ区，包括四川、云南、西藏等地；Ⅲ区，包括河北、山西、陕西、东北等地.

$$Y_k = \sum_{j=1}^{s_1} \left[G_k^{(j)} - \frac{\nu_k + \frac{1}{3}\sum_{k=1}^{3}\nu_k}{3\sum_{k=1}^{3}\nu_k} \right] + \sum_{s=1}^{s_2} \left[H_k^{(s)} - \frac{\nu_k + \frac{1}{3}\sum_{k=1}^{3}\nu_k}{3\sum_{k=1}^{3}\nu_k} \right]$$

$$= \sum_{j=1}^{s_1} G_k^{(j)} + \sum_{s=1}^{s_2} H_k^{(s)} - (s_1 + s_2)\frac{\nu_k + \frac{1}{3}\sum_{k=1}^{3}\nu_k}{3\sum_{k=1}^{3}\nu_k}$$

$$(k = 1, 2, 3). \tag{6.1}$$

对前述 1900～1965 年用公式报中的 58 次地震，每次分别算出 Y_1, Y_2, Y_3 三个值. 其中

Y_1——Ⅰ区的判别量，

Y_2——Ⅱ区的判别量，

Y_3——Ⅲ区的判别量.

s_1（或 s_2）——前 3 个月（或后 4 个月）中落入相关区的总次数.

$$G_k^{(j)} = \frac{e_k^{(j)}}{m_j},$$

m_j——前 3 个月第 j 次大震所在的相关小区中，7 个月内引起我国大震的次数.

$e_k^{(j)}$——该小区中转入我国大陆第 $k(1,2,3)$ 区，且经过时间在 $4\sim7$ 个月的大震次数．

$$H_k^{(s)} = \frac{f_k^{(s)}}{m_s},$$

m_s——后 4 个月第 s 次大震所在相关小区中，7 个月内引起我国大震的次数．

$f_k^{(s)}$——该区 m_s 次中转入我国大陆第 k 区，且经过时间小于 4 个月的大震次数．据地震目录得：

$\nu_1 = 20$——Ⅰ区大震发生次数，

$\nu_2 = 33$——Ⅱ区大震发生次数，

$\nu_3 = 8$——Ⅲ区大震发生次数．

(ii) 对 Y_1, Y_2, Y_3 进行比较，选其最大者所对应的区，作为预报区．

举例 1924 年 7 月 3 日新疆(36N，84E)发生 7.2 级大震，其前 7 个月内共有 9 次大震，该 9 次大震全部落入相关小区中，有

$$s_1 = 2, \quad s_2 = 7,$$

$$\begin{cases} G_1^{(1)} = \dfrac{2}{9}, \\ G_2^{(1)} = \dfrac{2}{9}, \\ G_3^{(1)} = \dfrac{0}{9}, \end{cases} \quad \begin{cases} G_1^{(2)} = \dfrac{2}{4}, \\ G_2^{(2)} = \dfrac{0}{4}, \\ G_3^{(2)} = \dfrac{0}{4}, \end{cases}$$

$$\begin{cases} H_1^{(1)} = \dfrac{5}{16}, \\ H_2^{(1)} = \dfrac{3}{16}, \\ H_3^{(1)} = \dfrac{2}{16}, \end{cases} \quad \begin{cases} H_1^{(2)} = \dfrac{3}{8}, \\ H_2^{(2)} = \dfrac{0}{8}, \\ H_3^{(2)} = \dfrac{2}{8}, \end{cases} \quad \begin{cases} H_1^{(3)} = \dfrac{1}{15}, \\ H_2^{(3)} = \dfrac{6}{15}, \\ H_3^{(3)} = \dfrac{1}{15}, \end{cases} \quad \begin{cases} H_1^{(4)} = \dfrac{3}{8}, \\ H_2^{(4)} = \dfrac{2}{8}, \\ H_3^{(4)} = \dfrac{2}{8}, \end{cases}$$

预测大地震的一种数学方法

$$\begin{cases} H_1^{(5)} = \dfrac{3}{16}, \\ H_2^{(5)} = \dfrac{3}{16}, \\ H_3^{(5)} = \dfrac{3}{16}, \end{cases} \begin{cases} H_1^{(6)} = \dfrac{3}{15}, \\ H_2^{(6)} = \dfrac{4}{15}, \\ H_3^{(6)} = \dfrac{1}{15}, \end{cases} \begin{cases} H_1^{(7)} = \dfrac{3}{16}, \\ H_2^{(7)} = \dfrac{3}{16}, \\ H_3^{(7)} = \dfrac{3}{16}, \end{cases}$$

代入公式(6.1)得

$$Y_1 = \frac{2}{9} + \frac{2}{4} + \frac{5}{16} + \frac{3}{8} + \frac{1}{15} + \frac{3}{8} + \frac{3}{16} + \frac{3}{15} + \frac{3}{16} - 9\frac{40}{183},$$

$$Y_2 = \frac{2}{9} + \frac{0}{4} + \frac{3}{16} + \frac{0}{8} + \frac{6}{15} + \frac{2}{8} + \frac{3}{16} + \frac{4}{15} + \frac{3}{16} - 9\frac{53}{183},$$

$$Y_3 = \frac{0}{9} + \frac{0}{4} + \frac{2}{16} + \frac{2}{8} + \frac{1}{15} + \frac{2}{8} + \frac{3}{16} + \frac{1}{15} + \frac{3}{16} - 9\frac{28}{183}.$$

显然 Y_1 最大,应报Ⅰ区,与实际相符.

公式(6.1)的来源如下:由于要预测的是下次大震发生的地区,此地区应只依赖于前 7 个月内落入相关小区中的地震,而不必考虑相关区外的地震,故(6.1)中只出现 s_i 而未出现 γ_i,这里起主要作用的是相关小区. (6.1)中的 $G_k^{(j)}$, $H_k^{(s)}$ 是相关小区对于大陆 3 个区的相关频率. (6.1)中第一项对应于前 3 个月,第二项对应于后 4 个月,第三项是由于考虑消除"自然概率"而引进的,表面上看来似乎引入

$$\sum_{j=1}^{s_1} \left(G_k^{(j)} - \frac{\nu_k}{\sum_{k=1}^{3} \nu_k} \right) + \sum_{s=1}^{s_2} \left(H_k^{(s)} - \frac{\nu_k}{\sum_{k=1}^{3} \nu_k} \right), \quad k = 1, 2, 3,$$

比较自然,但从实际效果考虑,改用修改后的"自然概率"即

$$\frac{\nu_k + \dfrac{1}{3}\sum_{k=1}^{3}\nu_k}{3\sum_{k=1}^{3}\nu_k}$$

更为适宜,既提高了资料数较少地区的预报效果,又保证了资

料数较多地区的预报效果基本不变,故最后采用了(6.1)的形式.

(iii) 采用上法效果:

Ⅰ区 17 个(另外 3 个在前面有关部分未报中,故这里不予考虑),全部报中,占 100%.

Ⅱ区 33 个,报中 26 个,占 80%.

Ⅲ区 8 个,报中 6 个,占 75%.

后验:

1969 年 7 月 18 日渤海地震,未报中(即未报上Ⅲ区);

1970 年 1 月 4 日云南通海地震,报中(即报上Ⅱ区);

1973 年 2 月 6 日四川甘孜地震,报中;

1973 年 7 月 14 日昆仑山地震,报中;

1974 年 5 月 11 日云南昭通地震,报中.

(1970 年 6 月 5 日中苏边界地震,也报中.)

华北地区地震的统计预报(一)[①]

本文目的是研究华北(34.5N—41N；109E—123E)及丹东附近地区近期有无 $4\frac{3}{4}(M_s)$ 以上地震的问题，所用的资料是 1880~1971 年：

华北 $M_s \geqslant 4\frac{3}{4}$ 地震记录；

大陆其他地区 $M_s \geqslant 5$（甘肃、宁夏等部分地区则取了 $M_s \geqslant 4.5$ 的资料）；

1900~1971 年我国台湾省 $M_s \geqslant 6$；

日本、千岛、阿留申等地 $M_s \geqslant 7$.

至于 1972~1975 年的记录则留作后验及试报.

基本思想 外区地震对华北地震有一定影响，有的触发华北地震，有的则起抵消作用，具体表现在间隔时间(外区震至华北震之间的间歇时间)的分布上呈现不同的规则. 根据间隔时间的统计特性及地质构造特点，可将外区分成 15 个大区，而每一个大区由若干小区组成. 利用随机过程的理论，可以构造一个阶梯函数 $H_{t,u,s}$，用它来判断时刻 t 以后一段时间内华北是否发

[①] 本文与吴荣合作.

生 $M_s \geqslant 4\frac{3}{4}$ 的地震.

数学模型 根据华北地震定义一个随机点过程 y_t

$$y_t = \begin{cases} 1, & t \text{ 时华北发震,} \\ 0, & \text{反之.} \end{cases}$$

$$t \geqslant 0$$

定义 $x_{t,s} = \max\limits_{t \leqslant u \leqslant t+s} y_u = \begin{cases} 1, & \overline{t, t+s} \text{ 时段内华北发震,} \\ 0, & \text{反之.} \end{cases}$

对 15 个外区也分别定义 15 个点过程 $y_t^{(k)}$

$$y_t^{(k)} = \begin{cases} 1, & t \text{ 时第 } k \text{ 区发震,} \\ 0, & \text{反之.} \end{cases}$$

$$k = 1, 2, \cdots, 15, \quad t \geqslant 0.$$

由于在我们所研究的震级范围内,地震是可数的故可以定义

$$\{\sigma_n\}, \quad n \in \mathbf{N}^*,$$

它表示事件 $(y_t = 1)$ 各次出现的时间(即华北发震时间). 同样以

$$\{\tau_n^{(k)}\} \quad n = 1, 2, 3, \cdots, k = 1, 2, \cdots, 15$$

表示事件 $(y_t^{(k)} = 1)$ 各次出现的时间.

令 $$r_n^{(k)} = \min_1(\sigma_1 : \sigma_1 > \tau_n^{(k)}) - \tau_n^{(k)}$$

$(n \in \mathbf{N}^*, k = 1, 2, \cdots, 15)$ 称为第 K 区的间隔时间(即第 K 区发震到华北发震的间歇时间). 根据具体资料的分析,对每一个 k,可以认为 $\{r_n^{(k)}\}(n \in \mathbf{N}^*)$ 是独立同分布的,并以 $F_k(x)(k=1, 2, \cdots, 15)$ 表示其经验分布函数,称为第 K 区的间隔时间分布.

所要研究的问题是:对任一时刻 t,在已知前一段时间 $\overline{t-u, t}$ 内,外区已发生的地震的条件下,预测 t 后一段时间 $\overline{t, t+s}$ 内华北发生地震的可能性有多大?

即考虑

$$P(X_{t,s} = 1 \mid y_v^{(1)}, y_v^{(2)}, \cdots, y_v^{(15)}, t-u \leqslant v \leqslant t)$$

$$= f_{t,u,s}(y_{t_1}^{(1)},\ y_{t_1}^{(2)},\ \cdots,\ y_{t_1}^{(15)};\ y_{t_2}^{(1)},\ y_{t_2}^{(2)},\ \cdots, \quad (1)$$
$$y_{t_2}^{(15)};\ \cdots;\ y_{t_m}^{(1)},\ y_{t_m}^{(2)},\ \cdots,\ y_{t_m}^{(15)};\ \cdots)$$
$$= f_{t,u,s}(\cdot), \quad t-u \leqslant t_m \leqslant t,\ m \in \mathbf{N}^*$$

$f_{t,u,s}$ 是一个无穷维的 Borel 可测函数.

若在 $y_t^{(k)}$ 的值域空间考虑问题, 由于 $y_t^{(k)}$ 只取 0 或 1 为值, 故 $f_{t,u,s}$ 是 0, 1 序列的函数, 其函数值代表了在不同情况下华北发生地震的条件概率, 这正是所需要的量.

事实上, 如果在 $\overline{t-u,t}$ 时段内, 落在间隔时间短的区域的地震比落在间隔时间长的区域的地震多(间隔时间短、长指分布集中于短或长的时间内), 那么可以认为在 $\overline{t,t+s}$ 时段内触发华北发震的可能性比抵消的可能性大, 反之则相反. (当然 s, u 需要适当的选取, 否则上述意义不存在).

因而 $f_{t,u,s}$ 的值应该依赖于 $\overline{t-u,t}$ 时段内出现的各地震所在区域的间隔时间的联合分布.

不失去一般性, 设在 $\overline{t-u,t}$ 时段内 15 个点过程 $y_t^{(k)}$ 共取 n 个 1 值, 对应的过程符标是 $1, 2, \cdots, l$ (其中这些过程取 1 值的次数是 $k_1, k_2, \cdots, k_l, \sum_{m=1}^{l} k_m = n$), 它们间隔时间 $\gamma^{(1)}$, $\gamma^{(2)}, \cdots, \gamma^{(l)}$ 的联合分布函数是 $F_{1,2,\cdots,l}(x_1, x_2, \cdots, x_l)$ 则 $f_{t,u,s}$ 可以表示成

$$f_{t,u,s} = \varphi(F_{1,2,\cdots,l}(x_1, x_2, \cdots, x_l)). \quad (2)$$

符标 $1, 2, \cdots, l$ 是与区间 $\overline{t-u,t}$ 内出现的 0, 1 序列有关, 而 $x_1, x_2, \cdots, x_l$ 还应该与 s 有关(下面将看到, 在我们所研究的 s 范围内 $x_1, x_2, \cdots, x_l$ 不依赖于 s).

但要直接构造 $f_{t,u,s}$ 比较困难, 事实上必要性也不大, 只要构造出一种函数, 能在某种程度上刻画它即可.

显然这样的函数应具备下述两条性质:

(i) 与间隔时间的分布有关.

(ii) 与 $f_{t,u,s}$ 具有相同的增减性.

我们认为, 具备上述两条性质的函数, 在某种程度上能刻画 $f_{t,u,s}$.

当然可以构造出很多这样的函数, 为简化起见, 试按下面一种方式来构造.

仍在公式(2)的假设条件下考虑问题, 定义 0, 1 序列的函数

$$H_{t,u,s} = \frac{\sum_{m=1}^{l} k_m \alpha_m F_m(x_m)}{\sum_{m=1}^{l} k_m \beta_m (1 - F_m(x_m))}, \quad (3)$$

其中 α_m, β_m 是非负权数, F_m 是第 m 区的间隔时间分布.

适当选取 u, s, x_m, α_m, β_m 这种形式的函数可以满足性质 1 和 2.

此公式的概率意义是明显的: 设在 $\overline{t-u, t}$ 时段内外区发生的 n 个地震发震时间分别是 $t_1, t_2, \cdots, t_u$, 其中任一地震的发震时刻设为 t_k, 该地震对未来 x_k 天内华北发生地震产生的概率是 $F(x_k)$, 不发地震产生的概率为 $(1-F(x_k))$, 将这些发震或不发震的概率分别加权的比.

$H_{t,u,s}$ 的精确化

我们的目的是: 对任何时刻 t 用公式(3)计算 $H_{t,u,s}$ 的值, 用它判断 t 以后 s 时间内华北发生地震的可能性大小, 所用资料是 t 以前 $\overline{t-u, t}$ 时段内外区发震的信息. 下面我们将具体说明 u, s 及公式(3)右边各量如何取等问题.

i) u 的取法: 我们自然希望 $\overline{t-u, t}$ 时段内外区发生的地震很多, 另一方面 u 要尽可能的短, 为此我们按下述方式定 u.

根据历史资料的统计分析, 若取 $u \geqslant 100$ 天, 则外区发生的地震较多(一般 $\geqslant 5$ 个), 因而对任一时刻 t 若它恰好是外区地震的发震时刻(今后为区别起见一般用 t_0 表示外区发震时刻), 则

取 $u=100$ 天若 t 不正好是外区地震的发震时刻, 但 $t_{10}<t<t_{20}$, t_{10}, t_{20} 外区地震的相邻两个发震时刻则取 $u=100+t-t_{10}$ (由于 $\overline{t_{10}\ t}$ 时段内外区未发震, 故 $t-(100+t-t_{10})$, t 时段内外区的信息与 $\overline{t_{10}-100,\ t_{10}}$ 时段内外区信息一样).

ii) x_m, $m=1, 2, \cdots, 15$ 的确定, 一般说来 x_m 的取法应该与 s 等有关, 而我们是采取先确定它然后再研究 s 取多长比较合适的办法.

对 15 个间隔时间分布函数 $F_k(x)$ 进行分析, 发现其分布多集中于 130 天内或 180 天外, 因而我们就取 $x_m = 130$ 天或 180 天而称前者为第一类分布函数, 后者为第二类分布函数.

iii) α_m, β_m 的确定: 自然要求公式(3)的分子上 $F(x_m)$ 大的起的作用大, 分母上 $(1-F(x_m))$ 大的起的作用大, 为此可取
$$\alpha_m = \log_2(1+F_m(x_m)),$$
$$\beta_m = \log_2(2-F_m(x_m)).$$

iv) s 的确定:

前面 i)~iii) 中各项规定与 s 无关, 但它已将公式(3)的右方具体化了, 即是说对任一时刻 t 我们可以利用 $\overline{t-u, t}$ 时段内外区发生的地震按公式(3)的右方计算出一个值来, 为区别于 $H_{t,u,s}$ 我们用 $\overline{H}_{t,u}$ 来表示它. 这个值如何刻画 t 以后华北发生地震的可能性呢? 即它与 $H_{t,u,s}$ 的关系如何? 这里应说明两层意思: ①如何用 $\overline{H}_{t,u}$ 确定 $H_{t,u,s}$ 中的 S? ②当 S 确定后, 如何用 $H_{t,u,s}$ 来刻画 $f_{t,u,s}$ 即说明 $\overline{t, t+S}$ 时段内华北发生地震的可能性大小问题. 下面我们将逐步说明这些问题.

对 t 在 1914~1971 年期间, 计算 $\overline{H}_{t,u}$, 分析 $\overline{H}_{t,u}$ 值, 有如下事实:

$\overline{A}$) 当 $\overline{H}_{t,u} \geqslant 0.9$ 时: 若 $t=t_0$ (即是外区发震时刻), 则在 t_0 后 110 天内华北发震可能性很大(仅个别例外), 若 $t_{10} < t < t_{20}$

(t_{10}, t_{20} 是外区相邻两个发震时刻),则在 t 后 $110-(t-t_{10})$ 天内华北发震可能性很大.

$\overline{\text{B}}$) 当 $\overline{H}_{t,u} \leqslant 0.5$ 时:若 $t=t_0$,则在 t_0 后 100 天内华北不发震的可能性很大(仅个别例外),若 $t_{10}<t<t_{20}$,则在 t 后 $100-(t-t_{10})$ 天内华北不发震的可能性很大.

$\overline{\text{C}}$) $0.5<\overline{H}_{t,u}<0.9$,$t$ 后华北发震的可能性不确定(当然指 t 后不太长的时段内如 100 天等)基于上述 $\overline{H}_{t,u}$ 值的分析,前面提出的①②问题也就迎刃而解了.

A) 当 $\overline{H}_{t,u}>0.9$ 时:

若 $t=t_0$; $H_{t,u,s}=\overline{H}_{t,u}$, $s=110$.

若 $t_{10}<t<t_{20}$: $H_{t,u,s}=\overline{H}_{t,s}$, $s=110-(t-t_0)$.

B) 当 $\overline{H}_{t,u} \leqslant 0.5$ 时,

若 $t=t_0$: $H_{t,u,s}=\overline{H}_{t,u}$, $s=100$.

若 $t_{10}<t<t_{20}$: $H_{t,u,s}=\overline{H}_{t,u}$, $s=100-(t-t_0)$.

C) 当 $0.5<\overline{H}_{t,u}<0.9$ 时:

若 $t=t_0$: $H_{t,u,s}=\overline{H}_{t,u}$, $s=100$.

若 $t_{10}<t<t_{20}$: $H_{t,u,s}=\overline{H}_{t,u}$, $s=100-(t-t_{10})$.

到此我们已具体构造出函数 $H_{t,u,s}$ 了,并由 $\overline{\text{A}}$),$\overline{\text{B}}$),$\overline{\text{C}}$),及 A),B),C),可以见到 $H_{t,u,s}$ 可以很好地刻画 $f_{t,u,s}$ 即 $H_{t,u,s}$ 能很好地说明在 t,$\overline{t+s}$ 时段内华北发震的可能性大小问题.

具体步骤

(i) **划区** 根据地质构造的特性及间隔时间的统计规律将外区划分成几个区,每一个地区间隔时间分布尽可能地集中,我们这里共分 15 个区.(部分地区由于资料太少,暂不归入).

(ii) 统计间隔时间分布密度 f_k(以 10 天为 1 个单位),$k=1,2,\cdots,15$.

例1 第5区(主要是新疆、青海部分地区).

考虑到中华人民共和国成立前后资料的准确性有差异,因而作了分段统计,以便在处理时有所区别(表1).

表1

	1	2	3	4	5	6	7	8	9	10	11	12	13	14	15	16	17	18	19	20	$t \geq 20$
1972~1975 (8月底)	1	2		1					1	1			1				1				1
1952~1971	10	1	2	2	5	4	4	6	3	1	3	1	2			2	1				9
1916~1951		2		5				3	4	2	1	1									14
1800~1915			1					1													3

(iii) 确定 x_m 对于所划分的 15 个可取 $x_m = 130$ 或 180(即分布集中于 130 天内或 180 天外)

(iv) 计算 $\alpha_i F_i$; $\beta_i(1-F_i)$ ($i=1, 2, 3, \cdots, 15$)

$$\alpha_i F_i = \begin{cases} \log_2(1+F_i^*(130)) \cdot F_i(130), & \text{若分布集中于 130 天内,} \\ \log_2(1+F_i^*(180)) \cdot F_i(180), & \text{若分布集中于 180 天外,} \end{cases}$$

$$\beta_i(1-F_i) = \begin{cases} \log_2(2-F_i^*(130)) \cdot (1-F_i(130)), & \text{若分布集中于 130 天内,} \\ \log_2(2-F_i^*(180)) \cdot (1-F_i(180)), & \text{若分布集中于 180 天外,} \end{cases}$$

F_i^* 表示所用资料取自 1952 年至 1971 年,这是为了加强资料的准确性. 而 F_i 是取自 1880 年至 1971 年,其中 1880 年至 1916 年作为参考.

并将 $\alpha_i F_i$, $\beta_i(1-F_i)$ 列成表 $i=1, 2, \cdots, 15$ 称为表 *.

(v) 在所研究的震级、区域范围内将外地各次地震按时间顺序(1914~1973 年)及区域位置排列成表.

(vi) 对任一时刻 t 计算 $\overline{H}_{t,u}$ 的值,(注意到 u 的取法是与 t 有关的),并作图.

例2 $t=$ 1967 年 1 月 10 日，由"震目"看出在 1967 年 1 月 10 日 5 区发生一个地震，由 $t=t_0$ 按前段(i)规定取 100 天，即 $\overline{t-u}, t=$ 1966 年 10 月 1 日至 1967 年 1 月 1 日，由"震目"可见在该时段内外区共发 7 个地震分别落在 1，2，4，7，8，5 区然后由表查得 $\alpha_1 F_1(130)$，$\alpha_2 F_2(180)$，$\alpha_4 F_4(130)$，$\alpha_7 F_7(180)$，$\alpha_8 F_8(130)$，$\alpha_5 F_5(130)$ 及 $\beta_1(1-F_1(130))$，$\beta_2(1-F_2(180))$，$\beta_4(1-F_4(130))$，$\beta_7(1-F_7(180))$，$\beta_8(1-F_8(130))$，$\beta_5(1-F_5(130))$，$K_1=1$，$K_2=2$，$K_4=1$，$K_7=1$，$K_8=1$，$K_5=1$.

利用公式(3)右端计算 $\overline{H}_{t,u}=0.78$.

例3 $t=$ 1967 年 1 月 20 日由图 1 看出在 1967 年 1 月 20 日外区未发震，按前段(i)规定取 $u=100+10=110$，即：$\overline{t-u}, t=$ 1966 年 10 月 1 日至 1967 年 1 月 20 日，该时段内外区发震情况与例 2 同，因而 $\overline{H}_{t,u}=0.78$.

仿前两例作法利用公式(3)右方可求出

t 在 1966 年 10 月 1 日至 1967 年 3 月 30 日时段内，函数 $\overline{H}_{t,u}$ 的值，并作出图 1.

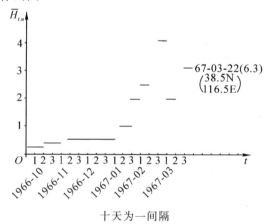

图 1　1966 年 10 月 1 日至 1967 年 3 月 27 日地震情况

(vii) 预报有无区间

分析 $\overline{H}_{t,u}$ 的值与 t 后华北有无地震的关系,确定规律性的指标及 S 的长度,进行预报,根据历史资料的分析我们将按如下规则进行预报.

A) $\overline{H}_{t,u} \geqslant 0.9$

i) $t=$ 某 t_0(外区发震时刻),则取 $S=110$ 天,且报 t 后 110 天内华北有 $M_s \geqslant 4\frac{3}{4}$ 级地震.

ii) $t_{10}<t<t_{20}$(t_{10},t_{20} 是外区某相邻两个发震时刻),则取 $S=110-(t-t_{10})$,预报 t 后,$110+t_0-t$ 天内,华北有 $M_s \geqslant 4\frac{3}{4}$ 级地震.

B) $\overline{H}_{t,u} \leqslant 0.5$

i) $t=$ 某 t_0,则取 $S=100$ 天,且报 t 后一百天内华北无 $M_s \geqslant 4\frac{3}{4}$ 级以上地震.

ii) $t_{10}<t<t_{20}$,取 $S=100-(t-t_{10})$,预报 t 后 $100-t+t_{10}$ 天内华北无 $M_s \geqslant 4\frac{3}{4}$ 级以上地震.

C) $0.5 < \overline{H}_{t,u} < 0.9$ 不作预报.

D) 若在一段时间内,$\overline{H}_{t,u}$ 从小于 0.5 ↑ 0.9 以上,(一般在华北有震情况下是如此),则可以进一步确定华北发震区间,以图 1 为例,以 $\overline{H}_{t,u} \leqslant 0.5$ 的最后一个时间区间的左端点为起点确定

无震区间 I_1:1966 年 11 月 11 日至 1967 年 2 月 20 日,

以 $\overline{H}_{t,u} \geqslant 0.9$ 的第一个时间区间的左端点为起点确定

有震区间 I_2:1967 年 2 月 1 日至 5 月 10 日,

除去 I_1 与 I_2 的共同部分即为

华北发震的时间区间:1967 年 2 月 21 日至 1967 年 5 月 10 日.

显然,若 I_1,I_2 的共同部分多,则预报时段就短.

(viii) 效果

1916～1971 年年底，华北 $M_s \geqslant 4\frac{3}{4}$ 地震共计 49 个（1966 年邢台大震只取主震，1966 年后取 5 级以上的余震），其中间隔时间不足 3 个月的 10 个，由于方法所限只将其作为 39 个资料处理（关于 3 个月内华北有 1 个以上地震的预报问题另作处理）．

若仅就外区发震时刻 t_0 的 $\bar{H}_{t,u}$ 值计算，约有 1 000 个数据，其中 $\geqslant 0.9$ 的约 140 个，$\leqslant 0.5$ 的约 800 个，若按上述规则预报则：虚报 7 次，漏报 1 次．

该 39 个地震（除去一个漏报的不计）所确定的预报区间长度 I，

$$|I| \leqslant 70 \text{ 天}, \qquad 28 \text{ 个},$$
$$70 < |I| \leqslant 110 \text{ 天}, \qquad 10 \text{ 个}.$$

1972 年初至 1975 年 3 月华北共有 8 个地震（间隔时间不足 3 个月的 3 个），按 5 个计算．在此期间，按上法预报，无虚报、漏报，确定预报区间长度：

$$|I| \leqslant 70 \text{ 天}, \qquad 3 \text{ 个},$$
$$70 < |I| < 110 \text{ 天}, \qquad 2 \text{ 个}.$$

实报：1975 年 8 月 28 日．$\begin{pmatrix} 40.8 \\ 122.7 \end{pmatrix}$ 5.2 级，报中．$\bar{H}_{t,u}$ 图形如图 2．

重点地区：$\begin{pmatrix} 39\text{—}41 \\ 109\text{—}105 \end{pmatrix}$ 或渤海湾（包括辽东半岛南部）

附注 本文仅是方法简介，在数学模型中提到许多问题，并未在此深入涉及，以后另作探讨，关于进一步的问题：华北发震地点，具体时间，3 个月内有 1 个以上地震等问题，尚在探究中，本文未作介绍．

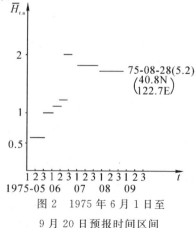

图 2 1975 年 6 月 1 日至 9 月 20 日预报时间区间

华北地区地震的统计预报(二)[①]

§1. 引 言

我们已在前文(一)中讨论了华北地区在某一时段内有否 $M_s \geqslant 4\frac{3}{4}$ 的地震发生. 本文在此基础上对发震的震级作进一步分类, 即研究震级 $4\frac{3}{4} \leqslant M_s < 5.4$ 或 $M_s \geqslant 5.4$.

为了对这两类震级作较正确的判断和区分, 关键在于寻求分辨这两类震级显著的预报因子, 这里给出了提取地震震级预报因子的方法, 在此基础上利用 Bayes(贝叶斯)聚合(集成)过程的概率模型作出震级的分类预报.

[①] 本文与钱尚玮合作.

§2. 预报因子提取的方法

二类震级分辨效果的好坏，主要取决于所提取的因子，我们通过对已有外区报华北地区的 15 个小区(见本文附注)资料的分析，二类震级对应于起报时间之前 120 天左右，外区的发震情况反映在地区的分布上是有区别的，但这种区别需要进行适当的处理后才能呈现出来．不仅如此二类震级有时也依赖于小区内发震的时间，发震的震级或发震的次数，因此从这种颇为复杂的关系中，提炼出区分二类震级的有用信息是我们选取预报因子的出发点．具体作法如下：

记预报的震级为 y，现分成两类：

第一类为 $4\frac{3}{4} \leqslant M_s < 5.4$，其取值为 0；

第二类为 $M_s \geqslant 5.4$，其取值为 1．
同样预报因子 x 也分成两类，其取值为 0 者规定为第一类，取值为 1 者规定为第二类，共提取了三个预报因子，现列述如下．

x_1：满足以下各项之一者编作取值为 1，否则取值为 0．

(i) 11，13，14 三区中至少出现二区有震，但 15 区要无震(21.5N 以下例外)；

(ii) 如果 11，13，14 三区中只有一区有震，但该震发生在离起报时间要近(约 20 天以内)．

x_2：满足以下各项之一者编作取值为 1，否则取值为 0．

(i) 8 区中 $\begin{pmatrix} 30-32\text{N} \\ 78-82\text{E} \end{pmatrix}$ 发震，但 9，10 区要无震；

(ii) 8 区无震，但 10 区有震；

(iii) 8 区有 $\geqslant 3$ 次的地震；

(iv) 8，9 区同时有震，且其一发生 7 级以上地震者.

x_3：满足以下各项之一者编作取值为 1，否则取值为 0.

(i) 4，5 区同时发震，但 3，6 区无震；

(ii) 3，4，5，6 区同时发震，但 1 区无震；

(iii) 1，2，3，4 区无震，但 8，9 区同时发震且其一为 7 级以上或 8 区内 $\begin{pmatrix} 30\text{—}32\text{N} \\ 79\text{—}82\text{E} \end{pmatrix}$ 有震.

华北地区震级 $\geqslant 5.4$ 的地震资料取自 1922 年至 1969 年共 17 次(不包括 1967 年 7 月 28 日 $\begin{pmatrix} 40.7\text{N} \\ 115.8\text{E} \end{pmatrix}$—5.5 级地震和 1967 年 12 月 3 日 $\begin{pmatrix} 37.7\text{N} \\ 115.2\text{E} \end{pmatrix}$—5.7 级地震，因为在 1967 年 7 月 28 日之前，1967 年 6 月 8 日于 $\begin{pmatrix} 39\text{N} \\ 123.7\text{E} \end{pmatrix}$ 发生了一次 $4\frac{3}{4}$ 地震，1967 年 12 月 13 日之前 1967 年 11 月 18 日于 $\begin{pmatrix} 35\text{N} \\ 123\text{E} \end{pmatrix}$ 发生一次 5 级地震，在预报时只对早先一个地震有效)，余下的 1975 年 2 月 4 日于 $\begin{pmatrix} 40.9\text{N} \\ 122.7\text{E} \end{pmatrix}$ 发生的 7.3 级地震作为后验资料. 震级 $4\frac{3}{4} \leqslant M_s < 5.4$ 的资料，取自 1918 年至 1972 年共 21 次地震(实际发生 24 次地震)①，余下 1972 年 4 月 14 日—1974 年 6 月 6 日共发震 4 次(实际发生 5 次)作为后验资料.

---

① 为了避免两类不同震级的资料数相差过大，故在取 $M_s \geqslant 5.4$ 的地震资料时，若两次地震相隔不到一月算作两次，而在取 $4\frac{3}{4} \leqslant M_s < 5.4$ 的地震资料时，若两次地震相隔不到一月算作一次.

§3. 分类预报的 Bayes(贝叶斯) 聚合过程的概率模型

设 $E_1, E_2, \cdots, E_n$ 为 n 个互不相容的完备事件组，它表示所考查的预报对象的分类.

记 $P(E_i)$ $i=1, 2, \cdots, n$ 表示各类预报对象出现的概率，通常称为先验概率，由历史资料确定，如果通过考察第一信息源后产生的信息记为 I_1，利用 Bayes 定理得

$$P(E_i \mid I_1) = \frac{P(E_i)P(I_1 \mid E_i)}{\sum_{j=1}^{n} P(E_j)P(I_1 \mid E_j)}, \quad i=1,2,\cdots,n, \quad (3.1)$$

其中 $P(E_i \mid I_1)$ 表示观察第一信息源产生的信息为 I_1 时预报事件 E_i 出现的概率，称为对 I_1 而言的后验概率.

$P(I_1 \mid E_i)$ 为事件 E_i 出现的条件下，信息为 I_1 时的条件概率.

倘若又有第二个信息源，经考察后产生的信息记为 I_2，对 I_1, I_2 而言的后验概率为

$$P(E_i \mid I_1, I_2) = \frac{P(E_i \mid I_1)P(I_2 \mid E_i, I_1)}{\sum_{j=1}^{n} P(E_j \mid I_1)P(I_2 \mid E_j I_1)}, \quad i=1,2,\cdots,n. \quad (3.2)$$

一般情形，若有 k 个信息源，考察后分别产生的信息为 $I_1, I_2, \cdots, I_k$，同样可得对 $I_1, I_2, \cdots, I_k$ 而言的后验概率为

$$P(E_i \mid I_1, I_2, \cdots, I_k)$$
$$= \frac{P(E_i \mid I_1, I_2, \cdots, I_{k-1})P(I_k \mid E_i, I_1, I_2, \cdots, I_{k-1})}{\sum_{j=1}^{n} P(E_j \mid I_1, I_2, \cdots, I_{k-1})P(I_k \mid E_j, I_1, I_2, \cdots, I_{k-1})},$$
$$i=1,2,\cdots,n. \quad (3.3)$$

或

$$P(E_i \mid I_1, I_2, \cdots, I_k)$$
$$= \frac{P(E_i)P(I_1 \mid E_i)P(I_2 \mid E_i, I_1)P(I_3 \mid E_i, I_1, I_2)\cdots}{\sum_{j=1}^{n} P(E_j)P(I_1 \mid E_j)P(I_2 \mid E_j, I_1)P(I_3 \mid E_j, I_1, I_2)\cdots}$$
$$\frac{P(I_k \mid E_i, I_1, I_2, \cdots, I_{k-1})}{P(I_k \mid E_j, I_1, I_2, \cdots, I_{k-1})}, \quad i=1,2,\cdots,n. \quad (3.4)$$

为简单起见，通常假定下式成立

$$P(I_k \mid E_i, I_1, I_2, \cdots, I_{k-1}) = P(I_k \mid E_i), \quad i=1, 2, \cdots, n \quad (3.5)$$

它表示各信息源产生的信息相互之间是无关的．即对任何 n 个事件它们是条件独立的．由此可见，(3.4)可改用下式计算

$$P(E_i \mid I_1, I_2, \cdots, I_k)$$
$$= \frac{P(E_i)P(I_1 \mid E_i)P(I_2 \mid E_i)P(I_3 \mid E_i)\cdots P(I_k \mid E_i)}{\sum_{j=1}^{n} P(E_j)P(I_1 \mid E_j)P(I_2 \mid E_j)P(I_3 \mid E_j)\cdots P(I_k \mid E_j)},$$
$$i=1,2,\cdots,n. \quad (3.6)$$

利用(3.6)可给出 n 类事件发生的后验概率，取其概率最大者为预报类．

§4. 分类预报的具体过程

(i) 资料的 0，1 取值处理

首先对历史资料中两类预报震级 y 与其对应的三个因子按因子提取的方法作 0，1 取值处理.

(ii) 求先验概率 $P(E_i)$

由于震级只分成两类，故 E_i 的下标 i 只取 1，2 两个数，$P(E_1)$ 表示震级 y 出现第一类（此时 y 取值为 0）的概率，$P(E_2)$ 表示震级 y 出现第二类（此时 y 取值为 1）的概率. 其计算公式为

$$P(E_i) = \frac{N_i}{N}, \quad i=1, 2$$

其中 N_1，N_2 分别为所取华北地震资料 1，2 两类震级发震的次数，$N=N_1+N_2$（这里 $N_1=21$，$N_2=17$，$N=38$）.

(iii) 求各因子 x_k 对两类预报对象 y 的条件概率 $P(I_k \mid E_i)$ ($k=1, 2, 3; i=1, 2$) 此处 $I_k(k=1, 2, 3)$ 为因子 $x_k(k=1, 2, 3)$ 的 1，2 两类与 y 某类共同出现的次数记作 I_{k_1}, I_{k_2} ($k=1, 2, 3$). 在实际中，如此给出的 I_{k_1}, I_{k_2} 之间是互斥的，且对 k 而言它们相互之间可以认为是无关的.

于是经 0，1 取值处理后的资料中容易求得各种情况的条件概率.

例如对 X_1 而言，只需从资料数 y 与 X_1 同取值为 0 的次数现共有 19 次，故

$$P(I_{11} \mid E_1) = \frac{P(I_{11}E_1)}{P(E_1)} = \frac{19}{21},$$

同样数 y 取值为 0 同时 X_1 取值为 1 的次数共 2 次，故

$$P(I_{12} \mid E_1) = \frac{P(I_{12}E_1)}{P(E_1)} = \frac{2}{21};$$

数 y 与 X_1 同取值为 1 的次数共 14 次，故

$$P(I_{12} \mid E_2) = \frac{P(I_{12}E_2)}{P(E_2)} = \frac{14}{17},$$

数 y 取值为 1 同时 X_1 取值为 0 的次数共 3 次，故

$$P(I_{11} \mid E_2) = \frac{P(I_{11}E_2)}{P(E_2)} = \frac{3}{17}.$$

为计算时查阅方便，将所求出的先验概率与各条件概率列述表 4-1 内.

由于只考虑两类事件，信息只取 I_1, I_2, I_3，故由(3.6)可知预报的后验概率只需用下式计算

$$P(E_i \mid I_1, I_2, I_3)$$
$$= \frac{P(E_i)P(I_1 \mid E_i)P(I_2 \mid E_i)P(I_3 \mid E_i)}{\sum_{j=1}^{n} P(E_j)P(I_1 \mid E_j)P(I_2 \mid E_j)P(I_3 \mid E_j)},$$
$$j = 1, 2. \tag{4.1}$$

表 4-1

	$P(E_i)$	因子 x_1 $P(I_{11} \mid E_i)$ $P(I_{12} \mid E_i)$	因子 x_2 $P(I_{21} \mid E_i)$ $P(I_{22} \mid E_i)$	因子 x_3 $P(I_{31} \mid E_i)$ $P(I_{32} \mid E_i)$
E_1	$\frac{21}{38} = 0.55$	$\frac{19}{21} = 0.90$ $\frac{2}{21} = 0.10$	$\frac{21}{21} = 1$ $\frac{0}{21} = 0$	$\frac{18}{21} = 0.86$ $\frac{3}{21} = 0.14$
E_2	$\frac{3}{17} = 0.45$	$\frac{3}{17} = 0.18$ $\frac{14}{17} = 0.82$	$\frac{5}{17} = 0.30$ $\frac{12}{17} = 0.70$	$\frac{4}{17} = 0.24$ $\frac{14}{17} = 0.76$

注 表 4-1 中的数字随着资料的增加，要作相应的调整.

§5. 效果检验

(i) 历史资料拟合检验

将逐个地震资料根据各因子的取值情况，经(4.1)计算结果表明震级 $M_s \geq 5.4$ 共 17 次，其中报对 15 次，2 次错报，历史拟合为 $\frac{15}{17}$，震级 $4\frac{3}{4} \leq M_s < 5.4$ 共 21 次其中报对 20 次，错报 1 次，历史拟合为 $\frac{20}{21}$．

(ii) 后验资料的检验

取作后验的地震资料共 5 次，其中 4 次震级为 $4\frac{3}{4} \leq M_s < 5.4$，1 次震级为 $M_s \geq 5.4$ 均报对．现以较近的两次为例说明：

例1 1973 年 12 月 31 日河北沧州地区 $\begin{pmatrix} 38.4N \\ 116.6E \end{pmatrix}$ 发生 5.1 级地震，根据(一)文中的方法预报，1974 年 12 月 21 日至 1975 年 1 月 20 日华北某区有 $\geq 4\frac{3}{4}$ 级地震发生，现按起报时间为 12 月 21 日之前 120 天内所选 15 个小区落震的分布情况（见表 5-1）作出震级分类预报．其余未列入的各区均无震落入．

表 5-1

地区号	1		4	
	①	②	①	②
时间	1973-08-11	1973-09-30	1973-08-02	1973-09-09
震级(M_s)	$6\frac{1}{4}$	5	5.5	$5\frac{3}{4}$
经纬度$\left(\frac{N}{E}\right)$	$\begin{pmatrix}32.8\\104\end{pmatrix}$	$\begin{pmatrix}37.9\\106.3\end{pmatrix}$	$\begin{pmatrix}27.5\\104.5\end{pmatrix}$	$\begin{pmatrix}31.5\\99.7\end{pmatrix}$

续表

地区号	8	12
时间	1973-09-08	1973-09-29
震级(M_s)	6	8
经纬度$\left(\dfrac{N}{E}\right)$	$\begin{pmatrix}33.1\\85.5\end{pmatrix}$	$\begin{pmatrix}42\\135.3\end{pmatrix}$

按 §2 中提取因子的方法根据表 5-1 可以判断 x_1, x_2, x_3 均取值为 0.

故由表 4-1 查得

$$P(E_1)=0.55, \quad P(I_{11}\mid E_1)=0.90,$$
$$P(I_{21}\mid E_1)=1, \quad P(I_{31}\mid E_1)=0.86,$$
$$P(E_2)=0.45, \quad P(I_{11}\mid E_2)=0.18,$$
$$P(I_{21}\mid E_2)=0.30, \quad P(I_{31}\mid E_2)=0.24,$$

所以

$$P(E_1)P(I_{11}\mid E_1)P(I_{21}\mid E_1)P(I_{31}\mid E_1)$$
$$=0.55\times0.90\times1\times0.86=0.425\ 7,$$
$$P(E_2)P(I_{11}\mid E_2)P(I_{21}\mid E_2)P(I_{31}\mid E_2)$$
$$=0.45\times0.18\times0.30\times0.24=0.005\ 8.$$

代入公式(4.1)得

$$P(E_1\mid I_{11}, I_{21}, I_{31})=\frac{0.425\ 7}{0.425\ 7+0.005\ 8}=\frac{0.425\ 7}{0.431\ 5}=0.987,$$

$$P(E_2\mid I_{11}, I_{21}, I_{31})=\frac{0.005\ 8}{0.425\ 7+0.005\ 8}=\frac{0.005\ 8}{0.431\ 5}=0.013,$$

故预报该震震级应为 $4\dfrac{3}{4}\leqslant M_s<5.4$，与实际相符.

例2 1975 年 2 月 4 日辽宁海城 $\begin{pmatrix}40.9\text{N}\\122.7\text{E}\end{pmatrix}$ 发生 7.3 级地震，根据 §1 中方法预报 1975 年 2 月 1 日至 1975 年 3 月 20 日华北某区有 $\geqslant 4\dfrac{3}{4}$ 级地震发生，现按起报时间 2 月 1 日之前 120

天内所选 15 个小区落震的分布情况（见表 5-2）作出震级分类预报．其余各区均无震落入，按 §2 中提取因子方法，根据表 5-2 可以判断 x_1，x_2，x_3，均取值为 1，事实上，对因子 x_1 而言它满足其第(ii)项；对因子 x_2 而言它满足其第(i)项；对因子 x_3 而言它满足其第(i)项．

表 5-2

地区号	2	4		
		①	②	③
时　　间	1975-01-04	1975-01-12	1975-01-15	1975-01-22
震级(M_s)	5.4	5	6.2	5
经纬度$\left(\dfrac{N}{E}\right)$	$\begin{pmatrix}38.8\\97.3\end{pmatrix}$	$\begin{pmatrix}24.8\\101.5\end{pmatrix}$	$\begin{pmatrix}29\\101.6\end{pmatrix}$	$\begin{pmatrix}26.6\\103\end{pmatrix}$

地区号	5	8	13
时　　间	1975-01-14	1975-01-19	1975-02-02
震级(M_s)	5	7.1	7.2
经纬度$\left(\dfrac{N}{E}\right)$	$\begin{pmatrix}43.5\\86.8\end{pmatrix}$	$\begin{pmatrix}31.9\\79.2\end{pmatrix}$	$\begin{pmatrix}52\\175\end{pmatrix}$

故由表 4-1 查得

$P(E_1)=0.55$，$P(I_{12}\mid E_1)=0.10$，$P(I_{22}\mid E_1)=0$，
$$P(I_{32}\mid E_1)=0.14,$$
$P(E_2)=0.45$，$P(I_{12}\mid E_2)=0.82$，$P(I_{22}\mid E_2)=0.70$，
$$P(I_{32}\mid E_2)=0.76,$$

所以　　　$P(E_1)P(I_{12}\mid E_1)P(I_{22}\mid E_1)P(I_{32}\mid E_1)$
$$=0.55\times 0.10\times 0\times 0.14=0,$$
$\quad\quad\quad P(E_2)P(I_{12}\mid E_2)P(I_{22}\mid E_2)P(I_{32}\mid E_2)$
$$=0.45\times 0.82\times 0.70\times 0.76=0.196\,3.$$

代入公式(4.1)得

$\quad\quad P(E_1\mid I_{12},I_{22},I_{32})=0,\quad\quad P(E_2\mid I_{12},I_{22},I_{32})=1.$

故预报该震震级应为 $M_s \geqslant 5.4$，与实际相符.

(iii) 实报一例

1975 年 8 月 28 日辽宁海城 $\begin{pmatrix} 40.8\text{N} \\ 122.7\text{E} \end{pmatrix}$ 发生 5.2 级地震，之前我组曾利用多种统计预报方法综合会商对该震作出了预报. 现仅以本文提出的预报方法说明如何对该震的震级作出分类预报的.

根据 §1 中方法提出 1975 年 5 月 30 日至 1975 年 9 月 11 日华北有 $M_s \geqslant 4\frac{3}{4}$ 地震发生，同样查起报时间之前 120 天内 15 个小区落震的分布情况（见表 5-3）其余各区因无震故未列入. 对照 §2 中提取因子的方法考察表 5-3，容易得出因子 x_1, x_2, x_3，均取值为 0. 仿例 1 计算即可预报将要发生的地震的震级在 $4\frac{3}{4}$ 至 5.4 之间，结果与实际吻合.

表 5-3

地区号	3	4	
时　　间	1975-03-18	1975-03-08	1975-03-16
震级(M_s)	5	5	5
经纬度$\left(\frac{\text{N}}{\text{E}}\right)$	$\begin{pmatrix} 25.5 \\ 99.4 \end{pmatrix}$	$\begin{pmatrix} 28.2 \\ 104.2 \end{pmatrix}$	$\begin{pmatrix} 29.4 \\ 101.9 \end{pmatrix}$
地区号	5			
	①	②	③	④
时　　间	1975-03-31	1975-04-28	1975-05-27	1975-06-04
震级(M_s)	6.0	6.2	5	6.1
经纬度$\left(\frac{\text{N}}{\text{E}}\right)$	$\begin{pmatrix} 46.5 \\ 96.9 \end{pmatrix}$	$\begin{pmatrix} 35.7 \\ 78.4 \end{pmatrix}$	$\begin{pmatrix} 40.5 \\ 77.6 \end{pmatrix}$	$\begin{pmatrix} 37.2 \\ 79.9 \end{pmatrix}$

续表

地区号	7	8	
		①	②
时　　间	1975-03-19	1975-05-05	1975-05-30
震级(M_s)	6.4	6.6	6.3
经纬度$\left(\dfrac{N}{E}\right)$	$\begin{pmatrix}35.3\\86.3\end{pmatrix}$	$\begin{pmatrix}32.8\\92.9\end{pmatrix}$	$\begin{pmatrix}26.5\\96\end{pmatrix}$
地区号	11	12	15
时　　间	1975-06-14	1975-06-19	1975-03-23
震级(M_s)	6.9	7.1	7
经纬度$\left(\dfrac{N}{E}\right)$	$\begin{pmatrix}48.9\\149.5\end{pmatrix}$	$\begin{pmatrix}44\\149.5\end{pmatrix}$	$\begin{pmatrix}22.4\\121.8\end{pmatrix}$

附注　1,2 区——甘肃、宁夏、内蒙古等部分地区．

　　　3,4 区——四川、云南大部分地区．

　　　5,6 区——新疆、青海等部分地区．

　　　7,8 区——西藏部分地区．

　　　9 区——华中、华南个别部分．

　　　10 区——东北部分地区．

　　11,12 区——日本、千岛群岛部分地区．

　　　13 区——阿留申、白令海峡部分地区．

　　14,15 区——我国台湾部分地区．

参考文献

[1] 王梓坤,吴荣. 华北地区地震的统计预报（一）. 南开大学学报（自然科学版），1977，(1)：8-15.

[2] WinKler R L, Murphy A H. Probability forecasting：The aggregation of information. International Symposium on Probability and Statistics in the Atmospheric Sciences，1971：83-89.

中断生灭过程的构造[①]

§1.

本文较[2]深入之处在于: 允许生灭过程中断的情形下, 对 $S<+\infty$ 和 $S=+\infty$ 两种情形的构造问题作了统一处理, 构造了全部(包括中断)生灭过程.

设 $E=\mathbf{N}$, $X(\omega)=\{x(t,\omega), t<\sigma(\omega)\}$ 或简记 $X=\{x(t), t<\sigma\}$ 是定义在完备概率空间 $(\Omega, \mathscr{F}, p)$ 上, 以 E 为其状态空间的可分 Borel 可测右下半连续(在 $E\cup\{+\infty\}$ 中)的时间齐次连续参数马氏过程, 其转移概率矩阵 $p(t)=\{p_{ij}(t), i,j\in E\}$ $(t\geqslant 0)$ 是一组满足下列条件的实值函数. 对任意 $S, t\geqslant 0$,

$$p_{ij}(t)\geqslant 0, \quad \sum_j p_{ij}(t)\leqslant 1, \tag{1.1}$$

$$p_{ik}(s+t)=\sum_j p_{ij}(s)p_{jk}(t), \tag{1.2}$$

[①] 收稿日期: 1974-02-20; 收简报稿日期: 1977-03-31.
本文与杨向群合作.

$$\lim_{t\downarrow 0} p_{ij}(t) = p_{ij}(0) = \delta_{ij} = \begin{cases} 1, & i=j, \\ 0, & i\neq j. \end{cases} \quad (1.3)$$

求和 $j \in E$. 当两过程有相同转移概率矩阵时，我们视它们为同一过程，故又称 $p(t)$ 为过程. 因 (1.1)~(1.3)，存在 Q 矩阵 $\boldsymbol{Q}=(q_{ij})$：

$$\lim_{t\downarrow 0} \frac{p_{ij}(t)-p_{ij}(0)}{t} = q_{ij}, \quad (1.4)$$

又称 X 或 $p(t)$ 为 Q 过程. 当 $p(\sigma=+\infty)=1$ 或 (1.1) 第二式对一切 i 成立等号时，称过程为不中断的，否则称为中断的.

Q 矩阵满足

$$0 \leqslant q_{ij} < +\infty, (i\neq j), \quad \sum_{j\neq i} q_{ij} \leqslant -q_{ii} \equiv q_i \leqslant +\infty.$$

$$(1.5)$$

特别当

$$q_{i,i+1}=b_i>0, \quad q_{i,i-1}=a_i>0, \quad (1.6)$$
$$q_i=a_i+b_i, \quad q_{ij}=0, \quad (|i-j|>1)$$

(补定义 $a_0=0$) 时 Q 过程称为生灭过程. Q 过程的构造问题是：给定满足 (1.5) 的 Q，求出一切 Q 过程. 本文对形如 (1.6) 的 Q，求出了一切 Q 过程.

对于中断过程 X，我们补定义 $x(t)=-1 (t\geqslant\sigma)$，则 $\hat{X}=\{x(t), t<+\infty\}$ 是状态空间为 $\hat{E}=\{-1\}\cup E$ 的过程 [1, P134 定理 3].

§ 2.

设给定 Q 如 (1.6). 引进下列特征数:

$$z_0 = 0, \quad z_n = 1 + \sum_{k=1}^{n-1} \frac{a_1 a_2 \cdots a_k}{b_1 b_2 \cdots b_k}, \quad z = \lim_{n \to +\infty} z_n;$$

$$y_0 = \frac{1}{b_0}, \quad y_n = \frac{b_1 b_2 \cdots b_{n-1}}{a_1 a_2 \cdots a_{n-1} a_n};$$

$$m_i = \frac{1}{b_i} + \sum_{k=0}^{i-1} \frac{a_i a_{i-1} \cdots a_{i-k}}{b_i b_{i-1} \cdots b_{i-k} b_{i-k-1}} = (z_{i+1} - z_i) \sum_{k=0}^{i} y_k, \quad (i \geqslant 0);$$

$$e_i = \frac{1}{a_i} + \sum_{k=0}^{+\infty} \frac{b_i b_{i+1} \cdots b_{i+k}}{a_i a_{i+1} \cdots a_{i+k} a_{i+k+1}}, (i > 0);$$

$$N_i = \sum_{j=i}^{+\infty} m_j = (z - z_i) \sum_{j=0}^{i} y_j + \sum_{j=i+1}^{+\infty} (z - z_j) y_j;$$

$$R = N_0 = \sum_{i=0}^{+\infty} (z - z_i) y_i, \quad S = \sum_{i=1}^{+\infty} e_i = \sum_{j=1}^{+\infty} z_j y_j.$$

$$C_{kj} = \begin{cases} 1, & k \leqslant j, \\ \dfrac{z - z_k}{z - z_j}, & k > j. \end{cases}$$

设 X 为 Q 过程, $\tau_1^{(n)}$ 为其第一个飞跃点, 令

$$\beta_1^{(n)} = \begin{cases} \inf\{t: \tau_1^{(n)} \leqslant t < \sigma, \ x(t) \leqslant n\} \\ \sigma, \qquad \text{上集合为空集} \end{cases} \quad (2.1)$$

定理 1 设 $R < +\infty$. X 为非最小 Q 过程. 则存在非负数列 $p, q, r_n (n \geqslant -1)$ 满足

$$\begin{cases} p + q = 1 \text{ 且 } S = +\infty \text{ 时 } q = 0; \\ \text{若 } p = 0, \text{ 则 } r_n = 0 (n \geqslant 0); \\ \text{若 } p > 0, \text{ 则 } 0 < \sum_{n=0}^{+\infty} r_n N_n < +\infty. \end{cases} \quad (2.2)$$

使得
$$v_1^{(n)} = p(x(\beta_1^{(n)}) = j), \quad -1 \leqslant j \leqslant n \tag{2.3}$$
可以表示为
$$\begin{cases} v_j^{(n)} = \dfrac{X_n}{A_n} r_j, & -1 \leqslant j < n \\ v_n^{(n)} = Y_n + \dfrac{X_n}{A_n} \sum_{l=n}^{+\infty} r_l C_{ln}. \end{cases} \tag{2.4}$$

其中 $A_n = \sum_{l=0}^{+\infty} r_l C_{ln}$,

$$X_n = \frac{A_n C_{n0}}{(r_{-1} + A_n) C_{n0} + d}, \quad Y_n = \frac{d}{(r_{-1} + A_n) C_{n0} + d},$$

$$d = \begin{cases} \dfrac{q}{p} A_0, & p > 0, \\ 1, & p = 0. \end{cases} \quad \frac{X_n}{A_n} = \frac{C_{n0}}{r_{-1} C_{n0} + 1}, \quad p = 0.$$

p, q 由 X 唯一决定. 若 $p > 0$, 则 $r_n (n \geqslant -1)$ 除常数因子外由 X 唯一决定. 若 $p = 0$, 则 $r_n (n \geqslant -1)$ 由 X 唯一决定.

我们称非负数列 $p, q, r_n (n \geqslant -1)$ 为过程 X 的特征数列.

定理 2 设 X 是具特征数列 $p, q, r_n (n \geqslant -1)$ 的 Q 过程, $p_{ij}(t)$ 为其转移概率, 则对任何 $\lambda > 0$,

$$\int_0^{+\infty} e^{-\lambda t} p_{ij}(t) dt$$
$$= f_{ij}(\lambda) + \xi_i(\lambda) \frac{\sum_{k=0}^{+\infty} r_k f_{kj}(\lambda) + dz \xi_j(\lambda) y_j}{r_{-1} + \sum_{k=0}^{+\infty} r_k (1 - \xi_k(\lambda)) + dz \sum_{k=0}^{+\infty} \xi_k(\lambda) y_k}, \tag{2.5}$$

其中 $\quad f_{ij}(\lambda) = \int_0^{+\infty} e^{-\lambda t} \overline{p}_{ij}(t) dt, \quad \xi_i(\lambda) = 1 - \lambda \sum_k f_{ik}(\lambda),$
$$\tag{2.6}$$

而 $\overline{p}(t) = \{\overline{p}_{ij}(t)\}$ 是最小 Q 过程.

基本定理 设 $R<+\infty$.

(i) 设已给可分 Borel 可测右连续(在 $E\cup\{+\infty\}$ 中)非最小 Q 过程 X, 则其特征数列 p, q, $r_n(n\geqslant -1)$ 满足(2.2).

(ii) 设给定一列非负数列 p, q, $r_n(n\geqslant -1)$ 满足(2.2), 则存在唯一的非最小 Q 过程 X, 它以此给数列为其特征数列. 过程 X 的转移概率 $p_{ij}(t)=\lim\limits_{n\to+\infty}p_{ij}^{(n)}(t)$, 这里 $p_{ij}^{(n)}(t)$ 是 $(Q, V^{(n)})$ Doob 过程的转移概率. $V^{(n)}=(V_j^{(n)}:0\leqslant j\leqslant n)$ 由 (2.4) 确定, 且(2.5)成立. X 不中断的充分必要条件是 $r_{-1}=0$; X 满足柯氏向前方程的充分必要条件是 $p=0$.

实际上, 我们得到的结果比上面基本定理更强. 将过程 X 变成 Doob 过程 $X_n\equiv g_n(X)$ 的变换 g_n 见[2].

定理 3 设 $R<+\infty$. X 是可分 Borel 可测右连续非最小 Q 过程, 则过程 $X_n=g_n(X)=\{x_n(t), t<\sigma_n\}$ 是 $(Q, V^{(n)})$ Doob 过程, 满足 $g_n(X_{n+1})=X_n$. 这里 $V^{(n)}=(v_j^{(n)}:0\leqslant j\leqslant n)$ 由 (2.3) 确定, 满足

$$\begin{cases} v_j^{(n)}=\dfrac{v_j^{(n)}}{\Delta_n}, \ (-1\leqslant j<n) \\ v_n^{(n)}=\dfrac{v_n^{(n+1)}+v_{n+1}^{(n+1)}C_{n+1,n}}{\Delta_n}, \\ \sum\limits_{j=-1}^n v_j^{(n)}=1, \ \sum\limits_{j=0}^n v_j^{(n)}>0, \ \Delta_n=\sum\limits_{k=-1}^n v_k^{(n+1)}+v_{n+1}^{(n+1)}C_{n+1,n}. \end{cases} \quad (2.7)$$

X 不中断的充分必要条件是 $v_{-1}^{(n)}=0(n\geqslant 0)$.

定理 4 设 $R<+\infty$; 任给非负数列 p, q, $r_n(n\geqslant -1)$ 满足(2.2), 按(2.4)可定义 $V_j^{(n)}(-1\leqslant j\leqslant n)$. 则

(i) 存在概率空间 $(\Omega, \mathscr{F}, p)$, 在其上可定义一列 $(Q, V^{(n)})$ Doob 过程 $X_n=\{x_n(t), t<\sigma_n\}$ 使适合 $g_n(X_{n+1})=X_n$.

(ii) 存在 $\Omega_0\in\mathscr{F}$, $p(\Omega_0)=1$. 对任 $\omega\in\Omega_0$, 当 $n\to+\infty$ 时

$\sigma_n(\omega)\uparrow\sigma(\omega)$ 存在，而且对几乎一切 $t\in[0,\sigma(\omega))$ [勒贝格测度]，$x_n(t,\omega)$ 存在有穷极限.

(iii) 令 $x(t,\omega)=\lim\limits_{n\to+\infty}x_n(t,\omega)$；若极限不存在，令 $x(t,\omega)=+\infty$. 则 $X=\{x(t),t<\sigma\}$ 是满足(2.3)的唯一 Q 过程，其转移概率 $p_{ij}(t)=\lim\limits_{n\to+\infty}p_{ij}^{(n)}(t)$，其中 $p_{ij}^{(n)}(t)$ 是 X_n 的转移概率.

定理5 设 $R<+\infty$，若令 η 为 $\beta_1^{(0)}$ 前的最后一个飞跃点 [2，§4]，则

$$p(x(\eta)=j)=\begin{cases}\dfrac{r_{-1}}{r_{-1}+A_0+d}, & j=-1,\\[6pt] \dfrac{r_j C_{j0}}{r_{-1}+A_0+d}, & 0\leqslant j<+\infty, \\[6pt] \dfrac{d}{r_{-1}+A_0+d}, & j=+\infty.\end{cases}\quad(2.8)$$

其中 $p,q,r_n(n\geqslant-1)$ 为 X 的特征数列.

定理6 设 $R<+\infty$. τ_n 为 Q 过程 X 的第 n 次跳跃时刻，τ 为第一个飞跃点. ξ_j 为 X 首达 j 的时刻. 对 $0\leqslant\varepsilon\leqslant+\infty$ 作函数

$$f_\varepsilon(x)=\begin{cases}x, & 0\leqslant x\leqslant\varepsilon,\\ \varepsilon, & \varepsilon<x<+\infty.\end{cases}$$

对 $k\geqslant i\geqslant 0$，令

$$H_{ki}^{(\varepsilon)}=E_k\Big\{\sum_{0\leqslant\tau_j<\min(\xi_j,\tau)}f_\varepsilon(\tau_{j+1}-\tau_j)\Big\},\quad(2.9)$$

其中 $E_k\{\ \}$ 表数学期望. 特别，

$$H_{ki}^{(+\infty)}=E_k\{\min(\xi_j,\tau)\},$$

则

$$H_{ki}^{(\varepsilon)}=\frac{z-z_k}{z-z_i}\sum_{j=i+1}^{k}(z_j-z_i)(1-e^{-(a_j+b_j)\varepsilon})y_j+$$

$$\frac{z_k-z_i}{z-z_i}\sum_{j=k+1}^{+\infty}(z-z_j)(1-e^{-(a_j+b_j)\varepsilon})y_j\leqslant N_k,$$

$$\lim_{\varepsilon\downarrow 0}H_{ki}^{(\varepsilon)}=0.$$

又如果 $S<+\infty$, 那么

$$\lim_{k \to +\infty} \frac{H_{ki}^{(\varepsilon)}}{C_{k0}} = \frac{1}{C_{i0}} \sum_{j=i+1}^{+\infty} (z_j - z_i)(1 - e^{-(a_j+b_j)\varepsilon}) y_j,$$

$$\lim_{i \to +\infty} \lim_{k \to +\infty} \frac{H_{ki}^{(\varepsilon)}}{C_{k0}} = 0,$$

$$\lim_{\varepsilon \downarrow 0} \lim_{k \to +\infty} \frac{H_{ki}^{(\varepsilon)}}{C_{k0}} = 0.$$

定理 7 对任何 Q 过程 $X = \{x(t), t<\sigma\}$, $p(\sigma<+\infty) = 0$ 或 1.

§3.

为了证明基本定理,我们用到过程的延拓,它也有独立的意义. 此节的矩阵 Q 不必如(1.6).

设 $\widetilde{p}(t) = \{\widetilde{p}_{ij}(t), i, j \in E\}$,$(t \geqslant 0)$满足(1.1)~(1.3)且过程中断. 任给分布 $\pi = \{\pi_j, j \in E\}$:
$$\pi_j \geqslant 0, \quad 0 < \sum_j \pi_j \leqslant 1,$$
令
$$\widetilde{L}_i(t) = 1 - \sum_j \widetilde{p}_{ij}(t), \quad \widetilde{L} = \sum_i \pi_i \widetilde{L}_i,$$
$$\widetilde{\pi}_j(t) = \sum_i \pi_i \widetilde{p}_{ij}(t), \quad \widetilde{L}^{(0)*}(t) = \begin{cases} 0, & t \leqslant 0, \\ 1, & t > 0. \end{cases}$$
$$\widetilde{L}^{(n+1)*} = \widetilde{L}^{(n)*} * \widetilde{L}, \quad (*\text{表示卷积})$$
$$\widetilde{k}_i = \sum_{n=0}^{+\infty} \widetilde{L}_i * \widetilde{L}^{(n)*},$$
$$p_{ij}(t) = \widetilde{p}_{ij}(t) + \int_0^t \widetilde{\pi}_j(t-s) \mathrm{d}\widetilde{k}_i(s). \quad (3.1)$$

引理 存在概率空间$(\Omega, \mathscr{F}, p)$,在其上可以定义一列过程 $X^{(n)} = \{x^{(n)}(t), t < \sigma^{(n)}\}(n \geqslant 0)$,具有性质:

(i) $X^{(n)}$ 的转移概率为 $\widetilde{p}(t)$;

(ii) 以概率 1,$X^{(n)}$ 可分 Borel 可测右下半连续;

(iii) $(\sigma^{(n)} = 0) \bigcup (\sigma^{(n)} = \infty) \subset (\sigma^{(n+1)} = 0)(n \geqslant 0)$,其中 $A \subset B$ 表示 $p(A \setminus B) = 0$;

(iv) $p(x^{(n+1)}(0) = j \mid 0 < \sigma^{(n)} < +\infty) = \pi_j$,$p(\sigma^{(n+1)} = 0 \mid 0 < \sigma^{(n)} < +\infty) = 1 - \sum_j \pi_j$;

(v) 在条件$(0 < \sigma^{(n)} < +\infty)$或$(x^{(n+1)}(0) = i)$之下,$X^{(m)}$

$(m \leqslant n)$ 与 $X^{(m)}$ $(m > n)$ 条件独立.

定理 8 设定义于 $(\Omega, \mathscr{F}, p)$ 上的一列过程 $X^{(n)} = \{x^{(n)}(t), t < \sigma^{(n)}\}$ $(n \geqslant 0)$ 具有性质 (i)~(v). 令

$$\tau^{(0)} = 0, \quad \tau^{(n+1)} = \sum_{m=0}^{n} \sigma^{(m)}, \quad \sigma = \lim_{n \to +\infty} \tau^{(n)}.$$

对 $0 \leqslant t < \sigma$, 令

$$x(t) = x^{(n)}(t - \tau^{(n)}), \quad \tau^{(n)} \leqslant t < \tau^{(n+1)}, \tag{3.2}$$

那么 $X = \{x(t), t < \sigma\}$ 是可分 Borel 可测右下半连续过程, 其转移概率由 (3.1) 给出. X 不中断的充分必要条件是 $\sum_j \pi_j = 1$.

注 1 定理 8 中的过程 X 的前面一部分是 $X^{(0)}$, 而且满足 $p(x(\tau^{(1)}) = j \mid 0 < \tau^{(1)} < +\infty) = \pi_j$. 因此我们称过程 X (或 $p(t)$) 是中断过程 $X^{(0)}$ (或 $\widetilde{p}(t)$) 的 πD-型延拓过程. 过程与其 D-型延拓过程的 Q 矩阵未必一致.

注 2 假定过程 $\widetilde{p}(t)$ 的 Q 矩阵中一切 $\widetilde{q}_i < +\infty$. 称 σ 为函数的 S-尾, 如 $0 < \sigma < +\infty$ 且存在 $u < \sigma$ 使对一切 $u \leqslant t < \sigma$ 有 $x(t) = x(u)$. 否则称 σ 为函数的 P-尾. 显然若过程 $\widetilde{X} = \{\widetilde{x}(t), t < \widetilde{\sigma}\}$ 有转移概率 $\widetilde{p}(t)$, 则

$M_{ij}(t) = p_i(\widetilde{x}(t) = j, t < \widetilde{\sigma}, \widetilde{\sigma}$ 为 $\widetilde{X}$ 的 P-尾),

$N_{ij}(t) = P_i(\widetilde{x}(t) = j, t < \widetilde{\sigma}, \widetilde{\sigma}$ 为 $\widetilde{X}$ 的 S-尾),

$M_i(t) = \sum_j M_{ij}(t), \quad N_i(t) = \sum_j N_{ij}(t), \quad R_i(t) = 1 - M_i(t)$

等由 $\widetilde{p}(t)$ 唯一决定.

我们将条件 (i)~(v) 修改成 (i*)~(v*).

(i*) 同 (i)(ii) 中将 "下半" 两字去掉得 (ii*) (iii*): $(\sigma^{(n)} = 0) \cup (\sigma^{(n+1)} = +\infty) \cup (\sigma^{(n)}$ 是 $X^{(n)}$ 的 S-尾$) \subset (\sigma^{(n+1)} = 0)$. (iv*) 和 (v*) 只需在 (iv)(v) 中用 $(0 < \sigma^{(n)} < +\infty, \sigma^{(n)}$ 是 $X^{(n)}$ 的 P-尾) 代之以 $(0 < \sigma^{(n)} < +\infty)$ 而得.

定理 8′ 设一切 $\widetilde{q}_i < +\infty$. 定义于同一概率空间的 $\widetilde{Q}$ 过程

具性质$(i^*) \sim (v^*)$. 则按(3.2)确定的 X 是可分 Borel 可测右连续 $\widetilde{Q}$ 过程, 其转移概率由(3.1)(在其中用 $R_i(t)$ 代替 $\widetilde{L}_i(t)$)给出. X 不中断的条件是 $\sum_j \pi_j = 1$ 并且

$$\sum_{j \neq i} \widetilde{q}_{ij} = \widetilde{q}_i (< +\infty). \tag{3.3}$$

称定理 $8'$ 中的过程 X(或 $p(t)$)是过程 $X^{(0)}$, 或 $\widetilde{p}(t)$ 的 πD^*-型延拓过程. 过程与其 D^*-延拓过程的 Q 矩阵相同. 当 $\widetilde{Q}$ 满足(3.3)时, D-型延拓与 D^*-型延拓一致.

注 3 当 $\widetilde{p}(t)$ 是熟知的最小 Q 过程 $\overline{p}(t)$ 时, $\overline{p}(t)$ 的 πD^*-型延拓过程 $p(t)$, 我们称为 (Q, π) Doob 过程, 它与通常的 Doob 过程(例如[1] II §19; [2] 及 [3] 中讨论的 Doob 过程)一致.

参考文献

[1] Chung K L. Markov Chains with Stationary Transition Probabilities. Springer-Verlag, 1960.

[2] 王梓坤. 生灭过程构造论. 数学进展, 1962, 5: 137-170.

[3] Reuter G E H. Denumerable Markov processes and the associated semigroup on I. Acta. Math., 1957, 97: 1-46.

数学物理学报，1978，21(3)

随机激发过程对地极移动的作用

§1. 地极移动的随机微分方程模型

长期的观察证实，地球自转轴的位置相对于地球本体并非固定不变，极点（北极或南极）在地球表面作随机飘移，构成所谓"地极移动"。大致地可将它分解为三种分量：长期极移、周年分量与钱德勒摆动。1891年钱德勒发现极移有一个1.2年的周期。地极移动（特别是钱德勒摆动）与大地震的关系引起了许多研究，但目前还无定论。我国在这些方面也做了不少工作（参看文献[1]）。正因为如此，就更增加了研究地极移动的必要性。在文献[3][4]中提出了钱德勒摆动所满足的随机微分方程，并研究了方程中系数 λ 与 ω 的统计估值问题。本篇的目的是求出这组方程的解，并以此解作为极移的数学模型。然后研究此模型的一些性质，并在此基础上提出了最佳预测公式。

以下所谓地极移动指的都是钱德勒摆动。

以 (ξ_t, η_t) 表示时刻 t 上地极的坐标，并简写为 (ξ, η)。地极移动的方程为

$$d\xi = -\lambda\xi dt - \omega\eta dt + bd\varphi,$$
$$d\eta = \omega\xi dt - \lambda\eta dt + bd\psi, \qquad (1.1)$$

其中 φ, ψ 是时间 t 的随机函数,称为激发函数,$b>0$ 为常数,它表示激发的强度,ω 是钱德勒摆动的角速度而 λ 为阻尼系数. 我们假定 φ_t, ψ_t 是两相互独立的布朗运动(即维纳(Wiener)随机过程),满足条件:

$$Ed\varphi = Ed\psi = 0; \quad E(d\varphi)^2 = E(d\psi)^2 = dt \qquad (1.2)$$

(E 表示数学期望). 因而 $d\varphi(:=\varphi_{t+dt}-\varphi_t)$ 有数学期望为 0、方差为 dt 的正态分布. $d\psi$ 也是如此. 工程或物理上常把方程 (1.1) 直观地理解为

$$\dot{\xi} = -\lambda\xi - \omega\eta + b\dot{\varphi},$$
$$\dot{\eta} = \omega\xi - \lambda\eta + b\dot{\psi},$$

其中 "·" 表示对 t 的随机微商,而 $\dot{\varphi}$ 与 $\dot{\psi}$ 则理解为独立的正态白噪声.

改写方程组 (1.1) 为矩阵形式

$$\begin{pmatrix} d\xi \\ d\eta \end{pmatrix} = \begin{pmatrix} -\lambda & -\omega \\ \omega & -\lambda \end{pmatrix} \begin{pmatrix} \xi \\ \eta \end{pmatrix} dt + \begin{pmatrix} b & 0 \\ 0 & b \end{pmatrix} \begin{pmatrix} d\varphi \\ d\psi \end{pmatrix}, \qquad (1.3)$$

或简写为

$$d\boldsymbol{X} = \boldsymbol{A}\boldsymbol{X}dt + \boldsymbol{B}d\boldsymbol{W}, \qquad (1.4)$$

其中 $\boldsymbol{X} = \boldsymbol{X}_t = \begin{pmatrix} \xi_t \\ \eta_t \end{pmatrix}$, $\boldsymbol{W} = \boldsymbol{W}_t = \begin{pmatrix} \varphi_t \\ \psi_t \end{pmatrix}$, 而 $\boldsymbol{A}$, $\boldsymbol{B}$ 分别为 (1.3) 中两矩阵. 这是二元常系数线性随机微分方程组(在 Ito 意义下). 为解此方程,需先给出起始条件:

$$\boldsymbol{X}_a = \boldsymbol{c}. \qquad (1.5)$$

这里 a 是开始时间,$\boldsymbol{c}$ 为已给的二维随机向量. 我们假定 $\boldsymbol{c}$ 与 $\boldsymbol{W}_t - \boldsymbol{W}_s$ ($a \leqslant s \leqslant t$) 独立. 容易证明:(1.4)(1.5) 的唯一解为

$$\boldsymbol{X}_t = e^{\boldsymbol{A}(t-a)}\left[\boldsymbol{c} + \int_a^t e^{-\boldsymbol{A}(s-a)}\boldsymbol{B}d\boldsymbol{W}_s\right]. \qquad (1.6)$$

事实上，微分(1.6)得

$$d\boldsymbol{X}_t = \boldsymbol{A}e^{\boldsymbol{A}(t-a)}\left[\boldsymbol{c} + \int_a^t e^{-\boldsymbol{A}(s-a)}\boldsymbol{B}d\boldsymbol{W}_s\right]dt + e^{\boldsymbol{A}(t-a)}e^{-\boldsymbol{A}(t-a)}\boldsymbol{B}d\boldsymbol{W}_t$$
$$= \boldsymbol{A}\boldsymbol{X}_t dt + \boldsymbol{B}d\boldsymbol{W}_t.$$

这就是(1.4)，故(1.6)满足(1.4)；又(1.6)满足(1.5)是明显的.

注意(1.6)中的 $\boldsymbol{A}$ 是一矩阵. 为了讨论解的性质，需要求出 $e^{\boldsymbol{A}t}$，这可用矩阵论中的一般方法. 但由于这里的 $\boldsymbol{A}$ 非常简单，故不如用初等方法把它直接计算出来，以便于阅读. 由定义

$$e^{\boldsymbol{A}t} = \boldsymbol{I} + \sum_{n=1}^{+\infty}\frac{(\boldsymbol{A}t)^n}{n!}, \tag{1.7}$$

其中 $\boldsymbol{I} = \begin{pmatrix} 1 & 0 \\ 0 & 1 \end{pmatrix}$ 为恒等矩阵. 因为

$$\boldsymbol{A}t = \begin{pmatrix} -\lambda t & -\omega t \\ \omega t & -\lambda t \end{pmatrix} = -\lambda t \boldsymbol{I} + \omega t \begin{pmatrix} 0 & -1 \\ 1 & 0 \end{pmatrix}.$$

令 $\boldsymbol{D} = \begin{pmatrix} 0 & -1 \\ 1 & 0 \end{pmatrix}$，得

$$e^{\boldsymbol{A}t} = e^{-\lambda t \boldsymbol{I}}e^{\omega t \boldsymbol{D}} = e^{-\lambda t}e^{\omega t \boldsymbol{D}} = e^{-\lambda t}\left(\boldsymbol{I} + \sum_{n=1}^{+\infty}\frac{(\omega t \boldsymbol{D})^n}{n!}\right). \tag{1.8}$$

但容易看出：

$$\boldsymbol{D}^{4n} = \boldsymbol{I}, \qquad \boldsymbol{D}^{4n+1} = \boldsymbol{D},$$
$$\boldsymbol{D}^{4n+2} = -\boldsymbol{I}, \quad \boldsymbol{D}^{4n+3} = -\boldsymbol{D} \quad (n \in \boldsymbol{N}^*).$$

代入(1.8)，即得

$$e^{\boldsymbol{A}t} = e^{-\lambda t}\sum_{n=0}^{+\infty}\left[\frac{(\omega t)^{4n}}{(4n)!}\boldsymbol{I} + \frac{(\omega t)^{4n+1}}{(4n+1)!}\boldsymbol{D} - \frac{(\omega t)^{4n+2}}{(4n+2)!}\boldsymbol{I} - \frac{(\omega t)^{4n+3}}{(4n+3)!}\boldsymbol{D}\right]$$

$$= e^{-\lambda t} \begin{pmatrix} \sum_{n=0}^{+\infty} (-1)^n \dfrac{(\omega t)^{2n}}{(2n)!} & -\sum_{n=0}^{+\infty} (-1)^n \dfrac{(\omega t)^{2n+1}}{(2n+1)!} \\ \sum_{n=0}^{+\infty} (-1)^n \dfrac{(\omega t)^{2n+1}}{(2n+1)!} & \sum_{n=0}^{+\infty} (-1)^n \dfrac{(\omega t)^{2n}}{(2n)!} \end{pmatrix}$$

$$= e^{-\lambda t} \begin{pmatrix} \cos \omega t & -\sin \omega t \\ \sin \omega t & \cos \omega t \end{pmatrix}. \tag{1.9}$$

这就是所要求的结果. 由(1.9)得下列各式：

$$e^{A^T t} = (e^{At})^T = e^{-\lambda t} \begin{pmatrix} \cos \omega t & \sin \omega t \\ -\sin \omega t & \cos \omega t \end{pmatrix}, \tag{1.10}$$

$$e^{(A+A^T)t} = e^{-2\lambda t I} = e^{-2\lambda t} \boldsymbol{I}, \tag{1.11}$$

$$e^{-At} = e^{\lambda t} \begin{pmatrix} \cos \omega t & \sin \omega t \\ -\sin \omega t & \cos \omega t \end{pmatrix}, \tag{1.12}$$

其中 $\boldsymbol{B}^T$ 表示矩阵 $\boldsymbol{B}$ 的转置矩阵. 以(1.9)及(1.12)代入(1.6)展开，并利用三角函数的和角公式，最后得(1.4)(1.5)的解为

$$\xi_t = e^{-\lambda(t-a)}[c_1 \cos \omega(t-a) - c_2 \sin \omega(t-a)] +$$
$$b\int_a^t e^{-\lambda(t-s)}[\cos \omega(t-s) d\varphi_s - \sin \omega(t-s) d\psi_s],$$
$$\eta_t = e^{-\lambda(t-a)}[c_1 \sin \omega(t-a) + c_2 \cos \omega(t-a)] +$$
$$b\int_a^t e^{-\lambda(t-s)}[\sin \omega(t-s) d\varphi_s + \cos \omega(t-s) d\psi_s], \tag{1.13}$$

其中 c_1, c_2 是开始向量的两个分量, $\boldsymbol{c} = \begin{pmatrix} c_1 \\ c_2 \end{pmatrix}$.

§2. 地极移动模型的概率性质

我们既然求出了地极坐标(ξ_t, η_t)的表达式(1.6)或(1.13),就可通过此式来讨论地极移动模型的一些性质.

(i) 由(1.6)或(1.13)可见,此题由两部分构成,第一部分即$e^{A(t-a)} \cdot c$系由开始向量c所引起,它随$t \to +\infty$而指数型地趋于0,即c的影响逐渐消失;第二部分即$e^{A(t-a)} \int_a^t e^{-A(s-a)} B dW_s$,它是由时间$[a, t]$中的布朗运动$W_s$所激励而产生的. 由假定$c$与$W_u - W_v (a \leqslant v \leqslant u)$独立,故这两部分也是独立的.

(ii) 由于(1.1)是常系数线性方程组,它的解只依赖于$t=a$时的位置c,而不依赖于时刻a以前地极的位置. 因而地极移动模型是无后效的. 或者说,无记忆的. 即已知地极现在的位置c时,它将来的位置与过去的位置是独立的. 从数学上说,$X_t \equiv \begin{pmatrix} \xi_t \\ \eta_t \end{pmatrix}$构成二维齐次马尔可夫随机过程,所谓齐次马尔可夫过程的定义是:对任意$n+1$个时刻$0 \leqslant s_1 \leqslant s_2 < \cdots < s_n < s_n + t$,任意$n$个二维向量$x_1, x_2, \cdots, x_n$,在已知$X_{s_i} = x_i (i=1, 2, \cdots, n)$的条件下,$X_{s_n+t}$的条件概率分布,重合于在$X_0 = x_n$的条件下,$X_t$的条件概率分布.

(iii) 如果取开始向量c有二维正态分布,或取c为非随机的二维常值向量(因而可视c为有退化的二维正态分布),那么解的第一部分是正态分布的. 由于$W_u - W_v (a \leqslant v < u)$有正态分布,作为它们的线性组合的极限,第二部分也有正态分布. 由于这两部分独立,故此时解X_t也有正态分布,这意味着地极移动模型是二维正态过程. 此过程的分布显然依赖于c的分布.

下面指出应如何合理地选取 c 的分布.

(iv) 试求出 $\boldsymbol{X}_t$ 的转移概率分布. 设 $X_0 = x$, $\boldsymbol{x} = \begin{pmatrix} x_1 \\ x_2 \end{pmatrix}$ 为任意二维常值向量. 在此条件下, 由(1.6), $\boldsymbol{X}_t$ 可表示为

$$\boldsymbol{X}_t = e^{At}\left(x + \int_0^t e^{-As}\boldsymbol{B}d\boldsymbol{W}_s\right), \tag{2.1}$$

而(2.1)中的 $\boldsymbol{X}_t$ 的分布即所要求的转移概率分布. 由(iii)中所述, 此 $\boldsymbol{X}_t$ 有二维正态分布, 故为求出此分布, 只要求出数学期望 $E\boldsymbol{X}_t$ 及方差矩阵 $D\boldsymbol{X}_t$. 由(2.1)

$$E\boldsymbol{X}_t = e^{At}x + e^{At} \cdot E\int_0^t e^{-As}\boldsymbol{B}d\boldsymbol{W}_s, \tag{2.2}$$

由 Itô 积分的性质, 后一项为 0, 故在 $X_0 = x$ 的条件下,

$$Ex_t = e^{At}x. \tag{2.3}$$

利用(1.9), 上式的分量表示为

$$E\xi_t = e^{-\lambda t}(x_1 \cos \omega t - x_2 \sin \omega t),$$
$$E\eta_t = e^{-\lambda t}(x_1 \sin \omega t + x_2 \cos \omega t). \tag{2.4}$$

其次, 由(2.1)(2.3)及 Itô 积分的性质(见[4, P410])得

$$D\boldsymbol{X}_t = E(\boldsymbol{X}_t - E\boldsymbol{X}_t)(\boldsymbol{X}_t - E\boldsymbol{X}_t)^T$$
$$= E\left[\left(\int_0^t e^{A(t-s)}\boldsymbol{B}d\boldsymbol{W}_s\right)\left(\int_0^t e^{A(t-s)}\boldsymbol{B}d\boldsymbol{W}_s\right)^T\right]$$
$$= E\int_0^t e^{(A+A^T)(t-s)}\boldsymbol{B}\boldsymbol{B}^T ds (\text{由积分性质})$$
$$= b^2\int_0^t e^{(A+A^T)(t-s)}ds = b^2\int_0^t e^{-2\lambda(t-s)\boldsymbol{I}}ds \quad (\text{由}(1.11))$$
$$= b^2\int_0^t e^{-2\lambda(t-s)}ds \cdot \boldsymbol{I} = b^2\frac{1-e^{-2\lambda t}}{2\lambda}\cdot\boldsymbol{I}, \tag{2.5}$$

其中 $\boldsymbol{I} = \begin{pmatrix} 1 & 0 \\ 0 & 1 \end{pmatrix}$. 故自 (x_1, x_2) 出发, 经时间 t 后, 地极转移到点 (y_1, y_2) 附近的转移概率密度为

$$p(x_1, x_2; t, y_1, y_2)$$
$$= \frac{\lambda}{\pi b^2 (1-e^{-2\lambda t})} \exp \frac{-\lambda}{b^2(1-e^{-2\lambda t})} \cdot \{[y_1 - e^{-\lambda t}(x_1 \cos \omega t - x_2 \sin \omega t)]^2 + [y_2 - e^{-\lambda t}(x_1 \sin \omega t + x_2 \cos \omega t)]^2\}. \quad (2.6)$$

从 $\boldsymbol{X}_t$ 的两个分量来看

$$D\xi_t = D\eta_t = b^2 \frac{1-e^{-2\lambda t}}{2\lambda}.$$

而 ξ_t 与 η_t 的相关系数则等于 0，故在 $\xi_0 = x_1$，$\eta_0 = x_2$ 的条件下，两分量 ξ_t, η_t 在同一时刻 t 上是独立的.

(v) 在(2.6)中令 $t \to +\infty$，得

$$\lim_{t \to +\infty} p(x_1, x_2; t, y_1, y_2) = \frac{\lambda}{\pi b^2} \exp \frac{-\lambda}{b^2}(y_1^2 + y_2^2). \quad (2.7)$$

右边与起点 (x_1, x_2) 无关，这说明极移模型具有遍历性：当 $t \to +\infty$ 时，它来到 (y_1, y_2) 附近的概率，与开始时从那一点出发是无关的. (2.7)中的极限分布是数学期望为 $\boldsymbol{0}$、方差矩阵为 $\frac{b^2}{2\lambda}\boldsymbol{I}$ 的二维正态分布，记为 $N\left(\boldsymbol{0}, \frac{b^2}{2\lambda}\boldsymbol{I}\right)$，并称它为平稳分布.

(vi) 由于地极移动已进行了漫长的时间，故可认为已到达此平稳分布状态. 因此，为研究今后的地极移动，可合理地取开始分布(即上述开始向量 $\boldsymbol{c}$ 的分布)为 $N\left(\boldsymbol{0}, \frac{b^2}{2\lambda}\boldsymbol{I}\right)$.

今证明，若设 $\boldsymbol{X}_0 = \boldsymbol{c}$ 的分布为 $N\left(\boldsymbol{0}, \frac{b^2}{2\lambda}\boldsymbol{I}\right)$，而且 $\boldsymbol{c}$ 与 $\boldsymbol{W}_u - \boldsymbol{W}_v (0 \leqslant v < u)$ 独立，则极移模型是平稳过程. 为此，只要证明 $\boldsymbol{X}_t$ 的一、二阶矩不随 t 而变. 事实上，由(1.6)得

$$E\boldsymbol{x}_t = e^{\boldsymbol{A}t} \cdot E\boldsymbol{c} = 0. \quad (2.8)$$

其次，由(1.6)得

$$E\boldsymbol{X}_t \boldsymbol{X}_t^{\mathrm{T}} = E\left[\left(e^{\boldsymbol{A}t}\boldsymbol{c} + \int_0^t e^{\boldsymbol{A}(t-s)} \boldsymbol{B} d\boldsymbol{W}_s\right)\left(e^{\boldsymbol{A}t}\boldsymbol{c} + \int_0^t e^{\boldsymbol{A}(t-s)} \boldsymbol{B} d\boldsymbol{W}_s\right)^{\mathrm{T}}\right]$$
$$= \boldsymbol{E}_1 + \boldsymbol{E}_2 + \boldsymbol{E}_3 + \boldsymbol{E}_4, \quad (2.9)$$

其中

$$E_1 = E[(e^{At}c)(e^{At}c)^T] = e^{At} \cdot Ecc^T \cdot e^{A^T t}$$
$$= e^{At} \cdot \frac{b^2}{2\lambda}I \cdot e^{A^T t} = \frac{b^2}{2\lambda} \cdot e^{-2\lambda t}I \quad (\text{用到}(1.11));$$

$$E_2 = E\Big[(e^{At}c)\Big(\int_0^t e^{A(t-s)}B dW_s\Big)^T\Big],$$

但由假设，c 与 $W_u - W_v (0 \leq v < u)$ 独立，故方括号中两因子也独立. 又因 $Ee^{At}c = e^{At}Ec = 0$，故

$$E_2 = e^{At}Ec \cdot E\Big(\int_0^t e^{A(t-s)}B dW_s\Big)^T = \mathbf{0}.$$

同理

$$E_3 = E\Big(\int_0^t e^{A(t-s)}B dW_s\Big)(e^{At}c)^T = \mathbf{0}.$$

$$E_4 = E\Big(\int_0^t e^{A(t-s)}B dW_s\Big)\Big(\int_0^t e^{A(t-s)}B dW_s\Big)^T$$
$$= \int_0^t e^{A(t-s)} \cdot b^2 I \cdot e^{A^T(t-s)} ds \quad (\text{由积分性质})$$
$$= b^2 I e^{(A+A^T)t} \int_0^t e^{-(A+A^T)s} ds$$
$$= b^2 \cdot e^{-2\lambda t} \cdot \int_0^t e^{2\lambda s} ds \cdot I \quad (\text{由}(1.11))$$
$$= \frac{b^2}{2\lambda}(1 - e^{-2\lambda t}) \cdot I. \qquad (2.10)$$

以 E_1, E_2, E_3, E_4 的值代入(2.9)得

$$EX_t X_t^T = \frac{b^2}{2\lambda}I. \qquad (2.11)$$

下面计算相关函数. 对 $\tau > 0$，由(1.6)得

$$X_{t+\tau} = e^{A\tau}\Big(X_t + \int_t^{t+\tau} e^{-A(s-t)} dW_s\Big),$$

故

$$EX_t X_{t+\tau}^T = EX_t X_t^T e^{A^T \tau} + EX_t\Big(e^{A\tau}\int_t^{t+\tau} e^{-A(s-t)} dW_s\Big)^T,$$

但 X_t 与 $\int_t^{t+\tau} e^{-A(s-t)} dW_s$ 独立, 而且 $E\int_t^{t+\tau} e^{-A(s-t)} dW_s = 0$, 故由此及 (2.11)(1.10) 得

$$EX_t X_{t+\tau}^T = EX_t X_t^T e^{A^T \tau} = \frac{b^2}{2\lambda} e^{-\lambda \tau} \begin{pmatrix} \cos\omega\tau & \sin\omega\tau \\ -\sin\omega\tau & \cos\omega\tau \end{pmatrix} \quad (\tau > 0).$$

两边取转置矩阵, 得

$$EX_{t+\tau} X_t^T = \frac{b^2}{2\lambda} e^{-\lambda\tau} \begin{pmatrix} \cos\omega\tau & -\sin\omega\tau \\ \sin\omega\tau & \cos\omega\tau \end{pmatrix} = EX_t X_{t-\tau}^T \quad (\tau > 0).$$

综合此两式, 即知不论 τ 的正负, 恒有

$$EX_t X_{t+\tau}^T = \frac{b^2}{2\lambda} e^{-\lambda|\tau|} \begin{pmatrix} \cos\omega\tau & \sin\omega\tau \\ -\sin\omega\tau & \cos\omega\tau \end{pmatrix}. \quad (2.12)$$

右边与 t 无关, 故 $\{X_t, t \geq 0\}$ 是平稳过程; 由于它是正态的, 故它的有穷维分布也是平稳的.

(vii) 强相关性与钱德勒周期.

由 (2.12) 可见坐标分量 ξ_t (或 η_t) 的自相关函数 $R(\tau)$:

$$R(\tau) \equiv E\xi_t \xi_{t+\tau} = \frac{b^2}{2\lambda} e^{-\lambda|\tau|} \cdot \cos\omega\tau$$

$$= E\eta_t \eta_{t+\tau}. \quad (2.13)$$

注意 $E\xi_t = E\eta_t = 0$, 如图 2-1. 当 $\tau = \frac{2k\pi}{\omega}(k \in \mathbf{N})$ 时, $R(\tau)$ 达到局部极大.

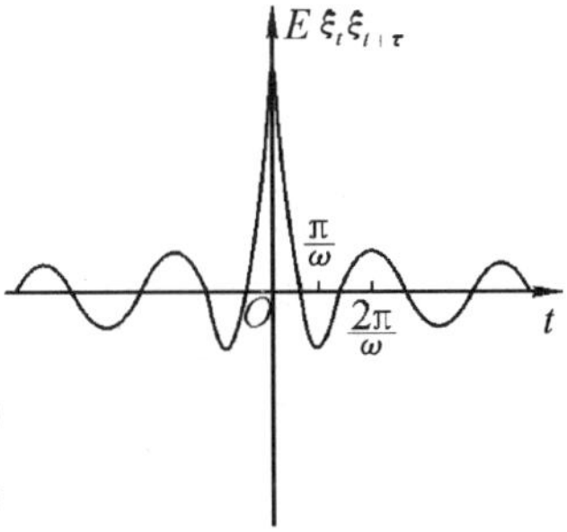

图 2-1

这表示有一 "相关" 周期为 $\frac{2\pi}{\omega}$, 每过一周期, 强正相关一次. 又当 $\tau = \frac{(2k+1)\pi}{\omega}(k \in \mathbf{N})$ 时, $R(\tau)$ 达到局部极小, 其周期也是 $\frac{2\pi}{\omega}$, 每过一周期, 强负相关一次.

在 [4] 中利用 60 年间极移的观察资料, 在除去长期极移与

周年分量后，求得 ω 与 λ 的极大似然估值为
$$\omega\approx 5.274, \quad \lambda\approx 0.06.$$
从而 $\frac{2\pi}{\omega}\approx 1.2$ 年(约 14 个月)，这与钱德勒周期符合.

(viii)"独立性"周期. 从理论上分析还可以发现另一新的周期 $\frac{\pi}{\omega}$，它是上述相关周期的一半. 如果采用上面 ω 的估值 5.274，那么它约等于 0.6 年，即约 7 个月，也是钱德勒周期的一半. 每过此周期，坐标分量间即具有概率独立性. 事实上，由(2.13)可见当
$$\tau=\frac{(2k+1)\pi}{2\omega}\ (k\in\mathbf{N}) \text{时}, R(\tau)=E\xi_t\xi_{t+\tau}=0.$$
这表示诸随机变量 $\xi_t, \xi_{t+\frac{\pi}{2\omega}}, \xi_{t+\frac{3\pi}{2\omega}}, \xi_{t+\frac{5\pi}{2\omega}}, \cdots$ 是两两不相关的(t 任意固定)，由于它们有正态分布，故也是两两独立的. 注意这些 τ 值中相邻两个的距离是 $\frac{\pi}{\omega}$，这就是上述独立性周期. 对 $\{\eta_t\}$ 也有同样结果.

类似地，考虑互相关函数
$$E\xi_t\eta_{t+\tau}=\frac{b^2}{2\lambda}\mathrm{e}^{-\lambda|\tau|}\sin\omega\tau=-E\eta_t\xi_{t+\tau}. \quad (2.14)$$
与上同样推理，可见：当 $\tau=\frac{(4k+1)\pi}{2\omega}\ (k\in\mathbf{N})$ 时 ξ_t 与 $\eta_{t+\tau}$ 强正相关，周期为 $\frac{2\pi}{\omega}$；当 $\tau=\frac{(4k-1)\pi}{2\omega}\ (k\in\mathbf{N}^*)$ 时，ξ_t 与 $\eta_{t+\tau}$ 强负相关，周期也是 $\frac{2\pi}{\omega}$；最后，当 $\tau=\frac{k\pi}{\omega}\ (k\in\mathbf{N})$ 时，ξ_t 与 $\eta_{t\pm\tau}$ 两两独立，这里 t 为任意固定的实数.

两坐标分量在不同的 t 上有上述种种独立性，周期都是 $\frac{\pi}{\omega}$，是颇令人感到奇怪的.

§3. 地极移动模型的预测问题

试根据时刻 t 以前的全部观察值 $\boldsymbol{X}_s \equiv \begin{pmatrix} \xi_s \\ \eta_s \end{pmatrix}$ $(s \leqslant t)$, 以最佳地预测将来的值 $\boldsymbol{X}_{t+\tau}$ $(\tau > 0)$. 最佳的标准是要求均方误差最小. 由[5, P396]知, $\boldsymbol{X}_{t+\tau}$ 的最佳估值为

$$\overline{\boldsymbol{X}}_{t+\tau} = E(\boldsymbol{X}_{t+\tau} \mid \boldsymbol{X}_s, \ s \leqslant t). \tag{3.1}$$

它是在已知 $(\boldsymbol{X}_s, s \leqslant t)$ 的条件下, $\boldsymbol{X}_{t+\tau}$ 的条件数学期望. 但由于上节所述, $\boldsymbol{X}_t$ 是齐次马尔可夫过程, 上式右方等于 $E(\boldsymbol{X}_{t+\tau} \mid \boldsymbol{X}_t)$. 为求后者, 由(1.6)得

$$\boldsymbol{X}_{t+\tau} = e^{\boldsymbol{A}\tau} \left(\boldsymbol{X}_t + \int_\tau^{t+\tau} e^{-\boldsymbol{A}(s-t)} \boldsymbol{B} d\boldsymbol{W}_s \right), \tag{3.2}$$

与(2.3)的推理一样, 当 $\boldsymbol{X}_t$ 已知时,

$$E(\boldsymbol{X}_{t+\tau} \mid \boldsymbol{X}_t) = e^{\boldsymbol{A}\tau} \boldsymbol{X}_t.$$

由此式及(3.1), 得最佳预测为

$$\overline{\boldsymbol{X}}_{t+\tau} = e^{\boldsymbol{A}\tau} \boldsymbol{X}_t. \tag{3.3}$$

以(1.9)代入此式, 即得分量的最佳预测为

$$\begin{aligned} \bar{\xi}_{t+\tau} &= e^{-\lambda\tau}(\xi_t \cos \omega\tau - \eta_t \sin \omega\tau), \\ \bar{\eta}_{t+\tau} &= e^{-\lambda\tau}(\xi_t \sin \omega\tau + \eta_t \cos \omega\tau). \end{aligned} \tag{3.4}$$

故预测值只指数型地依赖于最近期的一组观察值 (ξ_t, η_t). 试求出预测的均方误差

$$\begin{aligned} \sigma^2(\tau) &\equiv E \mid \boldsymbol{X}_{t+\tau} - \overline{\boldsymbol{X}}_{t+\tau} \mid^2 \\ &= E(\xi_{t+\tau} - \bar{\xi}_{t+\tau})^2 + E(\eta_{t+\tau} - \bar{\eta}_{t+\tau})^2. \end{aligned} \tag{3.5}$$

由(3.2)(3.3)得

$$E(\boldsymbol{X}_{t+\tau} - \overline{\boldsymbol{X}}_{t+\tau})(\boldsymbol{X}_{t+\tau} - \overline{\boldsymbol{X}}_{t+\tau})^\top$$

$$= E\left(\int_t^{t+\tau} e^{A(\tau-s+t)} B dW_s\right)\left(\int_t^{t+\tau} e^{A(\tau-s+t)} B dW_s\right)^T$$

$$= \frac{b^2}{2\lambda}(1 - e^{-2\lambda\tau}) \cdot I. \tag{3.6}$$

由于(3.5)右边等于(3.6)中对角线上的元素之和,故

$$\sigma^2(\tau) = \frac{b^2}{\lambda}(1 - e^{-2\lambda\tau}). \tag{3.7}$$

注意(3.3)或(3.4)不但是最佳线性预测公式,而且在非线性意义下也是最佳的. 这里我们充分地利用了二维过程(ξ_t, η_t)的马尔可夫性. 但若只考虑一个分量如ξ_t,并取其开始分布为期望为0、方差为$\frac{b^2}{2\lambda}$的正态分布,则它也构成一维平稳正态过程,但它不是一维马尔可夫过程. 当然,对$\{\xi_t, t \geq 0\}$也可应用平稳过程的线性预测理论(见[6, P184]),但公式变得相当麻烦.

§4. 小　结

我们从运动方程组(1.1)出发，用数学的方法，证明了极移的随机激发数学模型 X_t 具有二维正态性、无后效性和遍历性. 如果选取(2.7)中的右边为开始分布，那么它还是平稳的. 这就是说，X_t 的二级矩及有穷维分布都不随时间的推移而改变. 此外，研究相关函数后，可以发现 X_t 的两坐标间的关系，例如强相关性或独立性. 在研究了这些性质的基础上，导出了地极移动模型 X_t 的最佳预测公式.

上述结论的基础是方程组(1.1)，因此，问题在于(1.1)与极移运动符合到什么程度. 为此，我们可以检验观察资料是否的确具有上述的独立性周期. 任何数学模型只可能在若干主要方面符合实际，绝不可能穷尽实际的无限丰富性. 我们深信，方程组(1.1)也只能是地极移动的一个近似模型.

从逻辑上看，激发函数应是比较一般的随机过程 φ_t, ψ_t. 但文献[3][4]中从与实际的观察资料对比出发，认为假定它们是独立的布朗运动是有根据的.

在确定了模型以后，接着就是(1.1)中参数的估计问题. 如上所述，[4]中给出 $\lambda \approx 0.06$，$\omega \approx 5.274$，[7]中为 $\lambda \approx 0.067$，$\omega \approx 5.267$，但另一些人则得到显著不同的估值. 在[3]中，每十年计算一次，而十年内则认为它们是常数，例如 1940～1950 年 $\lambda \approx 0.054$，$\omega \approx 5.256$；1950～1960 年及 1960～1970 年，$\lambda \approx 0.022$，0.008；$\omega \approx 5.279$，5.250. 可见 λ 的估值波动较大而 ω 则较稳定. 求得 λ 与 ω 后，利用 $b^2 = \lambda(E\xi_t^2 + E\eta_t^2)$（见(2.13)），即可由观察资料求出 b^2 的估值. 从而看出激发强度是如何变化

的. 如果 λ 每十年而变，那么 b^2 也如此.

我们求出了极移模型的最佳预测公式，是否可通过极移的预测来预测大地震，这是一个值得深入研究的问题.

参考文献

[1] 傅承义. 地球十讲. 北京：科学出版社，1976.

[2] 巴特 M. 地震学的数学问题（郑治真，译）. 北京：科学出版社，1976.

[3] Naosuke Sekiguchi. On some natures of the exitation and damping of the polar motion. Rotation of the Earth, 1972：221-223.

[4] Арато М，Колмогоров А Н，Синай Я Г. Оδ оценке параметров комплексного спационарного Гауссовского марковского процесса. Доклады. Акад. Наук СССР，1962，146(4)：747-750.

[5] 王梓坤. 随机过程论. 北京：科学出版社，1965.

[6] Казакевиу Д И. 随机函数论原理及其在水文气象学中的应用. 章基嘉，译. 北京：科学出版社，1974.

[7] Munk W H，Macdonald G J F. The Rotation of the Earth(地球自转. 李启斌，等译)，北京：科学出版社，1976.

中断生灭过程构造中的概率分析方法[①]

§1. 引　言

设 $X(\omega)=\{x(t,\omega),\ t<\sigma(\omega)\}$ 或简记 $X=\{x(t),\ t<\sigma\}$ 为定义于完备概率空间 $(\Omega,\mathscr{F},P)$ 上的具可列多个状态的时间齐次的连续参数马氏链,或简称过程,相空间为 $E\in\mathbf{N}$,其转移概率矩阵 $\boldsymbol{P}(t)=\{P_{ij}(t),\ i,j\in E\}(t\geqslant 0)$ 是一组满足下列条件的实值函数：对任意 $t,s\geqslant 0$ 及 $i,k\in E$,

$$p_{ij}(t)\geqslant 0,\quad \sum_{j}p_{ij}(t)\leqslant 1; \tag{1.1}$$

$$p_{ik}(s+t)=\sum_{j}p_{ij}(s)p_{jk}(t); \tag{1.2}$$

$$\lim_{t\downarrow 0}p_{ij}(t)=p_{ij}(0)=\delta_{ij}; \tag{1.3}$$

其中求和号遍历一切 $j\in E$,又 $\delta_{ii}=1,\ \delta_{ij}=0(i\neq j).$

由 (1.1)～(1.3) 推知 [1,P126～127],存在极限

[①] 本文与杨向群合作.

$$\lim_{t\downarrow 0}\frac{p_{ij}(t)-p_{ij}(0)}{t}=q_{ij}, \tag{1.4}$$

因此又称 X 为 Q 过程，以表示 X 的转移概率矩阵 $\boldsymbol{P}(t)$ 与矩阵 $\boldsymbol{Q}=(q_{ij})$ 有关系 (1.4)，如果两个 Q 过程有相同的转移概率矩阵，我们视它们为同一 Q 过程。

当 $\sum_j p_{ij}(t)=1$ 对一切 i 成立时，Q 过程称为不中断的，否则称为中断的。

一般地，(1.4) 中 $\boldsymbol{Q}=(q_{ij})$ 满足

$$0\leqslant q_{ij}<+\infty, (i\neq j), \sum_{j\neq i}q_{ij}\leqslant -q_{ii}\equiv q_i\leqslant +\infty, \tag{1.5}$$

但以后我们恒假定

$$q_{ij}\geqslant 0, (i\neq j), \sum_{j\neq i}q_{ij}=-q_{ii}\equiv q_i<+\infty. \tag{1.6}$$

特别地，当

$$\begin{cases} q_{i,i+1}=b_i>0, \ q_{i,i-1}=a_i>0, \\ q_i=-q_{ii}=(a_i+b_i), \ q_{ij}=0, \ |i-j|>1, \end{cases} \tag{1.7}$$

(补定义 $a_0=0$) 时，Q 过程称为生灭过程。

Q 过程的构造问题是：给定满足 (1.5) 的 Q，试求出一切 Q 过程。

为了解决这个问题，不少作者采用各种分析方法 [4，6，8，10~15] 和半群方法 [14，16] 对某些特定的 Q 求出了全部 Q 过程。1958 年王梓坤 [2，3] 研究了解决构造 Q 过程的一种新的概率方法——样本函数逼近法。其基本思想是：用结构比较简单的 Q 过程 (Doob 过程) 来逼近任一 Q 过程，"逼近"是通过过程的样本函数来完成的。这种方法较之分析方法和半群方法的优点是：构造出来的 Q 过程的样本函数结构清楚，概率意义明显，为了研究 Q 过程的性质，可对简单 Q 过程 (Doob 过程) 的

中断生灭过程构造中的概率分析方法

性质进行研究，然后过渡到极限即可．因此，虽然这一方法目前还限于对生灭过程的研究，但它在理论上和实践上都有很重要的潜在价值，例如[5，7，9]．

[2]已用样本函数逼近法构造了一切不中断的生灭过程．在构造时使用了 Doob 过程的两种不同变换来分别处理 $S<+\infty$ 和 $S=+\infty$（见(2.8)）的情形．[2]最后指出，两种情况是否可用同一方法来处理？

本文仍用样本函数逼近法研究生灭过程的构造．在两方面较[2]深入：一方面解决了[2]最后提出的问题，另一方面允许过程中断．就是说，我们对 $s<+\infty$ 和 $S=+\infty$ 同时允许过程中断的情形作了统一处理，从而大大简化了[2]的证明，进一步展示出样本函数逼近法的作用．本文的摘要见[3_1]．

§2. 两个引理

关于中断马氏链 $X=\{x(t), t<\sigma\}$ 的定义见[1, P244]. 显然当且仅当

$$\sum_j p_{ij}(t) = 1 \qquad (2.1)$$

时，$p_i(\sigma=+\infty)=1$. 其中 $p_i(\cdot)=p(\cdot\mid x(0)=i)$，而以后用 E_i 表示关于 p_i 而取的数学期望. 对于中断马氏链 $X=\{x(t), t<\sigma\}$，我们可以任取 $\theta\notin E$，例如取 $\theta=-1$，当 $\sigma<+\infty$ 时，补充定义

$$x(t)=-1, \; t\geqslant\sigma. \qquad (2.2)$$

因此 $x(t)$ 对一切 $t\geqslant 0$ 有定义，并且

$$\sigma=\inf\{t: x(t)=-1\}.$$

[1, P245]指出，$\hat{X}=\{x(t), t<+\infty\}$ 是过程，相空间 $\hat{E}=\{-1\}\cup E$，转移概率矩阵 $\hat{P}(t)=\{\hat{P}_{ij}(t), i, j\in\hat{E}\}(t\geqslant 0)$ 由 [1, P134 定理 3] 给出，即

$$\begin{cases} \hat{p}_{ij}(t)=p_{ij}(t), \; \hat{p}_{-1,j}(t)=0, \; \hat{p}_{-1,-1}(t)=1, \\ \hat{p}_{i,-1}(t)=1-\sum_{j\in E}p_{ij}(t); i,j\in E. \end{cases} \qquad (2.3)$$

以后对任何中断过程 X，我们恒默认 $x(t)$ 按(2.2)而对一切 $t\geqslant 0$ 有定义，因此我们可以应用[1]的结果. 以后我们也恒设 X 因而 $\hat{X}$ 的初始分布是集中于 E 的概率分布，即

$$P(x(0)\in E)=P(0<\sigma)=1 \qquad (2.4)$$

以后如无特殊声明，Q 过程恒指生灭过程. 本文中我们将基本上采用[2]的术语和记号.

设给定形如(1.7)的矩阵 $Q=(q_{ij})$，引进下列特征数：

$$\begin{cases} z_0 = 0, z_n = 1 + \sum_{k=1}^{n-1} \frac{a_1 a_2 \cdots a_k}{b_1 b_2 \cdots b_k}, \\ z_1 = \lim_{n\to\infty} z_n = 1 + \sum_{k=1}^{+\infty} \frac{a_1 a_2 \cdots a_k}{b_1 b_2 \cdots b_k}, \\ y_0 = \frac{1}{b_0}, \quad y_n = \frac{b_1 b_2 \cdots b_{n-1}}{a_1 a_2 \cdots a_{n-1} a_n}; \end{cases} \quad (2.5)$$

$$\begin{cases} m_i = \frac{1}{b_i} + \sum_{k=0}^{i-1} \frac{a_i a_{i-1} \cdots a_{i-k}}{b_i b_{i-1} \cdots b_{i-k} b_{i-k-1}} \\ \qquad = (z_{i+1} - z_i) \sum_{k=0}^{i} y_k, \; i \geqslant 0, \\ e_i = \frac{1}{a_i} + \sum_{k=0}^{+\infty} \frac{b_i b_{i+1} \cdots b_{i+k}}{a_i a_{i+1} \cdots a_{i+k} a_{i+k+1}} \\ \qquad = (z_i - z_{i-1}) \sum_{k=i}^{+\infty} y_k, \; i > 0; \end{cases} \quad (2.6)$$

$$N_i = \sum_{j=i}^{+\infty} m_j = (z - z_i) \sum_{j=0}^{i} y_i + \sum_{j=i+1}^{+\infty} (z - z_j) y_j; \quad (2.7)$$

$$\begin{cases} R = N_0 = \sum_{j=0}^{+\infty} m_j = \sum_{i=0}^{+\infty} (z - z_i) y_i; \\ S = \sum_{i=1}^{+\infty} e_i = \sum_{j=1}^{+\infty} z_j y_j. \end{cases} \quad (2.8)$$

上述特征数的概率意义见[2，§2]，今考虑可分 Borel 可测 Q 过程 $X=\{x(t), t<\sigma\}$。由于可分性的需要，$x(t)$ 可能取值"$+\infty$"，但对任一固定的 $t\geqslant 0$，$p(x(t)=+\infty)=0$。根据[1，Ⅱ §7]，还可设 X 右连续，即在 $E\cup\{+\infty\}$ 中有

$$x(t) = \lim_{s\downarrow t} x(s), \quad \text{对一切 } t<\sigma. \quad (2.9)$$

注意我们默认了 $x(t)$ 按(2.2)而对一切 $t\geqslant 0$ 都有定义，$\hat{X}=\{x(t), t<+\infty\}$ 是 $\hat{Q}$ 过程，$\hat{Q}=(\hat{q}_{ij})(i, j\in \hat{E}=\{-1\}\cup E)$ 为

$$\hat{q}_{ij}=q_{ij}, \quad \hat{q}_{i,-1}=\hat{q}_{-1,j}=\hat{q}_{-1,-1}=0, \quad (i, j\in E). \quad (2.10)$$

实际上，由(2.3)，除 $\hat{q}_{i,-1}=0$ 之外，其余都是明显的. 但由于 $\hat{Q}=(\hat{q}_{ij})$ 应满足(1.5)，即 $\hat{q}_{i,-1}\geq 0$, $\sum\limits_{\substack{j\neq i\\j\in E}}\hat{q}_{ij}+\hat{q}_{i,-1}\leq -\hat{q}_{ii}$，注意(1.6)及(1.7)，必定 $\hat{q}_{i,-1}=0$.

设 ζ 为马尔可夫时间①，满足

$$P(\zeta\leq\sigma)=1, \quad p(x(\zeta)=+\infty)=0. \quad (2.11)$$

设 $\tau_\zeta^{(n)}$ 为 ζ 后的第 n 个跳跃点，$\tau_\zeta=\lim\limits_{n\to+\infty}\tau_\zeta^{(n)}$ 为 ζ 后的第一个飞跃点. 由[1, P125 定理6]，对 $i, j\in E$,

$$P(x(\zeta)=i, \ x(\tau_\zeta^{(1)})=j)=P(x(\zeta)=i)\frac{(1-\delta_{ij})q_{ij}}{q_i}, \quad (2.12)$$

对 $i, j\in E$ 求和便得

$$P(\zeta<\sigma, \ \zeta<\tau_\zeta^{(1)}<\sigma)=P(\zeta<\sigma)$$

将上面的讨论依次应用于马尔可夫时间 $\min(\tau_\zeta^{(1)}, \sigma)$, $\min(\tau_\zeta^{(2)}, \sigma)\cdots\cdots$ 我们便证得下面引理的前一部分.

引理 1 设 $X=\{x(t), t<\sigma\}$ 为可分 Borel 可测右连续 Q 过程.

(i) 设 ζ 为马尔可夫时间，满足(2.11)，则对几乎一切 $\omega\in(\zeta<\sigma)$ 有

$$\zeta(\omega)<\tau_\zeta^{(1)}(\omega)<\cdots<\tau_\zeta^{(n)}(\omega)<\cdots<\lim_{n\to+\infty}\tau_\zeta^{(n)}(\omega)$$
$$=\tau_\zeta(\omega)\leq\sigma(\omega). \quad (2.13)$$

其中 $\tau_\zeta^{(n)}$, τ_ζ 分别为 ζ 后的第 n 个跳跃点和第一个飞跃点.

---

① 即[1, P160]的 Optional random variable. 注意 ζ 以及前 ζ 域 $\mathscr{F}_\zeta$ 和后 ζ 域 $\mathscr{F}_\zeta[1,\text{P163}, \text{P168}]$ 等概念都是关于由 X 按(2.2)而得到的 $\hat{Q}$ 过程 $\hat{X}=\{x(t), t<+\infty\}$ 而言的.

特别地，取 $\zeta\equiv 0$，则对几乎一切 ω[①]，
$$0<\tau_1(\omega)<\cdots<\tau_n(\omega)<\cdots<\lim_{n\to+\infty}\tau_n(\omega)$$
$$=\tau(\omega)\leqslant\sigma(\omega), \qquad (2.14)$$
其中 τ_n，τ 为 X 的第 n 个跳跃点和第一个飞跃点.

（ii）对几乎一切 ω，对任意 $t<\sigma(\omega)$，如果 $x(t-0,\omega)=\lim_{s\uparrow t}x(s,\omega)\in E$，那么 $x(t,\omega)=\lim_{s\downarrow t}x(s,\omega)\in E$[②]，且 $|x(t,\omega)-x(t-0,\omega)|=1$，即 t 是跳跃点，且跃度为 1. 若 $\sigma(\omega)<+\infty$，则 $x(\sigma(\omega)-0,\omega)\equiv\lim_{s\uparrow\sigma(\omega)}x(s,\omega)=+\infty$，即 $\sigma(\omega)$ 是飞跃点.

往证(ii) $\hat{X}=\{x(t),t<\infty\}$ 是 $\hat{Q}$ 过程，$\hat{Q}$ 如 (2.10)(1.7). 设 Λ 为不具(ii)所述性质的全体，则对 $\omega\in\Lambda$，存在 $0<t<+\infty$，使 $x(t-0,\omega)\in E$ 而 $x(t,\omega)=-1$ 或 $+\infty$，或者 $x(t,\omega)\in E$ 但 $|x(t,\omega)-x(t-0,\omega)|>1$. 由于 $x(t-0,\omega)=\lim_{s\uparrow t}x(s,\omega)\in E$，故存在 $\varepsilon>0$，使 $x(r,\omega)=x(t-0,\omega)$ 对一切 $r\in(t-\varepsilon,t)$. 因此若 r 后的第一个跳跃点为 $\tau_r^{(1)}$，则必定

$$\Lambda\subset\bigcup_{\substack{r\geqslant 0\\r\in Q}}\{x(r)\in E \text{ 而 } x(\tau_r^{(1)})=-1 \text{ 或 } +\infty,$$
$$\text{或者 } x(\tau_r^{(1)})\in E \text{ 但 } |x(\tau_r^{(1)})-x(r)|>1\}.$$

依[1，P215 定理6]，上式右边含 r 的项的概率

$$\pi_r=\sum_{i\in E}P(x(r)=i)\left\{\frac{\hat{q}_{i,-1}}{\hat{q}_i}+\left(1-\sum_{\substack{j\neq i\\j\in\hat{E}}}\frac{\hat{q}_{ij}}{\hat{q}_i}\right)+\sum_{\substack{|j-i|>1\\j\in E}}\frac{\hat{q}_{ij}}{\hat{q}_i}\right\}=0,$$

---

① 注意 $P(0<\sigma)=1$，见(2.4)式.

② 依[1，II §6]，以概率 1，在 $E\cup\{\infty\}$ 中的极限 $\lim_{s\uparrow t}x(s)$ 对一切有穷的 $t\leqslant\sigma$ 存在.

故 $P(\Lambda) \leqslant \sum\limits_{\substack{r \geqslant 0 \\ r \in \mathbf{Q}}} \pi_r = 0$,从而 $P(\Lambda) = 0$. 引理 1 得证.

由引理 1,可确定 Q 过程 $X = \{x(t), t < \sigma\}$ 的第 n 个跳跃点 τ_n 和第一个飞跃点 τ,它们满足 (2.14),则 $\bar{X} = \{x(t), t < \tau\}$ 是最小 Q 过程,即它的转移概率矩阵 $\bar{P}(t) = \{\bar{P}_{ij}(t)\}(i, j \in E, t \geqslant 0)$ 是熟知的最小解 [1,Ⅱ§18 及 P245]. 又 [2, P168 定理 1] 指出 $N_i = E_i \tau$,而且 $P(\tau = +\infty) = 1$ 的充分必要条件是 $R = +\infty$.

令

$$\xi_i = \begin{cases} \inf\{t: x(t) = i\} \\ \sigma, \text{ 上面集合是空集}. \end{cases} \quad (2.15)$$

则由引理 1(ii) 易得

$$P_k(\xi_j \uparrow \tau(j \uparrow +\infty)) = 1. \quad (2.16)$$

又 [2,引理 2.2] 指出

$$C_{kj} \equiv P_k(\xi_j < \tau) = \begin{cases} 1, & k \leqslant j, \\ \dfrac{z - z_k}{z - z_j}, & k > j. \end{cases} \quad (2.17)$$

对任 $0 \leqslant \varepsilon \leqslant +\infty$,令

$$f_\varepsilon(x) = \begin{cases} x, & 0 \leqslant x < \varepsilon, \\ \varepsilon, & x \geqslant \varepsilon. \end{cases} \quad (2.18)$$

引理 2 设 $R < +\infty$. 对 $k \geqslant i \geqslant 0$,令

$$H_{ki}^{(\varepsilon)} = E_k \left\{ \sum_{0 \leqslant \tau_j < \min(\xi_i, \tau)} f_\varepsilon(\tau_{j+1} - \tau_j) \right\}. \quad (2.19)$$

特别

$$H_{ki}^{(\varepsilon)} = E_k\{\min(\xi_i, \tau)\}. \quad (2.20)$$

则

$$H_{ki}^{(\varepsilon)} = \frac{z - z_k}{z - z_i} \sum_{j=i+1}^{k} (z_j - z_i)(1 - e^{-(a_j + b_j)\varepsilon}) y_j +$$

$$\frac{z_k - z_i}{z - z_i} \sum_{j=k+1}^{+\infty} (z - z_j)(1 - e^{-(a_j + b_j)\varepsilon}) y_j \leqslant N_k; \quad (2.21)$$

$$\lim_{\varepsilon \downarrow 0} H_{ki}^{(\varepsilon)} = 0. \tag{2.22}$$

又如果 $S < +\infty$, 那么

$$\lim_{k \to +\infty} \frac{H_{ki}^{(\varepsilon)}}{C_{k0}} = \frac{1}{C_{i0}} \sum_{j=i+1}^{+\infty} (z_j - z_i)(1 - e^{-(a_j+b_j)\varepsilon}) y_j ; \tag{2.23}$$

$$\lim_{i \to +\infty} \lim_{k \to +\infty} \frac{H_{ki}^{(\varepsilon)}}{C_{k0}} = 0. \tag{2.24}$$

$$\lim_{\varepsilon \downarrow 0} \lim_{k \to +\infty} \frac{H_{ki}^{(\varepsilon)}}{C_{k0}} = 0. \tag{2.25}$$

证 设 $0 \leqslant k \leqslant n$. 令

$$H_{kin}^{(\varepsilon)} = E_k \left\{ \sum_{0 \leqslant \tau_i < \min(\xi_i, \xi_n)} f_\varepsilon(\tau_{j+1} - \tau_j) \right\}.$$

由 (2.16), $H_{kin}^{(\varepsilon)} \uparrow H_{ki}^{(\varepsilon)}$ $(n \uparrow +\infty)$. 熟知, 关于测度 P_k, τ_1 有以 $(a_k + b_k)$ 为参数的指数分布, 因而

$$E_k f_\varepsilon(\tau_1) = \frac{1}{a_k + b_k}(1 - e^{(a_k + b_k)\varepsilon}),$$

又 $x(\tau_1) = k-1, k+1$ 的概率分别为 $\dfrac{a_k}{a_k + b_k}$, $\dfrac{b_k}{a_k + b_k}$. 利用强马尔可夫性易得

$$H_{iin}^{(\varepsilon)} = 0,$$

$$H_{kin}^{(\varepsilon)} = \frac{a_k}{a_k + b_k} \left(\frac{1 - e^{-(a_k+b_k)\varepsilon}}{a_k + b_k} + H_{k-1,i,n}^{(\varepsilon)} \right) +$$

$$\frac{b_k}{a_k + b_k} \left(\frac{1 - e^{-(a_k+b_k)\varepsilon}}{a_k + b_k} + H_{k+1,i,n}^{(\varepsilon)} \right), \quad (i < k < n),$$

$$H_{nin}^{(\varepsilon)} = 0.$$

即

$$\begin{cases} H_{iin}^{(\varepsilon)} = 0, \\ a_k H_{k-1\,in}^{(\varepsilon)} - (a_k + b_k) H_{kin}^{(\varepsilon)} + b_k H_{k+1\,in}^{(\varepsilon)} + \\ \quad (1 - e^{-(a_k+b_k)\varepsilon}) = 0, \quad (i < k < n), \\ H_{nin}^{(\varepsilon)} = 0. \end{cases} \tag{2.26}$$

解此方程便得

$$H_{kin}^{(\varepsilon)} = \frac{z_n - z_k}{z_n - z_i}\sum_{j=i+1}^{k}(z_j - z_i)(1 - e^{-(a_j+b_j)\varepsilon})y_j +$$

$$\frac{z_k - z_i}{z_n - z_i}\sum_{j=k+1}^{n-1}(z_n - z_j)(1 - e^{-(a_j+b_j)\varepsilon})y_i.$$

令 $n \to +\infty$ 便得(2.21)中等号,注意(2.7)(2.21)中不等号容易验证. 由于 $R < +\infty$,利用控制收敛定理,从(2.21)得(2.22).

由(2.8)可见若 $S < +\infty$,则 $\sum_{j=0}^{+\infty} y_j < +\infty$. 由于

$$\frac{1}{C_{k0}}\frac{z_k - z_i}{z - z_i}\sum_{j=k+1}^{+\infty}(z - z_j)(1 - e^{-(a_j+b_j)\varepsilon})y_j$$

$$\leqslant z\sum_{j=k+1}^{+\infty} y_j \to 0, \ k \to +\infty;$$

$$\frac{1}{C_{i0}}\sum_{j=i+1}^{+\infty}(z_j - z_i)(1 - e^{-(a_j+b_j)\varepsilon})y_j$$

$$\leqslant z\sum_{j=i+1}^{+\infty} y_j \to 0, \ i \to +\infty.$$

故从(2.21)便得(2.23),从而得(2.24). 由 $R < +\infty$,利用控制收敛定理,从(2.23)得(2.25). 引理证完.

我们指出解方程(2.26)的方法. 显然只需解方程

$$\begin{cases} u_i = f_i, \\ a_k u_{k-1} - (a_k + b_k)u_k + b_k u_{k+1} + f_k = 0, \ (i < k < n) \\ u_n = f_n. \end{cases} \quad (2.27)$$

若令

$$u_k^+ = \frac{u_{k+1} - u_k}{z_{k+1} - z_k},$$

则(2.27)变为

$$\begin{cases} u_i = f_i, \\ u_k^+ - u_{k-1}^+ = -f_k y_k, \ i < k < n, \\ u_n = f_n. \end{cases}$$

由上式得

$$u_k^+ - u_i^+ = \sum_{l=i+1}^{k}(u_l^+ - u_{l-1}^+) = -\sum_{l=i+1}^{k} f_l y_l,$$

$$u_k^+ = u_i^+ - \sum_{l=i+1}^{k} f_l y_l,$$

由此

$$u_k = u_i + \sum_{l=i}^{k-1}(u_{l+1} - u_i) = u_i + \sum_{l=i}^{k-1} u_l^+(z_{i+1} - z_l)$$

$$= u_i + \sum_{l=i}^{k-1} u_i^+(z_{l+1} - z_i) - \sum_{l=i}^{k-1}\Big(\sum_{j=i+1}^{l} f_j y_j\Big)(z_{l+1} - z_l)$$

$$= f_i + u_i^+(z_k - z_i) - \sum_{j=i+1}^{k-1}\sum_{l=j}^{k-1} f_j y_j(z_{l+1} - z_l),$$

故

$$u_k = f_i + u_i^+(z_k - z_i) - \sum_{j=i+1}^{k-1}(z_k - z_i) f_i y_i, \quad (2.28)$$

特别 $k=n$ 时得

$$f_n = f_i + u_i^+(z_n - z_i) - \sum_{j=i+1}^{n-1}(z_n - z_j) f_j y_j,$$

从而

$$u_i^+ = \frac{f_n - f_i}{z_n - z_i} + \frac{1}{z_n - z_i}\sum_{j=i+1}^{n-1}(z_n - z_j) f_j y_j,$$

代入(2.28)并经整理后得

$$u_k = f_i \frac{z_n - z_k}{z_n - z_i} + f_n \frac{z_k - z_i}{z_n - z_i} + \frac{z_n - z_k}{z_n - z_i}\sum_{j=i+1}^{k-1}(z_j - z_i) f_j y_j +$$

$$\frac{z_k - z_i}{z_n - z_i}\sum_{j=k}^{n-1}(z_n - z_i) f_j y_j.$$

§3. 中断过程的延拓过程

设 $\widetilde{P}(t)=\{\widetilde{p}_{ij}(t), i,j\in E\}$, $t\geq 0$ 满足(1.1)~(1.3)及
$$\sum_j \widetilde{P}_{ij}(t)<1 \quad 对某 i. \tag{3.1}$$
$\widetilde{P}(t)$ 所对应的 $\widetilde{Q}$ 可以不必满足(1.6). 任给 $\pi=\{\pi_i, j\in E\}$ 满足
$$\pi_j\geq 0, \quad 0<\sum_j \pi_i\leq 1. \tag{3.2}$$
令
$$\widetilde{L}_i(t)=1-\sum_j \widetilde{P}_{ij}(t), \quad \widetilde{L}=\sum_i \pi_i \widetilde{L}_i,$$
$$\widetilde{\pi}_j(t)=\sum_i \pi_i \widetilde{P}_{ij}(t), \quad \widetilde{L}^{(0)*}(t)=\begin{cases}0, & t\leq 0,\\ 1, & t>0,\end{cases}$$
$$\widetilde{L}^{(n+1)*}=\widetilde{L}^{(n)*}*\widetilde{L}, \quad (*\text{表示卷积})$$
$$\widetilde{K}_i=\sum_{n=0}^{+\infty}\widetilde{L}_i*\widetilde{L}^{(n)*},$$
$$P_{ij}(t)=\widetilde{P}_{ij}(t)+\int_0^t \widetilde{\pi}_j(t-s)\mathrm{d}\widetilde{K}_i(s). \tag{3.3}$$

引理 3 存在概率空间 $(\Omega, \mathscr{F}, P)$, 在其上可以定义一列过程 $X^{(n)}=\{x^{(n)}(t), t<\sigma^{(n)}\}$[①] $(n\geq 0)$, 它们具有下列性质:

(i) $X^{(n)}$ 的转移概率矩阵 $\widetilde{P}(t)$;

(ii) 以概率 1, $x^{(n)}(t,\omega)$, $t<\sigma^{(n)}(\omega)$ 是 t 的右下半连续函数[②]; $X^{(n)}$ 可分 Borel 可测;

(iii) $(\sigma^{(n)}=0)\bigcup(\sigma^{(n)}=+\infty)\subset(\sigma^{(n+1)}=0)$[③], $(n\geq 0)$;

[①] 除 $X^{(0)}$ 外, $X^{(n)}(n\geq 1)$ 可以不满足(2.4).

[②] 即在 $E\cup\{+\infty\}$ 中, $\lim_{s\downarrow t}x^{(n)}(s,\omega)=x^{(n)}(t,\omega)$, $t<\sigma^{(n)}(\omega)$.

[③] $A\subset B$ 即 $P(A-B)=0$.

(iv)　$P(x^{(n+1)}(0)=j \mid 0<\sigma^{(n)}<+\infty)=\pi_j$,
$$P(\sigma^{(n+1)}=0 \mid 0<\sigma^{(n)}<+\infty)=1-\sum_j \pi_j;$$

(v) 在条件 $(0<\sigma^{(n)}<+\infty)$ 或 $(x^{(n+1)}(0)=i)$ 之下，$X^{(m)}(m\leqslant n)$ 与 $X^{(m)}(m>n)$ 条件独立，即设
$$\Lambda_m=\{x^{(m)}(t_{mk})=j_{mk},\ 1\leqslant k\leqslant l_m\},$$
$$(0\leqslant t_{m1}<\cdots<t_{ml_m},\ j_{m1},\cdots,j_{ml_m}\in E)$$

则对任 $l\geqslant 1$，$n\geqslant 0$ 有
$$P\Big(\bigcap_{a=0}^{n+l}\Lambda_a\Big|\Delta\Big)=P\Big(\bigcap_{a=0}^{n}\Lambda_a\Big|\Delta\Big)P\Big(\bigcap_{a=n+1}^{n+l}\Lambda_a\Big|\Delta\Big),$$

其中，　$\Delta=(0<\sigma^{(n)}<+\infty)$ 或 $\Delta=(x^{(n+1)})(0)=i$.

证　利用作独立乘积空间的技巧，不难证明：存在概率空间 $(\Omega, \mathscr{F}, P)$，在其上可以定义一列过程 $X^{(0)}=\{x^{(0)}(t), t<\sigma^{(0)}\}$，$X_i^{(n)}=\{x_i^{(n)}(t), t<\sigma_i^{(n)}\}(n\geqslant 1, i\in E)$ 及取值于 $E\cup\{+\infty\}$ 的随机变量族 $f^{(n)}(n\geqslant 0)$，它们具有下列性质 (i′)～(iv′).

(i′) $X^{(0)}$，$X_i^{(n)}$ 的转移概率矩阵都是 $\widetilde{P}(t)$；

(ii′) 以概率 1，$x^{(0)}(t,\omega)$，$t<\sigma^{(0)}(\omega)$ 及 $x_i^{(n)}(t,\omega)$，$t<\sigma_i^{(n)}(\omega)(n\geqslant 1, i\in E)$ 是 t 的右下半连续函数；$X^{(0)}$，$X_i^{(n)}$ 可分 Borel 可测；

(iii′)　$P(x^{(0)}(0)\in E)=P(\sigma^{(0)}>0)=1$,
$$P(x_i^{(n)}(0)=i)=1,\quad (n\geqslant 1, i\in E),$$
$$P(f^{(n)}=i)=\pi_i,\quad (i\in E),$$
$$P(f^{(n)}=+\infty)=1-\sum_i \pi_i;$$

(iv′) 诸 $X^{(0)}$，$X_i^{(n)}(n\geqslant 1, i\in E)$，$f^{(n)}(n\geqslant 0)$ 相互独立.

设 $C(t)$ 为集 $\{t: 0<t<+\infty\}$ 的示性函数. 对于 $\omega\in(f^{(0)}\in E, C(\sigma^{(0)})=1)$，令 $x^{(1)}(t,\omega)=x_{f^{(0)}(\omega)}^{(1)}(t,\omega)$，$t<\sigma^{(1)}(\omega)=\sigma_{f^{(0)}(\omega)}^{(1)}(\omega)$；否则令 $\sigma^{(1)}(\omega)=0$. 对于 $\omega\in(f^{(1)}\in E, C(\sigma^{(1)})=$

1)，令 $x^{(2)}(t,\omega) = x^{(2)}_{f^{(1)}(\omega)}(t,\omega)$，$t < \sigma^{(2)}(\omega) = \sigma^{(2)}_{f^{(1)}(\omega)}(\omega)$；否则令 $\sigma^{(2)}(\omega) = 0$。如此继续，我们得到一列 $X^{(n)} = \{x^{(n)}(t), t < \sigma^{(n)}\}$ $(n \geqslant 0)$。易见 $X^{(n)}$ 是过程，并且由于(i')~(iv')，不难证明(i)~(v)满足。实际上(i)~(iii)是易证的，今往证(iv)(v)。

因为 $\sigma^{(n)}$ 只依赖于 $X^{(0)}$，$X_i^{(m)}(m \leqslant n, i \in E)$，$f^{(m)}(m < n)$，由(iii')(iv')，

$$P(x^{(n+1)}(0) = j, 0 < \sigma^{(n)} < +\infty)$$
$$= P(f^{(n)} = j, 0 < \sigma^{(n)} < +\infty)$$
$$= P(f^{(n)} = j) P(0 < \sigma^{(n)} < +\infty)$$
$$= \pi_j P(0 < \sigma^{(n)} < +\infty),$$

此即(iv)。

为证（v），我们固定 n 及 $j \in E$。令 $\overline{X}^{(0)} = X_j^{(n+1)}$，$\overline{X}_i^{(m)} X_i^{(n+1+m)}$，$\overline{f}^{(m)} = f^{(n+m)}$。像刚才按照 $X^{(0)}$，$X_i^{(m)}(m \geqslant 1, i \in E)$，$f^{(m)}(m \geqslant 0)$ 定义 $X^{(m)} = \{x^{(m)}(t), t < \sigma^{(m)}\}(m \geqslant 0)$ 一样，按照 $\overline{X}^{(0)}$，$\overline{X}_i^{(m)}(m \geqslant 1, i \in E)$，$\overline{f}^{(m)}(m \geqslant 0)$，我们可以定义 $\overline{X}^{(m)} = \{\overline{X}^{(m)}(t), t < \overline{\sigma}^{(m)}\}(m \geqslant 0)$。显然 $\overline{X}^{(m)}(m \geqslant 0)$ 只依赖于 $X_i^{(m)}(m > n, i \in E)$ 及 $f^{(m)}(m > n)$。并且如令

$$\overline{\Lambda}_{n+1+m} = \{\overline{x}^{(m)}(t_{n+1+m,k}) = j_{n+1+m,k}, 1 \leqslant k \leqslant l_{n+1+m}\},$$

$$\overline{N} = \bigcap_{a=n+1}^{n+l} \overline{\Lambda}_a, \quad M = \bigcap_{a=0}^{n} \Lambda_a, \quad N = \bigcap_{a=n+1}^{n+l} \Lambda_a,$$

则易见

$$(x^{(n+1)}(0) = j) \bigcap \overline{N} = (x^{(n+1)}(0) = j) \bigcap N,$$

且

$$P(\overline{N}) = P(N \mid x^{(n+1)}(0) = j).$$

于是由(iii')(iv')及(iii)以及对 $\Delta = (0 < \sigma^{(n)} < +\infty)$ 有

$$\Delta \bigcap (x^{(n+1)}(0) \in E) = \Delta,$$

故

$$P(MN\Delta) = \sum_j P(M\Delta, x^{(n+1)}(0) = j, N)$$
$$= \sum_j P(M\Delta, f^{(n)} = j, \bar{N})$$
$$= \sum_j P(M\Delta) P(f^{(n)} = j) P(\bar{N})$$
$$= \sum_j P(M\Delta) \pi_j P(N \mid x^{(n+1)}(0) = j)$$
$$= P(M\Delta) \sum_j P(x^{(n+1)}(0) = j \mid \Delta) P(N \mid x^{(n+1)}(0) = j)$$
$$= p(M\Delta) P(N \mid \Delta).$$

由此可知(v)对 $\Delta = (0 < \sigma^{(n)} < +\infty)$ 成立，类似可证对 $\Delta = (x^{(n+1)}(0) = i)$ 也成立. 引理于是证毕.

定理 1 设定义于同一概率空间 $(\Omega, \mathscr{F}, P)$ 上的一列过程 $X^{(n)} = \{x^{(n)}(t), t < \sigma^{(n)}\}$ $(n \geq 0)$ 具有引理 3 性质 (i)~(v). 令 $\tau^{(0)} = 0$, $\tau^{(n+1)} = \sum_{m=0}^n \sigma^{(m)}$, $\sigma = \lim_{n \to +\infty} \tau^{(n)} = \sum_{m=0}^{+\infty} \sigma^{(m)}$. 对 $0 \leq t < \sigma$, 令

$$x(t) = x^{(n)}(t - \tau^{(n)}), \quad \tau^{(n)} \leq t < \tau^{(n+1)}. \tag{3.4}$$

则 $X = \{x(t), t < \sigma\}$ 是可分 Borel 可测右下半连续过程，它的转移概率由 (3.3) 给出. X 不中断的充分必要条件是

$$\sum_j \pi_j = 1. \tag{3.5}$$

证 由 (ii) 可知，X 可分 Borel 可测右下半连续. 对于 X 是以 (3.3) 为转移概率的马氏链的严格证明，可仿 [6, 定理 7] 用离散逼近法证明. 这里我们只给出启发性的辅导.

设 $0 \leq t_1 < \cdots < t_l < t_{l+1}$, $j_1, j_2, \cdots, j_l, j_{l+1} \in E$. 令
$$\Lambda_k = \{x(t_a) = j_a, \ 1 \leq a \leq k\},$$
$$\Delta(m_1, m_2, \cdots, m_k) = \{x^{(m_a)}(t_a - \tau^{(m_a)}) = j_a,$$
$$\tau^{(m_a)} \leq t_a < \tau^{(m_a+1)}, \ 1 \leq a \leq k\},$$

则显然

$$P(\Lambda_{l+1}) = \sum_{0 \leq m_1 \leq m_2 \leq \cdots \leq m_l+1} P(\Delta(m_1, m_2, \cdots, m_{l+1})). \tag{3.6}$$

设 $m_l = m_{l+1} = m$，则存在 $k < l$ 使 $m_1 \leqslant m_2 \cdots \leqslant m_k \leqslant m_{k+1} = \cdots = m_l = m_{l+1} = m$，因而

$$P(\Delta(m_1, m_2, \cdots, m_{l+1}))$$
$$= \sum_i P(\Delta(m_1, m_2, \cdots, m_k), \tau^{(m)} \leqslant t_{k+1} \cdot x^{(m)}(0) = i,$$
$$x^{(m)}(t_a - \tau^{(m)}) = j_a, k+1 \leqslant a \leqslant l+1)$$
$$= \sum_i \int_0^{t_{k+1}} P(x^{(m)}(t_a - \tau^{(m)}) = j_a,$$
$$k+1 \leqslant a \leqslant l+1 \mid \Delta(m_1, m_2, \cdots, m_k),$$
$$\tau^{(m)} = s, x^{(m)}(0) = i) \mathrm{d}s P(\Delta(m_1, m_2, \cdots, m_k),$$
$$\tau^{(m)} \leqslant s, x^{(m)}(0) = i). \tag{3.7}$$

由于 $\Delta(m_1, m_2, \cdots, m_k), \tau^{(m)}$ 只依赖于 $X^{(n)}(n \leqslant m-1)$，故由性质(v)，上式被积表达式等于

$$P(x^{(m)}(t_a - s) = j_a, k+1 \leqslant a \leqslant l+1 \mid \Delta(m_1, m_2, \cdots, m_k), \tau^{(m)}$$
$$= s, x^{(m)}(0) = i) = P(x^{(m)}(t_a - s) = j_a, k+1 \leqslant a$$
$$\leqslant l+1 \mid x^{(m)}(0) = i) = p(x^{(m)}(t_a - s) = j_a, k+1 \leqslant a$$
$$\leqslant l \mid x^{(m)}(0) = i) \widetilde{P}_{j_l j_{l+1}}(t_{l+1} - t_l) = P(x^{(m)}(t_a - s)$$
$$= j_a, k+1 \leqslant a \leqslant l \mid \Delta(m_1, m_2, \cdots, m_k), \tau^{(m)} = s, x^{(m)}(0)$$
$$= i) \widetilde{P}_{j_l j_{l+1}}(t_{l+1} - t_l),$$

代入(3.7)并逆转刚才的计算得当 $m_l = m_{l+1}$ 时，

$$P(\Delta(m_1, m_2, \cdots, m_{l+1}))$$
$$= P(\Delta(m_1, m_2, \cdots, m_l)) \widetilde{P}_{j_l j_{l+1}}(t_{l+1} - t_l). \tag{3.8}$$

今设 $m_l = m < m_{l+1} = r$，则由性质(v)，

$$P(\Delta(m_1, m_2, \cdots, m_{l+1}))$$
$$= \sum_i P(\Delta(m_1, m_2, \cdots, m_l), x^{(r)}(0) = i, x^{(r)}(t_{l+1} - \tau^{(r)}) =$$
$$j_{l+1}, t_l < \tau^{(r)} < t_{l+1})$$
$$= \sum_i \int_{t_l}^{t_{l+1}} P(x^{(r)}(t_{l+1} - \tau^{(r)}) j_{i+1} \mid \Delta(m_1, m_2, \cdots, m_l),$$

$$\tau^{(r)} = s, \ x^{(r)}(0) = i) \mathrm{d}s P(\Delta(m_1, m_2, \cdots, m_l),$$
$$\tau^{(r)} = s, \ x^{(r)}(0) = i). \tag{3.9}$$

由于性质(v)，被积表达式等于
$$P(x^{(r)}(t_{l+1} - s) = j_{l+1} \mid x^{(r)}(0) = i) = \widetilde{P}_{ij_{l+1}}(t_{l+1} - s) \tag{3.10}$$

又由性质(iii)(iv)(v)
$$P(\Delta(m_1, m_2, \cdots, m_l), \ \tau^{(r)} \leqslant s, \ x^{(r)}(0) = i)$$
$$= P(\Delta(m_1, m_2, \cdots, m_l), \ \tau^{(r)} \leqslant s, \ 0 < \sigma^{(r-1)} < +\infty, \ x^{(r)}(0) = i)$$
$$= p(\Delta(m_1, m_2, \cdots, m_l), \ \tau^{(r)} \leqslant s, \ 0 < \sigma^{(r-1)} < +\infty) \pi_i \tag{3.11}$$

如果能够证明：对 $s \geqslant t_l$,
$$P(\Delta(m_1, m_2, \cdots, m_l), \ \tau^{(r)} \leqslant s, \ 0 < \sigma^{(r-1)} < +\infty)$$
$$= P(\Delta(m_1, m_2, \cdots, m_l)) \cdot (\widetilde{L}_{j_l} * \widetilde{L}^{(r-m-1)*})(s - t_l), \tag{3.12}$$

那么将(3.8)~(3.12)代入(3.6)中得

$$P(\Lambda_{l+1})$$
$$= \sum_{0 \leqslant m_1 \leqslant m_2 \leqslant \cdots \leqslant m_l} P(\Delta(m_1, m_2, \cdots, m_l)) \{ \widetilde{P}_{j_l j_{l+1}}(t_{l+1} - t_l) +$$
$$\sum_{m_{l+1} = m_l + 1}^{+\infty} \int_{t_l}^{t_{l+1}} \widetilde{\pi}_{j_{l+1}}(t_{l+1} - s) \mathrm{d}s (\widetilde{L}_{j_l} * \widetilde{L}^{(m_{l+1} - m_l - 1)*})(s - t_l) \}$$
$$= \sum_{0 \leqslant m_1 \leqslant m_2 \leqslant \cdots \leqslant m_l} P(\Delta(m_1, m_2, \cdots, m_l)) \Big\{ \widetilde{P}_{j_l j_{l+1}}(t_{l+1} - t_l) +$$
$$\int_0^{t_{l+1} - t_l} \widetilde{\pi}_{j_{l+1}}(t_{l+1} - t_l - s) \mathrm{d}s \Big(\sum_{m_{l+1} = m_l + 1}^{+\infty} \widetilde{L}_{j_l} * \widetilde{L}^{(m_{l+1} - m_l - 1)*} \Big)(s) \Big\}$$
$$= \Big\{ \sum_{0 \leqslant m_1 \leqslant m_2 \leqslant \cdots \leqslant m_l} P(\Delta(m_1, m_2, \cdots, m_l)) \Big\} \Big\{ \widetilde{P}_{j_l j_{l+1}}(t_{l+1} - t_l) +$$
$$\int_0^{t_{l+1} - t_l} \widetilde{\pi}_{j_{l+1}}(t_{l+1} - t_l - s) \mathrm{d} \widetilde{K}_{j_l}(s) \Big\}$$
$$= P(\Lambda_l) p_{j_l j_{l+1}}(t_{l+1} - t_l).$$

由此说明 X 是过程，而且它的转移概率为(3.3).

往证(3.12). 即要证

$$P(\Delta(m_1, m_2, \cdots, m_l), \tau^{(r)} \leqslant s+t_l, 0<\sigma^{(r-1)}<+\infty)$$
$$=P(\Delta(m_1, m_2, \cdots, m_l))(\widetilde{L}_{j_l} * \widetilde{L}^{(r-m-1)*})(s), (s\geqslant 0).$$
(3.13)

设 $r=m+1$. 上式左边等于

$$\sum_i P(\Delta(m_1,m_2,\cdots,m_{l-1}), \tau^{(m)}+\sigma^{(m)}\leqslant s+t_l, x^{(m)}(0)=i,$$
$$x^{(m)}(t_l-\tau^{(m)})=j_l, \tau^{(m)}\leqslant t_l)$$
$$=\sum_i \int_0^{t_l} P(x^{(m)}(t_l-\tau^{(m)})=j_l, \tau^{(m)}+\sigma^{(m)}\leqslant s+t_l \mid \Delta(m_1,m_2,\cdots,$$
$$m_{l-1}), \tau^{(m)}=u, x^{(m)}(0)=i) \mathrm{d}u P(\Delta(m_1,m_2,\cdots,m_{l-1}), x^{(m)}(0)=$$
$$i, \tau^{(m)}\leqslant u).$$

由性质(v)，被积表达式等于

$$P(x^{(m)}(t_l-u)=j_l, \sigma^{(m)}\leqslant s+t_l-u \mid x^{(m)}(0)=i)$$
$$=P(x^{(m)}(t_l-u)=j_l \mid x^{(m)}(0)=i)\widetilde{L}_{j_l}(s).$$

代回原来的式子并逆转刚才的计算得(3.13)的左边$=P(\Delta(m_1, m_2, \cdots, m_l))\widetilde{L}_{j_l}(s)$. 即(3.13)对 $r=m+1$ 正确.

用归纳法证(3.13)，设 $r>m+1$. (3.13)左边等于

$$\sum_i P(\Delta(m_1,m_2,\cdots,m_l), \tau^{(r-1)}+\sigma^{(r-1)}\leqslant s+t_l, x^{(r-1)}(0)=i)$$
$$=\sum_i \int_{t_l}^{s+t_l} P(\tau^{(r-1)}+\sigma^{(r-1)}\leqslant s+t_l \mid \Delta(m_1,m_2,\cdots,m_l),$$
$$\tau^{(l-1)}=u, x^{(r-1)}(0)=i)\mathrm{d}u P(\Delta(m_1, m_2, \cdots, m_l),$$
$$\tau^{(r-l)}\leqslant u, x^{(r-1)}(0)=i)$$
$$=\sum_i \int_{t_l}^{s+t_l} P(\sigma^{(r-1)}\leqslant s+t_l-u \mid x^{(r-1)}(0)=i)\mathrm{d}u$$
$$P(\Delta(m_1, m_2, \cdots, m_l), \tau^{(r-1)}\leqslant u, x^{(r-1)}(0)=i),$$

但

$$P(\sigma^{(r-1)}\leqslant s+t_l-u \mid x^{(r-1)}(0)=i)=\widetilde{L}_i(s+t_l-u),$$

又由性质(iii)(iv)(v)及归纳法假设，对 $u\geqslant t_l$ 有

$$P(\Delta(m_1, m_2, \cdots, m_l), \tau^{(r-1)} \leqslant u, x^{(r-1)}(0)=i)$$
$$= P(\Delta(m_1, m_2, \cdots, m_l), \tau^{(r-1)} \leqslant u, 0 < \sigma^{(r-2)} < +\infty, x^{(r-1)}(0)=i)$$
$$= P(\Delta(m_1, m_2, \cdots, m_l), \tau^{(r-1)} \leqslant u, 0 < \sigma^{(r-2)} < +\infty) \pi_i$$
$$= P(\Delta(m_1, m_2, \cdots, m_l))(\widetilde{L}_{j_l} * \widetilde{L}^{(r-1-m-1)*})(u-t_l)\pi_i.$$

因此(3.13)式左边等于

$$\sum_i \int_{t_l}^{s+t_l} \pi_i \widetilde{L}_i(s+t_l-u) P(\Delta(m_1,m_2,\cdots,m_l)) \mathrm{d}u (\widetilde{L}_{j_l} * \widetilde{L}^{(r-1-m-1)*})(u-t_l)$$
$$= P(\Delta(m_1,m_2,\cdots,m_l)) \int_0^s \widetilde{L}(s-v) \mathrm{d}v (\widetilde{L}_{j_l} * \widetilde{L}^{(r-1-m-1)*})(v)$$
$$= P(\Delta(m_1, m_2, \cdots, m_l))(\widetilde{L}_{j_l} * \widetilde{L}^{(r-m-1)*})(s).$$

得证(3.13). 从而得证 X 是过程，其转移概率由(3.3)得出.

在(3.3)中对 j 求和得

$$\sum_j p_{ij}(t) = 1 - \widetilde{L}_i(t) + \int_0^t \sum_k \pi_k(1-\widetilde{L}_k(t-s)) \mathrm{d}\widetilde{K}_i(s)$$
$$= 1 - \widetilde{L}_i(t) + \int_0^t \Big(\sum_k \pi_k - \widetilde{L}(t-s)\Big) \mathrm{d}\widetilde{K}_i(s)$$
$$= 1 - \widetilde{L}_i(t) + \Big(\sum_k \pi_k\Big) \widetilde{K}_i(t) - \sum_{n=0}^{+\infty}(\widetilde{L}_i * \widetilde{L}^{(n+1)*})(t)$$
$$= 1 - \widetilde{K}_i(t)\Big(1 - \sum_k \pi_k\Big).$$

由(3.1)，至少有一个 i 使 $\widetilde{K}_i \neq 0$，故 X 不中断即(2.1)对一切 i 成立当且只当 $\sum_k \pi_k = 1$.

定理证完.

注1 由于定理1中的过程 $X = \{x(t), t < \sigma\}$ 的前面一部分是中断过程 $X^{(0)} = \{x(t), t < \tau^{(1)}\}$，而且满足 $P(x(\tau^{(1)}) = j \mid 0 < \tau^{(1)} < +\infty) = \pi_i$，而它们的转移概率矩阵 $P(t)$ 和 $\widetilde{P}(t)$ 有关系(3.3). 因此我们称过程 X 是中断过程 $X^{(0)}$ 的 $\pi = \{\pi_i, j \in E\}$ D 型延拓过程，或称由(3.3)确定的过程 $P(t)$ 是中断过程 $\widetilde{P}(t)$ 的 $\pi = \{\pi_j, j \in E\}$ D 型延拓过程.

在(3.3)两边取拉氏变换得：对 $\lambda>0$,

$$\psi_{ij}(\lambda) \equiv \int_0^{+\infty} e^{-\lambda t} p_{ij}(t) dt = \widetilde{f}_{ij}(\lambda) + \tilde{\xi}_i(\lambda) \frac{\sum_k \pi_k \widetilde{f}_{kj}(\lambda)}{1 - \sum_k \pi_k \tilde{\xi}_k(\lambda)}$$

$$= \widetilde{f}_{ij}(\lambda) + \tilde{\xi}_i(\lambda) \frac{\sum_k \pi_k \widetilde{f}_{kj}(\lambda)}{1 - \sum_k \pi_k + \sum_k \pi_k(1 - \tilde{\xi}_k(\lambda))},$$

其中

$$\widetilde{f}_{ij}(\lambda) = \int_0^{+\infty} e^{-\lambda t} \widetilde{p}_{ij}(t) dt, \tilde{\xi}_i(\lambda) = 1 - \lambda \sum_j \widetilde{f}_{ij}(\lambda).$$

注 2 当 $\widetilde{Q}$ 满足(1.6)时，由样本函数的性质易知，过程 $X^{(0)}$ (或 $\widetilde{P}(t)$)与其 D 型延拓过程 X (或 $P(t)$)有相同的 Q 矩阵. 但当 $\widetilde{Q}$ 不满足(1.6)时，上述结论未必成立. 为此我们引进 D^* 型延拓过程.

我们假定一切 $\tilde{q}_i < +\infty$.

称 σ 为函数 $X = \{x(t), t<\sigma\}$ 的 s-尾，如果 $0<\sigma<+\infty$, 并且存在 $u<\sigma$ 使对一切 $t: u \leqslant t<\sigma$, 有 $x(t) = x(u)$. 否则称 σ 为函数 X 的 p-尾.

显然，如果过程 $\widetilde{X} = \{\widetilde{x}(t), t<\tilde{\sigma}\}$ 有转移概率 $\widetilde{P}(t)$, 那么

$$M_{ij}(t) = p_i(\widetilde{x}(t) = j, t<\tilde{\sigma}, \tilde{\sigma} \text{ 为 } \widetilde{X} \text{ 的 } p\text{-尾}),$$
$$N_{ij}(t) = P_i(\widetilde{x}(t) = j, t<\tilde{\sigma}, \tilde{\sigma} \text{ 为 } \widetilde{X} \text{ 的 } s\text{-尾}),$$
$$M_i(t) = \sum_j M_{ij}(t), N_i(t) = \sum_j N_{ij}(t),$$
$$R_i(t) = 1 - M_i(t),$$

等由 $\widetilde{p}(t)$ 唯一确定.

我们将性质(i)~(v)修改为(i*)~(v*):

(i*) $\widetilde{Q}$ 过程 $X^{(n)}$ ($n \geqslant 0$)有相同的转移概率;

(ii*) 以概率 1, $X^{(n)}$ 是 t 的右连续函数即满足(2.9); $X^{(n)}$ 可分 Borel 可测;

(iii*) $(\sigma^{(n)}=0)\bigcup(\sigma^{(n)}=+\infty)\bigcup(\sigma^{(n)}$ 为 $X^{(n)}$ 的 s-尾$)$
$\subseteq (\sigma^{(n+1)}=0)$, $(n\geqslant 0)$;

(iv*) $P(x^{(n+1)}(0)=j\mid 0<\sigma^{(n)}<+\infty$ 且 $\sigma^{(n)}$ 为 $X^{(n)}$ 的 p-尾$)$
$=\pi_j$,
$P(\sigma^{(n+1)}=0\mid 0<\sigma^{(n)}<+\infty$ 且 $\sigma^{(n)}$ 为 $X^{(n)}$ 的 p-尾$)$
$=1-\sum_j \pi_j$;

(v*) 在条件 $(0<\sigma^{(n)}<+\infty$ 且 $\sigma^{(n)}$ 为 $X^{(n)}$ 的 p-尾,或 $(X^{(n+1)}(0)=i)$ 之下, $X^{(m)}(m\leqslant n)$ 与 $X^{(m)}(m>n)$ 条件独立.

依照定理 1 的证明,不难得到:

定理 1′ 假定一切 $\tilde{q}_i<\infty$. 设定义于同一概率空间 $(\Omega,\mathscr{F},P)$ 上的一列 $\tilde{Q}$ 过程 $X^{(n)}$ 具有性质 (i*)~(v*),则按 (3.4) 定义的 X 是可分 Borel 可测右连续 $\tilde{Q}$ 过程,其转移概率由 (3.3)(在其中用 $R_i(t)$ 代替 $\tilde{L}_i(t)$) 给出. X 不断的充分必要条件是 (1.6) 对 $\tilde{Q}$ 成立并且 $\sum_j \pi_j = 1$.

对定理 1′ 的最后一句话尚需证明.

由 (3.3) 可得

$$\sum_j p_{ij}(t) = 1-\left(1-\sum_j \pi_j\right)k_i(t) + N_i(t) + \int_0^t \sum_j \pi_j N_j(t-s)\mathrm{d}k_i(s).$$

(3.13′)

如果 (1.6) 对 $\tilde{Q}$ 成立并且 $\sum_j \pi_j = 1$,那么显然 $N_i(t)\equiv 0$,故由上式得 $\sum_j p_{ij}(t) = 1$ 即 X 不断.

如果 X 不断即 $\sum_j p_{ij}(t) = 1$,此时必定一切 $N_i(t) = 0$. 因不然设对某 i 有 $N_i(t) = p_i(t<\sigma^{(0)}, \sigma^{(0)}$ 为 $X^{(0)}$ 的 s-尾$) > 0$,根据 (ii*),在正概率集 $(t<\sigma^{(0)}, \sigma^{(0)}$ 为 $X^{(0)}$ 的 s-尾$)$ 上, $\sigma=$

$\sigma^{(0)} < +\infty$,此与 X 不断矛盾. 于是由 $N_i(t)=0$ 及 $(3.13')$ 可得 $\sum_j \pi_j = 1$;而且必定 (1.6) 对 $\widetilde{Q}$ 成立,因为例如对某 i 有 $q_i > \sum_{j\neq i} \widetilde{q_{ij}}$,则有 $p_i(0 < \sigma^{(0)} < +\infty$ 且 $\sigma^{(0)}$ 是 $X^{(0)}$ 的 s-尾)$\geqslant \dfrac{\bar{q}_i - \sum_{j\neq i}\widetilde{q}_{ij}}{q_i} > 0$,从而有 $N_i(t) = p_i(t < \sigma^{(0)},\ \sigma^{(0)}$ 是 $X^{(0)}$ 的 s-尾)> 0,此不可能.

我们称定理 $1'$ 中的过程 X(或 $P(t)$)是过程 $X^{(0)}$(或 $\widetilde{P}(t)$)的 $\pi = \{\pi_j, j \in E\} D^*$ 型延拓过程. 显然当 $\widetilde{Q}$ 满足 (1.6) 时, D^* 型延拓与 D 型延拓一致.

对于 D^* 型延拓过程, 我们有

$$\psi_{ij}(\lambda) \equiv \int_0^{+\infty} e^{-\lambda t} p_{ij}(t) dt = \widetilde{f}_{ij}(\lambda) + \eta_i(\lambda) \frac{\sum_k \pi_k \widetilde{f}_{kj}(\lambda)}{1 - \sum_k \pi_k \eta_k(\lambda)}$$

$$= \widetilde{f}_{ij}(\lambda) + \eta_i(\lambda) \frac{\sum_k \pi_k \widetilde{f}_{kj}(\lambda)}{1 - \sum_k \pi_k + \sum_k \pi_k(1 - \eta_k(\lambda))},$$

其中

$$\widetilde{f}_{ij}(\lambda) = \int_0^{+\infty} e^{-\lambda t} \widetilde{p}_{ij}(t) dt,$$

$$\eta_i(\lambda) = 1 - \lambda \sum_j \int_0^{+\infty} e^{-\lambda t} M_{ij}(t) dt.$$

注 3 特别地,当 $\widetilde{p}(t)$ 是熟知的最小 Q 过程 $\overline{p}(t)$ 时, 定理 $1'$ 化为

定理 2 设一切 $q_i < +\infty$, 并且 $\widetilde{p}(t)$ 是熟知的最小 Q 过程 $\overline{p}(t)$, 满足 (3.1), 设定义于同一概率空间 $(\Omega, \mathscr{F}, p)$ 上的一列最小 Q 过程 $X^{(n)}(n \geqslant 0)$ 具有性质 $(ii^*) \sim (v^*)$, 则按 (3.4) 定义的 X 是可分 Borel 可测右连续 Q 过程, 其转移概率由 (3.3)(在其中用 $\overline{p}(t)$ 代替 $\widetilde{p}(t)$ 并用 $R_i(t)$ 代替 $\widetilde{L}_i(t)$)给出. X 不断的充分必要条件是 (1.6) 成立并且 $\sum_j \pi_j = 1$.

最小 Q 过程 $X^{(0)}$（或 $\overline{p}(t)$）的 $\pi=\{\pi_j,\ j\in E\}D^*$ 型延拓过程 X（或 $p(t)$），我们称为 (Q,π) Doob 过程.

显然，当 (1.6) 成立时，对 (Q,π) Doob 过程 $p(t)$，我们有

$$\psi_{ij}(\lambda) \equiv \int_0^{+\infty} e^{-\lambda t} p_{ij}(t) dt = f_{ij}(\lambda) + \xi_i(\lambda) \frac{\sum_k \pi_k f_{kj}(\lambda)}{1 - \sum_k \pi_k \xi_k(\lambda)}$$

$$= f_{ij}(\lambda) + \xi_i(\lambda) \frac{\sum_k \pi_k f_{kj}(\lambda)}{1 - \sum_k \pi_k + \sum_k \pi_k(1-\xi_k(\lambda))}, \quad (3.14)$$

其中

$$f_{ij}(\lambda) = \int_0^{+\infty} e^{-\lambda t} \overline{p}_{ij}(t) dt, \quad \xi_i(\lambda) = 1 - \lambda \sum_j f_{ij}(\lambda). \quad (3.15)$$

§4. Q 过程的变换

设 $X=\{x(t), t<\sigma\}$ 是可分 Borel 可测右连续 Q 过程，ζ 为马尔可夫时间，满足 (2.11)，进一步设 $p(\zeta<\sigma)>0$. 由引理 1，除概率为零的集不计外，$(\zeta<\sigma)=(\tau_\zeta^{(n)}<\sigma)(n\geqslant 0)$. 前 $\tau_\zeta^{(n)}$ 域 $\mathscr{F}_{\tau_\zeta^{(n)}}$ 和后 τ_ζ 域 $\mathscr{F}'_{\tau_\zeta}$ 是 $(\zeta<\sigma)$ 上的 σ-代数 [1, P163, P168]. 令 $\mathscr{B}_{[0,\tau_\zeta]}$ 是含一切 $\mathscr{F}_{\tau_\zeta^{(n)}}(n\geqslant 0)$ 的 Ω 上的 σ-代数.

引理 4 设 $R<+\infty$. 在条件 $(\zeta<\sigma)$ 之下，$\mathscr{B}_{[0,\tau_\zeta]}$ 与 $\mathscr{F}'_{\tau_\zeta}$ 条件独立. 即对任意 $\Lambda\in\mathscr{B}_{[0,\tau_\zeta]}$, $M=\{x(\tau_\zeta+t_k)=j_k, 1\leqslant k\leqslant l\}$, $(0\leqslant t_1<t_2<\cdots<t_l, j_1, j_2, \cdots, j_l\in E)$,

$$P(M\mid \zeta<\sigma, \Lambda)=P(M\mid \zeta<\sigma) \quad (4.1)$$

而且右边的值与 ζ 的选择无关.

证 仿 [2, 定理 5.1] 的证明可得

$$P(M\mid \zeta<\sigma, \mathscr{B}_{[0,\tau_\zeta]})=c,$$ 对几乎一切 $\omega\in(\zeta<\sigma)$,

其中 c 是与 ζ 的选择无关的常数. 任取 $\Lambda\in\mathscr{B}_{[0,\tau_\zeta]}$, 则

$$P(M\Lambda\mid \zeta<\sigma)=\int_\Lambda P(M\mid \zeta<\sigma, \mathscr{B}_{[0,\tau_\zeta]})P(d\omega\mid \zeta<\sigma)$$
$$=cP(\Lambda\mid \zeta<\sigma).$$

特别取 $\Lambda=(\zeta<\sigma)\in\mathscr{B}_{[0,\tau_\zeta]}$ 而得 $P(M\mid \zeta<\sigma)=c$, 从而上式成为

$$P(M\Lambda\mid \zeta<\sigma)=P(M\mid \zeta<\sigma)P(\Lambda\mid \zeta<\sigma),$$

引理得证.

定义 1 设函数 $X=\{x(t), t<\sigma\}$ 取值于 $E\cup\{+\infty\}$. 称函数 $Y=\{y(t), t<\delta\}$ 自函数 X 经 $C\{\alpha_k, \beta_k\}$ 变换得来，如果存在两数列 $\{\alpha_k\}\{\beta_k\}$ 使

$$0=\beta_0\leqslant\alpha_1\leqslant\beta_1\leqslant\cdots\leqslant\alpha_k\leqslant\beta_k\leqslant\cdots\leqslant\sigma$$

而 $\delta = \sum_{k=0}^{+\infty}(\alpha_{k+1}-\beta_k)$，$Y=\{y(t), t<\delta\}$ 如下定义：令 $r_1=\alpha_1$，当 $0 \leqslant t < r_1$ 时，令 $y(t)=x(t)$；若 $y(t)$ 已在 $[0, r_k]$ 中有定义，则令 $r_{k+1}=r_k+(\alpha_{k+1}-\beta_k)$，在 $[r_k, r_{k+1})$ 中，令 $y(r_k+t)=x(\beta_k+t)$，当 $0 \leqslant t < \alpha_{k+1}-\beta_k$ 时。

设 $R<+\infty$。$X=\{x(t), t<\sigma\}$ 是非最小 Q 过程。令

$$\tau_1^{(n)} \text{ 为 } X \text{ 的第一个飞跃点}, \tag{4.2}$$

$$\beta_1^{(n)} = \begin{cases} \inf\{t: \tau_1^{(n)} \leqslant t<\sigma, x(t) \leqslant n\}, \\ \sigma, \text{上集合为空集}. \end{cases} \tag{4.3}$$

设 $\tau_{m-1}^{(n)}, \beta_{m-1}^{(n)}$ 已定义，若 $\beta_{m-1}^{(n)}=\sigma$，则令 $\tau_m^{(n)}=\beta_m^{(n)}=\sigma$，否则令

$$\tau_m^{(n)} \text{ 为 } \beta_{m-1}^{(n)} \text{ 后的第一个飞跃点}, \tag{4.4}$$

$$\beta_m^{(n)} = \begin{cases} \inf\{t: \tau_m^{(n)} \leqslant t<\sigma, x(t) \leqslant n\}, \\ \sigma, \text{上集合为空集}. \end{cases} \tag{4.5}$$

则由引理 1，对几乎一切 ω，若 $\beta_{m-1}^{(n)}(\omega)<\sigma(\omega)$，则 $\beta_{m-1}^{(n)}(\omega)<\tau_m^{(n)}(\omega) \leqslant \sigma(\omega)$，从而对几乎一切 ω 有

$$0 \equiv \beta_0(\omega) \leqslant \tau_1^{(n)}(\omega) \leqslant \beta_1^{(n)}(\omega) \leqslant \cdots \leqslant \tau_k^{(n)}(\omega) \leqslant \beta_k^{(n)}(\omega) \leqslant \cdots \leqslant \sigma(\omega).$$

对函数 $X(\omega)=\{x(t, \omega), t<\sigma(\omega)\}$ 施行 $C(\tau_k^{(n)}(\omega), \beta_k^{(n)}(\omega))$ 变换而得 $X_n(\omega)=\{x_n(t, \omega), t<\sigma_n(\omega)\}$。我们记此变换为

$$g_n(X(\omega))=X_n(\omega). \tag{4.6}$$

又令变换 $g_{n+1,n}=g_n g_{n+1}$，即 $g_{n+1,n}(X)=g_n[g_{n+1}(X)]$，则显然

$$g_{n+1,n}(X_{n+1}(\omega))=X_n(\omega). \tag{4.7}$$

特别，如果 X 是 (Q, π) Doob 过程，并且 $\pi_j=0(j>n)$，那么

$$g_n(X(\omega))=X(\omega) \tag{4.8}$$

定理 3 设 $R<+\infty$，$X=\{x(t), t<\sigma\}$ 是可分 Borel 可测右连续非最小 Q 过程。则由 (4.6) 确定的 $X_n=\{x_n(t), t<\sigma_n\}$ 是 $(Q, V^{(n)})$ Doob 过程，满足 (4.7)。这里 $V^{(n)}=(v_j^{(n)}, 0 \leqslant j \leqslant n)$：

$$v_j^{(n)}=P(x(\beta_1^{(n)})=j)(-1 \leqslant j \leqslant n) \tag{4.9}$$

满足

$$\begin{cases} v_j^{(n)} = \dfrac{v_j^{(n+1)}}{\sum\limits_{k=-1}^{n} v_k^{(n+1)} + v_{n+1}^{(n+1)} C_{n+1,n}}, (-1 \leqslant j < n), \\ v_n^{(n)} = \dfrac{v_n^{(n+1)} + v_{n+1}^{(n+1)} C_{n+1,n}}{\sum\limits_{k=-1}^{n} v_k^{(n+1)} + v_{n+1}^{(n+1)} C_{n+1,n}}, \\ \sum\limits_{j=-1}^{n} v_j^{(n)} = 1, \sum\limits_{j=0}^{n} v_j^{(n)} > 0. \end{cases} \quad (4.10)$$

X 不断的充分必要条件是 $v_{-1}^{(n)} = 0$，$(n \geqslant 0)$．

证 易证 $\beta_k^{(n)}$ 是马尔可夫时间，满足（2.11）．首先证由（4.9）确定的量满足（4.10）．

令 $\tilde{\beta}_1^{(n)}$ 为对 Q 过程 $\widetilde{X} = \{\tilde{x}(t), t < \tilde{\sigma}\} = \{x(\beta_1^{(n+1)} + t), t < \sigma - \beta_1^{(n+1)}\}$ 按（4.2）（4.3）方式确定的量，Λ_n 为过程 $\widetilde{X}$ 自 $n+1$ 出发经有穷步（包括零步）跳跃到达 n 的事件，$\overline{\Lambda}_n$ 为 Λ_n 的对立事件．易验证，$(x(\beta_1^{(n+1)}) = n+1)$，$\Lambda_n$，$\overline{\Lambda}_n \in \mathscr{B}_{[0, \tau_2^{(n+1)}]}$，$(x(\beta_1^{(n+1)}) = j) \in \mathscr{F}_{\tau_2}'^{(n+1)}$．由引理 4，对 $-1 \leqslant j \leqslant n$，

$$\Delta_j^{(n)} \equiv P(x(\beta_1^{(n+1)}) = n+1, \overline{\Lambda}_n, \tilde{x}(\tilde{\beta}_1^{(n)}) = j)$$
$$= P(x(\beta_1^{(n+1)}) = n+1, \overline{\Lambda}_n) P(\tilde{x}(\tilde{\beta}_1^{(n)}) = j \mid x(\beta_1^{(n+1)}) = n+1)$$
$$= P(x(\beta_1^{(n+1)}) = n+1) P(\overline{\Lambda}_n \mid x(\beta_1^{(n+1)}) = n+1) P(x(\beta_1^{(n)}) = j)$$
$$= v_{n+1}^{(n+1)} (1 - C_{n+1,n}) v_j^{(n)}.$$

由引理 1(ii)，

$$v_n^{(n)} = P(x(\beta_1^{(n)}) = n) = P(x(\beta_1^{(n+1)}) = n) + P(x(\beta_1^{(n+1)}) = n+1, \Lambda_n) + \Delta_n^{(n)}$$
$$= v_n^{(n+1)} + v_{n+1}^{(n+1)} C_{n+1,n} + v_{n+1}^{(n+1)} (1 - C_{n+1,n}) v_n^{(n)},$$

而对 $-1 \leqslant j < n$，则

$$v_j^{(n)} = P(x(\beta_1^{(n)}) = j) = P(x(\beta_1^{(n+1)}) = j) + \Delta_j^{(n)}$$
$$= v_j^{(n+1)} + v_{n+1}^{(n+1)} (1 - C_{n+1,n}) v_j^{(n)}.$$

于是

$$\begin{cases} v_j^{(n)} = \dfrac{v_j^{(n+1)}}{1-v_{n+1}^{(n+1)}(1-C_{n+1,n})}, & (-1 \leqslant j \leqslant n), \\ v_n^{(n)} = \dfrac{v_n^{(n+1)} + v_{n+1}^{(n+1)} C_{n+1,n}}{1-v_{n+1}^{(n+1)}(1-C_{n+1,n})}. & \end{cases} \quad (4.11)$$

由上式知或者对一切 n，$P(\beta_1^{(n)} < \sigma) > 0$，或者对一切 n，$P(\beta_1^{(n)} = \sigma) = 0$. 若后者成立，则由于 $P(\lim\limits_{n\to+\infty} \beta_1^{(n)} = \tau_1) = 1$[见 5，4°]，因此 $P(\tau_1 = \lim\limits_{n\to+\infty} \beta_1^{(n)} = \sigma) = 1$，$X$ 是最小 Q 过程，与定理假设不合，于是 $\sum\limits_{n=0}^{n} v_j^{(n)} = P(\beta_1^{(n)} < \sigma) > 0, (n \geqslant 0)$.

往证 $\sum\limits_{j=-1}^{n} v_j^{(n)} = 1$. 由此及(4.11)，便可得证(4.10).

当 X 不中断时，[2，定理 5.2]已证明 $P(\beta_1^{(n)} < +\infty) = 1$. 因此 $v_{-1}^{(n)} = 0$，$\sum\limits_{j=0}^{n} v_j^{(n)} = 1$.

设 X 中断. 由定理 1(取 $\pi_0 = 1$，$\pi_j = 0 (j > 0)$)，我们可构造 $X = \{x(t), t < \sigma\}$ 的不中断的延拓 Q 过程 $\widetilde{X} = \{\tilde{x}(t), t < +\infty\}$，即

$$\begin{cases} x(t) = \tilde{x}(t), & t < \sigma, \\ P(\tilde{x}(\sigma) = 0 \mid \sigma < +\infty) = 1. \end{cases} \quad (4.12)$$

令 $\tilde{\beta}_1^{(n)} = \inf\{t: t \geqslant \tau_1, \tilde{x}(t) \leqslant n\}$，显然 $\beta_1^{(n)} = \tilde{\beta}_1^{(n)}$. 既然 $\widetilde{X}$ 不中断，$P(\tilde{\beta}_1^{(n)} < +\infty) = 1$，因此 $\sum\limits_{j=-1}^{n} v_j^{(n)} = P(\beta_1^{(n)} < +\infty) = 1$.

设 $v_{-1}^{(n)} = 0$，即 $P(\beta_1^{(n)} < \sigma) = \sum\limits_{j=0}^{n} v_j^{(n)} = 1$. 由强马氏性，$P(\beta_1^{(n)} < \beta_2^{(n)} < \cdots \beta_k^{(n)} < \cdots < \sigma) = 1$. 此说明对几乎一切 ω，在[0, $\sigma(\omega)$)中有无穷多个 $j = j(\omega)(\leqslant n)$ 区间，由[1，P149 定理 7]，$P(\sigma = +\infty) = 1$，即 X 不中断.

往证 $X_n = g_n(X)$ 是 $(Q, V^{(n)})$ Doob 过程. 令 $X^{(m)} = \{x(\beta_m^{(n)} +$

$t)$, $t<\tau_{m+1}^{(n)}-\beta_m^{(n)}\}$, $(m\geqslant 0)$. $X^{(m)}(m\geqslant 0)$ 满足定理 2 中的性质 (i*)~(iii*) 是平凡的. 由于 $R<+\infty$ 及引理 1, $\beta_m^{(n)}<\sigma$ 当且只当 $0<\tau_{m+1}^{(n)}-\beta_m^{(n)}<+\infty$. 由引理 4,

$$P(x^{(m+1)}(0)=j \mid 0<\tau_{m+1}^{(n)}-\beta_m^{(n)}<+\infty)$$
$$=P(x(\beta_{m+1}^{(n)})=j \mid \beta_m^{(n)}<\sigma)=v_j^{(n)}, \quad (0\leqslant j\leqslant n).$$

因此性质(iv*)满足. 性质(v*)对 $\Delta=(x^{(m+1)}(0)=i)$ 满足可由 X 的强马氏性直接推出. 不难验证 $X^{(m)}(m\leqslant k)$ 是 $\mathscr{B}_{[0,\tau_{k+1}^{(n)}]}$ 可测的, 而 $X^{(m)}(m>k)$ 是 $\mathscr{F}'_{\tau_{k+1}^{(n)}}$ 可测的. 又已指出 $(0<\tau_{k+1}^{(n)}-\beta_k^{(n)}<+\infty)=(\beta_k^{(n)}<\sigma)$. 由引理 4, 性质(v*)是满足的. 于是按照定理 2 方式而得到的过程是 $(Q, V^{(n)})$ Doob 过程. 但这样得到的过程正是 $X_n=g_n(X)$.

定理证完.

引理 5 对任意 Q 过程 $X=\{x(t), t<\sigma\}$, $P(\sigma<+\infty)=0$ 或 1.

证 当 X 最小或不中断时, 引理显然正确. 不失一般性, 可设 X 可分 Borel 可测右连续中断的非最小 Q 过程.

由引理 3, $P(\beta_1^{(n)}=\sigma=+\infty)=1-\sum_{j=-1}^n v_j^{(n)}=0$, 即

$$P(\beta_1^{(n)}<\sigma=+\infty)+P(\sigma<+\infty)=1. \quad (4.13)$$

由强马氏性

$$P(\beta_1^{(n)}<\sigma=+\infty)=\sum_{j=0}^n v_j^{(n)} P_j(\sigma=+\infty),$$
$$P_j(\sigma=+\infty)=P_j(\sigma-\tau_1=+\infty)$$
$$=\frac{a_j}{a_j+b_j}P_{j-1}(\sigma=+\infty)+\frac{b_j}{a_j+b_j}P_{j+1}(\sigma=+\infty).$$

故 $P_j(\sigma=+\infty)=P(\sigma=+\infty)$ 与 j 无关, 因此(4.13)成为

$$\left(\sum_{j=0}^n v_j^{(n)}\right)P(\sigma=+\infty)+P(\sigma<+\infty)=1.$$

由定理 3，$0 < \sum_{j=0}^{n} v_j^{(n)} < 1$. 因此从上式必定有 $P(\sigma=+\infty)=0$，即 $P(\sigma<+\infty)=1$. 引理证完.

定理 4 设 $R<+\infty$. $X=\{x(t), t<\sigma\}$ 为可分 Borel 可测右连续非最小 Q 过程. 则存在非负数列 $p, q, r_n(n \geqslant -1)$ 满足

$$\begin{cases} p+q=1 \text{ 且 } S=+\infty \text{ 时 } q=0, \\ r_n=0(n \geqslant 0), \ p=0, \\ 0 < \sum_{n=0}^{+\infty} r_n N_n < +\infty, p > 0. \end{cases} \quad (4.14)$$

使得
$$v_j^{(n)} = P(x(\beta_1^{(n)})=j), \ -1 \leqslant j \leqslant n, \quad (4.15)$$

可表示为

$$\begin{cases} v_j^{(n)} = \dfrac{X_n}{A_n} r_j, \ -1 \leqslant j < n, \\ v_n^{(n)} = Y_n + \dfrac{X_n}{A_n} \sum_{l=n}^{+\infty} r_l C_{l_n}. \end{cases} \quad (4.16)$$

其中

$$\begin{cases} A_n = \sum_{l=0}^{+\infty} r_l C_{l_n}, \\ X_n = \dfrac{A_n C_{n0}}{(r_{-1}+A_n)C_{n0}+d}, \\ Y_n = \dfrac{d}{(r_{-1}+A_n)C_{n0}+d}, \\ d = \begin{cases} \dfrac{q}{p} A_0, p > 0 \\ 1, p = 0, \end{cases} \\ \dfrac{X_n}{A_n} = \dfrac{C_{n0}}{r_{-1}C_{n0}+1}, p=0. \end{cases} \quad (4.17)$$

如果令 η 为 $\beta_1^{(0)}$ 前的最后一个飞跃点[2, §4]，那么

$$P(x(\eta)=j)=\begin{cases}\dfrac{r_{-1}}{r_{-1}+A_0+d}, & j=-1,\\[2pt] \dfrac{r_jC_{j0}}{r_{-1}+A_0+d}, & 0\leqslant j<+\infty,\\[2pt] \dfrac{d}{r_{-1}+A_0+d}, & j=+\infty.\end{cases} \quad (4.18)$$

p, q 由 X 唯一决定. 若 $p>0$, 则 $r_n(n\geqslant-1)$ 除常数因子外由 X 唯一决定. 若 $p=0$, 则 $r_n(n\geqslant-1)$ 由 X 唯一决定.

证 分几步证明.

i) 设 $v_j^{(n)}(-1\leqslant j\leqslant n)$ 由 (4.15) 确定. 令

$$R_n=\sum_{i=0}^{n-1}v_j^{(n)}C_{i0},\ S_n=v_n^{(n)}C_{n0},\ \Delta_n=R_n+S_n,$$

则由 (4.10),

$$0<\Delta_n=\frac{\Delta_{n+1}}{\delta_{n+1}},$$

$$S_n=\frac{v_n^{(n+1)}C_{n0}+S_{n+1}}{\delta_{n+1}},$$

其中

$$\delta_{n+1}=\sum_{j=-1}^{n}v_j^{(n+1)}+v_{n+1}^{(n+1)}C_{n+1,n}.$$

故由 (4.10), $v_j^{(n)}/\Delta_n$ 与 $n>(j\geqslant-1)$ 无关, 而且极限

$$\frac{S_n}{\Delta_n}\downarrow q\geqslant0,\ \frac{R_n}{\Delta_n}\uparrow p\geqslant0 \quad (4.19)$$

存在. 如果 $p=0$, 取 $r=1$; 如果 $p>0$, 任取正数 $r>0$. 令

$$r_j=\frac{v_j^{(n)}}{\Delta n}r,\ n>j\geqslant-1. \quad (4.20)$$

这样我们得到非负数列 $p, q, r_n(n\geqslant-1)$.

ii) 显然 $p+q=1$; 又 $p=0$ 时, 显然 $r_n=0(n\geqslant0)$. 若 $p>0$, 则至少存在一个 $r_K>0(K\geqslant0)$, 故只需证明

$$\sum_{n=0}^{+\infty}r_nN_n<+\infty. \quad (4.21)$$

我们注意 $E\beta_1^{(0)} < +\infty$. 当 X 不中断时，此事实已在[2, 定理5.2]中证明；当 X 中断时，则由定理1，存在不中断的 Q 过程 $\widetilde{X}$ 满足(4.12)，故也有 $E\beta_1^{(0)} = E\widetilde{\beta}_1^{(0)} < +\infty$. 于是我们有

$$E\tau_2^{(0)} = E\beta_1^{(0)} + E(\tau_2^{(0)} - \beta_1^{(0)}) = E\beta_1^{(0)} + v_0^{(0)} E_0 \tau$$
$$= E_0 \beta_1^{(0)} + v_0^{(0)} R < +\infty.$$

令

$$M_i^{(n)} = \begin{cases} \tau_{i+1}^{(n)} - \beta_i^{(n)}, & \beta_i^{(n)} < \beta_1^{(0)}, \\ 0, & \text{反之}, \end{cases}$$

显然 $\tau_2^{(0)} \geqslant \sum_{i=1}^{+\infty} M_i^{(n)}$. 设 $C(\wedge)$ 为 $\wedge$ 的示性函数，则

$$+\infty > E\tau_2^{(0)} \geqslant \sum_{i=1}^{+\infty} EM_i^{(n)}$$
$$= \sum_{i=1}^{+\infty} E\{(\tau_{i+1}^{(n)} - \beta_i^{(n)}) C(\beta_i^{(n)} < \beta_1^{(0)})\},$$

但

$$E\{(\tau_2^{(n)} - \beta_1^{(n)}) C(\beta_1^{(n)} < \beta_1^{(0)})\} = \sum_{j=1}^{n} v_j^{(n)} E_j \tau_1^{(n)} = \sum_{j=1}^{n} v_j^{(n)} N_1,$$
$$E\{(\tau_{i+1}^{(n)} - \beta_i^{(n)}) C(\beta_i^{(n)} < \beta_i^{(0)})\}$$
$$= \sum_{j=1}^{n} P(\beta_{i-1}^{(n)} < \beta_1^{(n)}, x(\beta_i^{(n)}) = j) E_j \tau_1^{(n)}$$
$$= \sum_{j=1}^{n} \Big(\sum_{k=1}^{n} v_k^{(n)} (1 - C_{k0})\Big)^{i-1} v_j^{(n)} N_j,$$

故

$$+\infty > E\tau_2^{(0)} \geqslant \frac{\sum_{j=1}^{n} v_j^{(n)} N_j}{1 - \sum_{k=1}^{n} v_k^{(n)} (1 - C_{k0})} \geqslant \frac{\sum_{j=1}^{n-1} v_j^{(n)} N_j}{v_{-1}^{(n)} + \Delta_n},$$

由(4.20)即

$$+\infty > E\tau_2^{(0)} \geqslant \frac{\sum_{j=1}^{n-1} r_j N_j}{r_{-1} + r}.$$

令 $n \to +\infty$ 便得(4.21).

iii) 往证(4.16).

由于(4.19)(4.20),

$$P = \lim_{n \to +\infty} \frac{R_n}{\Delta_n} \lim_{n \to +\infty} \frac{1}{r} \sum_{i=1}^{n-1} r_i C_{i0} = \frac{A_0}{r},$$

因此

$$r = \begin{cases} \dfrac{A_0}{p}, & p > 0, \\ 1, & p = 0 \end{cases} \tag{4.22}$$

由(4.10)用归纳法易得：$m > n$,

$$\begin{cases} v_j^{(n)} = \dfrac{v_j^{(m)}}{\delta_{m,n}}, & -1 \leqslant j < n, \\[2mm] v_n^{(n)} = \dfrac{\sum\limits_{j=n}^{m} v_j^{(m)} C_{jn}}{\delta_{m,n}}, \\[2mm] 0 < \Delta_n = \dfrac{\Delta_m}{\delta_{m,n}}, \\[2mm] \delta_{m,n} = \sum\limits_{j=-1}^{n} v_j^{(m)} + \sum\limits_{i=n+1}^{m} v_j^{(m)} C_{jn}. \end{cases} \tag{4.23}$$

由此及(4.19)(4.20),

$$\frac{v_n^{(n)}}{\Delta_n} = \frac{\sum\limits_{j=n}^{m} v_j^{(n)} C_{jm}}{\Delta_m} = \frac{1}{r} \sum_{j=n}^{m-1} r_j c_{jn} + \frac{S_m}{\Delta_m C_{n0}},$$

从而当 $m \to +\infty$ 时有

$$\frac{v_n^{(n)}}{\Delta_n} = \frac{1}{r} \sum_{j=n}^{+\infty} r_j c_{jn} + \frac{q}{C_{n0}}. \tag{4.24}$$

由(4.20)(4.24),

$$1 = \sum_{j=-1}^{n} v_j^{(n)} = \Delta_n \left(\frac{\sum\limits_{j=-1}^{n-1} r_j}{r} + \frac{\sum\limits_{j=n}^{+\infty} r_j C_{jn}}{r} + \frac{q}{C_{n0}} \right).$$

因此
$$\Delta_n = \frac{rC_{n0}}{(r_{-1}+A_n)C_{n0}+qr},$$
代回(4.20)(4.24)中并注意(4.22)便得(4.16).

iv) 往证(4.17).

若 X 不中断,则 $1 = \sum_{j=0}^{n} v_j^{(n)} = P(\beta_1^{(n)} < +\infty)$. 由强马氏性,$P(\beta_1^{(n)} < \beta_2^{(n)} < \cdots < +\infty) = 1$,故对几乎一切 ω,在 $[0, \lim_{l \to +\infty} \beta_l^{(n)})$ 中有无穷多个 $j(\omega)(\leqslant n)$ 区间,按[1, P149 定理 5],$p(\lim_{l \to +\infty} \beta_l^{(n)} = +\infty) = 1$. 若 X 中断,由引理 5,$P(\sigma < +\infty) = 1$;又由引理 1,对几乎一切 ω,若 $\beta_l^{(n)}(\omega) < \sigma(\omega)$,则 $\beta_l^{(n)}(\omega) < \tau_{l+1}^{(n)}(\omega) \leqslant \beta_{l+1}^{(n)}(\omega)$,因此按[1, P149 定理 5],$P(存在 l 使 \beta_l^{(n)} = \sigma) = 1$. 这样,不论 X 中断与否,恒有
$$P(\lim_{l \to +\infty} \beta_l^{(n)} = \sigma) = 1. \tag{4.25}$$

而且对几乎一切 ω,存在唯一的 l 使 $\beta_{l-1}^{(n)} < \eta < \beta_l^{(n)} \leqslant \beta_l^{(0)}$,记此 l 为 l_n,即
$$l_n = \min(l : \beta_l^{(n)} \geqslant \eta),$$
$$\beta_{l_n}^{(n)} = \begin{cases} \inf\{t : \eta \leqslant t < \sigma, \ x(t) \leqslant n\}, \\ \sigma, \quad 上集合是空集. \end{cases}$$

如果对某 $n > j \geqslant -1$ 有 $x(\beta_{l_n}^{(n)}) = j$,那么必定 $\beta_{l_n}^{(n)} = \eta$. 因为如果 $\eta < \beta_{l_n}^{(n)}$,按定义对 $t \in (\eta, \beta_{l_n}^{(n)})$ 有 $n < x(t) < +\infty$,而 $x(\beta_{l_n}^{(n)}) = j < n$. 由引理 1(ii),此不可能. 那么对 $n > j \geqslant -1$,
$$(x(\beta_{l_n}^{(n)}) = j) \subset (\beta_{l_n}^{(n)} = \eta, \ x(\eta) = j) \subset (x(\eta) = i). \tag{4.26}$$
又显然,如果 $x(\eta) = j$,那么对 $n > j$ 有 $\beta_{l_n}^{(n)} = \eta$,即
$$(x(\eta) = j) \subset (\beta_{l_n}^{(n)} = \eta, \ x(\eta) = j) \subset (x(\beta_{l_n}^{(n)}) = j),$$
故
$$(x(\eta) = j) = (x(\beta_{l_n}^{(n)}) = j), \quad -1 \leqslant j < n.$$

往证

$$(x(\eta)=+\infty)=\lim_{n\to+\infty}(x(\beta_{l_n}^{(n)})=n).$$

设 $\omega\in$ 右边，则必定 $\beta_{l_n}^{(n)}\downarrow\eta\geqslant\eta$，从而由 X 的右连续性，$x(\eta)=\lim_{n\to+\infty}x(\beta_{l_n}^{(n)})=\lim_{n\to+\infty}n=+\infty$，但由 η 的定义，对任 $\eta<t<\beta_1^{(0)}$，$x(t)\neq+\infty$. 因此 $\eta=\eta$，$x(\eta)=+\infty$，即 $\omega\in$ 左边. 设 $\omega\in$ 左边，则对任意 n 必定 $x(\beta_{l_n}^{(n)})=n$，因为不然由 (4.26) 将导致

$$x(\eta)=j\neq+\infty.$$

这样

$$P(x(\eta)=j)=P(x(\beta_{l_n}^{(n)})=j),\quad(-1\leqslant j<n)$$

$$P(x(\eta)=+\infty)=\lim_{n\to+\infty}P(x(\beta_{l_n}^{(n)})=n)$$

但对 $0\leqslant j\leqslant n$,

$$P(x(\beta_{l_n}^{(n)})=j)=\sum_{l=1}^{+\infty}P(x(\beta_l^{(n)})=j,l_n=l)$$

$$=\sum_{l=1}^{+\infty}P(\beta_{l-1}^{(n)}<\beta_1^{(0)},x(\beta_l^{(n)})=j,\beta_1^{(n)}\leqslant\beta_1^{(0)}<\tau_{l+1}^{(n)})$$

$$=\sum_{l=1}^{+\infty}(\sum_{k=1}^{n}v_k^{(n)}(1-C_{k0}))^{l-1}v_j^{(n)}C_{j0}$$

$$=\frac{v_j^{(n)}C_{j0}}{v_{-1}^{(n)}+\Delta_n},$$

$$P(x(\beta_{l_n}^{(n)})=-1)=\sum_{l=1}^{+\infty}p(x(\beta_l^{(n)})=-1,l_n=l)$$

$$=\sum_{l=1}^{+\infty}P(\beta_{l-1}^{(n)}<\beta_1^{(0)},x(\beta_l^{(n)})=-1)$$

$$=\sum_{l=1}^{+\infty}\sum_{k=1}^{n}v_k^n(1-C_{k0})^{l-1}v_{-1}^{(n)}$$

$$=\frac{v_{-1}^{(n)}}{v_{-1}^{(n)}+\Delta_n},$$

注意 (4.19)(4.20)，我们得

$$P(x(\eta)=j)=\begin{cases}\dfrac{r_{-1}}{r_{-1}+r}, & j=-1,\\[2mm]\dfrac{r_jC_{j0}}{r_{-1}+r}, & 0\leqslant j<+\infty,\\[2mm]\dfrac{rq}{r_{-1}+r}, & j=+\infty.\end{cases}$$

再注意(4.22)，上式即(4.18).

v) 设 $S=+\infty$. 令 τ_t 为 t 之前的最后一个飞跃点[2, P147]. 令

$$\xi_0(\omega)=\begin{cases}\inf\{t: x(\tau_t,\omega)=+\infty, x(t,\omega)=0\},\\ \sigma, \text{上集合是空集}.\end{cases} \quad (4.27)$$

若 X 不中断，则[2, §4]已证明 $P(\xi_0(\omega)<+\infty)=0$. 若 X 中断，则由定理1，存在 X 的延拓过程 $\widetilde{X}$ 满足(4.12). 既然对 $\widetilde{X}$ 按(4.27)方式定义的 $\widetilde{\xi}_0$ 有 $P(\widetilde{\xi}_0<+\infty)=0$, 又 $(\xi_0<\sigma)\subset(\widetilde{\xi}_0<+\infty)$, 故 $P(\xi_0<\sigma)=0$. 但 $(x(\eta)=+\infty)\subset(\xi_0<\sigma)$, 因此 $P(x(\eta)=+\infty)=0$. 由(4.16)(4.18)，$q=0$.

vi) p, q 由 X 唯一决定，当 $p>0$ 时 $r_n(n\geqslant-1)$ 除常数因子外由 X 唯一决定的证明可仿[2, 定理7.1]进行. 当 $p=0$ 时，$r_n(n\geqslant-1)$ 由 X 唯一决定亦可类似证明.

定理证完.

定义 2 称 $p, q, r_n(n\geqslant-1)$ 为 Q 过程 X 的特征数列.

§5. Q 过程的构造

设 $R<+\infty$. 任给非负数列 p, q, $r_n(n\geqslant -1)$ 满足(4.14). 按(4.16)(4.17)可定义 $v_j^{(n)}(-1\leqslant j\leqslant n)$. 仿[2,定理 7.1]可证 $v_j^{(n)}(-1\leqslant j\leqslant n)$ 满足(4.10). 因此, 仿[2,引理 3.3]的证明可得: 存在概率空间 $(\Omega, \mathscr{F}, P)$, 在其上可定义一列 $(Q, V^{(n)})$ Doob 过程 $X_n=\{x_n(t), t<\sigma_n\}(n\geqslant 0)$ 满足(4.7). 这里
$$V^{(n)}=(V_j^{(n)}, 0\leqslant j\leqslant n).$$

定理 5 存在 $\Omega_0\in\mathscr{F}$, $P(\Omega_0)=1$. 对任 $\omega\in\Omega_0$, 当 $n\to+\infty$ 时, $\sigma_n(\omega)\uparrow\sigma(\omega)$ 存在, 而且对几乎一切 $t\in[0, \sigma(\omega))$①, $x_n(t,\omega)$ 收敛.

证 由(4.7), $\sigma_n(\omega)\leqslant\sigma_{n+1}(\omega)$, 故可令 $\sigma(\omega)=\lim\limits_{n\to+\infty}\sigma_n(\omega)$.

设按(4.2)~(4.5)方式对 X_k 而定义的量记为 $\tau_m^{(k,n)}$, $\beta_m^{(k,n)}$. 由(4.7), $\beta_m^{(n,0)}\leqslant\beta_m^{(n+1,0)}\leqslant\sigma_{n+1}$, 故存在极限 $\beta_m^{(0)}=\lim\limits_{n\to+\infty}\beta_m^{n,0}\leqslant\sigma$. 由(4.25), $\lim\limits_{m\to+\infty}\beta_m^{(0)}\geqslant\lim\limits_{m\to+\infty}\beta_m^{(n,0)}=\sigma_n$. 因此
$$P(\lim_{m\to+\infty}\beta_m^{(0)}=\sigma)=1. \tag{5.1}$$

仿[2, P154~155]定义 T_ε 及 L_m. 即对 $n>m$, 令
$$L_{n,m}^{(i)}=\begin{cases}\beta_i^{(n,m)}-\tau_i^{(n,m)}, & \beta_i^{(n,m)}<\beta_1^{(n,m)},\\ 0, & \text{反之},\end{cases}$$
$$T_\varepsilon^{(n,m)}=\sum_{\tau_1^{(n)}\leqslant\tau_{ij}^{(n)}<\beta_1^{(n,m)}}f_\varepsilon(\tau_{i,j+1}^{(n)}-\tau_{i,j}^{(n)})$$

这里 $\tau_{i,j}^{(n)}$ 为 X_n 过程的第 i 个飞跃点后的第 j 个跳跃点, $f_\varepsilon(x)$ 如 (2.18). 由(4.7)易得

---

① 对勒贝格测度而言.

中断生灭过程构造中的概率分析方法

$$L_{n,m} \equiv \sum_{i=1}^{+\infty} L_{(n,m)}^{(i)} \leqslant L_{n+1,m} \equiv \sum_{i=1}^{+\infty} L_{n+1,m}^{(i)},$$

$$L_{n,m+1} \leqslant L_{n,m},$$

$$T_{\varepsilon}^{(n,0)} \leqslant T_{\varepsilon}^{(n+1,0)},$$

$$T_{\varepsilon_1}^{(n,0)} \leqslant T_{\varepsilon_2}^{(n,0)}, \ \varepsilon_1 < \varepsilon_2.$$

故可令

$$L_{n,m} \uparrow L_m, \ (n \uparrow +\infty);$$

$$L_m \uparrow L, \ (m \uparrow +\infty);$$

$$T_{\varepsilon}^{(n,0)} \uparrow T_{\varepsilon}, \ (n \uparrow +\infty);$$

$$T_{\varepsilon} \downarrow T, \ (\varepsilon \downarrow 0)$$

如果能够证明

$$P(L=0) = P(T=0) = 1, \qquad (5.2)$$

由此及(5.1),逐字重复[2,定理 6.1]的证明便可得证定理.

为证(5.2),令

$$\bar{\beta}_i^{(n,m)} = \inf\{t: t \geqslant \tau_{i0}^{(n)}, \ x_n(t) \leqslant m\},$$

考虑

$$\bar{L}_{n,m}^{(i)} = \begin{cases} \min(\tau_{i+1,0}^{(n)}, \bar{\beta}_i^{(n,m)}) - \tau_{i,0}^{(n)}, & \tau_{i0}^{(n)} < \beta_1^{(n,0)} \text{ 且 } m < x_n(\tau_{i0}^{(n)}), \\ 0, & \text{反之}. \end{cases}$$

$$\bar{T}_{\varepsilon_i}^{(n,0)} = \begin{cases} \sum_{\tau_{i0}^{(n)} \leqslant \tau_{ij}^{(n)} < \min(\tau_{i+1,0}^{(n)}, \bar{\beta}_i^{(n,0)})} f_{\varepsilon}(\tau_{i,j+1}^{(n)} - \tau_{ij}^{(n)}), & \tau_{i0}^{(n)} < \beta_1^{(n,0)}, \\ 0, & \text{反之}. \end{cases}$$

则

$$\sum_{i=1}^{+\infty} \bar{L}_{n,m}^{(i)} = L_{n,m},$$

$$\sum_{i=1}^{+\infty} \bar{T}_{\varepsilon_i}^{(n,0)} = T_{\varepsilon}^{(n,0)}.$$

但由引理 2 及(4.16)(4.17),以及定理 1 中的性质(v),

$$EL_{n,m} = \sum_{i=1}^{+\infty} E\bar{L}_{n,m}^{(i)} = \sum_{i=1}^{+\infty} \sum_{j=m+1}^{n} P(\tau_{i0}^{(n)} < \beta_1^{(n,0)}, x_n(\tau_{i0}^{(n)}))$$

$$= j)E\{\min(\tau_{i+1,0}^{(n)}, \bar{\beta}_i^{(n,m)}) - \tau_{i,0}^{(n)} \mid x_n(\tau_{i0}^{(n)}) = j\}$$

$$= \sum_{i=1}^{+\infty} \sum_{j=m+1}^{+\infty} \Big(\sum_{k=1}^{+\infty} v_k^{(n)}(1-C_{k0})\Big)^{i-1} v_j^{(n)} H_{jm}^{(+\infty)}$$

$$= \frac{\sum_{j=m+1}^{n} v_j^{(n)} H_{jm}^{(m)}}{v_{-1}^{(n)} + \sum_{k=0}^{n} v_k^{(n)} C_{k0}}$$

$$= \frac{\sum_{j=m+1}^{n-1} r_i H_{jm}^{(m)} + \Big(\dfrac{d}{C_{n0}} + \sum_{l=n}^{+\infty} r_l C_{ln}\Big) H_{nm}^{(+\infty)}}{r_{-1} + A_0 + d},$$

$$ET_\varepsilon^{(n,0)} = \sum_{i=1}^{+\infty} E\overline{T}_{\varepsilon_i}^{(n,0)} = \sum_{i=1}^{+\infty} \sum_{k=1}^{n} P(\tau_{i0}^{(n)} < \beta_1^{(n,0)}, x_n(\tau_{i0}^{(n)})$$

$$= K)E\Big\{\sum_{\tau_{i0}^{(n)} \leqslant \tau_{ij}^{(n)} < \min(\tau_{i+1,0}^{(n)}, \bar{\beta}_i^{(n,0)})} f_\varepsilon(\tau_{i,j+1}^{(n)} - \tau_{ij}^{(n)}) \mid x_n(\tau_{i0}^{(n)}) = K\Big\}$$

$$= \sum_{i=1}^{+\infty} \sum_{k=1}^{n} \Big(\sum_{j=1}^{n} v_j^{(n)}(1-C_{j0})\Big)^{i-1} v_k^{(n)} H_{k0}^{(\varepsilon)}$$

$$= \frac{\sum_{k=1}^{n} v_k^{(n)} H_{k0}^{(\varepsilon)}}{v_{-1}^{(n)} + \sum_{j=1}^{n} v_j^{(m)} C_{j0}} = \frac{\sum_{k=1}^{n-1} r_k H_{k0}^{(\varepsilon)} + \Big(\dfrac{d}{c_{n0}} + \sum_{e=n}^{+\infty} r_e C_{en}\Big) H_{n0}^{(\varepsilon)}}{r_{-1} + A_0 + d}.$$

由 (2.7) 可验证 $C_{en} \leqslant N_e (e \geqslant n)$,由引理 2 及 (4.14)(4.17),

$H_{jm}^{(\varepsilon)} \leqslant N_j$, $\sum_{j=0}^{+\infty} r_j N_j < +\infty$, $\lim_{\varepsilon \downarrow 0} H_{k0}^{(\varepsilon)} = 0$; 若 $S = +\infty$, 则 $d = 0$; 若 $S < +\infty$, 则 $\lim_{m \to +\infty} \lim_{n \to +\infty} \dfrac{H_{nm}^{(+\infty)}}{C_{n_0}} = \lim_{\varepsilon \downarrow 0} \lim_{n \to +\infty} \dfrac{H_{n0}^{(\varepsilon)}}{C_{n_0}} = 0.$ 因此由上两式得

$$EL = \lim_{m \to +\infty} \lim_{n \to +\infty} EL_{nm}$$

$$= \frac{\lim_{m \to +\infty} \sum_{j=m+1}^{+\infty} r_j H_{jm}^{(+\infty)} + \lim_{m \to +\infty} \lim_{n \to +\infty} \Big(\dfrac{d}{C_{no}} + \sum_{l=n}^{+\infty} r_l C_{ln}\Big) H_{nm}^{(+\infty)}}{r_{-1} + A_0 + d} = 0,$$

$$ET = \lim_{\varepsilon \downarrow 0} \lim_{n \to +\infty} ET_\varepsilon^{(n,0)}$$

$$= \frac{\lim_{\varepsilon \downarrow 0}\sum_{k=1}^{+\infty} r_k H_{ko}^{(\varepsilon)} + \lim_{\varepsilon \downarrow 0}\lim_{n\to+\infty}\left(\dfrac{d}{c_{no}} + \sum_{l=n}^{+\infty} r_l C_{ln}\right) H_{no}^{(\varepsilon)}}{r_{-1} + A_0 + d} = 0.$$

由此得(5.29)，于是定理证完．

记 $\sigma(\omega) = \lim\limits_{n\to+\infty}\sigma_n(\omega)$，$x(t,\omega)$ 为 $x_n(t,\omega)$ 的极限函数；若极限不存在，则令 $x(t,\omega) = +\infty$．

定理 6 $X = \{x(t), t < \sigma\}$ 是 Q 过程，它的转移概率 $P_{ij}(t) = \lim\limits_{n\to+\infty} P_{ij}^{(n)}(t)$，这里 $P_{ij}^{(n)}(t)$ 是 $(Q, V^{(n)})$ Doob 过程的转移概率．X 是满足

$$P(x(\beta_1^{(n)}) = i) = v_i^{(n)}, \quad (-1 \leqslant i \leqslant n)$$

的唯一 Q 过程．

证 仿[2，定理 6.2，定理 6.4]证明．

定理 7 设 X 是具特征数列 P，p，$r_n(n \geqslant -1)$ 的 Q 过程，$P_{ij}(t)$ 为其转移概率，则对任 $\lambda > 0$，

$$\int_0^{+\infty} e^{-\lambda t} p_{ij}(t) dt = f_{ij}(\lambda) + \xi_i(\lambda) \frac{\sum_{k=0}^{+\infty} r_k f_{kj}(\lambda) + dz \xi_i(\lambda) y_j}{r_{-1} + \sum_{k=0}^{+\infty} r_k (1 - \xi_k(\lambda)) + dz \sum_{k=0}^{+\infty} \xi_k(\lambda) y_k}.$$

(5.3)

其中 $f_{ij}(\lambda)$，$\xi_i(\lambda)$ 由(3.15)确定，y_k，z 由(2.5)确定，而

$$d = \begin{cases} \dfrac{q}{p} A_0, & p > 0, \\ 1, & p = 0. \end{cases} \quad (5.4)$$

证 利用(3.14)，仿[5，定理 1，定理 2]证明．

总结定理 3~7，我们得本文的基本结果．

基本定理 设 $R < +\infty$．

(i) 设已给可分 Borel 可测右连续非最小 Q 过程 $X = \{x(t), t < \sigma\}$，则它的特征数列 p，q，$r_n(n \geqslant -1)$ 满足(4.14)．

(ii) 设给定一列非负数列 p,q,$r_n(n\geqslant -1)$ 满足 (4.14)，则存在唯一的非最小 Q 过程 $X=\{x(t),t<\sigma\}$，其特征数列重合于此已给数列，此时过程的转移概率 $P_{ij}(t)=\lim\limits_{n\to +\infty}P_{ij}^n(t)$，这里 $P_{ij}^{(n)}(t)$ 是 $(Q,V^{(n)})$ Doob 过程的转移概率，$V^{(n)}=(V_j^{(n)},0\leqslant j\leqslant n)$ 由 (4.16)(4.17) 确定，又 (5.3) 成立. X 不断的充分必要条件是 $r_{-1}=0$. X 满足柯氏向前方程的充分必要条件是 $P=0$.

注 因 $P=0$ 等价于 $r_{n=0}(n\geqslant 0)$，由 (5.3)，上面最后一句话是 [12，定理 5.1] 的平凡推论，用概率方法的直接证明也是可能的，证明同 [2，定理 7.2].

参考文献

[1] Chung K L. Markov Chains with Stationary Transition Probabilites. Springer-Verlag, 1960.

[2] 王梓坤. 生灭过程构造论. 数学进展, 1962, 5: 137-170.

[3] 王梓坤. Классификация всех процессов размножения и гибели, Научные доклады высшей щколы. Физ Матем. Науки, 1958, 4: 19-25.

[3_1] 王梓坤, 杨向群. 中断生灭过程的构造. 数学学报, 1978, 21: 66-71.

[4] 杨超群. 一类生灭过程. 数学学报, 1965, 15: 9-31.

[5] 杨超群. 关于生灭过程构造论的注记. 数学学报, 1965, 15: 174-187.

[6] 杨超群. 柯氏向后微分方程组的边界条件. 数学学报, 1966, 16: 429-452.

[7] 杨超群. 生灭过程的性质. 数学进展, 1966, 9: 365-380.

[8] 杨超群. 双边生灭过程. 南开大学学报(自然科学版),

1964, 5(5): 9-40.

[9] 侯振挺. 齐次可列马尔可夫过程中的概率——分析法. 科学通报, 1973, 3: 115-118.

[10] 孙振祖. 一类马氏过程的一般表达式. 郑州大学学报, 1962, 21: 17-23.

[11] Feller W. On boundaries and lateral conditions for the Kolmogorov's differential equations. Ann. of Math., 1957, 65: 527-570.

[12] Feller W. The birth and death processes as diffusion processes. J. Math. Pures Appl., 1959, 38(9): 301-345.

[13] Reuter G E H. Denumerable Markov processes and the associated semigroup on Ⅰ. Acta Math., 1957, 97: 1-46.

[14] Reuter G E H. Denumerable Markov processes Ⅱ. J. London Math. Soc., 1959, 34: 81-91.

[15] Karlin S, Mcgregor J. The differential equations of birth and death processes and the stieltjes moment problem. Trans. Am. Math. Soc., 1957, 85(2): 489-546.

[16] Дынкин Е Б. Скачкообразные Марковские процессы. Теория Вероят. и ее Прмм. Ⅲ, 1958, 1: 41-60.

Scientia Sinica, 1980, 13(3)

Sojourn Times and First Passage Times for Birth and Death Processes[①]

Abstract In this paper, the Laplace transforms of integral functionals for birth and death processes are found. Particularly, it is proved that the distributions of sojourn times before hitting a state are the mixed exponentials; by appropriate normalizations, their limit distributions can be only of two types: the exponential or the mixed exponential. The parameters involved in these distributions are found. The first passage time can be considered as a special case of sojourn time.

§ 1. Introduction

Let $X = \{x(t, \omega), t \geqslant 0\}$ be a homogeneous denumerable Markov process, defined on probability space $(\Omega, \mathscr{F}, \mathscr{P})$, with state space $E = \mathbf{N}$. We say that it is a birth and death process if

---

① Received: 1979-01-15.

Sojourn Times and First Passage Times for Birth and Death Processes

its transition probabilities $p_{ij}(t)$ satisfy the following conditions:

$$p_{ii}(t) = 1 - (a_i + b_i)t + o(t), \quad t \to 0,$$
$$p_{i,i+1}(t) = b_i t + o(t), \tag{1.1}$$
$$p_{i,i-1}(t) = a_i t + o(t),$$

where $a_0 = 0$, $a_i > 0$ $(i > 0)$, $b_i > 0$ $(i \geq 0)$. Suppose that X is a strong Markov process, separable and measurable, with right lower semicontinuous path[1]. Let

$$\tau_n(\omega) = \inf(t: t \geq 0, x(t, w) = n) \tag{1.2}$$

be the first passage time to state n by X. We shall suppress ω, for example $x(t, \omega) = x(t)$ or x_t, etc. Let $V(i)$ $i \in E$ be a non-negative function. Our aim is to find the distribution of the stochastic integral functional

$$\xi^{(n)} = \int_0^{\tau_n} V(x_t) dt. \tag{1.3}$$

A general method of difference equations was suggested in [2], and was developed by [3]~[5]. This method involves a solution for a system of equations which are not always easy. Here we shall suggest another method of recurrence. We believe it is more simple in many cases.

Specially, if $V(i) = U(i)$,

$$U(i) = 1 \quad (0 \leq i < n); \quad U(i) = 0, \ i \geq n, \tag{1.4}$$

we denote $\xi^{(n+k)}$ in (3) by τ_{nk}, i.e.

$$\tau_{nk} = \int_0^{\tau_{n+k}} U(x_t) dt, \tag{1.5}$$

which is the sojourn time in $(0, 1, 2, \cdots, n-1)$ by X before hitting $n+k$; and τ_{n0} is the first passage time to n, i.e. $\tau_{n0} = \tau_n$.

Therefore, the first passage time is a special case of sojourn time.

In [7], the τ_n's distribution is expressed by an integral, which can be deduced from a general theorem in this paper (see Corollary 2). When the upper limit in integral (1.5) is a constant, the sojourn time was investigated in [6].

§ 2. Distributions of integral functionals

Consider $\xi^{(n)}$ in (3). Let P_k be the law for X starting at k, and denote E_k by the corresponding expectation. Put

$$F_{kn}(x) = P_k(\xi^{(n)} \leqslant x), \tag{2.1}$$

$$\varphi_{kn}(\lambda) = E_k \exp(-\lambda \xi^{(n)}) = \int_0^{+\infty} e^{-\lambda x}\, dF_{kn}(x). \tag{2.2}$$

Lemma 1

$$\varphi_{kn}(\lambda) = h_k(\lambda) h_{k+1}(\lambda) \cdots h_{n-1}(\lambda),\ k<n, \tag{2.3}$$

$h_m(\lambda)$ being defined by the recurrent formula

$$\begin{cases} h_0(\lambda) = \dfrac{b_0}{\lambda V(0) + b_0}, \\ h_m(\lambda) = \dfrac{b_m}{\lambda V(m) + c_m - a_m h_{m-1}(\lambda)},\ m \geqslant 1, \end{cases} \tag{2.4}$$

where $c_m = a_m + b_m$.

Proof Let

$$h_m(\lambda) = E_m \exp(-\lambda \xi^{(m+1)}). \tag{2.5}$$

For simplicity, denote $\int_u^v V(x_t)\,dt$ by $\int_u^v$. For optional time τ, let $\mathscr{F}_\tau$ be the pre-τ σ-algebra[1], and θ_τ be the shift operator for X[8]. We have

$$\varphi_{kn}(\lambda) = E_k \exp\left[-\lambda\left(\int_0^{\tau_{k+1}} + \int_{\tau_{k+1}}^{\tau_n}\right)\right]$$

$$= E_k\left\{E_k\left[\exp-\lambda\left(\int_0^{\tau_{k+1}} + \theta_{\tau_{k+1}}\int_0^{\tau_n}\right)\Big|\mathscr{F}_{\tau_{k+1}}\right]\right\}$$

$$= E_k \exp\left(-\lambda\int_0^{\tau_{k+1}}\right) E_{k+1} \exp\left(-\lambda\int_0^{\tau_n}\right)$$

$$= h_k(\lambda) \varphi_{k+1,n}(\lambda) = h_\lambda(\lambda) h_{k+1}(\lambda) \cdots h_{n-1}(\lambda),$$

which is (2.3). Now, we proceed to prove (2.4). If β is the first jump of X, them we have

$$P_m(\beta \leqslant x) = 1 - e^{-c_m x} \equiv F_m(x),$$
$$E_m e^{-\lambda V(m)\beta} = \int_0^{+\infty} e^{-\lambda V(m)x} dF_m(x) = \frac{c_m}{\lambda V(m) + c_m}. \quad (2.6)$$

If $m=0$, (2.6) is reduced to the first equality of (2.4). Moreover,

$$h_m(\lambda) = E_m\left\{E_m\left[\exp\left(-\lambda \int_0^{\tau_{m+1}}\right)\Big|\mathscr{F}_\beta\right]\right\}$$
$$= E_m\left\{E_m\left[\exp\left(-\lambda \int_0^{\beta} - \lambda \int_\beta^{\tau_{m+1}}\right)\Big|\mathscr{F}_\beta\right]\right\}$$
$$= E_m\left\{\exp\left(-\lambda \int_0^{\beta}\right) E_m\left[\exp\left(-\lambda \int_\beta^{\tau_{m+1}}\right)\Big|\mathscr{F}_\beta\right]\right\}$$
$$= E_m\left\{\exp\left(-\lambda \int_0^{\beta}\right) E_{x(\beta)} \exp\left(-\lambda \int_0^{\tau_{m+1}}\right)\right\}$$
$$= E_m e^{-\lambda V(m)\beta}\left[P_m(x(\beta) = m-1) E_{m-1} \exp\left(-\lambda \int_0^{\tau_{m+1}}\right) + P_m(x(\beta) = m+1) E_{m+1} \exp\left(-\lambda \int_0^{\tau_{m+1}}\right)\right]$$
$$= \frac{c_m}{\lambda V(m) + c_m}\left\{\frac{a_m}{c_m}\varphi_{m-1,m+1} + \frac{b_m}{c_m}\right\}.$$

Substituting (2.3) into $\varphi_{m-1,m+1}$, we obtain the second equality of (2.4). The lemma is immediate.

Theorem 1

$$\varphi_{kn}(\lambda) = b_k b_{k+1} \cdots b_{n-1} \frac{L_k(\lambda)}{L_n(\lambda)}, \quad k < n, \quad (2.7)$$

where $L_m(\lambda)$ is a polynomial of order $l \leqslant m$. $L_m(\lambda)$ is defined by the following recurrent formulas:

$$\begin{cases} L_0(\lambda) = 1, \quad L_1(\lambda) = \lambda V(0) + b_0, \\ L_m(\lambda) = [\lambda V(m-1) + c_{m-1}] L_{m-1}(\lambda) - a_{m-1} b_{m-2} L_{m-2}(\lambda), \quad m > 1. \end{cases}$$

$$(2.8)$$

Sojourn Times and First Passage Times for Birth and Death Processes

Proof By (2.4) and induction, we see that $h_m(\lambda)$ can be represented as

$$h_m(\lambda) = \frac{U_{m+1}(\lambda)}{L_{m+1}(\lambda)}, \qquad (2.9)$$

where U_{m+1} and L_{m+1} are polynomials of order not higher than m and $m+1$ respectively. Substituting (2.9) into (2.4), we get

$$h_m(\lambda) = \frac{b_m L_m(\lambda)}{[\lambda V(m) + c_m] L_m(\lambda) - a_m U_m(\lambda)}. \qquad (2.10)$$

Thus, we can put

$$\begin{cases} U_{m+1}(\lambda) = b_m L_m(\lambda), \\ L_{m+1}(\lambda) = [\lambda V(m) + c_m] L_m(\lambda) - a_m U_m(\lambda), \end{cases} \qquad (2.11)$$

then

$$h_m(\lambda) = \frac{b_m L_m(\lambda)}{L_{m+1}(\lambda)}. \qquad (2.12)$$

Substituting (2.12) into (2.3), we obtain (2.7). From (2.11) we get the last formula of (2.8). The first two equalities of (2.8) follow from (2.12) and the first formula of (2.4). The proof is completed.

By (2.7), there exist

$$\varphi_{0n}(\lambda) = \frac{b_0 b_1 \cdots b_{n-1}}{L_n(\lambda)}, \qquad (2.13)$$

$$\varphi_{kn}(\lambda) = \frac{\varphi_{0n}(\lambda)}{\varphi_{0k}(\lambda)}, \quad k < n. \qquad (2.14)$$

Therefore, without loss of generality we may suppose that $X(0) = 0$; and for finding $\varphi_{0n}(\lambda)$ it is sufficient to find $L_n(\lambda)$. The latter either follows from (2.8), or can be determined directly by the following method. Suppose

$$L_n(\lambda) = d_{n0} + d_{n1}\lambda + \cdots + d_{nn}\lambda^n. \qquad (2.15)$$

Evidently, we have

$$d_{n0} = L_n(0) = b_0 b_1 \cdots b_{n-1}, \tag{2.16}$$

$$d_{nn} = V(0)V(1)\cdots V(n-1), \tag{2.17}$$

$$d_{n1} = L'_n(0) = b_0 b_1 \cdots b_{n-1} M_1, \tag{2.18}$$

where M_i is the i-th moment of $\xi^{(n)}$:

$$M_i = E_0 [\xi^{(n)}]^i. \tag{2.19}$$

Corollary 1

(i) $d_{nm} = \sum_{i=1}^{m} (-1)^{i+1} \dfrac{M_i}{i!} d_{n,m-i}$, $1 \leq m \leq n$; (2.20)

(ii) $d_{nm} = (-1)^{\left[\frac{n}{2}\right]} \Delta_m$, $1 \leq m \leq n$, (2.21)

where $[a]$ is the greatest integer $\leq a$, and Δ_m is a determinant obtained by substituting the row vector

$$d_{n0} \left(M_1, \frac{M_2}{2!}, \frac{M_3}{3!}, \cdots, \frac{M_n}{n!} \right)^{\mathrm{T}}$$

into the m-th row of Δ:

$$\Delta = \begin{vmatrix} 1 & 0 & 0 & \cdots & 0 \\ M_1 & -1 & 0 & \cdots & 0 \\ \dfrac{M_2}{2!} & -M_1 & 1 & \cdots & 0 \\ \dfrac{M_3}{3!} & -\dfrac{M_2}{2!} & M_1 & \cdots & 0 \\ \vdots & \vdots & \vdots & & \vdots \\ \dfrac{M_{n-1}}{(n-1)!} & -\dfrac{M_{n-2}}{(n-2)!} & \dfrac{M_{n-3}}{(n-3)!} & \cdots & (-1)^{n+1} \end{vmatrix}.$$

Proof From (2.13), we see

$$\varphi_{0n}(\lambda) L_n(\lambda) = b_0 b_1 \cdots b_{n-1}.$$

Differentiating both sides for m times, and denoting

$$\varphi^{(i)} = \varphi_{0n}^{(i)}(0), \quad \varphi^{(0)} = \varphi_{0n}(0),$$

$$L^{(i)} = L_n^{(i)}(0), \quad L^{(0)} = L_n(0), \ i \geq 1,$$

we get
$$\sum_{i=0}^{m} \frac{m!}{i!(m-i)!} \varphi^{(m-i)} L^{(i)} = 0.$$

Substituting $\varphi^{(i)} = (-1)^i M_i$ and $L^{(i)} = i! \ d_{ni}$ into the above equation, we have

$$\sum_{i=0}^{m} (-1)^{m-i} \frac{M_{m-i}}{(m-i)!} d_{ni} = 0, \qquad (2.22)$$

or

$$d_{nm} = \sum_{i=0}^{m-1} (-1)^{m-i+1} \frac{M_{m-i}}{(m-i)!} d_{ni}, \ 1 \leqslant m \leqslant n,$$

that is (2.20). Set $m = 1, 2, \cdots, n$ in (2.22), and we obtain

$$M_1 d_{n0} = d_{n1},$$

$$\frac{M_2}{2!} d_{n0} = M_1 d_{n1} - d_{n2},$$

$$\cdots$$

$$\frac{M_n}{n!} d_{n0} = \frac{M_{n-1}}{(n-1)!} d_{n1} - \frac{M_{n-2}}{(n-2)!} d_{n2} + \cdots + (-1)^{n+1} d_{nn}.$$

Solving these equations of unknown variables $d_{n1}, d_{n2}, \cdots, d_{nn}$ by Cramer's method, and using $\Delta = (-1)^{\left[\frac{n}{2}\right]}$, we get (2.21). The proof is completed.

In [2] the moments $M_l = E_0[\xi^{(n)}]^l$ and the more general moments

$$M_{kn}^{(l)} = E_k[\xi^{(n)}]^l, \qquad M_l = M_{0n}^{(l)},$$

were found. It was proved there

$$M_{kn}^{(l)} = \sum_{i=k}^{n-1} G_{in}^{(l)}, \qquad (2.23)$$

$$G_{in}^{(i)} = \frac{lV(i) M_{in}^{(l-1)}}{b_i} + \sum_{k=0}^{i-1} \frac{a_i a_{i-1} \cdots a_{i-k} lV(i-k-1) M_{i-k-1,n}^{(i-1)}}{b_i b_{i-1} \cdots b_{i-k} b_{i-k-1}},$$

$$(2.24)$$

$$G_{in}^{(i)} = \frac{V(i)}{b_i} + \sum_{k=0}^{i-1} \frac{a_i a_{i-1} \cdots a_{i-k} V(i-k-1)}{b_i b_{i-1} \cdots b_{i-k} b_{i-k-1}}. \quad (2.25)$$

$G_{in}^{(1)}$ can be obtained by (2.25), $M_{kn}^{(1)}$ by (2.23), $M_{kn}^{(2)}$ by (2.24) (2.23), $\cdots$. It is seen that the higher moments of $\xi^{(n)}$ may be represented by the lower moments, so that we get all $M_{kn}^{(l)}$ including M_l.

By (2.20) or (2.21), we know that the value of d_{nm} depends only on the moments of $\xi^{(n)}$ with order $k \leqslant m$, and hence the coefficients of $L_n(\lambda)$ depend at most on the moments of $\xi^{(n)}$ with order $k \leqslant n$. Therefore, the distribution of $\xi^{(n)}$ is uniquely determined by its moments of order $k \leqslant n$. More exactly, if there are just m values of $V(i)$ ($0 \leqslant i < n$) greater than 0, then the order of $L_n(\lambda)$ is m, so that $F_{0n}(x)$ is uniquely determined by its moments of order $k \leqslant m$.

Equality (2.20) establishes relations among the coefficients of $L_n(\lambda)$. For different n, we can find a similar formula. By (2.7),

$$\varphi_{kn}(\lambda) L_n(\lambda) = b_k b_{k+1} \cdots b_{n-1} L_k(\lambda), \quad k < n.$$

Repeating the proof of Corollary 1, we have

$$d_{nm} = \sum_{i=1}^{m} \frac{(-1)^{i+1}}{i!} M_{kn}^{(i)} d_{n,m-i} + b_k b_{k+1} \cdots b_{n-1} d_{km}, \quad 1 \leqslant m \leqslant n \quad (2.26)$$

(Put $d_{km} = 0$ if $k < m$). In particular, (2.26) reduces to (2.20) when $k = 0$.

§3. Distributions of Sojourn times and first passage times

Consider the sojourn time τ_{nk} in (1.5), where n and $n+k$ are fixed. Let $S_m(\lambda)$ be the corresponding polynomials which by (1.4) and (2.8) satisfy the following relations:

$$S_0(\lambda)=1, \quad S_1(\lambda)=\lambda+b_0, \tag{3.1}$$

$$S_i(\lambda)=(\lambda+c_{i-1})S_{i-1}(\lambda)-a_{i-1}b_{i-2}S_{i-2}(\lambda), \quad 1<i\leqslant n, \tag{3.2}$$

$$S_{n+i}(\lambda)=c_{n+i-1}S_{n+i-1}(\lambda)-a_{n+i-1}b_{n+i-2}S_{n+i-2}(\lambda), \quad 1\leqslant i\leqslant k. \tag{3.3}$$

Lemma 2 The roots of $S_m(\lambda)=0$ $(1\leqslant m\leqslant n+k)$ are negative and simple.

Proof If $m\leqslant n$[10], the conclusion is well-known for $S_m(\lambda)$ satisfying (3.1) and (3.2), so we prove it when $n<m\leqslant n+k$. Consider

$$\mathscr{S}_l(\lambda)=\frac{S_l(\lambda)}{b_0b_1\cdots b_{l-1}}=\frac{1}{\varphi_{0l}(\lambda)}, \quad 1\leqslant l\leqslant n+k. \tag{3.4}$$

Clearly, $\mathscr{S}_l(\lambda)$ and $S_l(\lambda)$ have the same nulls. By (2.16), the constant term of $\mathscr{S}_l(\lambda)$ is $\mathscr{S}_l(0)=1$. By (3.3),

$$\mathscr{S}_{n+i}(\lambda)=\frac{c_{n+i-1}}{b_{n+i-1}}\mathscr{S}_{n+i-1}(\lambda)-\frac{a_{n+i-1}}{b_{n+i-1}}\mathscr{S}_{n+i-2}(\lambda), \quad 1\leqslant i\leqslant k. \tag{3.5}$$

Let us suppress λ in the following. From (3.5) follows

$$\mathscr{S}_{n+i}-\mathscr{S}_{n+i-1}=\frac{a_{n+i-1}}{b_{n+i-1}}(\mathscr{S}_{n+i-1}-\mathscr{S}_{n+i-2})=\cdots$$

$$=\frac{a_{n+i-1}a_{n+i-2}\cdots a_n}{b_{n+i-1}b_{n+i-2}\cdots b_n}(\mathscr{S}_n-\mathscr{S}_{n-1}).$$

Summing up both sides from $i=1$ to k, we have
$$\mathscr{S}_{n+k} = \mathscr{S}_n + A_{nk}(\mathscr{S}_n - \mathscr{S}_{n-1}), \qquad (3.6)$$
where
$$A_{nk} = \sum_{j=0}^{k-1} \frac{a_n a_{n+1} \cdots a_{n+j}}{b_n b_{n+1} \cdots b_{n+j}} > 0.$$

For fixed n, the polynomials $\mathscr{S}_n$ and $\mathscr{S}_{n-1}$ are of order n and $n-1$ respectively by (3.1) and (3.2). The coefficients of the highest power are positive by (2.17) so that $\mathscr{S}_{n+k}$ is also a polynomial of order n. If $\mu_i^{(l)}$ are the nulls of $\mathscr{S}_l(\lambda)$, it is known that[10]
$$0 > \mu_1^{(n)} > \mu_1^{(n-1)} > \mu_2^{(n)} > \mu_2^{(n-1)} > \cdots > \mu_{n-1}^{(n-1)} > \mu_n^{(n)}.$$
Clearly
$$\mathscr{S}_{n+k}(\mu_i^{(n)}) = -A_{nk}\mathscr{S}_{n-1}(\mu_i^{(n)}).$$
Since $\lim_{\lambda \to \infty} \mathscr{S}_{n-1}(\lambda) = +\infty$ and $\mu_1^{(n)} > \mu_1^{(n-1)}$, we have $\mathscr{S}_{n-1}(\mu_1^{(n)}) > 0$ hence $\mathscr{S}_{n+k}(\mu_1^{(n)}) < 0$. Similarly, $\mathscr{S}_{n+k}(\mu_i^{(n)})$ has the same sign as $(-1)^i$, so that $\mathscr{S}_{n+k}(\lambda)$ has a null in $(\mu_{i+1}^{(n)}, \mu_i^{(n)})$, $i = 1, 2, \cdots$, $n-1$. Since $\mathscr{S}_{n+k}(0) = 1$, another null is in $(\mu_1^{(n)}, 0)$. Therefore
$$0 > \mu_1^{(n+k)} > \mu_1^{(n)} > \mu_2^{(n+k)} > \mu_2^{(n)} > \cdots > \mu_n^{(n+k)} > \mu_n^{(n)}. \qquad (3.7)$$
The lemma is proved.

Remark If A_{nk} in (3.6) is replaced by another positive number, (3.7) still holds.

For convenience, let $\lambda_i^{(m)} = -\mu_i^{(m)} > 0$. By Lemma 2
$$S_m(\lambda) = \begin{cases} \prod_{i=1}^{m}(\lambda + \lambda_i^{(m)}), & m \leq n, \\ \prod_{i=1}^{n}(\lambda + \lambda_i^{(m)}), & n < m \leq n+k, \end{cases} \qquad (3.8)$$
where $0 < \lambda_1^{(m)} < \lambda_2^{(m)} < \cdots$. Define
$$F_{mnk}(x) = P_m(\tau_{nk} \leq x).$$

Theorem 2 By starting at $m (m < n)$, the distribution

Sojourn Times and First Passage Times for Birth and Death Processes

function $F_{mnk}(x)$ of sojourn time τ_{nk} is a mixedly exponential, which has a density given by

$$f_{mnk}(x) = \sum_{i=1}^{n} \frac{b_m b_{m+1} \cdots b_{n+k-1} S_m(-\lambda_i^{(n+k)})}{S'_{n+k}(-\lambda_i^{(n+k)})} e^{-\lambda_i^{(n+k)} x}, \quad (3.9)$$

where $S_i(\lambda)$ ($i = m, n+k$) are given by (3.1)~(3.3), and S'_{n+m} denotes the derivative of S_{n+m},

$$S'_{n+k}(-\lambda_i^{(n+k)}) = \prod_{j=1, j \neq i}^{n} (\lambda_j^{(n+k)} - \lambda_i^{(n+k)}). \quad (3.10)$$

Proof By (2.7),

$$\varphi_{m,n+k}(\lambda) = E_m e^{-\lambda \tau_{nk}} = \frac{b_m b_{m+1} \cdots b_{n+k-1} S_m(\lambda)}{S_{n+k}(\lambda)}.$$

Substituting (3.8) into the above equation, and taking inverse Laplace transform, we obtain the required results.

The conclusion is apparent.

Since the first passage time to n is $\tau_n = \tau_{n0}$, we have

Corollary 2 To begin with m ($m < n$), the first passage time τ_n to n has density

$$f_{mn}(x) = \sum_{i=1}^{n} \frac{b_m b_{m+1} \cdots b_{n-1} S_m(-\lambda_i^{(n)})}{S'_n(-\lambda_i^{(n)})} e^{-\lambda_i^{(n)} x}, \quad (3.11)$$

where $S_i(\lambda)$ ($i = m, n$) are given by (3.1) and (3.2).

In the following, we shall discuss briefly the sojourn time at a high level. Put

$$h(i) = 0 \quad (0 \leq i < n); \quad h(i) = 1, \quad i \geq n. \quad (3.12)$$

Then $\zeta_{nk} = \int_0^{\tau_{n+k}} h(x_t) dt$ is the sojourn time in $(n, n+1, \cdots, n+k-1)$, before hitting $n+k$. By (2.6), the corresponding polynomial $H_m(\lambda)$ must satisfy

$$H_0(\lambda) = 1, \quad H_i(\lambda) = b_0 b_1 \cdots b_{i-1} \quad (1 \leq i \leq n),$$
$$H_{n+1}(\lambda) = b_0 b_1 \cdots b_{n-1} (\lambda + b_n),$$

$$H_{n+i}(\lambda) = (\lambda + c_{n+i-1}) H_{n+i-1}(\lambda) - a_{n+i-1} b_{n+i-2} H_{n+i-2}(\lambda),$$
$$1 \leqslant i \leqslant k.$$

We see that $H_i(\lambda)$ ($1 \leqslant i \leqslant n$) are constants, and $H_{n+i}(\lambda)$ is a polynomial in λ of order i. The latter depends on λ just like (3.2); hence it has i unlls, negative and simple. By (2.7)

$$E_m \exp(-\lambda \zeta_{nk}) = \frac{b_m b_{m+1} \cdots b_{n+k-1} H_m(\lambda)}{H_{n+k}(\lambda)}, \quad 0 \leqslant m < n+k.$$

Therefore, $P_m(\zeta_{nk} \leqslant x)$ has a density as follows:

$$h_{mnk}(x) = \sum_{i=1}^{k} \frac{b_m b_{m+1} \cdots b_{n+k-1} H_m(-\gamma_i^{(n+k)})}{H'_{n+k}(-\gamma_i^{(n+k)})} e^{-\gamma_i^{(n+k)} x},$$

where $-\lambda_i^{(n+k)}$ are nulls of $H_{n+k}(\lambda)$, $i = 1, 2, \cdots, k$.

§ 4. The limit distribution

Introduce

$$Z = 1 + \sum_{k=1}^{+\infty} \frac{a_1 a_2 \cdots a_k}{b_1 b_2 \cdots b_k}. \qquad (4.1)$$

It was proved in [9][①] that all states in X are recurrent if $Z = +\infty$. Consider the polynomial $\mathcal{S}_l(\lambda)$ in (3.4). Let

$$\mathcal{S}_n(\lambda) = 1 + c_{n1}\lambda + c_{n2}\lambda^2 + \cdots + c_{nn}\lambda^n. \qquad (4.2)$$

By (2.20) and (2.23)~(2.25), we note

$$c_{n1} = E_0 \tau_n = E_0 \tau_{n0} = \sum_{i=0}^{n-1}\left(\frac{1}{b_i} + \sum_{j=0}^{i-1}\frac{a_i a_{i-1}\cdots a_{i-j}}{b_i b_{i-1}\cdots b_{i-j}b_{i-j-1}}\right). \qquad (4.3)$$

In general, we have

$$E_0 \tau_{nk} = \sum_{i=0}^{n+k-1}\left(\frac{U(i)}{b_i} + \sum_{j=0}^{i-1}\frac{a_i a_{i-1}\cdots a_{i-j} U(i-j-1)}{b_i b_{i-1}\cdots b_{i-j}b_{i-j-1}}\right), \qquad (4.4)$$

where $U(i)$ is defined by (1.4). Let

$$A_n = \lim_{k \to +\infty} A_{nk} = \sum_{j=0}^{+\infty}\frac{a_n a_{n+1}\cdots a_{n+j}}{b_n b_{n+1}\cdots b_{n+j}}. \qquad (4.5)$$

Clearly,

$$A_n = \left(Z - 1 - \sum_{i=1}^{n-1}\frac{a_1 a_2 \cdots a_i}{b_1 b_2 \cdots b_i}\right)\frac{b_1 b_2 \cdots b_{n-1}}{a_1 a_2 \cdots a_{n-1}}.$$

Hence $A_n < +\infty$ if and only if $Z < +\infty$.

---

① In fact, there is a deeper result that all states of the imbedded Markov chain (with discrete time parameter) in X are recurrent if and only if $Z = +\infty$.

We now proceed to find the limit of $P_0\left(\dfrac{\tau_{nk}}{E_0\tau_{nk}}\leqslant x\right)$ when $k\to+\infty$.

Theorem 3 There are only two possibilities:

(i) if $Z=+\infty$, then
$$\lim_{k\to+\infty} P_0\left(\frac{\tau_{nk}}{E_0\tau_{nk}}\leqslant x\right)=1-e^{-x}; \tag{4.6}$$

(ii) if $Z<+\infty$, then
$$\lim_{k\to+\infty} P_0\left(\frac{\tau_{nk}}{E_0\tau_{nk}}\leqslant x\right)=\int_0^x \sum_{j=1}^n \frac{B_n}{Q'_n(-\beta_j^{(n)})} e^{-\beta_j^{(n)}t}\,dt, \tag{4.7}$$

where $\beta_j^{(n)} = d_n \alpha_j^{(n)} > 0$, $d_n = A_n(c_{n1}-c_{n-1,1})+c_{n1}>0$, $-\alpha_j^{(n)}$ ($j=1, 2, \cdots, n$) are n different roots of

$$\mathscr{S}(\lambda)\equiv 1+\sum_{i=1}^n [A_n(c_{ni}-c_{n-1,i})+c_{ni}]\lambda^i = 0$$

($c_{n-1,n}=0$). And

$$B_n=\frac{(d_n)^n}{(A_n+1)c_{nn}}, \qquad Q_n(\lambda)=B_n\mathscr{S}\left(\frac{\lambda}{d_n}\right).$$

Proof From (3.6) and (4.2), we easily know

$$\mathscr{S}_{n+k}(\lambda) = 1+\sum_{i=1}^n [A_{nk}(c_{ni}-c_{n-1,i})+c_{ni}]\lambda^i, \tag{4.8}$$

and from (3.4),

$$E_0 e^{-\lambda\tau_{nk}}\equiv\varphi_{0,n+k}(\lambda)=\frac{1}{\mathscr{S}_{n+k}(\lambda)},$$

$$E_0\tau_{nk}=-\varphi'_{0,n+k}(0)=A_{nk}(c_{n1}-c_{n-1,1})+c_{n1}.$$

From these equalities and (4.8), we have

$$E_0\exp\left(-\lambda\frac{\tau_{nk}}{E_0\tau_{nk}}\right)=\frac{1}{\mathscr{S}_{n+k}\left(\dfrac{\lambda}{E_0\tau_{nk}}\right)}$$

$$=\frac{1}{1+\lambda+\displaystyle\sum_{i=2}^n \frac{A_{nk}(c_{ni}-c_{n-1,i})+c_{ni}}{[A_{nk}(c_{n1}-c_{n-1,1})+c_{n1}]^i}\lambda^i}. \tag{4.9}$$

(i) Suppose $Z = +\infty$, then $\lim\limits_{k \to \infty} A_{nk} = +\infty$. For any $\lambda > 0$,

$$\left| \mathscr{S}_{n+k}\left(\frac{\lambda}{E_0 \tau_{nk}}\right) - 1 - \lambda \right| \leqslant \sum_{i=2}^{n} \left| \frac{A_{nk}(c_{ni} - c_{n-1,i}) + c_{ni}}{[A_{nk}(c_{n1} - c_{n-1,1}) + c_{n1}]^i} \lambda^i \right|.$$

The right side will approach 0 when $k \to +\infty$, therefore

$$\lim_{k \to +\infty} E_0 \exp\left(-\lambda \frac{\tau_{nk}}{E_0 \tau_{nk}}\right) = \frac{1}{1+\lambda},$$

which is equivalent to (4.6).

(ii) Suppose $Z < +\infty$, then $A_n = \lim\limits_{k \to +\infty} A_{nk} < +\infty$. In virtue of (3.6) and (4.8), it leads to

$$\mathscr{S}(\lambda) = \lim_{k \to +\infty} \mathscr{S}_{nk}(\lambda) = \mathscr{S}_n(\lambda) + A_n(\mathscr{S}_n(\lambda) - \mathscr{S}_{n-1}(\lambda))$$

$$= 1 + \sum_{i=1}^{n} [A_n(c_{ni} - c_{n-1,i}) + c_{ni}]\lambda^i.$$

According to the remark in Lemma 2, the equation $\mathscr{S}(\lambda) = 0$ has n different negative roots, say $-\alpha_j^{(n)}$, $j = 1, 2, \cdots, n$. Since $c_{n1} = E_0 \tau_n$, we have

$$d_n = A_n(c_{n1} - c_{n-1,1}) + c_{n1} > 0.$$

If $-\beta_j^{(n)}$ is the roots of $\mathscr{S}\left(\frac{\lambda}{d_n}\right) = 0$, it is clear that $\beta_j^{(n)} = d_n \alpha_j^{(n)} > 0$. According to (4.9)

$$\lim_{k \to +\infty} E_0 \exp\left(-\lambda \frac{\tau_{nk}}{E_0 \tau_{nk}}\right) = \left\{1 + \sum_{i=1}^{n}[A_n(c_{ni} - c_{n-1,i}) + c_{ni}]\left(\frac{\lambda}{d_n}\right)^i\right\}^{-1}$$

$$= \left[\mathscr{S}\left(\frac{\lambda}{d_n}\right)\right]^{-1} = \frac{B_n}{B_n \mathscr{S}\left(\frac{\lambda}{d_n}\right)} = \frac{B_n}{Q_n(\lambda)}.$$

The coefficient of λ^n in $\mathscr{S}\left(\frac{\lambda}{d_n}\right)$ is

$$\frac{(A_n + 1)c_{nn}}{(d_n)^n} = \frac{1}{B_n},$$

so that the coefficient of λ^n in $Q_n(\lambda)$ is 1. Since $\lim\limits_{\lambda \to 0} \mathscr{S}\left(\frac{\lambda}{d_n}\right) = 1$, we have $\lim_{\lambda\to 0}\dfrac{B_n}{Q_n(\lambda)}=1$. By the continuity Theorem [11, P451] and inverse formula of Laplace transform, we complete the proof of (4.7).

It is interesting to note that the limit distribution does not depend on the number n of states in sojourn set $(0, 1, 2, \cdots, n-1)$ if $Z=+\infty$ by Theorem 3; it is not so in the case $Z<+\infty$.

References

[1] Chung K L. Markov Chains with Stationary Transition Probabilities. Springer, 1976.

[2] Wang Tzu-kwen. On distributions of functionals of birth and death processes and their applications in the theory of queues. Scientia Sinica, 1961, 10(2): 160-170.

[3] 吴立德. 齐次可列马尔可夫过程积分型泛函的分布. 数学学报, 1963, 13: 86-93.

[4] 杨超群. 可列马氏过程的积分型泛函和双边生灭过程的边界性质. 数学进展, 1964, 7: 397-424.

[5] 侯振挺, 郭青峰. 齐次可列马尔可夫过程. 北京: 科学出版社, 1978.

[6] Karlin S, Mcgregor J L. Occupation time laws for birth and death processes. Proceedings of the Fourth Berkeley Symp. on Math. Statistics and Probability, 1961: 249-272.

[7] Soloviev A D. Asymptotic distribution of the moment of first crossing of high level by a birth and death process. Proceedings of the Sixth Berkeley Symp. on Math. Statistics and Probability, 1972: 71-86.

[8] Дынкин Е В. Марковские процессы. Москва, 1963.

[9] 王梓坤. 生灭过程构造论. 数学进展, 1962, 5: 137-179.

[10] Ledermann W, Reuter G E H. Spectral theory for the differential equations of simple birth and death processes. Phil. Trans. Roy. Soc., London, Ser. A, 1954, 246: 321-369.

[11] Feller W. An Introduction to Probability Theory and Its Applications. New York: John Wiley and Sons, 1971, 2.

生灭过程停留时间与首达时间的分布

摘要 本文求出了生灭过程积分型泛函的分布的拉普拉斯变换. 证明了: 停留时间的精确分布是混合指数型的; 适当规范化后, 其极限分布只能是指数型的或混合指数型的两种; 各分布中的参数也已求出. 有关首中时的结果可作为停留时的特殊情况而推出.

Scientia Sinica, 1981, 24(3)

Last Exit Distributions and Maximum Excursion for Brownian Motion*

Abstract[①] This paper discusses the last exit time and place of a sphere, maximum excursion before last exit and the time for first attaining the maximum. Distributions and moments of the above four random variables are found. The properties of moments distinguish the Brownian motion in various dimensions. In particular, it is shown that the distributions of last exit place and first hitting place of a sphere with center 0 are the same, i. e. the uniform distribution on the sphere, if $X_0 = 0$.

§ 1. Distributions of last exit place

Let $X = \{X_t(\omega),\ t \geqslant 0\}$ be Brownian motion in $n(\geqslant 3)$ di-

---

* Received: 1979-12-28.

[①] Some results of Parts Ⅰ and Ⅱ were published in [5] in the form of correspondence.

Last Exit Distributions and Maximum Excursion for Brownian Motion

mensional Euclidean space $\mathbf{R}^n$ with Borel σ-field $\mathscr{B}^n$. Define the first hitting time and last exit time of $B \in \mathscr{B}^n$ by

$$T_B = \begin{cases} \inf(t>0, X_t \in B), & \text{if the set in the right hand side is} \\ +\infty, & \text{nonempty, otherwise;} \end{cases}$$

$$\gamma_B = \begin{cases} \sup(t>0, X_t \in B), & \text{if the set in the right hand side is} \\ 0, & \text{nonempty, otherwise,} \end{cases}$$

respectively. Distributions of first hitting place $X(T_B)$ and last exit place $X(\gamma_B)$ are denoted as

$$H_B(x, A) = P_x(X(T_B) \in A),$$
$$L_B(x, A) = P_x(X(\gamma_B) \in A, \gamma_B > 0),$$

respectively, where P_x is the probability for $X_0 = x$, and E_x is the expectation relative to P_x. The transition density is

$$p(t, x, y) = (2\pi t)^{-\frac{n}{2}} \exp\left(-\frac{|y-x|^2}{2t}\right)$$

with potential kernel

$$u(x,y) = \int_0^{+\infty} p(t,x,y) \, dt = \frac{k}{|x-y|^{n-2}}, \quad (1.1)$$

$$k = \frac{\Gamma\left(\frac{n}{2} - 1\right)}{2\pi^{\frac{n}{2}}}.$$

Potential of measure ν is defined by

$$U\nu(x) = \int_{\mathbf{R}^n} u(x,y) \nu(dy).$$

Theorem 1 For any relatively compact set B, we have

$$L_B(x, dy) = u(x, y) \lim_{|z| \to +\infty} \frac{H_B(z, dy)}{u(z, y)}. \quad (1.2)$$

Proof Consider $P_B I(x) \equiv P_x(T_B < +\infty)$. Let ∂B be the boundary of B. By [1] there exists σ-finite measure μ_B with support in ∂B, and for any $x \in \mathbf{R}^n$, we have

$$P_B I(x) = U_{\mu_B}(x), \tag{1.3}$$

$$\mu_B(\mathrm{d}y) = \frac{L_B(x, \mathrm{d}y)}{u(x, y)}. \tag{1.4}$$

On the other hand, by [2] there exists measure ν_B with support in ∂B, and for any $x \in \mathbf{R}^n$, we have

$$P_B I(x) = U \nu_B(x), \tag{1.5}$$

$$\nu_B(\mathrm{d}y) = \lim_{|z| \to +\infty} \frac{H_B(z, \mathrm{d}y)}{u(z, 0)}. \tag{1.6}$$

Noting $P_B I \leqslant I$, by (1.3) (1.5) and the uniqueness of charge, we have

$$\mu_B(\mathrm{d}y) = \nu_B(\mathrm{d}y). \tag{1.7}$$

Substituting (1.4) and (1.6) into (1.7), and using

$$\lim_{|z| \to +\infty} \frac{u(z, y)}{u(z, 0)} = 1,$$

we get (1.2). The proof is complete.

Theorem 1 means that the distribution of last exit place can be expressed by that of first hitting place. In the following we consider the ball S_r and sphere ∂S_r:

$$S_r = (x: |x| \leqslant r); \quad \partial S_r = (x: |x| = r) \quad (r > 0).$$

By continuity of the path,

$$P_x(\gamma_{S_r} = \gamma_{\partial S_r}) = 1, \quad P_x(\gamma_{S_r} < +\infty) = 1 \ (x \in S_r, \ n \geqslant 3).$$

Corollary 1 For any x, we have

$$L_{\partial S_r}(x, \mathrm{d}y) = \frac{r^{n-2}}{|x-y|^{n-2}} U_r(\mathrm{d}y), \tag{1.8}$$

$$L_{\partial S_r}(0, \mathrm{d}y) = U_r(\mathrm{d}y), \tag{1.9}$$

where $U_r(\mathrm{d}y)$ is the uniform distribution on ∂S_r, i.e.

$$U_r(\mathrm{d}y) = \frac{1}{|\partial S_r|} L_{n-1}(\mathrm{d}y). \tag{1.10}$$

$|\partial S_r|$ is the area of ∂S_r, and L_{n-1} is Lebesgue measure on $\mathbf{R}^{n-1}$.

Proof By solving Dirichlet problem [3, Chap. 13], we find that

$$H_{\partial S_r}(z, dy) = r^{n-2} ||z|^2 - r^2| |y-z|^{-n} U_r(dy). \tag{1.11}$$

It follows from (1.1) and (1.11) that in the sense of strong convergence we have

$$\lim_{|z| \to \infty} \frac{H_{\partial S_r}(z, dy)}{u(z, y)} = \frac{r^{n-2}}{k} U_r(dy). \tag{1.12}$$

Substituting (1.12) and (1.1) into (1.2) we obtain (1.8). Putting $x = 0$ in (1.8) and noting $|y| = r$ when $y \in \partial S_r$, we get (1.9). This proves the corollary.

It is well-known that $H_{\partial S_r}(0, dy) = U_r(dy)$. Therefore, by starting at 0, the first hitting place and last exit place of ∂S_r have the same distribution-uniform distribution (1.10) on ∂S_r.

Remark From (1.8) (1.10), and identity

$$\frac{1}{|\partial S_r|} \int_{\partial S_r} \frac{L_{n-1}(dz)}{|y-z|^{n-2}} = \begin{cases} |y|^{2-n}, & \text{if } |y| > r, \\ r^{2-n}, & \text{if } |y| \leq r, \end{cases} \tag{1.13}$$

we can derive a well-known result [3, Chap. 13, §3]:

$$P_x(T_{\partial S_r} < +\infty) = L_{\partial S_r}(x, \partial S_r)$$

$$= \frac{r^{n-2}}{|\partial S_r|} \int_{\partial S_r} \frac{L_{n-1}(dy)}{|x-y|^{n-2}}$$

$$= \begin{cases} \left|\dfrac{r}{x}\right|^{n-2}, & \text{if } |x| > r, \\ 1, & \text{if } |x| \leq r, \end{cases} \tag{1.13$_1$}$$

As to the joint distribution of last exit place and time, we have

Theorem 2 For $D \subset \partial S_r$, $D \in \mathscr{B}^n$, $t > 0$,

$$P_x(X(\gamma_{\partial S_r}) \in D, \ \gamma_{\partial S_r} > t)$$
$$= \frac{r^{n-2}}{(2\pi t)^{\frac{n}{2}} |\partial S_r|} \int_{\mathbf{R}^n} e^{-\frac{|y-x|^2}{2t}} \int_D \frac{L_{n-1}(dz)}{|y-z|^{n-2}} dy. \quad (1.14)$$

Proof Let $C = (X(\gamma_{\partial S_r}) \in D, \ \gamma_{\partial S_r} > 0)$, then
$$P_y(C) = L_{\partial S_r}(y, D).$$

By Markov property
$$P_x(X(\gamma_{\partial S_r}) \in D, \ \gamma_{\partial S_r} > t) = E_x P_{X(t)}(C)$$
$$= \int_{\mathbf{R}^n} P_y(C) p(t, x, y) dy$$
$$= (2\pi t)^{-\frac{n}{2}} \int_{\mathbf{R}^n} e^{-\frac{|y-x|^2}{2t}} L_{\partial S_r}(y, D) dy.$$

Substituting (1.8) and (1.10) into the above equation, we obtain (1.14).

Last Exit Distributions and Maximum Excursion for Brownian Motion

§ 2. Distributions of last exit time

Let
$$F_{0r}(t) \equiv P_0(\gamma_{\partial S_r} \leqslant t).$$

Theorem 3[①] $F_{0r}(t) = \int_0^t f(s)ds$ has density

$$f(s) = \frac{r^{n-2}}{2^{\frac{n}{2}-1}\Gamma\left(\frac{n}{2}-1\right)} s^{-\frac{n}{2}} e^{-\frac{r^2}{2s}}. \tag{2.1}$$

Proof Taking $D = \partial S_r$ in (1.14), we have

$$P_0(\gamma_{\partial S_r} > t)$$
$$= \frac{r^{n-2}}{(2\pi t)^{\frac{n}{2}} |\partial S_r|} \times \Bigg[\int_{|y| \leqslant r} e^{-\frac{|y|^2}{2t}} \int_{\partial S_r} \frac{L_{n-1}(dz)}{|y-z|^{n-2}} dy +$$
$$\int_{|y| > r} e^{-\frac{|y|^2}{2t}} \int_{\partial S_r} \frac{L_{n-1}(dz)}{|y-z|^{n-2}} dy \Bigg].$$

Substituting (1.13) into the above equation, transforming the integral to polar coordinates, we get

$$P_0(\gamma_{\partial S_r} > t)$$
$$= \frac{1}{(2\pi t)^{\frac{n}{2}}} \Bigg[\int_{|y| \leqslant r} e^{-\frac{|y|^2}{2t}} dy + \int_{|y| > r} e^{-\frac{|y|^2}{2t}} \left|\frac{r}{y}\right|^{n-2} dy\Bigg]$$
$$= \frac{G}{(2\pi t)^{\frac{n}{2}}} \Bigg[\int_0^r a^{n-1} e^{-\frac{a^2}{2t}} da + r^{n-2} \int_r^{+\infty} a e^{-\frac{a^2}{2t}} da\Bigg]$$
$$= \frac{1}{\Gamma\left(\frac{n}{2}\right)} \int_0^{\frac{r^2}{2t}} v^{\frac{n}{2}-1} e^{-v} dv + \frac{r^{n-2}}{(2t)^{\left(\frac{n}{2}\right)-1} \Gamma\left(\frac{n}{2}\right)} e^{-\frac{r^2}{2t}}, \tag{2.2}$$

[①] When this paper was in print, Professor K. L. Chung informed the author that this theorem was also obtained independently by Professor R. K. Getoor in [6] not long ago.

where $G = \dfrac{2\pi^{\frac{n}{2}}}{\Gamma\left(\dfrac{n}{2}\right)}$. Hence

$$\frac{d}{dt}P_0(\gamma_{\partial S_r}>t) = \frac{-r^{n-2}}{2^{(\frac{n}{2})-1}\Gamma\left(\dfrac{n}{2}-1\right)}t^{-\frac{n}{2}}e^{-\frac{r^2}{2t}},$$

which is equivalent to (2.1). The proof is complete.

Let $\gamma \equiv \gamma_{\partial S_r}$.

Theorem 4 For $n(\geqslant 3)$ dimensional Brownian motion, we have the moment of m-th order $E_0(\gamma^m)<+\infty$ if and only if $m<\dfrac{n}{2}-1$. Moreover

$$E_0(\gamma^m) = \frac{r^{2m}}{(n-4)(n-6)\cdots(n-2m-2)},\quad (n>4). \quad (2.3)$$

Proof By (2.1), there exists

$$E_0(\gamma^m) = \frac{r^{n-2}}{2^{\frac{n}{2}-1}\Gamma\left(\dfrac{n}{2}-1\right)}\int_0^{+\infty} s^{m-\frac{n}{2}}e^{-\frac{r^2}{2s}}ds$$

$$= \frac{r^{2m}}{2^m\Gamma\left(\dfrac{n}{2}-1\right)}\int_0^{+\infty} u^{\frac{n}{2}-m-2}e^{-u}du.$$

The last integral converges if and only if $\dfrac{n}{2}>m+1$; then its value is $\Gamma\left(\dfrac{n}{2}-m-1\right)$. Since

$$\Gamma\left(\frac{n}{2}-1\right) = \Gamma\left(\frac{n}{2}-m-1\right)\prod_{i=1}^{m}\left(\frac{n}{2}-i-1\right),$$

(2.3) follows immediately.

To emphasize the dimension n of $\mathbf{R}^n$, we denote γ by $\gamma(n)$. The first hitting time $T_{\partial S_r}$ of ∂S_r is abbreviated to $T(n)$. The sojourn time in the ball S_r of $\mathbf{R}^n$ is denoted by $J(n)$:

Last Exit Distributions and Maximum Excursion for Brownian Motion

$$J(n) = \int_0^{+\infty} \chi_{s_r}(X_t) dt,$$

where $\chi_A(x) = 1$ or 0, according as $x \in A$ or not.

Corollary 2

$$E_0[T(n)] = E_0[J(n+2)] = E_0[\gamma(n+4)] = \frac{r^2}{n} \quad (n \geq 1).$$

(2.4)

Proof

$$E_0[J(n)] = E_0\left[\int_0^{+\infty} \chi_{s_r}(X_t) dt\right]$$

$$= \int_0^{+\infty} \frac{1}{(2\pi t)^{\frac{n}{2}}} \int_{|y| \leq r} e^{-\frac{|y|^2}{2t}} dy dt$$

$$= k \int_{|y| \leq r} \frac{dy}{|y|^{n-2}} = k \frac{2\pi^{\frac{n}{2}}}{\Gamma\left(\frac{n}{2}\right)} \int_0^r a \, da$$

$$= \frac{r^2}{n-2}.$$

Hence $E_0[J(n+2)] = \frac{r^2}{n}$. It is proved in [4] that if $X_0 = 0$, then $T(n)$ and $J(n+2)$ have the same distribution. Therefore $E_0[T(n)] = \frac{r^2}{n}$. Finally, by (17) $E_0[\gamma(n+4)] = \frac{r^2}{n}$.

Corollary 2 means that the larger the dimension n is, the faster the Brownian particle leaves S_r.

§ 3. Maximum excursion

Define $M_r = \max\limits_{0 \leqslant t \leqslant \gamma} |X_t|$, which is the maximum excursion of the Brownian particle before last exit from ∂S_r. Clearly, Mr is the maximum displacement from the origin. In the following we write

$$\gamma = \gamma_r = \gamma_{\partial S_r}, \quad T_r = T_{\partial S_r}.$$

Theorem 5 For x, $|x| \leqslant r$,

$$P_x(M_r \leqslant a) = \begin{cases} 0, & \text{if } a \leqslant r, \\ 1 - \left(\dfrac{r}{a}\right)^{n-2}, & \text{if } a > r. \end{cases} \quad (3.1)$$

Proof Suppose $a > r$. By (13_1),

$$P_x(M_r \geqslant a) = P_x(T_a < \gamma_r < +\infty)$$

$$= \int_{\partial S_a} P_x(X(T_a) \in db) P_b(0 < \gamma_r < +\infty)$$

$$= \int_{\partial S_a} P_x(X(T_a) \in db) L_{\partial S_r}(b, \partial S_r)$$

$$= \int_{\partial S_a} P_x(X(T_a) \in db) \left(\dfrac{r}{a}\right)^{n-2}$$

$$= \left(\dfrac{r}{a}\right)^{n-2} \int_{\partial S_r} P_x(X(T_a) \in db) = \left(\dfrac{r}{a}\right)^{n-2},$$

so that

$$P_x(M_r > a) = \lim_{\varepsilon \downarrow 0} P_x(M_r \geqslant a + \varepsilon)$$

$$= \lim_{\varepsilon \downarrow 0} \left(\dfrac{r}{a+\varepsilon}\right)^{n-2} = \left(\dfrac{r}{a}\right)^{n-2},$$

which proves the second conclusion in (3.1).

Suppose $a < r$. By definition of M_r, $P_x(M_r \leqslant a) = 0$.

Suppose $a=r$. We have

$$P_x(M_r=r)=\lim_{\varepsilon\downarrow 0}P_x(r-\varepsilon<M_r\leqslant r+\varepsilon)$$
$$=\lim_{\varepsilon\downarrow 0}[P_x(M_r\leqslant r+\varepsilon)-P_x(M_r\leqslant r-\varepsilon)]$$
$$=\lim_{\varepsilon\downarrow 0}\left[1-\left(\frac{r}{r+\varepsilon}\right)^{n-2}\right]=0.$$

Hence

$$P_x(M_r\leqslant r)=\lim_{\varepsilon\downarrow 0}P_x(M_r\leqslant r-\varepsilon)+P_x(M_r=r)=0.$$

The proof is complete.

(3.1) shows that $P_x(M_r\leqslant a)$ does not depend on x, $|x|\leqslant r$. It has a density

$$g_r(a)=\begin{cases}0, & \text{if } a\leqslant r,\\ \dfrac{(n-2)r^{n-2}}{a^{n-1}}, & \text{if } a>r.\end{cases} \quad (3.2)$$

The moment of order m is

$$E_x(M_r^m)=(n-2)r^{n-2}\int_r^{+\infty}a^m a^{1-n}\,da$$
$$=\begin{cases}+\infty, & \text{if } m\geqslant n-2,\\ \dfrac{n-2}{n-m-2}r^m, & \text{if } m<n-2.\end{cases} \quad (|x|\leqslant r) \quad (3.3)$$

Now we introduce two characteristic numbers:

$$C_\gamma=\max(\text{Integers } m\geqslant 0,\ E_0(\gamma_r^m)<+\infty), \quad (3.4)$$
$$C_M=\max(\text{Integers } m\geqslant 0,\ E_0(M_r^m)<+\infty). \quad (3.5)$$

By (2.3) and (3.3) it is seen that they depend on n, but not on $r>0$. We have the following table 3-1:

Table 3-1

n	3	4	5	6	...	$2k-1$	$2k$
C_γ	0	0	1	1	...	$k-2$	$k-2$
C_M	0	1	2	3	...	$2k-4$	$2k-3$

We see that although the $(2k-1)$-and $2k$-dimensional Brownian motions have same $C_Y = k-2$, yet they have different $C_M = 2k-4$ and $2k-3$ respectively. Therefore, each of the Brownian motions of various dimensions has different (C_Y, C_M).

By (3.3), the mean and variance of M_r are: when $|x| \leqslant r$,

$$E_x(M_r) = \frac{n-2}{n-3}r, \quad D_x(M_r) = \frac{n-2}{(n-3)^2(n-4)}r^2. \quad (3.6)$$

Let N_r be the modified variable of M_r:

$$N_r = \frac{M_r - r}{\sqrt{D_x(M_r)}} \quad (n > 4), \quad (3.7)$$

which depends on n.

Theorem 6 If $|x| \leqslant r$, then

$$\lim_{n \to +\infty} P_x(N_r \leqslant a) = \begin{cases} 0, & \text{if } a \leqslant 0, \\ 1 - e^{-a}, & \text{if } a > 0. \end{cases} \quad (3.8)$$

Proof $P_x(N_r > a) = P_x\left(\dfrac{M_r - r}{\dfrac{r}{n-3}\sqrt{\dfrac{n-2}{n-4}}} > a \right)$

$$= P_x\left(M_r > \frac{ar}{n-3}\sqrt{\frac{n-2}{n-4}} + r \right).$$

When $\dfrac{ar}{n-3}\sqrt{\dfrac{n-2}{n-4}} + r \leqslant r$, i.e., when $a \leqslant 0$, by Theorem 5 we have $P_x(N_r > a) = 1$, from which the first conclusion of (3.8) follows. If $a > 0$, by Theorem 5

$$P_x(N_r > a) = \left(\frac{r}{\dfrac{ar}{n-3}\sqrt{\dfrac{n-2}{n-4}} + r} \right)^{n-2}$$

$$= \frac{1 + \dfrac{a}{n-3}\sqrt{\dfrac{n-2}{n-4}}}{\left[1 + \dfrac{a}{n-3}\sqrt{\dfrac{n-2}{n-4}} \right]^{n-3}} \to e^{-a} \quad (n \to +\infty).$$

The proof is complete.

§ 4. Time for first attaining maximum

Define the time for first attaining maximum M_r as
$$\alpha_r = \min_t(\ |X_t|\ = M_r,\ t \leqslant \gamma_r). \qquad (4.1)$$
Recalling $\gamma_r \equiv \gamma_{\partial s_r}$, $T_r \equiv T_{\partial s_r}$, we know from [4] that
$$P_0(T_r > t) = \sum_{i=1}^{+\infty} \xi_{ni} \exp\left(-\frac{q_{ni}^2}{2r^2}t\right), \qquad (4.2)$$
where q_{ni} are the positive roots of $J_\nu(z) = 0$, and $J_\nu(z)$ $\left(\nu = \frac{n}{2} - 1\right)$ is Bessel function. Moreover,
$$\xi_{ni} = \frac{q_{ni}^{\nu-1}}{2^{\nu-1}\Gamma(\nu+1)J_{\nu+1}(q_{ni})}.$$

Theorem 7
$$P_0(\alpha_r > t) = (n-2)r^{n-2}\sum_{i=1}^{+\infty}\xi_{ni}\int_r^{+\infty}\frac{1}{a^{n-1}}e^{-\frac{q_{ni}^2 t}{2a^2}}da. \qquad (4.3)$$

Proof Since α_r is the first hitting time of random sphere ∂S_{M_r}, i.e., $\alpha_r = T_{M_r}$, by (3.2) we get
$$P_0(\alpha_r > t) = P_0(T_{M_r} > t)$$
$$= \int_r^{+\infty} P_0(T_a > t) P_0(M_r \in da)$$
$$= \int_r^{+\infty} P_0(T_a > t)\frac{(n-2)r^{n-2}}{a^{n-1}}da. \qquad (4.4)$$

Substituting (4.2) into the above equation we obtain (4.3).

Corollary 3 $E_0 \alpha_r = \dfrac{n-2}{n(n-4)}r^2 \quad (n > 4)$.

Proof By (4.4), there exists
$$E_0\alpha_r = \int_0^{+\infty} P_0(\alpha_r > t)dt$$

$$= \int_r^{+\infty} \left[\int_0^{+\infty} P_0(T_a > t) dt \right] \frac{(n-2)r^{n-2}}{a^{n-1}} da,$$

but by (2.4), there exists

$$\int_0^{+\infty} P_0(T_a > t) dt = E_0(T_a) = \frac{a^2}{n},$$

so that we get

$$E_0 \alpha_r = \frac{(n-2)r^{n-2}}{n} \int_r^{+\infty} \frac{da}{a^{n-3}} = \frac{n-2}{n(n-4)} r^2.$$

This proves the corollary.

We have shown that for the same n and r,

$$E_0 T_r = \frac{r^2}{n} < E_0,$$

$$J_r = \frac{r^2}{n-2} < E_0 \alpha_r = \frac{(n-2)r^2}{n(n-4)} < E_0 r_r = \frac{r^2}{n-4}$$

(J_r is the sojourn time in S_r), where $E_0 J_r < E_0 \alpha_r$ is not intuitively obvious.

References

[1] Chung K L. Probabilistic approach in potential theory to the equilibrium problem. Ann. Inst. Fourier, Grenoble, 1973, 23: 313-322.

[2] Port S C & Stone C J. Classical potential theory and Brownian motion. Proceedings of the Sixth Berkeley Symposium on Mathematical Statistics and Probability, 1972, 3: 143-176.

[3] Дынкин Е Б. Марковские Ироцессы, 1963.

[4] Ciesielski Z & Taylor S J. First passage times and sojourn times for Brownian motion in space and the exact

Hausdorff measure of the sample path. Trans. Amer. Math. Soc., 1962, 103: 434-450.

[5] Wang Zikun. Last exit distributions for n-dimensional Brownian motion. Kexue Tongbao, 1980, 25: 446.

[6] Getoor R K. The Brownian escape process. The Annals of Probability, 1979, 7: 864-867.

高维布朗运动的末遇时间与位置

摘要 集 B 的首中时与末遇时由(1)(2)定义. 首中位置与末遇位置有分布为(3)(4). 我们证明了：

(i) 对相对紧集 B，末遇位置的分布由(8)给出；

(ii) 若 $X(0)=0$，$B=\partial S_r$ (9)，由此分布是 ∂S_r 上的均匀分布(见(11)(12))；

(iii) $X(\gamma_{\partial S_r})$ 与 $\gamma_{\partial S_r}$ 的联合分布由(13)给出；

(iv) 如 $X(0)=0$，∂S_r 的末遇时有密度(16)；

(v) 对 n 维 ($n\geqslant 3$) 布朗运动，$E_0(\gamma^m)<+\infty$ 的充分必要条件是 $m<\dfrac{n}{2}-1$，$E_0(\gamma^m)$ 由(19)给出，$\gamma=\gamma_{\partial S_r}$；

(vi) 令 $\gamma(n)$，$T(n)$，$J(n)$ 分别表 n 维布朗运动首中、末遇 ∂S_r 的时间及在 S_r 中的停留时间，则等式(21)成立.

我们的目的是研究布朗运动末遇时间与末遇位置的分布. 设 $\{X_t(\omega), t\geqslant 0\}$ 为 n 维欧氏空间 $\mathbf{R}^n$ 中的布朗运动，$n\geqslant 3$，$\mathscr{B}^n$ 为 $\mathbf{R}^n$ 中全体 Borel 集所成的 σ-代数. 对 $B\in\mathscr{B}^n$，定义 B 的首中与末遇时间分别为

$$T_B=\begin{cases}\inf(t>0,\ X_t\in B), & \text{右方集非空,}\\ +\infty, & \text{反之;}\end{cases} \quad (1)$$

$$\gamma_B = \begin{cases} \sup(t>0, X_t \in B), & \text{右方集非空}, \\ 0, & \text{反之}. \end{cases} \quad (2)$$

首中位置与末遇位置分别是 $X(T_B)$ 与 $X(\gamma_B)$，它们的分布是

$$H_B(x, A) = P_x(X(T_B) \in A), \quad (3)$$

$$L_B(x, A) = P_x(X(\gamma_B) \in A, \gamma_B > 0), \quad (4)$$

P_x 表开始分布集中在点 x 上的条件概率，E_x 表对应的数学期望. 转移概率密度为

$$p(t, x, y) = \frac{1}{(2\pi t)^{\frac{n}{2}}} \exp\left(-\frac{|y-x|^2}{2t}\right), \quad (5)$$

势核为

$$u(x, y) = \int_0^{+\infty} p(t, x, y) dt = k |x-y|^{2-n}, \quad (6)$$

$$k = \frac{\Gamma\left(\frac{n}{2} - 1\right)}{2\pi^{\frac{n}{2}}}, \quad (7)$$

$\Gamma(x)$ 表 Gamma 函数.

定理 1 对任一相对紧集 B，有

$$L_B(x, dy) = u(x, y) \lim_{|z| \to +\infty} \frac{H_B(z, dy)}{u(z, y)}. \quad (8)$$

此定理表示：末遇位置分布 L_B 可通过首中位置分布 H_B 表达出来；只要 H_B 已知，即可求出 L_B.

考虑中心为 O，半径为 $r(>0)$ 的球与球面

$$S_t = (x: |x| \leqslant r), \quad \partial S_r = (x: |x| = r), \quad (9)$$

以 $U_r(dy)$ 表球面 ∂S_r 上的均匀分布，即

$$U_r(dy) = \frac{1}{|\partial S_r|} L_{n-1}(dy), \quad (10)$$

其中 $|\partial S_r|$ 表 ∂S_r 的面积，L_{n-1} 表 R^{n-1} 上 Lebesgue 测度.

系 1 对一切 x，有

$$L_{\partial S_r}(x, \mathrm{d}y) = \frac{r^{n-2}}{|x-y|^{n-2}} U_r(\mathrm{d}y), \tag{11}$$

特别地

$$L_{\partial S_r}(0, \mathrm{d}y) = U_r(\mathrm{d}y) = H_{\partial S_r}(0, \mathrm{d}y), \tag{12}$$

于是我们得到了一个有趣的结果：自 O 出发，首中与末遇 ∂S_r 的位置有相同的分布，即球面上的均匀分布.

由系 1 可推出一熟知的结果：

$$P_X(T_{\partial S_r} < +\infty) = P_X(0 < \gamma_{\partial S_r} < +\infty) =$$
$$= L_{\partial S_r}(x, \partial S_r) = \frac{r^{n-2}}{|\partial S_n|} \int_{\partial S_r} \frac{L_{n-1}(\mathrm{d}y)}{|x-y|^{n-2}} =$$
$$= \begin{cases} \left|\dfrac{r}{x}\right|^{n-2}, & |x| > r, \\ 1, & |x| \leqslant r. \end{cases}$$

关于 ∂S_r 的末遇位置与末遇时间的联合分布，有

定理 2 对任意 $D \subset \partial S_r$, $D \in \mathscr{B}^{n-1}$, $t > 0$,

$$P_0(X(\gamma_{\partial S_r}) \in D, \gamma_{\partial S_r} > t) =$$
$$= \frac{r^{n-2}}{(2\pi t)^{\frac{n}{2}} |\partial S_r|} \int_{\mathbf{R}^n} e^{-\frac{|y|^2}{2t}} \int_D \frac{L_{n-1}(\mathrm{d}z)}{|y-z|^{n-2}} \mathrm{d}y, \tag{13}$$

由此可以求出 ∂S_r 的末遇时间的分布. 令

$$F_{0r}(t) = p_0(\gamma_{\partial S_r} \leqslant t). \tag{14}$$

定理 3 $F_{0r}(t) = \int_0^t f(s)\mathrm{d}s,$ \tag{15}

其中密度函数为

$$f(s) = \frac{r^{n-2}}{2^{\frac{n}{2}-1} \Gamma\left(\frac{n}{2}-1\right)} s^{-\frac{n}{2}} e^{-\frac{r^2}{2s}}, \tag{16}$$

简写 $\gamma = \nu_{\partial S_r}$.

定理 4 对 $n(\geqslant 3)$ 维布朗运动，当且仅当 $m < \dfrac{n}{2} - 1$ 时，m

阶矩 $E_0(\gamma^m)<+\infty$，而且

$$E_0(\gamma^m)=\frac{\gamma^{2m}}{(n-4)(n-6)\cdots(n-2m-2)}. \quad (17)$$

于是我们得到 $2k-1$ 与 $2k$ 维布朗运动的特点，它是其他维的布朗运动所没有的：

对 1，2 维， $P_0(\gamma=+\infty)=1$；

对 3，4 维， $P_0(\gamma<+\infty)=1$，但 $E_0\gamma=+\infty$；

对 5，6 维， $E_0\gamma<+\infty$，但一级以上矩不存在；

一般地，对 $n=4+2l-1$，$n=4+2l$ 维，($l\geqslant 1$)，存在不超过 l 阶的矩 $E_0(\gamma^m)<+\infty$，$m\leqslant l$，而更高的整数阶矩不存在.

为了强调维数，记 γ 为 $\gamma(n)$，首中时 $T_{\partial S_r}$ 为 $T(n)$. n 维布朗运动在 S_T 中的停留时间定义为

$$J(n)=\int_0^{+\infty}\chi_{S_T}(X_t)\mathrm{d}t \quad (18)$$

其中 $\chi_A(x)=1$，如 $x\in A$，否则为 0.

定理 5 对 $n\geqslant 1$，有

$$E_0[T(n)]=E_0[J(n+2)]=E_0[\gamma(n+4)]=\frac{r^2}{n}. \quad (19)$$

参考文献

[1] 王梓坤. 随机过程论. 北京：科学出版社，1978.

[2] Chung K L. Ann. Inst, Fourier, Grenoble，1973，23(3)：313-322.

[3] Fort S C, Stone C J. Proceedings of the Sixth Berkeley Symp. on Math. Statistics and Probability，1972，3：143-176.

[4] Ciesielski Z, Taylci S J. Trans，Amer. Math. Soc.，1962，103：434-450.

Last Exit Time and Position for N-dimensional Brownian Motion

Abstract The hitting time and last exit time of a set B are defined by (1)(2). The hitting position and last exit position have distribution (3)(4). We have proved that:

(i) for relatively compact set B, the distribution of last exit position is given by (8);

(ii) if $X(0)=0$, $B=\partial S_r$, (9), then this distribution is the uniform distribution on ∂S_r, (see (11)(12));

(iii) the joint distribution of $X(\gamma_{\partial S_r})$ and $\gamma_{\partial S_r}$ is given by (13);

(iv) if $X(0)=0$, the last exit time of ∂S_r has a density (16);

(v) for n-dimensional ($n \geqslant 3$) Brownian motion, $E_0(\gamma^m) < +\infty$ iff $m < \frac{n}{2} < -1$, $E_0(\gamma^m)$ is given by (19), where $\gamma = \gamma_{\partial S_r}$;

(vi) let $\gamma(n)$, $T(n)$, $J(n)$ be the last exit time, the hitting time of ∂S_r, the sojourn time of S_r, for n-dimensional Brownian motion respectively, then the equility (21) holds.

自然杂志，1980，3(3)

概率论的若干新进展

客观世界中许多现象具有必然性，它们在一定条件下必然出现，例如，"同性的电互相排斥"，"在标准大气压下，水到100℃会沸腾"，等．但另一些现象却不如此，它们是偶然的，即使在同样的条件下，可能出现也可能不出现，例如，"掷硬币得正面""6月1日是晴天"，等．虽然在一次试验中，我们不能准确预言某种偶然现象是否出现，但通过多次试验，却往往可以指出它出现的可能性的大小，即指出它出现的概率．例如多次掷硬币，正面出现的次数大约占一半，因此我们可以说出现正面的概率为$\frac{1}{2}$．概率论是数学的一个重要分支，它研究的对象，正是偶然现象的数量关系．在数学上，偶然性也叫作随机性．

由于现代科学技术的精确化程度愈来愈高，必须充分估计到各种偶然因素的作用．实践的需要和科学工作者的共同努力，使得概率论的发展非常迅速．近年来它的发展具有下列特点：一是随机过程理论的高度抽象化，这主要是由于法国学派的工作；二是特殊随机过程研究的深化与精化，例如布朗运动，虽然它是最早被研究的过程之一，但深刻的新结果却每年都在出现；三是由于与其他学科（如统计物理、位势理论等）的相互渗透，研究领域迅速扩展；四是应用更为广泛而深入.

我国数学工作者在概率论的研究中做了不少工作，其中有些成绩显著. 例如 1978 年侯振挺由于非保守 Q 过程唯一性的研究而荣获英国戴维森（Davison）奖，引起了广泛的注意. 本文的目的是回顾一下主要是 1978 年内公开发表的论文，而以国内工作为重点. 即使如此，由于文献浩繁而资料又短缺不全，肯定会有不少重要成果未被提到，这是需要请求读者原谅的.

§1. 关于马尔可夫过程

自然界有些随机过程具有无后效性，就是说，如已知它现在的状态，那么，它将来的状态便不依赖于过去的状态. 这种过程由俄国数学家马尔可夫首先研究. 因而通常称为马尔可夫过程. 作为例子，可以举出布朗运动，因为当已知在液体中做布朗运动的粒子现在所处的状态（位置）时，将来的状态是不依赖于它过去所曾走过的路径的，所以布朗运动是一种马尔可夫过程.

马尔可夫过程的理论相当丰富而深刻，它始终是概率论最活跃的分支之一. 近 20 年来，由于发现它与位势理论的深刻联系而得到更迅速的发展. 本年内的部分工作见[1~3]. 在[1]中研究对标准过程 X，何时它的过分函数（Excessive functions）重合于正的超调和函数. [3]则刻画了具平稳独立增量过程的半极集（Semipolar sets）.

结合统计物理中的"细致平衡"问题，在北京大学举办的讨论班上，钱敏、侯振挺、汪培庄、陈木法等系统地研究了马尔可夫过程的可逆性问题，其中已正式发表的部分结果见[4，54].

非保守 Q 过程唯一性的问题如下：设已给矩阵 $Q=(q_{ij})$，其中 i,j 遍历某可列集 E，q_{ij} 满足条件

$$0 \leqslant q_{ij} < +\infty\ (i \neq j),\quad -\infty < q_{ii} \leqslant 0, \tag{1.1}$$

$$\sum_j q_{ij} \leqslant 0, \tag{1.2}$$

若 (1.2) 中取等号（一切 i），则称 Q 为保守的，否则称为非保守的。现在问：是否存在齐次可列标准马尔可夫过程 $(p_{ij}(t))$，使它的转移密度矩阵重合于 Q？亦即满足

$$\lim_{t \to 0} \frac{p_{ij}(t) - \delta_{ij}}{t} = q_{ij},\quad (i, j \in E) \tag{1.3}$$

其中 $\delta_{ij} = 1$ 或 0，视 $i=j$ 或 $i \neq j$ 而定。如存在，何时唯一？如不唯一，试找出全部满足 (1.3) 的过程 $(p_{ij}(t))$。人们称这种过程为 Q 过程。

远在 1940 年，费勒 (W. Feller) 证明，Q 过程总是存在的；而且他还具体地造出了一个所谓最小的 Q 过程，记为 $\{f_{ij}(t)\}$。1945 年，杜布 (J. L. Doob) 证明，只有两种可能：或者只有一个 Q 过程，或者有无穷多个 Q 过程。1974 年，侯振挺解决了对一般非保守的 Q，Q 过程的唯一性问题，他所得到的定理如下：

定理 设已给任一满足 (1.1)(1.2) 的 Q，则存在唯一 Q 过程的充分必要条件是：

(i) 存在常数 $c_\lambda > 0$，使

$$\lambda \boldsymbol{\Phi}(\lambda) \boldsymbol{I} \geqslant c_\lambda \boldsymbol{I},$$

这里 $\boldsymbol{\Phi}(\lambda) = (\varphi_{ij}(\lambda))$，$\varphi_{ij}(\lambda) = \int_0^{+\infty} e^{-\lambda t} f_{ij}(t) dt$，而 $\boldsymbol{I}$ 是分量皆为 1 的向量；

(ii) 若向量 $\boldsymbol{X} = (x_i)$ 满足

$$\boldsymbol{X}(\lambda \boldsymbol{I} - \boldsymbol{Q}) = \boldsymbol{0},\quad x_i \geqslant 0,\quad \sum_i x_i < +\infty,$$

则 $\boldsymbol{X} = \boldsymbol{0}$。

关于此定理的详细讨论见[5].

对一类非保守的所谓生灭 Q 矩阵(那里只有状态 0 是非保守的)，Q 过程的唯一性已由杨向群研究过[6].

找出全部 Q 过程的问题尚未彻底解决，只是对一些特殊的 Q 有了完全的解答. 在[7]中找出了全部(可以中断)的生灭过程，用的是概率方法，即用较简单的过程以逼近任一 Q 过程的极限过渡法，此方法的进一步发展见[5，8]. 在[8]中建立了 W 变换，通过样本函数的截舍把一个可列马尔可夫过程变为另一过程.

在[9，58]中，墨文川求出了这类过程的齐次可加泛函的表示及击中分布. 关于更一般的马尔可夫系统的可加泛函的研究见[10].

胡迪鹤研究了一般空间的 Q 过程问题以及间断型过程的性质[11,12,58]，对 Q 过程的唯一性、存在性等有所阐述.

此外，李志阐在寻求全体 Q 过程，吴让泉在研究首达时间的性质，侯振挺、陈木法在研究马尔可夫过程与场论、强马尔可夫性等问题上，都有较好的进展.

§2. 关于鞅、随机积分与随机微分方程

鞅也是一种特殊的随机过程。称随机过程 $\{X_t, -\infty < t < +\infty\}$ 为鞅，如果对任意 $s<t$，有
$$E(X_t \mid X_u, u \leqslant s) = X_s,$$
这里 $E(X_t \mid X_u, u \leqslant s)$ 表示在已知过去 X_u, $u \leqslant s$ 的条件下，将来的变量 X_t 的平均值。而上式则表示此平均值等于现在的值 X_s。例如，如果 X_t 代表某商店第 t 日的收入，那么上式意味着，当已知它在第 s 日及以前各日的收入时，将来第 $t(t>s)$ 日的平均收入，等于它现在（即第 s 日）的收入。这反映收入的某种稳定性（或公平性）。这里，我们把第 s 日视为"现在"，第 s 日以前为"过去"，第 s 日以后为"将来"。

1976 年，梅耶(P. A. Meyer)研究了可选过程对局部鞅的随机积分。在[13]中，严加安将此结果推广到被积过程为可测过程的情形，方法上也作了改进。陈木法在[55]中作了有关解析集的研究。陈培德在[14]中讨论了对平方可积鞅的随机积分的性质。近年来随机微分方程受到普遍重视，它对自动控制等实际问题有了较多的应用。[15]研究了方程
$$X_t = X_0 + \sum_{i=1}^{mn} \int_0^t F_i X_s - dZ_s^i,$$
当驱动项 Z^i（它是半鞅）受到扰动时，讨论了解的稳定性问题。关于随机微分方程与决定性微分方程间的关系见[16]，那里试图把基本事件 ω 固定，以便化前一类方程为后者。关于半鞅的随机微分方程的解的唯一性与存在性见[17]。在[18]中，用随机微分方程作为地极移动的模型，通过解此方程来探讨地极移动的性质。其他有关研究见[59, 64]。

§3. 关于随机场及概率论在统计物理中的应用

1964～1965年，由于杜布鲁申(Р. Л. Добрушин)等人的开创性工作，开始了对无穷多个具有相互作用的质点所组成的随机场的研究，此后，有关的论文大量涌现．关于这方面的一般性介绍见[19，20]．本年度的一些进展见[22～23]．与统计物理有关，出现了值得注意的著作[24，25]．在国内，严士健、李占柄研究了非平衡系统中的概率问题[26]．

§4. 关于极限定理

概率论研究大量的偶然现象，由于量大，常常涉及极限情况. 通俗地说，在极限情况下所获得的定理便是极限定理. 例如，独立地掷 n 次硬币，所得正面的次数记为 $K(n)$，可以证明
$$\lim_{n\to+\infty}\frac{K(n)}{n}=\frac{1}{2},$$
这就是一种最简单的极限定理.

极限定理的研究虽然历史悠久，但新结果仍然层出不穷，或讨论相关随机变量，或研究多维情况，或改进误差估计，等等. 作为一例，见[27].

刘文在[28]中提出了一种新的纯分析方法，以研究马尔可夫链的强大数定理，这种方法别开生面，饶有兴趣.

陆传荣、林正炎、陆传赉在[29～31]中，讨论了项数随机的极限定理. 设 $v(t)$ 为取正整数值的随机变量，$\{x_i\}$ 为平稳序列，他们研究了何时有
$$\lim_{t\to+\infty}P\left(\frac{1}{\sigma\sqrt{v(t)}}\sum_{i=1}^{v(t)}x_i\leqslant x\right)=\frac{1}{\sqrt{2\pi}}\int_0^x e^{-y^{\frac{3}{2}}}dy,$$
并将此结果推广到可加随机函数及鞅的情形. 上式中 $\sigma>0$ 为某常数.

在[32]中，张惟明研究了在什么条件下，关于 Levy 过程 $X(t)$，$a\leqslant t\leqslant b$，下式成立：
$$\lim_{K\to+\infty}\sum_{K=1}^{2^n}\left|X\left(\frac{K}{2^n}\right)-X\left(\frac{K-1}{2^n}\right)\right|^2=\int_0^1 f(t)dt,$$
其中 $f(t)$ 是某一非随机的函数，它与过程的协方差函数有关.

§5. 关于点过程与排队论

点过程也是一种特殊的随机过程. 例如, 以 X_t 表 t 时以前天空的流星个数(或交换台 t 时以前打电话的次数), 那么 $\{X_t, t \geqslant 0\}$ 便是一个点过程. 流星出现的时刻可用直线上一个点来表示.

近年来关于点过程的论文也较多. 厉则治在[33]中研究了点过程的构造. 戴永隆在[34, 35]中对一种特殊的点过程即所谓重随机泊松过程进行了研究.

与点过程密切相关的是排队论问题, 顾客流即是用某种点过程来描述的. 近年来国内关于排队论的部分工作见复旦大学等的论文[36~39, 53, 56].

§6. 其 他

余家荣在[40]中对随机狄利克雷级数的性质进行了深入的研究，得到了很有意义的结果，近年来出现了不分明(Fuzzy)数学(Fuzzy也译作"模糊"或"弗晰". ——编者注)这一分支，在[41，42]中介绍了不分明概率空间.

把概率统计方法应用于预报中，取得了较多成果，有关气象预报见[43，44]. 福建师范大学等单位在水文预报与病虫害预报方面进行了不少工作[45~48,57]. 在地震统计预报方面，南开大学统计预报组所研究的方法总结在[49]中. 有关平稳序列预报见[50]. 林忠民等还将概率统计方法应用于建筑结构设计中[51]. 其他的一些应用或研究见[60~64].

1978年12月，由中山大学主办，在广州召开了全国概率论学术会议，会上交流了大量科研成果. 完全可以预言，我国的概率论研究、应用与教学，必将取得更大的成绩，定能更好地为我国的四个现代化服务.

参考文献

[1] Bliedtner J, Hansen W. Z. W. (即 Z. Wahrschienlichkeistheorie Verw. Gebiēte), 1978, 42: 309.

[2] Jeulin T. Z. W., 1978, 42: 229.

[3] Kanda M. Z. W., 1978, 42: 141.

[4] 钱敏平. 北京大学学报, 1978, 4: 1.

[5] 侯振挺, 郭青峰. 齐次可列马尔可夫过程. 北京: 科学出版社, 1978.

[6] 杨超群. 数学学报, 1965, 15: 9.

[7] 王梓坤, 杨向群. 数学学报, 1978, 21: 66.

[8] 杨向群. 湘潭大学学报, 1978, (1): 29.

[9] 墨文川. 山东大学学报, 1978, (2): 1.

[10] Dynkin E B. Ann. Prob., 1977, 5(5): 653.

[11] 胡迪鹤. 数学学报, 1978, 21(2): 190.

[12] 胡迪鹤. 数学学报, 1978, 21(3): 285.

[13] 严加安. 数学学报, 1978, 21(1): 18.

[14] 陈培德. 数学学报, 1978, 21(4): 363.

[15] Protter P. Z. W., 1978, 44: 337.

[16] Sussmann H J. Ann Prob., 1978, 6(1): 19.

[17] Гадьчук Л И. Teop. Бероям и Npum., 1978, 23: 782.

[18] 王梓坤. 地球物理学报, 1978, 21(3): 225.

[19] Spitzer F. Ann. Math. Monthly, 1971, 78: 142.

[20] Spitzer F. Adv. Math., 1975, 16: 139.

[21] Holley A. Liggett T M. Ann. Prob., 1978, 6: 198.

[22] Liggett T M. Ann. Prob., 1978, 6(4): 629.

[23] Harris T E. Ann. Prob., 1978, 6(3): 355.

[24] Добрушин Р Л, Минлюс Р А. Ycnexu Matem.

Иаук，1977，32(2)：67.

[25] Добрушин Р Л，Минлюс Р А. Теория Бероям. и Npим.，1978，23(4)：715.

[26] 严士健，李占柄. 北京师范大学学报，1979，(1)：1.

[27] Ellis R S，Newman C M. Z. W.，1978，44：117.

[28] 刘文. 数学学报，1978，21(3)：231.

[29] 林正炎，陆传荣，陆传赘. 杭州大学学报，1978，(2)：79.

[30] 陆传荣，林正炎. 杭州大学学报，1978，(3)：27.

[31] 陆传荣，林正炎. 杭州大学学报，1978，(4)：15.

[32] 张惟明. 郑州大学学报，1978，(1)：29.

[33] 厉则治. 厦门大学学报，1979，(2)：8.

[34] 戴永隆. 中山大学学报，1978，(3)：34.

[35] 戴永隆. 中山大学学报，1978，(3)：47.

[36] 郑祖康. 应用数学学报，1977，(4)：30.

[37] 韩继业. 应用数学学报，1978，1(1)：59.

[38] 吴立德，汪嘉冈，李贤平，倪重匡. 复旦学报，1977，(2)：1.

[39] 信息论教研组造机组. 复旦学报，1978，(3)：52.

[40] 余家荣. 数学学报，1978，21(2)：97.

[41] 潘雪明. 计算机应用与应用数学，1978，(8)：56.

[42] 仲崇骥. 吉林师大学报，1978，(1)：5.

[43] 安徽大学数学系应用数学专业，安徽省气象局研究所. 数学的实践与认识. 1978，(2)：7.

[44] 刘为纶，等. 杭州大学学报，1978，(4)：23.

[45] 数学系应用数学教研组，福建水文总站中长期预报研究组. 福建师大学报，1975，(1)：5.

[46] 卢正勇，庄兴元，卢衍风，戴桂珍. 福建师大学报，

1976，(2)：87.

[47] 王明义. 数学的实践与认识. 1978，(3)：7.

[48] 浙江师院数学系病虫害预报小组，浙江金华县病虫测报组. 数学的实践与认识. 1978，(4)：1.

[49] 南开大学数学系统计预报组. 概率与统计预报及在地震与气象中的应用. 科学出版社，1978.

[50] 罗乔林. 应用数学学报，1978，1(1)：79.

[51] 林忠民，等. 福建师大学报，1978，(1)：51.

[52] 胡迪鹤. 武汉大学学报，1978，(4)：1；1979，(1)：15.

[53] 张福基. 应用数学学报，1979，2(1)：44.

[54] 侯振挺，汪培庄. 北京师范大学学报，1979，(1)：23.

[55] 陈木法. 北京师范大学学报，1979，(1)：45.

[56] 杨德庄，颜基义，刘奇志. 中国科学技术大学学报，1978，8(1)：37.

[57] 曹鸿兴，陈国范. 数学的实践与认识，1978，(4)：5.

[58] 墨文川. 山东大学学报，1978，(3～4)：1.

[59] 张炳根. 山东海洋学院学报，1979，(1)：1.

[60] 张文修. 西安交通大学学报，1978，(2)：1.

[61] 余宙文，蒋德才，张大错. 山东海洋学院学报，1978，(1)：1.

[62] 余宙文，陈敦隆. 山东海洋学院学报，1978，(2)：16.

[63] 徐秉舒. 华南工学院学报，1978，(1)：1.

[64] 杜海传. 东北工学院学报，1978，(2)：62.

Scientia Sinica(Series A), 1983, 26(1)

Stochastic Waves for Symmetric Stable Processes and Brownian Motion[1]

Abstract The concepts of stochastic first waves and last waves are introduced for right continuous strong Markov processes. For symmetric stable processes and the Brownian motion, we find the distributions of these waves and that of the time difference between their arrivals, including the distributions of last exit points for balls in $\mathbf{R}^d$ (see (2.16)).

§ 1.

A special kind of stochastic waves was investigated in [6] but made no suggestion for a formal definition. The concept of stochastic first waves was introduced recently by E. B. Dynkin and R. Vanderbei for continuous strong Markov processes to

---

[1] Received: 1981-09-18.

study the harmonic functions in connection with several Markov processes. In this paper, we extend the definition of stochastic waves to right continuous strong Markov processes. We introduce the last waves besides the first ones so that the concept of stochastic waves becomes more complete. For two kinds of processes, the symmetric stable process and the Brownian motion, the distributions of the first waves and the last waves (including the last exit points for balls) and those of the time difference between their arrivals are found. It is well known that the sample functions of separable symmetric stable processes are right continuous with left limits.

Let $X=\{x_t(\omega), t\geqslant 0\}$ be right continuous strong Markov process in d-dimensional Euclidean measurable space $(\mathbf{R}^d, \mathscr{B}^d)$ with Borel σ-algebra $\mathscr{B}^d$ in $\mathbf{R}^d$, $\varphi(s)(s\geqslant 0)$ be a continuous strictly increasing function with $\varphi(0)=0$. For any $r\geqslant 0$ we define

$$h_r^\varphi(\omega)=\inf(t>0, \varphi(|x_t(\omega)|)-\varphi(|x_0(\omega)|)>r) \tag{1.1}$$

with convention $\inf \varnothing =+\infty$. On $(\omega: h_r^\varphi<+\infty)$ we define

$$W^f(r, \omega)=x(h_r^\varphi, \omega). \tag{1.2}$$

The process $\{W^f(r, \omega), r\geqslant 0\}$ is called the stochastic first wave of X.

By (1), we obtain

$$h_r^\varphi(\omega)=\inf(t>0, |x_t|>\psi(|x_0|, r)), \tag{1.3}$$

where

$$\psi(|x_0|, r)=\varphi^{-1}(\varphi(|x_0|)+r)\geqslant |x_0|.$$

Hence h_r^φ can be considered as the first exit time of $\overline{B}_{\psi(|x_0|,r)}(0)$,

Stochastic Waves for Symmetric Stable Processes and Brownian Motion

which is a closed ball in $\mathbf{R}^d$ with center 0 and radius $\psi(|x_0|, r)$. (This fact simplifies some proofs hereafter: firstly we prove some conclusions for balls (or spheres) with radius) r, substitute r by $\psi(|x_0|, r)$, then we obtain the same conclusions for balls (or spheres) with radius $\psi(|x_0|, r)$.

Since X is right continuous, the first wave $w^f(r, \omega)$ at time r concentrates on $(y: |y| \geq \psi(|x_0|, r))$.

In particular, if $\varphi(s)=s$, then $\psi(|x_0|, r)=|x_0|+r$.

We shall suppose in addition that X has left limit. For $r > 0$, let

$$l_r^\varphi(\omega) = \sup(t>0, \ |x_{t-}|<\psi(|x_0|, r)), \quad (1.4)$$

with convention $\sup \emptyset = 0$. l_r^φ is the last exit time of the open ball $B_{\psi(|x_0|, r)}(0)$. On $(\omega: l_r^\varphi > 0)$, we define

$$w^l(r, \omega) = x(l_r^\varphi-, \omega). \quad (1.5)$$

The process $\{w^l(r, \omega), r>0\}$ is called the stochastic last wave of X, which concentrates on $(y: |y| \leq \psi(|x_0|, r))$ at time r.

Especially, if X is continuous, then $w^f(r, \omega)$ and $w^l(r, \omega)$ concentrate on the same sphere $(y: |y|=\psi(|x_0|, r))$, of which the radius increases with r. Hence w^f and w^l extend outwards of wave form and with random positions on sphere. That is why we call them stochastic waves.

Instead of (1.1), we may consider

$$\overline{h_r^\varphi}(\omega) = \inf(t>0, \ \varphi(|x_t-x_0|)>r)$$

etc, which will not be studied here.

§ 2.

Let X be symmetric stable process in $\mathbf{R}^d$ with index α, $0<\alpha<2$, that is X is a process with stationary independent increments $x_{s+t}-x_s$, of which the characteristic function is

$$E\exp[i\xi(x_{st}-x_s)]=\exp(-t|\xi|^\alpha), \quad \xi\in\mathbf{R}^d. \quad (2.1)$$

It is a Markov process with bounded continuous transition density $p(t, x, y)$ depending only on t and $y-x$, $p(t, x, y)= p(t, y, -x)$, where $p(t, x)$ is determined uniquely by

$$e^{-t|\xi|^\alpha} = \int_{\mathbf{R}^d} e^{i(x,\xi)} p(t,x)\,dx.$$

By inverse Fourier transformation, we have

$$p(t,x,y) = \frac{1}{(2\pi)^{\frac{d}{2}}|y-x|^{\frac{d}{2}-1}}\int_0^{+\infty} e^{-t\beta^\alpha}\beta^{\frac{d}{2}} J_{\frac{d}{2}-1}(|y-x|\beta)\,d\beta, \quad (2.2)$$

where $J_v(x)$ is the Bessel function with index v (see [2]).

We can suppose that the process is right continuous with left limits and strong Markov property.

According to (2.1), X has scale invariance: for any $r>0$, the process

$$\left\{\frac{x(r^\alpha t)}{r},\ t\geqslant 0\right\}$$

is also symmetric stable with index α and for any $B\in\mathscr{B}^d$,

$$P_x(x(t)\in B)=P_{rx}\left(\frac{1}{r}x(r^\alpha t)\in B\right). \quad (2.3)$$

We shall suppose $d>\alpha$ which is equivalent to the transience of X.

Let $rB = (rx: x \in B)$, $r > 0$. The first hitting time of rB is denoted by h_{rB}

$$h_{rB}(\omega) = \inf(t > 0, x_t(\omega) \in rB). \tag{2.4}$$

The distribution of the first hitting point $x(h_{rB})$ is

$$H_{rB}(x, dy) = P_x(x(h_{rB}) \in dy).$$

Let $h_B = h_{1B}$, $H_B = H_{1B}$.

Lemma 1 If $H_B(x, dy)$ has density $f(x, y)$ with respect to d-dimensional Lebesgue measure, then $H_{rB}(x, dy)$ has a density $f_r(x, y)$:

$$f_r(x, y) = \frac{f\left(\dfrac{x}{r}, \dfrac{y}{r}\right)}{r^d}. \tag{2.5}$$

Proof By scale invariance, there exist

$$\delta_B \equiv \inf\left(t > 0, \frac{x(r^a t)}{r} \in B\right)$$

$$= \inf(t > 0, x(r^a t) \in rB) = \frac{h_{rB}}{r^a},$$

$$P_{rx}(x(h_{rB}) \in rA) = P_{rx}\left(\frac{x(r^a \delta_B)}{r} \in A\right) = P_x(x(h_B) \in A).$$

Substituting x by $\dfrac{x}{r}$, we have

$$P_x(x(h_{rB}) \in rA) = P_{\frac{x}{r}}(x(h_B) \in A)$$

$$= \int_A f\left(\frac{x}{r}, z\right) dz = \int_{rA} f\left(\frac{x}{r}, \frac{y}{r}\right) r^{-d} dy.$$

The proof is complete.

Consider the last exit time l_B of B

$$l_B(\omega) = \sup(t > 0, x_{t-}(\omega) \in B).$$

Denote the distribution of the last exit point $x(l_B^-)$ by

$$L_B(x, dy) = P_x(l_B > 0, x(l_{B^-}) \in dy).$$

The Riesz potential kernel $g(x)$ is

$$g(x) = \int_0^{+\infty} p(t,x) dt = \frac{C_1}{|x|^{d-\alpha}}. \tag{2.6}$$

$$C_1 = \frac{\Gamma\left(\frac{d-\alpha}{2}\right)}{2^\alpha \pi^{\frac{d}{2}} \Gamma\left(\frac{\alpha}{2}\right)}. \tag{2.7}$$

Denote

$$g(x, y) = g(y-x).$$

The following lemma means that the distribution of the last exit point can be expressed by that of the first hitting point. This conclusion is also correct for $d(\geqslant 3)$ dimensional Brownian motion[6].

Lemma 2 If X is a symmetric stable process in $\mathbf{R}^d (d > \alpha)$, then for any relatively compact set $B \in \mathscr{B}^d$, $x \in \mathbf{R}^d$, we have in the sense of weak convergence of measures

$$L_B(x, dy) = g(x, y) \lim_{|z| \to +\infty} \frac{H_B(z, dy)}{g(z)}. \tag{2.8}$$

Proof Let $\mu_B(dy)$ be the equilibrium measure of B. In the weak sense[4,5] there

$$\lim_{|z| \to +\infty} \frac{H_B(z, dy)}{g(z)} = \mu_B(dy).$$

On the other hand by[3]

$$\frac{L_B(x, dy)}{g(x, y)} = \mu_B(dy),$$

(2.8) follows from these two equalities. The proof is complete.

Theorem 1 If X is a symmetric stable process, $d > \alpha$,

Stochastic Waves for Symmetric Stable Processes and Brownian Motion

then from any $x \in \mathbf{R}^d$, the distributions of stochastic first wave and last wave are

$$P_x(w^f(r,\omega) \in A)$$

$$= C_2 \int_{A_r} \frac{|\psi^2(|x|,r) - |x|^2|^{\frac{a}{2}} dy}{|\psi^2(|x|,r) - |y|^2|^{\frac{a}{2}} |x-y|^d}, \quad (2.9)$$

$$P_x(w^l(r,\omega) \in A)$$

$$= C_2 \int_{A'_r} \frac{dy}{|\psi^2(|x|,r) - |y|^2|^{\frac{a}{2}} |x-y|^{d-a}}, \quad (2.10)$$

respectively, where $r > 0$, $C_2 = \dfrac{\Gamma\left(\dfrac{d}{2}\right) \sin \dfrac{\pi a}{2}}{\pi^{1+\frac{d}{2}}}$, and

$$A_r = A \cap (y: |y| \geq \psi(|x|, r));$$
$$A'_r = A \cap (y: |y| \leq \psi(|x|, r)).$$

Proof First consider the ball with center 0 and radius r. Let

$$h_r^{(0)} = \inf(t > 0, \ |x_t| > r); \quad h_r^{(i)} = \inf(t > 0, \ |x_t| < r).$$

Then

$H_r^{(0)}(x, dy) = P_x(x(h_r^{(0)}) \in dy)$ concentrates on $(|y| \geq r)$;
$H_r^{(i)}(x, dy) = P_x(x(h_r^{(i)}) \in dy)$ concentrates on $(|y| \leq r)$.

We omit index r if $r = 1$, so that $h^{(0)} = h_1^{(0)}$, etc. It was shown in [1] that if $|x| < 1$, then with respect to d-dimensional Lebesgue measure, $H^{(0)}(x, dy)$ has density

$$h^{(0)}(x, y) = \frac{c_2 |1 - |x|^2|^{\frac{a}{2}}}{|1 - |y|^2|^{\frac{a}{2}} |x - y|^d}, \quad |y| \geq 1;$$

and if $|x| > 1$, $H^{(i)}(x, y)$ has density

$$h^{(i)}(x, y) = \frac{c_2 |1 - |x|^2|^{\frac{a}{2}}}{|1 - |y|^2|^{\frac{a}{2}} |x - y|^d}, \quad |y| \leq 1.$$

Note that these two expressions are the same but different in their domains of definition.

By Lemma 1, $H_r^{(0)}(x, \mathrm{d}y)$ has density

$$h_r^{(0)}(x, y) = h^{(0)}\left(\frac{x}{r}, \frac{y}{r}\right) r^{-d}$$

$$= \frac{C_2 \mid r^2 - \mid x \mid^2 \mid^{\frac{\alpha}{2}}}{\mid r^2 - \mid y \mid^2 \mid^{\frac{\alpha}{2}} \mid x-y \mid^d}, \quad \mid y \mid \geq r,$$

(2.11)

if $\mid x \mid < r$; and $H_r^{(i)}(x, \mathrm{d}y)$ has density

$$h_r^{(i)}(x, y) = \frac{C_2 \mid r^2 - \mid x \mid^2 \mid^{\frac{\alpha}{2}}}{\mid r^2 - \mid y \mid^2 \mid^{\frac{\alpha}{2}} \mid x-y \mid^d}, \quad \mid y \mid \leq r,$$

(2.12)

if $\mid x \mid > r$. We substitute r by $\psi(\mid x \mid, r)$, then (2.9) is proved.

To prove (2.10), put $B = B_r \equiv (y: \mid y \mid < r)$ in [13], then

$$\frac{H_{B_r}(z, \mathrm{d}y)}{g(z)} = \frac{h_r^{(i)}(z, y)\mathrm{d}y}{g(z)} = \frac{C_2 \mid r^2 - \mid z \mid^2 \mid^{\frac{\alpha}{2}} \mid z \mid^{d-\alpha}}{C_1 \mid r^2 - \mid y \mid^2 \mid^{\frac{\alpha}{2}} \mid z-y \mid^d} \mathrm{d}y.$$

(2.13)

Now we show in weak convergence

$$\lim_{\mid z \mid \to +\infty} \frac{H_{B_r}(z, \mathrm{d}y)}{g(z)} = \frac{C_2 \mathrm{d}y}{C_1 \mid r^2 - \mid y \mid^2 \mid^{\frac{\alpha}{2}}}. \quad (2.14)$$

In fact, since $0 < \alpha < 2$, we have

$$\int_{B_r} \frac{\mathrm{d}y}{\mid r^2 - \mid y \mid^2 \mid^{\frac{\alpha}{2}}} = \frac{2\pi^{\frac{d}{2}}}{\Gamma(\frac{d}{2})} \int_0^r \frac{s^{d-1}\mathrm{d}s}{(r^2-s^2)^{\frac{\alpha}{2}}} < +\infty.$$

(2.15)

For any bounded continuous function $f(y)$ on B_r, $\sup \mid f(y) \mid < M$, we have

Stochastic Waves for Symmetric Stable Processes and Brownian Motion

$$\left| \int_{B_r} \frac{H_{B_r}(z, dy) f(y)}{g(z)} - \int_{B_r} \frac{C_2 f(y) dy}{C_1 |r^2 - |y|^2|^{\frac{a}{2}}} \right|$$

$$\leq M \left\{ \int_{|y| \leq r-\delta} \left| \frac{H_{B_r}(z, dy)}{g(z)} - \frac{C_2 dy}{C_1 |r^2 - |y|^2|^{\frac{a}{2}}} \right| + \int_{r-\delta < |y| < r} \frac{H_{B_r}(z, dy)}{g(z)} + \int_{r-\delta < |y| < r} \frac{C_2 dy}{C_1 |r^2 - |y|^2|^{\frac{a}{2}}} \right\}.$$

For given $\varepsilon > 0$, and by (2.15) and (2.13), we can take $|z|$ sufficiently large and $\delta > 0$ sufficiently small such that the last two integrals are smaller than $\frac{\varepsilon}{3M}$. The first integral approaches to zero as $|z| \to +\infty$ by dominated convergence theorem. This proves (2.14).

Remember that $g(x, y) = \frac{C_1}{|y-x|^{d-a}}$. By Lemma 2 and (2.14), we know that at any x, the distribution $L_{B_r}(x, dy)$ of the last exit point for the ball B_r has density

$$l_{B_r}(x, y) = \frac{C_2}{|r^2 - |y|^2|^{\frac{a}{2}} |y-x|^{d-a}} \quad (|y| \leq r).$$

(2.16)

Substituting r by $\psi(|x|, r)$ in (2.16), we obtain (2.10). The proof is complete.

In order to find the distribution of time difference $l_r^\varphi - h_r^\varphi$ for the first and last wave, denote the integrand in (2.9) by $h_r^\varphi(x, y)$, i.e.

$$h_r^\varphi(x, y) = \frac{C_2 |\psi^2(|x|, r) - |x|^2|^{\frac{a}{2}}}{|\psi^2(|x|, r) - |y|^2|^{\frac{a}{2}} |x-y|^d}.$$

Let $e_r(z)$ be the probability that the process starting from z never hits the ball ($|x| < r$), then[1]

$$e_r(z) = P_z(h_r^{(i)} = +\infty)$$

$$= \begin{cases} C_3 \int_0^{|\frac{z}{r}|^2 - 1} \dfrac{u^{\frac{\alpha}{2}-1}}{(1+u)^{\frac{d}{2}}} du, & \text{if } |z| \geq r, \\ 0, & \text{if } |z| < r, \end{cases} \qquad (2.17)$$

$$C_3 = \frac{\Gamma\left(\dfrac{d}{2}\right)}{\Gamma\left(\dfrac{\alpha}{2}\right)\Gamma\left(\dfrac{d-\alpha}{2}\right)}.$$

Theorem 2 For any x,

$$P_x(l_r^\varphi - h_r^\varphi \leq t)$$
$$= \int_{|y| \geq \psi(|x|,r)} h_r^\varphi(x,y) \int_{|z| \geq \psi(|x|,r)} p(t,y,z) e_{\psi(|x|,r)}(z) dz dy.$$
$$(2.18)$$

Proof Define l_r to be the last exit time of ball B_r: $l_r = \sup(t > 0, |x_{t-}| < r)$.
Then

$$P_x(l_r - h_r^{(0)} \leq t) = P_x(\theta_{h_r^{(0)}} l_r \leq t)$$
$$= \int_{|y| \geq r} P_x(x(h_r^{(0)}) \in dy) P_y(l_r \leq t), \qquad (2.19)$$

where θ_t is the shift operator for X. Since $e_r(z) = 0$ on $(|z| < r)$,

$$P_y(l_r \leq t) = \int_{\mathbf{R}^d} p(t,y,z) e_r(z) dz = \int_{|z| \geq r} p(t,y,z) e_r(z) dz.$$
$$(2.20)$$

Substituting it into (2.19), we obtain

$$P_x(l_r - h_r^{(0)} \leq t)$$
$$= \int_{|y| \geq r} P_x(x(h_r^{(0)}) \in dy) \int_{|z| \geq r} p(t,y,z) e_r(z) dz. \qquad (2.21)$$

Substituting (2.11) into (2.21) and changing r by $\psi(|x|, r)$, we get (2.18).

§ 3.

Now let us study the stochastic waves for $d(\geqslant 3)$-dimensional Brownian motion. Let $S_a=(x: |x|=a)$ be the sphere in $\mathbf{R}^d$ and $U_a(dy)$ be the uniform distribution on S_a.

Theorem 3 Let X be $d \geqslant 3$ dimensional Brownian motion. For any $x \in \mathbf{R}^d$, $A \in \mathcal{B}^d$, $r>0$, we have

$$P_x(w^f(x,\omega) \in A)$$
$$= \int_A \frac{\psi(|x|,r)^{d-2} \, ||x|^2 - \psi^2(|x|,r)|}{|x-y|^d} U_{\psi(|x|,r)}(dy). \quad (3.1)$$

$$P_x(w^l(r,\omega) \in A) = \int_A \frac{[\psi(|x|,r)]^{d-2}}{|x-y|^{d-2}} U_{\psi(|x|,r)}(dy), \quad (3.2)$$

$$P_x(l_r^\varphi - h_r^\varphi \leqslant t) = \int_{|y|=\psi(|x|,r)} \frac{\psi(|x|,r)^{d-2} \, ||x|^2 - \psi^2(|x|,r)|}{|x-y|^d},$$

$$\int_{|z|\geqslant \psi(|x|,r)} \tilde{p}(t,y,z)\tilde{e}_{\psi(|x|,r)}(z) dz U_{\psi(|x|,r)}(dy), \quad (3.3)$$

where

$$\tilde{p}(t, y, z) = \frac{1}{(2\pi t)^{\frac{d}{2}}} \exp\left(-\frac{|z-y|^2}{2t}\right),$$

$$\tilde{e}_r(z) = \begin{cases} 1 - \left|\dfrac{r}{z}\right|^{d-2}, & |z| \geqslant r, \\ 0, & |z| < r. \end{cases} \quad (3.4)$$

Since the proof is similar to that given above, we give only a sketch of it. It was shown in [6] that the distributions of the first hitting point and the last exit point for S_r are

$$r^{d-2} \, ||x|^2 - r^2| \, |y-x|^{-d} U_r(dy),$$

and
$$r^{d-2} |x-y|^{2-d} U_r(\mathrm{d}y),$$
respectively.

Note that $\bar{e}_r(z)$ is the probability that the motion starting from z never hits S_r. Following the proof of Theorem 2, we obtain (3.3). It is interesting to point out that $e_r(z)$ reduces to $\bar{e}_r(z)$ when $\alpha=2$.

Theorem 4 For $d>4$, $\dfrac{d}{2}-1>m$, m being a positive integer, $x\in \mathbf{R}^d$, we have
$$E_x[(l_r^\varphi - h_r^\varphi)^m] \leqslant \frac{[2\psi(|x|,r)]^{2m}}{(d-4)(d-6)\cdots(d-2m-2)}. \quad (3.5)$$

Proof Let $S_r(x)=(y: |y-x|=r)$, $\bar{B}_r(x)=(y: |y-x|\leqslant x)$. Consider the first hitting time h_r and the last exit time l_r for $S_r(0)$. Under the given conditions[6] we know
$$E_0 l_r^m = \frac{r^{2m}}{(d-4)(d-6)\cdots(d-2m-2)}.$$

From a point x in the ball, the last exit time increases with the radius. For any $x\in \bar{B}_r(0)$, clearly $\bar{B}_r(0)\subset \bar{B}_{2r}(x)$; therefore, starting from such x, we have that the last exit time of $S_r(0)$ can not exceed that of $S_{2r}(x)$. Since the phase space is homogeneous, the latter has the same distribution as that of l_{2r}, which is the last exit time of $S_{2r}(0)$, starting from 0. Hence
$$E_x[(l_r-h_r)^m] \leqslant E_x l_r^m \leqslant E_0 l_{2r}^m$$
$$= \frac{(2r)^{2m}}{(d-4)(d-6)\cdots(d-2m-2)}. \quad (3.6)$$

Since $x\in \bar{B}_{\psi(|x|,r)}(0)$ for any $x\in \mathbf{R}^d$, we get (3.5) by substituting r in the right side of (3.6) by $\psi(|x|,r)$. The proof is complete.

References

[1] Blumenthal R M, Getoor R K, Ray D B. On the distribution of first hits for the symmetric stable process. Trans. Amer. Math. Soc., 1961, 99: 540-554.

[2] Bochner S, Chandrasekaran K C. Fourier Transforms, 1965.

[3] Chung K L. Remarks on equilibrium potential and energy. Ann. Inst. Fourier, Grenoble, 1975, 25 (3-4): 131-138.

[4] Port S C. On hitting place for stable processes. Ann. Math. Stat., 1967, 38: 1 021-1 026.

[5] Port S C. A remark on hitting places for transient stable processes. Ann. Math. Stat., 1968, 39: 365-371.

[6] Wang Zikun. Last exit distributions and maximum excursion for Brownian motion. Scientia Sinica, 1981, 24: 324-331.

国家地震局分析预报中心编. 地震统计预报论文集. 1982.

关于 1976 年四川松潘大地震的统计预报[①]

§1. 预报意见及根据

1976 年 6 月初,我们根据几种统计预报方法的计算,向四川省地震局和国家地震局发出了在四川省松潘—康定附近可能在 1976 年 7 月 9 日或 8 月 19 日前后发生 $M_s = 6.0$ 以上地震的参考意见. 稍后国家地震局于 6 月下旬在成都召开了地震会商会,我们在大会发言中进一步对可能发生的震级作了补充,提出了可能震级在 $M_s = 7.0$ 以上,很可能是 $M_s = 7\frac{1}{4}$,至少在 $M_s = 6.5$ 级以上. 并对上述关于地区、震级、时间预报的理论根据作了简略的介绍. 一共提了 6 项工作:

1.1 南北地震带发震地区[1]和时间[2]的马尔可夫链随机转移法. 此项工作的计算结果是:

1976 年 5 月 29 日云南龙陵 7.6 级地震发生后,下次南北地

---

① 本文与李漳南、吴荣、朱成熹合作.

带发生 $M_s=5.0$ 以上地震的地区最大可能是四川的松潘—康定—马边地区,可能时间是 7 月 9 日或 8 月 19 日前后.

1.2 全国大陆 $M_s=6\frac{1}{4}$ 以上地震的马尔可夫链随机转移法. 其计算结果是：云南龙陵 7.6 级地震以后,四川的松潘—康定和青海的霍布逊湖附近发震的可能性较大,可能时间是 7 月 9 日或 8 月 19 日前后.

1.3 环西太平洋及我国大陆 $(M_s \geqslant 6.5)$ 地震的随机转移法. 其计算结果是：龙陵地震后 3~4 个月内下次大地震 $(M_s = 6.5$ 以上) 在我国北部地区 (包括松潘) 发生的可能性较大.

1.4 预测南北带最大地震 (半年内) 震级的多因素综合密度缺震法[3]. 其计算结果是：1976 年下半年南北带的最大震级在 $M_s = 6.5 \sim 7.5$,最有可能 (精报值) 是 $M_s = 7\frac{1}{4}$.

1.5 南北带大震的弹跳迁移及其统计预测方法①. 其计算结果是：1974 年云南大关 7.1 级地震后,下次 $(M_s \geqslant 7)$ 将往北迁移,地点在川区的东北部 $\begin{pmatrix} 30.5°\sim 32.5°N \\ 101.5°\sim 104°E \end{pmatrix}$ 地区,发震时间为 1976 年 8~11 月.

1.6 预测大地震 $(M_s \geqslant 7)$ 的一种方法[4]. 由于 1976 年 5 月 29 日龙陵大地震的发生,此法必须过 7 个月后才能进行计算,但从龙陵及附近地区发生地震后的资料来看,几乎百分之百,在四个月内,我国还有 7 级以上大地震发生.

实际上,于 1976 年 8 月 16 日和 23 日在四川松潘先后发生两次 7.2 级地震.

---

① 李漳南：我国南北带大震的迁移规律及其统计预报.

§2. 方法的基本思想

2.1 首先谈谈马尔可夫链随机转移法预报地点和时间的基本思想,较详细的可参考文献[1]和[2]. 为此先介绍一下随机转移的概念. 设有一个随机地(即偶然地)运动的质点 A, 它每经过一段时间(可以是相等的, 也可以是随机的)作一次随机的转移. 它所可能处的状态(或称地区)假定为 $1, 2, \cdots, m$. 譬如说, 它现在处于状态 3, 那么下一次可能转移到 1, 也可能转移到 2, 也可能转移到第 m 个. 事先不能准确地预言它到底转移到哪里, 只能说它转移到某状态的可能性有多大. 表示这种可能性大小的数字叫概率.

在时刻 t, 设已观察到质点 A 处于状态 j, 在这个条件下, 经过了时间 s, 在时刻 $t+s$ 它转移到状态 K 的概率用 $_tP_{iK}(s)$ 来表示. 称 $_tP_{iK}(s)$ ($j, K=1, 2, \cdots, m, t \geq 0, s > 0$) 为一重转移概率. 如果 $_tP_{iK}(s) = P_{iK}(s)$ 与 $t \geq 0$ 无关, 那么称 ($_tP_{iK}(s)$) 是齐次的(或平稳的). 下面对于平稳情形, 用 $P_{iK}(s)$ 代替 ($_tP_{iK}(s)$). 进而对状态和时间的预报分别讨论之. 先讨论预报状态的情形, 这时我们取 $s=1, 2, \cdots$, (s 表示发震的次序数), 用 P_{iK} 代替 $P_{iK}(1)$. 如果对另一状态 i, 有 $P_{ji} \geq P_{jk}$, 那么下一次转移到 K 的可能性比转移到 i 的可能性大或相等. 若上面的不等式不仅对一个状态 i, 而是对一切 i 都成立, 即有

$$P_{jk} = \max P_{ji}, \qquad (2.1)$$

式中 max 表示最大值. 假若转移的统计规律完全由一重转移概率 ($P_{ji}, j, i=1, 2, \cdots, m$) 决定, 则下一次转移到状态 K 的可能性最大. 这个简单的想法是本方法的出发点.

但实际上影响转移统计规律的,除了一重转移概率外,还可能有其他的因素,如二重转移概率以及小地区因素等.

现在将上述概念应用到地震的迁移和预报中,把质点 A 设想为地震,把状态 $1,2,\cdots,m$,设想为可能发生地震的地区;A 在这些状态上转移就表示地震在这些地区转移.根据一重、二重地区和震级以及小地区等因素,构成一个预报测度 $M(k)$:$k=1,2,\cdots,m$,

$$M[(i,m_2),(j,m_1),(j,r);k]$$
$$=\alpha_2 P_{i\cdot k}+\beta_2 P_{m_2\cdot k}+\alpha_1 P_{jk}+\beta_1 P_{m_1 k}+\gamma P_{(j,r)}k, \quad (2.2)$$

其中,$P_{i\cdot k}$ 表示上次处于状态 i,这次任意的条件下,下次处于状态 k 的二重转移概率.

$P_{m_2\cdot k}$ 表示上次地震震级处于 m_2 状态,这次任意的条件下,下次处于状态 k 的二重转移概率.

$P_{m_1 k}$ 表示这次地震震级处于 m_1 状态的条件下,下次处于状态 k 的一重转移概率.

$P_{(j,r)k}$ 表示这次地震的震中处于 j 中的 r 小区条件下,下次处于状态 k 的转移概率.

$\alpha_1,\beta_1,\alpha_2,\beta_2,\gamma$ 是正的常数,它可用"记分法"[1]得到.

$M[(i,m_2),(j,m_1),(j,r);k]$ 除了依赖于最后二次地震的地区 i,j 和震级 m_2,m_1 以及 j 中的小地区 r 外,由于(2.2)中右端各转移概率及常数都以其相应的频率代替,因而它还与地震的总次数 n 有关.

为了要预报下一次(设为第 $n+1$ 次)发震于哪一状态,先从图中查明近二次发生地震的地区和震级状态 $(i,m_2),(j,m_1)$ 是什么,再查近一次发震的小地区 (j,r) 是什么,有了这些数据就可以从一直累积到第 n 次地震的有关表中得到(2.2)中右端各转移概率及常数的具体数字,由(2.2)就可以求得

$$M(1),M(2),\cdots,M(m),$$

然后从它们中找出最大的,譬如 $M(K_0)$ 最大,则认为下次地震出现于 K_0 状态的可能最大,其概率近似地为

$$\frac{M(K_0)}{M(1)+M(2)+\cdots+M(m)}.$$

为了保证可能性更大,我们往往需要预报最大可能 K_0 与次大可能 K_1 两个状态. 总可能性(概率)为

$$\frac{M(K_0)+M(K_1)}{M(1)+M(2)+\cdots+M(m)}.$$

根据上述方法,我们知道了下次地震($M_t \geqslant M_0$)将发生在状态 K,试问状态 K 到底将在什么时候发震?可以设想状态 K 发震的时间与下列因素有关.

首先,它和这次地震所处的状态 j 有关,这关系可以通过下面 3 个因素来反映:

(i) 状态 j 的停留时间 τ_j:状态 j 发震之日起到下一次发震(不论那一个状态)为止的这段时间间隔.

(ii) 由状态 j 转移到状态 K 的转移时间 τ_{jk}:j 区发震到紧接着 K 区发震的时间间隔.

上面两个因素部分地反映了状态 K 的发震时间所受到的外部地区的影响. 显然,τ_{jK} 是 τ_j 的一部分,但为了突出 j 与 K 两地区的关系,有必要把 τ_{jK} 单独提取处理.

(iii) 这次地震的震级大小所处的状态 m_1(将震级分为若干档,每档称为一个震级状态)的停留时间 $\tau(m_1)$:这次震级状态为 m_1 的地震到下次地震($M_s \geqslant M_0$)的时间间隔.

此外,K 的发震时间还应与 K 本身有关,故还要考虑:

(iv) 状态 K 的回转时间 $\tau_K^{(R)}$:由 K 区发震之日到 K 区再次发震之时的这段时间间隔.

显然上述各时间因素,在地震的活跃期和平静期是有明显差别的,因此还须考虑:

(v) 活跃期(或平静期)状态的停留时间 $\tau^{(a)}$(对应地,$\tau^{(0)}$).

以上 5 个因素都对预测 K 的发震时间提供一定信息,为了取长补短,就要把这些因素进行综合处理. 办法是取加权线性组合.

$$\rho_{(j,m_1,K,a)}(t) = \alpha_1 \rho_{\tau_j}(t) + \alpha_2 \rho_{\tau_{jK}}(t) + \alpha_3 \rho_{\tau^{(m_1)}}(t) + \alpha_4 \rho_{t^{(a)}}(t) + \alpha_5 \rho_{\tau_{K^{(K)}}}(t + L_{K^{(i)}}), \quad (2.3)$$

式中, $\rho_\tau(t)$ 表示时间随机变数 τ 的分布密度(用分布频率代替); $\tau_K(j)$ 表示这次发震于状态 j, 而以前最后一次发震于状态 K 的时间间隔; $\alpha_1, \alpha_2, \alpha_3, \alpha_4, \alpha_5$ 是正常数,可用记分法得到[2].

我们称 $\rho_{(I,m_1,K,a)}(t)$ 为活跃期的预报密度. 式(2.3)中右端以 $\tau(\theta)$ 代替 $\tau(\alpha)$ 时, 我们称 $\rho_{(I,m_1,K,\theta)}(t)$ 为平静期预报密度, 简记为 $\rho_1(t)$.

另一方面,从理论上可以证明马氏链的停留时间的分布密度是负指数曲线,预报密度(经验密度)曲线与理论密度 $\rho(t)$ 的相对"最低点"就是我们的最可能的预报值(如图 2-1 所示). 这一结论我们将在"马尔可夫泛函的一种预报方法"一文中证明之.

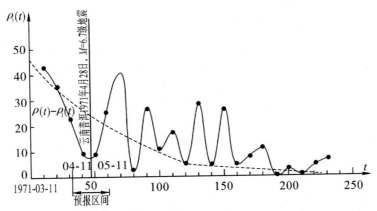

图 2-1 图中虚线 $\rho(t)$ 为理论分布密度. 实线 $\rho_1(t)$ 为预报分布密度

在 §1 中提到的前三项工作, 都是基本上根据上述思想, 结合不同大的地震区划的不同震级, 如南北地震带($M_s \geqslant 5.0$); 我国大陆及相邻地区($M_s \geqslant 6$); 西太平洋岛弧($M_s \geqslant 7$)及相邻大陆($M_s \geqslant 6.5$)等进行的.

2.2 其次，谈谈最大震级预报的一种方法．它的基本思想是用历史地震资料预测一个时段（如半年、一年、一个季度等）某地震带最大地震的震级．为此，我们首先分析了影响今后一个时间段（以半年为例）内最大震级的因素．大量的直观现象表明，它与过去半年的地震活动有关．所谓"地震活动情况"，大致可由地震频度、震级、较大地震发生的地区反映出来，由于历史地震资料对中小地震记录不全，因此对地震频度暂不考虑，对于震级和地区，只考虑主要地震的情况，即只考虑过去半年内最大地震的震级和发生的地区，这两个因素分别称为一重震级和一重地区．

然而，今后半年的地震活动不仅与过去半年有关，而且与前一年（二个半年）、前一年半（三个半年）和前两年（四个半年）等的地震活动有关．相应地我们称之为二重、三重、四重等．我们只取到四重，因为五重以上的分布密度，基本上接近于四重．由于每重都有地区和震级两个因素，故共有 8 个因素．对每个因素根据统计资料分布的集中情况和地质背景等，将其分为若干档，并统计出每个因素按档的分布频率表．根据前两年地震的实发情况，可列出相应的 8 个因素的分布密度．再利用线性叠加法（其系数可用前述记分法得到）构成一个综合分布密度，我们称它为预报密度，以 $\rho(\cdot, m)$ 表示．预报密度与对应的自然密度的相对比值，称为判别密度，以 $\Phi(\cdot, m)$ 表之，使 $\Phi(\cdot, m)$ 连续五个 m（$m = 4\frac{3}{4}$, 5, $5\frac{1}{4}$, $5\frac{1}{2}$, ..., $7\frac{1}{2}$, $7\frac{3}{4}$, 8, $8\frac{1}{4}$, $8\frac{1}{2}$）之和达最大值所对应的震级 m 的区间（正好 M_s 一级），则是最可能发生的震级范围．即是我们对下一个半年最大震级的粗报范围．然后，在粗报范围内，找出理论密度（可以证明它近似一个重指数分布密度）与预报密度 $\rho(\cdot, m)$ 正误差最大的 m 值（简称"相对低点"）作为最大震级的精报值

关于1976年四川松潘大地震的统计预报

(误差为 $\frac{1}{4}$ 级).

此项工作完成于1973年,1974年起至今为实际预报,其预报及实发情况,如图2-2所示.

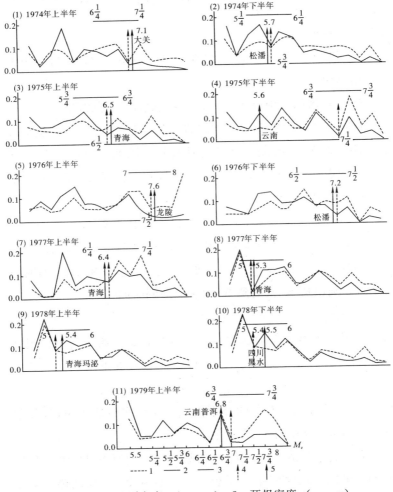

图 2-2　1—差别密度 $\Phi(\cdot, m)$;　2—预报密度 $\rho(\cdot, m)$;
3—粗报震级区间;　4—精报震级;　5—实发震级

2.3 下面谈谈利用南北带 $M_s \geq 7$ 的震中迁移及其统计预报的方法. 其基本思想如下:

南北带的南段川滇地区

$$\left. \begin{cases} 21.5°\sim 32.5°\text{N} \\ 100.0°\sim 104.0°\text{E} \end{cases} \right\} M_s \geq 7$$ 的震中迁移,自 1900 年至 1975 年有以下现象:

(i) 以 29°N 为界呈南北往返弹跳. 以下记 29°N 的北侧为 S_1,南侧为 S_2.

(ii) S_1 区的震中以 101.5°E 为界呈东西往返弹跳.

(iii) 在 S_1 区的 101.5°E 两侧的地震,各侧又呈上下往返弹跳.

(iv) 在 S_2 区的震中呈逆时针旋转. 因此,根据 1974 年 5 月大关地震在 S_2 区,运用(i)～(iii)即可判断下次 $M_s \geq 7$ 震中落入的大致区域.

发震时间先预测年份再进一步估计月份. 令 $t_1, t_2, \cdots, t_N$ 及 $r_1, r_2, \cdots, r_{N-1}$ 分别表示在 S_1 及 S_2 区内依次发震的时间,在我们这里 $t_1 < r_1 < t_2 < r_2 \cdots$. 问题是已知 $\{r_i\}_{i,n}$ 和 $\{t_i\}_{i,n}$ 如何预测 t_{n+1},利用从 S_2 发震到 S_1 发震的间隔时间

$$P_i = t_{i+1} - r_i,$$

对 $i \geq 2$ 有 $\Delta P_i \approx \Delta P_{i+1}$ ($\Delta P_i = P_{i+1} - P_{i-1}$) 这一特性,可得到利用 $\{r_i\}, \{t_i\}\ i=1, 2, \cdots, n$ 预测 t_{n+1} 的公式如下

$$t_{n+1} = r_n + (t_{n-1} - r_{n-2}) - \delta_n \quad (n \geq 3),$$

其中 δ_n 为只与 ΔP_i 有关的修正值. 计算结果只取年份(计算值得 $t=1976.4$ 故预测为 1976 年).

为了进一步估计在 1976 年的哪月,我们对 S_1 区 1900 年以来 $M_s \geq 6$ 的发震时间进行了统计,发现在 30.5°N 北侧,除了一个例外,全在春秋间即 2～4 月或 8～11 月发震,而在我们根据(i)～(iii)预测的未来震中的区域内,全在 8～11 月发震. 于

是推测下次的发震时间为 1976 年 8~11 月.

2.4 谈谈预测我国大地震($M_s \geqslant 7$)方法的基本思想:

(i) 找出与我国大陆大震相关密切的地区共 11 块,记为 $A_1, A_2, \cdots, A_{11}$. 这些地区大致分布在阿富汗、土耳其、缅甸、苏门答腊、苏拉威西、所罗门群岛、菲律宾、斐济、汤加、美国、墨西哥、中南美、阿留申、日本、千岛群岛等地. 每块地区又包含若干块小地区. 所谓相关密切, 是指每一块地区 A_i 发震后, 我国大陆在半年(或 7 个月)以内接着发震的频率较高. 这些地区称为**相关区**, 它所包含的每一小地区则为**相关小区**.

(ii) 在 1900~1965 年中, 在地区 A_i 内总共发生地震的次数设为 n_i, 其中有 m_i 次各在 7 个月内引起我国大震. 令

$$a_i = \frac{m_i}{n_i},$$

它是该地区的地震在 7 个月内转移到我国的频率. 这个数越大, 表示该区与我国大震的关系越密切, 将 a_i 称为 A_i 的**相关频率**.

(iii) 全球在某指定的 7 个月(例如 1950 年 1~7 月)中发生的大地震的总数设为 γ, 其中计有 S 次分别落在相关区 $A_{i1}, A_{i2}, \cdots, A_{is}$ 中(A_{i1} 可能重合于 A_{s2} 等), 它们对应的相关频率是: $a_{i1}, a_{i2}, \cdots, a_{is}$ 作线性组合, 给出预报判别函数

$$X = d_1 a_{i1} + d_2 g_{i2} + \cdots + d_s a_{is},$$

权数 d_i 的取法见[5]. 称 X 为这个月的**判别量**

(iv) 根据 X 的大小, 就可预报这 7 个月后的 4 个月中(在上例中为 1950 年 8~11 月)我国大陆是否有大震发生:

若 $X \geqslant 0.53$ 则报有, 若 $X \leqslant 0.5$ 则报无;

若 $0.5 < X < 0.53$ 则不作结论, 继续观察.

§3. 结束语

松潘地震的统计预报效果是较好的，其主要原因是：

3.1 南北带及与其有关的工作做得较多，探索的时间也比较长．我们从1972年与徐道一等同志合作以来，由于南北带地震资料较多，我们首先就从它着手，探索它的具体规律和一般处理的数学方法，我们的主要精力和兴趣都在南北带．

3.2 多种方法综合预报．这次我们对南北带及有关的工作都事先作了计算，如第一段所提到的6项工作，还有一些零星的工作，只作参考未列入根据．综合这些结果，得出一个比较适宜可靠的综合结论，似乎比用一种手段要好．

3.3 各方面的重视也很重要．当时国家地震局很重视，为此在成都召开了会商会，电告我们派三名代表参加．

我们的工作得到国家地震局和天津市地震局的领导和同志们的大力支持和帮助，在此表示深切的谢意．在工作中还存在不少缺点和错误，有的工作还做得很肤浅，还存在不少问题，欢迎同志们批评指正．

参考文献

[1] 王梓坤，朱成熹，李漳南，徐道一，等. 地震迁移的统计预报. 地质科学，1973，(4)：294-306；数学学报，1974，17(1)：5-19.

[2] 南开大学数学系统计预报组，天津地震队. 预测下次地震发震时间的一种方法. 数学学报，1975，18(2)：86-90.

[3] 朱成熹. 预测最大地震震级的一种方法. 地震科学研究，1980，(1)：33-38.

[4] 南开大学数学系统计预报组. 预测大地震的一种数学方法. 地球物理学报，1975，18(2)：118-127.

后　记

　　王梓坤教授是我国著名的数学家、数学教育家、科普作家、中国科学院院士。他为我国的数学科学事业、教育事业、科学普及事业奋斗了几十年，做出了卓越贡献。出版北京师范大学前校长王梓坤院士的 8 卷本文集（散文、论文、教材、专著，等），对北京师范大学来讲，是一件很有意义和价值的事情。出版数学科学学院的院士文集，是学院学科建设的一项重要的和基础性的工作。

　　王梓坤文集目录整理始于 2003 年。

　　北京师范大学百年校庆前，我在主编数学系史时，王梓坤老师很关心系史资料的整理和出版。在《北京师范大学数学系史（1915～2002）》出版后，我接着主编 5 位老师（王世强、孙永生、严士健、王梓坤、刘绍学）的文集。王梓坤文集目录由我收集整理。我曾试图收集王老师迄今已发表的全部散文，虽然花了很多时间，但比较困难，定有遗漏。之后《王梓坤文集：随机过程与今日数学》于 2005 年在北京师范大学出版社出版，2006 年、2008 年再次印刷，除了修订原书中的错误外，主要对附录中除数学论文外的内容进行补充和修改，其文章的题目总数为 147 篇。该文集第 3 次印刷前，收集补充散文目录，注意到在读秀网（http：//www.duxiu.com），可以查到王老师的

后 记

散文被中学和大学语文教科书与参考书收录的一些情况,但计算机显示的速度很慢。

出版《王梓坤文集》,原来预计出版 10 卷本,经过测算后改为 8 卷。整理 8 卷本有以下想法和具体做法。

《王梓坤文集》第 1 卷:科学发现纵横谈。在第 4 版内容的基础上,附录增加收录了《科学发现纵横谈》的 19 种版本目录和 9 种获奖名录,其散文被中学和大学语文教科书、参考书、杂志等收录的 300 多篇目录。苏步青院士曾说:在他们这一代数学家当中,王梓坤是文笔最好的一个。我们可以通过阅读本文集体会到苏老所说的王老师文笔最好。其重要体现之一,是王老师的散文被中学和大学语文教科书与参考书收录,我认为这是写散文被引用的最高等级。

《王梓坤文集》第 2 卷:教育百话。该书名由北京师范大学出版社高等教育与学术著作分社主编谭徐锋博士建议使用。收录的做法是,对收集的散文,通读并与第 1 卷进行比较,删去在第 1 卷中的散文后构成第 2 卷的部分内容。收录 31 篇散文,30 篇讲话,34 篇序言,11 篇评论,113 幅题词,20 封信件,18 篇科普文章,7 篇纪念文章,以及王老师写的自传。1984 年 12 月 9 日,王梓坤教授任校长期间倡议在全国开展尊师重教活动,设立教师节,促使全国人民代表大会常务委员会在 1985 年 1 月 21 日的第 9 次会议上作出决定,将每年的 9 月 10 日定为教师节。第 2 卷收录了关于在全国开展尊师重教月活动的建议一文。散文《增人知识,添人智慧》没有查到原文。在文集中专门将序言列为收集内容的做法少见。这是因为,多数书的目录不列序言,而将其列在目录之前。这需要遍翻相关书籍。题词定有遗漏,但数量不多。信件收集的很少,遗漏的是大部分。

《王梓坤文集》第 3~4 卷:论文(上、下卷)。除了非正式发表的会议论文:上海数学会论文,中国管理数学论文集论文,以及在《数理统计与应用概率》杂志增刊发表的共3篇论文外，其余数学论文全部选入。

《王梓坤文集》第5卷：概率论基础及其应用。删去原书第3版的4个附录。

《王梓坤文集》第6卷：随机过程通论及其应用（上卷）。第10章及附篇移至第7卷。《随机过程论》第1版是中国学者写的第一部随机过程专著（不含译著）。

《王梓坤文集》第7卷：随机过程通论及其应用（下卷）。删去原书第13～17章，附录1～2；删去内容见第8卷相对应的章节。《概率与统计预报及在地震与气象中的应用》列入第7卷。

《王梓坤文集》第8卷：生灭过程与马尔可夫链。未做调整。

王梓坤的副博士学位论文，以及王老师写的《南华文革散记》没有收录。

《王梓坤文集》第1～2卷，第3～4卷，第5～8卷，分别统一格式。此项工作量很大。对文集正文的一些文字做了规范化处理，第3～4卷论文正文引文格式未统一。

将数学家、数学教育家的论文、散文、教材（即在国内同类教材中出版最早或较早的）、专著等，整理后分卷出版，在数学界还是一个新的课题。

本套王梓坤文集列入北京师范大学学科建设经费资助项目（项目编号CB420）。本书的出版得到了北京师范大学出版社的大力支持，得到了北京师范大学出版社高等教育与学术著作分社主编谭徐锋博士的大力支持，南开大学王永进教授和南开大学数学科学学院党委书记田冲同志提供了王老师在《南开大学》（校报）上发表文章的复印件，同时得到了王老师的夫人谭得伶教授的大力帮助，使用了读秀网的一些资料，在此表示衷心的感谢。

李仲来

2016-01-18

图书在版编目（CIP）数据

论文. 上卷/王梓坤著；李仲来主编. —北京：北京师范大学出版社，2018.8（2019.12重印）
（王梓坤文集；第3卷）
ISBN 978-7-303-23662-6

Ⅰ. ①论… Ⅱ. ①王… ②李… Ⅲ. ①概率论－文集 Ⅳ. ①O211-53

中国版本图书馆CIP数据核字（2018）第090386号

营 销 中 心 电 话	010—58805072 58807651	
北师大出版社高等教育与学术著作分社	http://xueda.bnup.com	

Wang Zikun Wenji

出版发行：北京师范大学出版社 www.bnup.com
北京市海淀区新街口外大街19号
邮政编码：100875

印　　刷：	鸿博昊天科技有限公司
经　　销：	全国新华书店
开　　本：	890 mm×1240 mm　1/32
印　　张：	17.875
字　　数：	405千字
版　　次：	2018年8月第1版
印　　次：	2019年12月第2次印刷
定　　价：	88.00元

策划编辑：谭徐锋　岳昌庆	责任编辑：岳昌庆
美术编辑：王齐云	装帧设计：王齐云
责任校对：陈　民	责任印制：马　洁

版权所有　侵权必究

反盗版、侵权举报电话：010—58800697
北京读者服务部电话：010—58808104
外埠邮购电话：010—58808083
本书如有印装质量问题，请与印制管理部联系调换。
印制管理部电话：010—58805079